NITRENES

REACTIVE INTERMEDIATES IN ORGANIC CHEMISTRY

Edited by GEORGE A. OLAH and LESTER FRIEDMAN
Case Western Reserve University

A series of collective volumes and monographs on the chemistry of all the important species of organic reaction intermediates:

CARBONIUM IONS, 4 volumes
Edited by George A. Olah of Case Western Reserve University and Paul v. R. Schleyer of Princeton University (1968–)

RADICAL IONS
Edited by E. T. Kaiser of the University of Chicago and L. Kevan of the University of Kansas (1968)

NITRENES
Edited by W. Lwowski of New Mexico State University

Planned for the Series

CARBANIONS
Edited by R. Waack, Polaroid Corp.

CARBENES and CARBENOIDS
Edited by L. Friedman of Case Western Reserve University

FREE RADICALS
Edited by J. K. Kochi of Case Western Reserve University

ARYNES
Edited by R. M. Stiles of The University of Michigan

NITRENES

Edited by
WALTER LWOWSKI
Research Center
New Mexico State University
Las Cruces, New Mexico

INTERSCIENCE PUBLISHERS   A Division of John Wiley & Sons
New York · London · Sydney · Toronto

Library of Congress Catalogue Card Number: 76-97256

SBN 471 55710 2

Printed in the United States of America

10 9 8 7 6 5 4 3 2 1

Introduction to the Series

Reactive intermediates have always occupied a place of importance in the spectrum of organic chemistry. They were, however, long considered only as transient species of short life-time. With the increase in chemical sophistication many reactive intermediates have been directly observed, characterized, and even isolated. While the importance of reactive intermediates has never been disputed, they are usually considered from other points of view primarily relative to possible reaction mechanism pathways based on kinetic, stereochemical and synthetic chemical evidence. It was felt that it would be of value to initiate a series that would be primarily concerned with the reactive intermediates themselves and their impact and importance in organic chemistry. In each volume, critical, but not necessarily exhaustive coverage is anticipated. The reactive intermediates will be discussed from the points of view of: formation, isolation, physical characterization, and reactions.

The aim, therefore, is to create a forum wherein all the resources at the disposal of experts in the field could be brought together to enable the reader to become acquainted with the reactive intermediates in organic chemistry and their importance.

As the need arises, it is anticipated that supplementary volumes will be published to present new data in this rapidly developing field.

Cleveland, Ohio   GEORGE A. OLAH and LESTER FRIEDMAN

Preface

Nitrene chemistry has experienced a renaissance during the last decade, with many independent research groups taking up work in the field. The need has arisen to collect the wealth of new data and, perhaps more importantly, to gather in one place the conclusions that have been drawn and the hypotheses that have been advanced. At this time, one would not expect the research groups to be in complete agreement in their views on nitrene chemistry. Many crucial experiments have not yet been done, and the perspectives of the research groups differ, especially when they are working with different nitrenes. No attempt has been made here to unify these views. On the contrary, the authors were asked to speak in their own voices and were encouraged to put forth their working hypotheses and speculations, so that the reader will be exposed to the ideas that govern the course of present and future nitrene research. Consequently, the chapters vary widely in style, some being written in more lyrical prose than others. This seemed a small price to pay for preserving the authors' original views.

There is some overlap between the various chapters. This is believed to be desirable, because it exposes different interpretations of certain observations. Furthermore, few people will read all chapters at one time; rather, they will read one or two chapters and will expect these to be comprehensive.

Many colleagues have generously contributed unpublished data and ideas. My sincere thanks go to all of them.

Las Cruces, New Mexico   WALTER LWOWSKI
August 1969

Authors

A. G. Anastassiou (with Mrs. J. N. Shepelavy), *Department of Chemistry, Syracuse University, Syracuse, New York*

R. Stephen Berry, *Department of Chemistry and The James Franck Institute, The University of Chicago, Chicago, Illinois*

J. H. Boyer, *Department of Chemistry, University of Illinois, Chicago Circle Campus, Chicago, Illinois*

David S. Breslow, *Hercules Incorporated, Wilmington, Delaware*

O. E. Edwards, *Division of Pure Chemistry, National Research Council of Canada, Ottawa, Canada*

P. T. Lansbury, *Department of Chemistry, State University of New York at Buffalo, Buffalo, New York*

David M. Lemal, *Department of Chemistry, Dartmouth College, Hanover, New Hampshire*

Frederick D. Lewis, *Department of Chemistry, University of Rochester, Rochester, New York*

Walter Lwowski, *Research Center, New Mexico State University, Las Cruces, New Mexico*

F. D. Marsh, *Central Research Department, Experimental Station, E. I. du Pont de Nemours and Co., Inc., Wilmington, Delaware*

William H. Saunders, Jr., *Department of Chemistry, University of Rochester, Rochester, New York*

H. E. Simmons, *Central Research Department, Experimental Station, E. I. du Pont de Nemours and Co., Inc., Wilmington, Delaware*

Peter A. S. Smith, *University of Michigan, Ann Arbor, Michigan*

Contents

NITRENES

CHAPTER 1

Introduction

WALTER LWOWSKI

*Research Center, New Mexico State University,
Las Cruces, N.M. 88001*

I. DEFINITION

As used in this book, the name nitrene stands for the electron-deficient, electroneutral molecule NH and its derivatives. The derivatives are formally produced by substitution of the hydrogen in NH. Consequently, they are called (in one word) alkylnitrenes, arylnitrenes, and so on. The definition is based on the structure of the molecule, and not on any particular chemical property. Thus, the definition differs from others, proposed occasionally, which reserve the name nitrene for species of certain chemical properties. Such functional definitions are thought to be undesirable, because it is impossible to ascertain that all members of a given class of molecules will display a certain chemical property. For example, a series of structurally analogous nitrenes might lose the power to insert into a certain type of C—H bonds when the substituent on the nitrogen is gradually changed. Observing the insertion reaction would then depend on reaction conditions and sensitivity of analytical methods, and the border between nitrenes and non-nitrenes would become blurred.

1

II. SCOPE

This book, as part of the series "Reactive Intermediates in Organic Chemistry," discusses nitrenes insofar as they are involved in organic chemistry. The considerable body of knowledge on nitrenes in inorganic chemistry is not covered. However, Chapter 12 gives some references to the chemistry of NH, and to the azides (and potential nitrenes) derived from the metalloids.

III. HISTORICAL

Nitrenes were first proposed by Tiemann in 1891 (1) as short-lived intermediates in the Lossen rearrangement (see Chapter 6, Section V). Stieglitz (2) adopted the nitrene mechanism for the Curtius rearrangements, and Curtius himself used the nitrene hypothesis freely, to explain intermolecular reactions of sulfonyl and other azides. He pointed out (3) that these "residues from azides" (formulated R—N$\diagdown^{\diagup}$) "must be considered only as short-lived intermediary stages." In 1913, Curtius remarked (4) on the "unmistakable analogy" of the "residues from sulfonylazides" to the short-lived residues R—CH$\diagdown^{\diagup}$ which were even then assumed to be intermediates in the reactions of ethyl diazoacetate with aromatic hydrocarbons (5). Bertho (6) summarized, in 1928, the thoughts of Curtius and his collaborators on the mechanisms of azide reactions.

The electronic absorption band, at 3360 Å, of the parent nitrene, NH, was seen by Eder in 1892, and also noted in the solar atmosphere, the NH_3/O_2 flame, and the flame of wet cyanogen (7,8). It was specifically assigned to NH by Frank and Reichardt in 1936, and by Keyser in 1960 (9,10). (See Chapter 2, Section I). As an intermediate in reactions of chloroamine, NH was postulated by Raschig in 1924 (11), and as a product of the base-induced decomposition of hydroxylamine-O-sulfonic acid by Sommer in 1925 (12). It now seems that NH is not an intermediate in these reactions (see Chapter 12, Section I).

The interest in nitrenes was revived after 1950, as attested by the reviews that have appeared in 1959 (13), 1963 (14), and 1964 (15).

IV. NOMENCLATURE

The early workers in the nitrene field did not coin a name for the species, preferring to call it a "residue." No consensus developed in the recent period of renewed activity in the field, leaving one with the problem of choosing a name which is practical. Such a name should lead to a minimum of confusion, both in print and as a spoken word. It should also be as uncomplicated as possible, should indicate that the species is a nitrogen derivative, and should have analogies in nomenclature to gain a mnemonic advantage.

The names imene (16), imin (17), phenylstickstoff (for C_6H_5—N, analogous names could be formed for other nitrenes) (18), imin radical (19), nitrene (20), azacarbene (21), azene (22), azylene (23), and imido intermediate (15) have all been used. *Chemical Abstracts* prefers "imidogen," but also uses terms such as nitrene in its biweekly indexes. All these names are open to some objection or other. "Imen" and "imin" will lead

to confusion with imines, $\diagdown\!\!C\!\!=\!\!N\!\!-$, as will "imine radical." Azacarbene
$\diagup$

is somewhat derivative, Azene will lead to confusion with azines,

$\diagdown\!\!C\!\!=\!\!N\!\!-\!\!N\!\!=\!\!C\diagdown$ (especially when spoken). "Azylene" and "imidogen"
$\diagup$ $\diagdown$

have not found much acceptance, indicating perhaps a degree of difficulty in pronunciation, and "imidogen intermediate" is outright awkward. The term "nitrene" seems to have found widespread acceptance, as of the time of this writing. Objections can be raised against "nitrene": The term has been used before, by Staudinger (24), for what later turned out to be azomethine-ylides (25). However, the term "nitrene" is rarely if ever used now for Staudinger's compounds. Another disadvantage of the term "nitrene" is the "e" at its end—the ending "ene" being reserved for olefinic and aromatic compounds (15). Dropping the "e" (26) might overcome this objection. Because it meets the conditions outlined above, and because it is widely accepted, the term "nitrene" will be used in this book.

Kirmse (27) has discussed the difficulties associated with naming divalent carbon intermediates, commonly known as carbenes. Part of the appeal of the name "nitrene" is, of course, because of its analogy with "carbene." The latter word, however, also does not conform to IUPAC rules. As Kirmse points out, the IUPAC nomenclature leads to difficulties, and it would perhaps be best if a serious attempt to revise this nomenclature would be made.

V. NITRENE AND AZIDE MECHANISMS

Nitrenes, being univalent nitrogen derivatives, are usually generated from some derivative of trivalent nitrogen by some elimination or reduction process, in the presence of the substrate with which the nitrene is supposed to react. Thus, the possibility exists that reactions observed under these circumstances are not those of the nitrene, but those of the precursor, of an excited state of the precursor, of a partially reduced form of the precursor, and so on.

In the case of azides as nitrene precursors, loss of N_2 as the *first, unassisted*, reaction step leads to a nitrene. Reactions of excited azides (vibrationally "hot," or singlet or triplet states of electronically excited azides) with the substrate, followed by loss of nitrogen, are, of course, not nitrene reactions. Many examples for the duality of azide decomposition mechanisms are known, and can be found in the following chapters, wherever azides are used. For example, the reaction of an azide in the presence of an olefin, to give nitrogen and an aziridine, can be induced by heat or light. Is loss of nitrogen the first step, or is a triazoline intermediate formed first? The latter reaction, triazoline formation, is well documented. Furthermore, triazolines are known to decompose to give nitrogen and aziridines (a sampling of the literature is given in references 28–38). However, the direct addition of nitrenes to olefins is also a well-documented reaction. The decision between mechanisms 1 and 2 must be made for every reaction studied, as long as the azide is used for a nitrene precursor. Similar questions arise for practically all other presumed nitrene reactions.

From the structure of the product alone, the intervention of a nitrene intermediate can neither be proved nor disproved. (Note, however, the use of labeling to show the intervention of a symmetrical species, as in the work of Simmons and Anastassiou on cyanonitrene (39); see Chapter 9 Section III-A-1.) From experience in the field, a reaction indicative of a nitrene intermediate is insertion into C—H bonds. Even this reaction, however, can be written as an azide reaction, as in equation 3b. While no case of an azide insertion, followed by elimination of nitrogen, has yet been uncovered, one cannot be sure that such reactions will not be found.

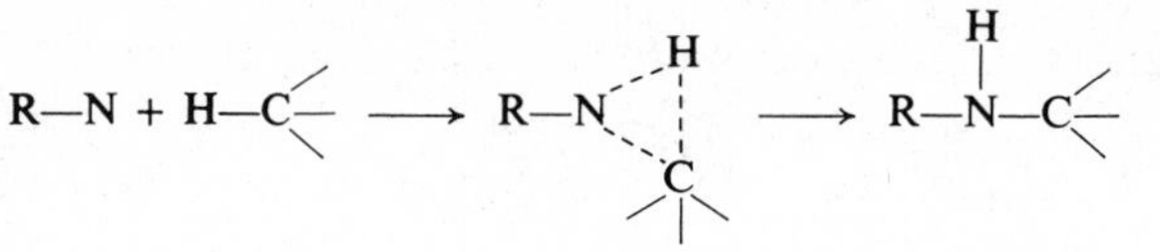

$$R\text{—}N + H\text{—}C\diagdown \longrightarrow R\text{—}N\diagup\overset{H}{\underset{C}{\diagdown}} \longrightarrow R\text{—}\overset{H}{\underset{|}{N}}\text{—}C\diagdown \qquad (3a)$$

Nitrene Insertion

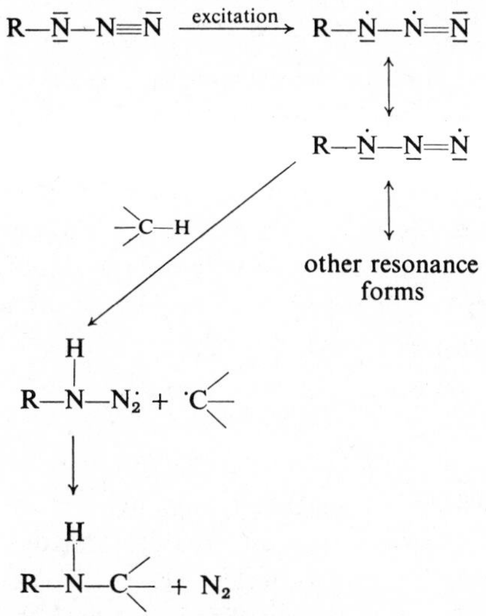

$$R\text{—}\bar{\underline{N}}\text{—}N\text{≡}\bar{N} \xrightarrow{\ \text{excitation}\ } R\text{—}\dot{\underline{N}}\text{—}\dot{N}\text{=}\bar{\underline{N}}$$

$$\updownarrow$$

$$R\text{—}\dot{\underline{N}}\text{—}N\text{=}\dot{N} \qquad (3b)$$

$$\updownarrow$$

other resonance forms

$$\diagup\!\!\!\text{C—H}$$

$$R\text{—}\overset{H}{\underset{|}{N}}\text{—}N_2^{\cdot} + {}^{\cdot}C\diagdown$$

$$\downarrow$$

$$R\text{—}\overset{H}{\underset{|}{N}}\text{—}C\diagdown + N_2$$

Azide Mechanism (Abstraction–Substitution)

A number of generally applicable methods have been used to ascertain (or to make very probable) the intervention of nitrenes in given reactions. These methods are briefly discussed in the following paragraphs. Most of the examples refer to differentiating between nitrene and azide mechanisms, but the problem is basically the same for other methods of making what might be a nitrene. In the deoxygenations of nitro and nitroso compounds (see Chapter 5), certain radicals or partially reduced precursors might react to give products which can, on paper, be easily interpreted as "nitrene products."

A. Kinetic Methods

In all known nitrene reactions, the rate-determining step is the formation of the nitrene. In the cases of azide or α-elimination precursors, the rate law is thus first-order in the disappearance of azide or precursor (such as the anion $Ar\text{—}SO_2O\text{—}\bar{N}\text{—}R$). This has long been recognized (cf. 40), and the first kinetic criterion applied to such nitrene reactions is that they must be first-order. For example, the decomposition of benzenesulfonyl azide is first-order in most solvents, but not in thiols (41). In the presence of triphenylphosphine, benzenesulfonyl azide decomposes in a first-order process, but benzoyl azide loses nitrogen according to a second-order rate law (42). The copper-catalyzed decomposition of benzenesulfonyl azide (43) cannot involve a free nitrene.

1. *Independence of the Rate on the Nature of the Solvent*

In many cases, the substance that is to react with the nitrene or the nitrene precursor is employed as the solvent, and the rate of disappearance of the precursor would always be first-order or pseudo-first-order. A nitrene mechanism demands that this rate should vary only little with changes of the solvent, since the latter is presumed to participate in the transition state only by solvation. Many studies have been made along this line. In typical nitrene reactions (with azides as the precursors), rates varied by a factor of 10 or less. In typical azide reactions, rate factors of many thousands were observed. Some examples may be found in references 44–52.

2. *Substituent Effects*

Appl and Huisgen (53) have studied the influence of substituents on the rate of decomposition of substituted phenyl azides. The lack of pronounced substituent effects indicates that the nitrogen is lost without a concerted change in the ring system. This is discussed in more detail in Chapter 4, Section I-B. Substituent effects are not always interpreted easily, as can be seen from the discussion of the mechanisms of the Curtius rearrangement, in Chapter 6, Section V.

3. *Volume Change of Activation*

Measuring the dependence of reaction rate on the pressure a (condensed phase) system is subject to has become an important tool, to determine whether the volume of the transition state is larger or smaller than that of the starting material(s) (54). The volume change of activation of benzazide (55,56) is positive but small, indicating that bond-breaking is important in the transition state. Le Noble found small, negative, activation volumes in the reaction of chloroamine with hydroxide ion. In accord with other evidence (see Chapter 12, Section I), this indicates that a nitrene is not formed (57):

$$H_2N\text{---}Cl + {}^-OH \longrightarrow HO\text{---}NH_2 + Cl^- \tag{4}$$

rather than the nitrene reaction

$$H_2N\text{---}Cl + {}^-OH \rightleftharpoons H\bar{N}\text{---}Cl + H_2O$$

$$\text{slow} \cancel{\Big|}$$

$$HN + Cl^- \tag{5}$$

$$H_2O \Big|$$

$$H_2N\text{---}OH$$

where, in the rate-determining step, the volume would increase by forming two species ($NH + Cl^-$) from one ($H\bar{N}Cl$).

In contrast to the chloroamine reaction, the hydrolysis of difluoroamine has a large, positive activation volume ($\Delta V^{\ddagger} = +14$ ml/mole) and gives difluorodiazine as the main product (58,59). Both these facts indicate a nitrene mechanism:

$$HNF_2 + {}^-OH \rightleftharpoons {}^-NF_2 + H_2O$$

$$\text{slow} \Big|$$

$$NF + F^- \tag{6}$$

$$\Big|$$

$$N_2F_2$$

B. Physical Verification

In the carbene field, Trozzolo (60) has recently generated the species in a frozen matrix, at very low temperatures, observed their ESR and UV spectra, and then allowed them to react by raising the temperature. He could thus correlate chemical reactions with the availability of carbenes in known electronic states. The technique seems to hold great promise for the nitrene field as well.

C. Chemical Methods

1. *Independent Routes to the same Nitrene*

The intervention of a nitrene intermediate has been made highly probable in several reactions by generating the species by two or more independent routes, and showing that the same products or product mixtures result. Carbethoxynitrene, for example, can be generated by photolysis (61) and pyrolysis (62,63) of ethyl azidoformate, and by α-elimination (64) of p-nitrobenzenesulfonic acid from $N(p$-nitrobenzenesulfonyloxy)-urethan, $(p)O_2N\text{---}C_6H_4SO_2ONHCOOEt$. The selectivity of the C—H insertion of the species generated in each instance has been compared, and found to be nearly the same (63–67). Phenylnitrene seems to be the only likely intermediate in a number of reactions, all leading to 2-dialkylamino-dihydroazepines: The pyrolysis (68) and photolysis (69) of phenyl azide the deoxygenation of nitrosobenzene (70) and nitrobenzene (71), and the photolyses of oxaziranes (72,73) such as **1** (72), all in the presence of a dialkylamine (eq. 7). The precise nature of the intermediates generated by

(1)

$$(7)$$

deoxygenation of aromatic nitro and nitroso compounds has been discussed on the basis of the product mixtures formed, compared to the azide decomposition reaction of the analogous azides (71,74–78). Other examples may be found in the individual chapters of this book, especially those on arylnitrenes (Chapter 4) and deoxygenation reactions (Chapter 5).

2. Analogy of Chemical Properties

Although analogies in chemical behavior might not be a good basis for nomenclature (see Section IV, above), they are obviously valuable for assigning the nature of intermediates. The ability of the intermediate formed by thermolysis of phenyl azide to insert into unactivated C—H bonds (79) strongly points to a nitrene, as do the properties of the intermediates formed in the deoxygenation of alkyl nitro- and nitrosobenzenes (77). The photolysis of certain nitrile oxides gives products like those observed in the photolyses of carbonyl azides (80), and the oxidation of 3-aminobenzoxazolin-2-one gave an intermediate which added to olefins stereospecifically, to give aziridines (81), in analogy with the additions of carbethoxynitrene (82).

3. Differences in Chemical Properties

Sometimes, a given precursor gives two distinct sets of products, depending on the mode of decomposition. This can force one to postulate two different mechanisms, only one of which can be a nitrene mechanism. Consideration of the nature of the products, the selectivity with which they are formed, and the mode of decomposition of the precursor, might then allow one to assign a nitrene mechanism to one of the reaction paths.

Saunders has applied this type of reasoning to the photolysis and thermolysis of triarylmethyl azides (83–85), and Edwards (86) used it considering the reactions of ethyl azidoformate with dihydropyran.

REFERENCES

1. F. Tiemann, *Ber.*, **24**, 4162 (1891).
2. J. Stieglitz, *Amer. Chem. J.*, **18**, 751 (1896).
3. T. Curtius and F. Schmidt, *Ber.*, **55**, 1571 (1922).
4. T. Curtius, *Z. Angew. Chem.*, **26**, III, 134 (1913).
5. E. Buchner and T. Curtius, *Ber.*, **18**, 2377 (1885).
6. A. Bertho, *J. Prakt. Chem. Ser. 2*, **120**, 89 (1928).
7. J. M. Eder, *Monatsh. Chem.*, **12**, 86 (1892).
8. A. Fowler and C. C. L. Gregory, *Phil. Trans. Roy. Soc. (London)*, Ser. *A*, **218**, 351 (1919).
9. H. H. Frank and H. Reichardt, *Naturwissenschaften*, **24**, 171 (1936).
10. L. F. Keyser, *J. Am. Chem. Soc.*, **82**, 5245 (1960).
11. F. Raschig, *Schwefel- und Stickstoffstudien*, Verlag Chemie, Berlin, 1924, p. 76.
12. F. Sommer, O. F. Schultz, and M. Nassau, *Z. Anorg. Allg. Chem.*, **147**, 142 (1925).
13. W. Kirmse, *Angew. Chem.*, **71**, 540 (1959).
14. L. Horner and A. Christmann, *Angew. Chem.*, **75**, 707 (1963).
15. R. A. Abramovitch and B. A. Davis, *Chem. Rev.*, **64**, 149 (1964).
16. A. Lüttringhaus, J. Jander, and R. Schneider, *Chem. Ber.*, **92**, 1756 (1959).
17. W. Kirmse, *Angew. Chem.*, **71**, 537 (1959).
18. M. Appl and R. Huisgen, *Chem. Ber.*, **92**, 2961 (1959).
19. F. O. Rice and T. A. Luckenbach, *J. Am. Chem. Soc.*, **82**, 2681 (1960); J. F. Heacock and M. T. Edmison, *J. Am. Chem. Soc.*, **82**, 3460 (1960).
20. G. Smolinsky, *J. Am. Chem. Soc.*, **82**, 4717 (1960).
21. I. L. Knunyants and E. G. Bykhovskaya, *Proc. Acad. Sci. USSR*, **132**, 513 (1960).
22. G. Smolinsky, *J. Am. Chem. Soc.*, **83**, 2489 (1961).
23. P. A. S. Smith, L. O. Krbecheck, and W. Resemann, *Abstracts*, 144th Natl. Meeting, ACS, Los Angeles, April 1963, p. 35M.
24. H. Staudinger and K. Miescher, *Helv. Chim. Acta*, **2**, 554 (1919).
25. C. H. Hassall and A. E. Lippman, *J. Chem. Soc.*, 1059 (1953).
26. J. S. McConaghy, Jr. and W. Lwowski, *J. Am. Chem. Soc.*, **89**, 2357 (1967).
27. W. Kirmse, *Carbene Chemistry*, Academic Press, New York, 1964, p. 3ff.
28. L. Wolff, *Liebigs Ann.*, **394**, 30, 68 (1912).
29. K. Alder and G. Stein, *Liebigs Ann.*, **485**, 211 (1931).
30. R. Huisgen, *Angew. Chem.*, **72**, 370 (1960); G. Szeimies and R. Huisgen, *Chem. Ber.*, **99**, 491 (1966).
31. R. Fusco, G. Bianchetti, D. Pocar, and R. Ugo, *Gazz. Chim. Ital.*, **92**, 1040 (1962); *Chem. Abstr.*, **58**, 12560 (1963).
32. R. Huisgen, G. Grashey, and J. Sauer, in *The Chemistry of Alkenes*, S. Patai, Ed., Interscience, New York, 1964, p. 835ff.
33. A. C. Oehlschlager, P. Tillman, and L. H. Zalkow, *Chem. Commun.*, **1**, 596 (1965).
34. A. S. Bailey and J. E. White, *Chem. Ind. (London)*, **1965**, 1628.
35. M. E. Hermes and F. D. Marsh, *J. Am. Chem. Soc.*, **89**, 4760 (1967).

36. K. R. Henry-Logan and R. A. Clark, *Tetrahedron Letters*, 801 (1968).
37. P. Scheiner, *Tetrahedron*, **24**, 2757 (1968).
38. P. Scheiner, *J. Am. Chem. Soc.*, **90**, 988 (1968).
39. A. G. Anastassiou, H. E. Simmons, and F. D. Marsh, *J. Am. Chem. Soc.*, **87**, 2296 (1965); A. G. Anastassiou and H. E. Simmons, *ibid.*, **89**, 3177 (1967).
40. J. Stieglitz and G. O. Curme, Jr., *Ber.*, **46**, 911 (1913).
41. M. Takebayashi and T. Shingaki, *Sci. Rept. Osaka U.*, **8**, 43 (1959); *Chem. Abstr.*, **54**, 17303f (1960).
42. J. E. Leffler and Y. Tsuno, *J. Org. Chem.*, **28**, 902 (1963).
43. H. Kwart and A. A. Khan, *J. Am. Chem. Soc.*, **89**, 1950, 1951 (1967).
44. K. E. Russell, *J. Am. Chem. Soc.*, **77**, 3487 (1955).
45. P. A. S. Smith and J. H. Hall, *J. Am. Chem. Soc.*, **84**, 480 (1962).
46. P. Walker and W. A. Waters, *J. Chem. Soc.*, **1962**, 1632.
47. L. Horner and A. Christmann, *Chem. Ber.*, **96**, 388 (1963).
48. M. F. Sloan, W. B. Renfrow, and D. S. Breslow, *Tetrahedron Letters*, **1964**, 2905.
49. R. Huisgen and H. Blaschke, *Chem. Ber.*, **98**, 2985 (1965).
50. D. S. Breslow, T. J. Prosser, A. F. Marcantonio, and C. A. Genge, *J. Am. Chem. Soc.*, **89**, 2384 (1967).
51. D. S. Breslow and E. I. Edwards, *Tetrahedron Letters*, **1967**, 2123.
52. A. S. Bailey and J. E. White, *J. Chem. Soc.* (*B*), **1966**, 819.
53. M. Appl and R. Huisgen, *Chem. Ber.*, **92**, 2961 (1959).
54. E. Whalley, in *Advances in Physical Organic Chemistry*, Vol. 2, V. Gold, Ed., Academic Press, New York, 1964, p. 93ff; W. J. le Noble, *Progr. Phys. Org. Chem.*, **5**, 207 (1967); *J. Chem. Ed.*, **44**, 729 (1967); *J. Am. Chem. Soc.*, **87**, 2434 (1965).
55. K. R. Brower, *J. Am. Chem. Soc.*, **85**, 1401 (1963).
56. K. R. Brower, *J. Am. Chem. Soc.*, **83**, 4370 (1961).
57. W. J. le Noble, *Tetrahedron Letters*, **1966**, 727.
58. W. J. le Noble and D. Skulnik, *Tetrahedron Letters*, **1967**, 5217.
59. G. A. Ward and C. M. Wright, *J. Am. Chem. Soc.*, **86**, 4333 (1964).
60. A. M. Trozzolo, *Accounts Chem. Res.*, **1**, 329 (1968).
61. W. Lwowski and T. W. Mattingly, Jr., *Tetrahedron Letters*, **1962**, 277.
62. R. J. Cotter and W. F. Beach, *J. Org. Chem.*, **29**, 751 (1964).
63. M. F. Sloan, T. J. Prosser, N. R. Newburg, and D. S. Breslow, *Tetrahedron Letters*, **1964**, 2945.
64. W. Lwowski, T. J. Maricich, and T. W. Mattingly, Jr., *J. Am. Chem. Soc.*, **85**, 1200 (1963).
65. D. S. Breslow, T. J. Prosser, A. F. Marcantonio, and C. A. Genge, *J. Am. Chem. Soc.*, **89**, 2384 (1967).
66. W. Lwowski and T. W. Mattingly, Jr., *J. Am. Chem. Soc.*, **87**, 1947 (1965).
67. W. Lwowski and T. J. Maricich, *J. Am. Chem. Soc.*, **87**, 3630 (1965).
68. R. Huisgen, D. Vossius, and M. Appl, *Chem. Ber.*, **91**, 1, 12 (1958).
69. W. v. E. Doering and R. A. Odum, *Tetrahedron*, **22**, 81 (1966).
70. R. A. Odum and M. Brenner, *J. Am. Chem. Soc.*, **88**, 2074 (1966).
71. R. J. Sundberg, W. G. Adams, R. H. Smith, and D. E. Blackburn, *Tetrahedron Letters*, **1968**, 777.
72. E. Meyer and G. W. Griffin, *Angew. Chem.*, **79**, 648 (1967).
73. J. S. Splitter and M. Calvin, *Tetrahedron Letters*, **1968**, 1445.
74. R. A. Abramovitch, Y. Ahmad, and D. Newman, *Tetrahedron Letters*, **1961**, 752.

75. G. Smolinsky and B. I. Feuer, *J. Org. Chem.*, **31**, 3882 (1966).
76. R. J. Sundberg, *Tetrahedron Letters*, **1966**, 477.
77. R. J. Sundberg, *J. Am. Chem. Soc.*, **88**, 3781 (1966).
78. R. A. Abramovitch and B. A. Davis, *J. Chem. Soc.* (*C*), **1968**, 119.
79. J. H. Hall, J. W. Hill, and H-C. Tsai, *Tetrahedron Letters*, **1965**, 2211.
80. G. Just and W. Zehetner, *Tetrahedron Letters*, **1967**, 3389.
81. R. S. Atkinson and C. W. Rees, *Chem. Commun.*, **1967**, 1230.
82. J. S. McConaghy and W. Lwowski, *J. Am. Chem. Soc.*, **89**, 2357 (1967).
83. W. H. Saunders and J. C. Ware, *J. Am. Chem. Soc.*, **80**, 3328 (1958).
84. W. H. Saunders and E. A. Caress, *J. Am. Chem. Soc.*, **86**, 861 (1964).
85. F. D. Lewis and W. H. Saunders, *J. Am. Chem. Soc.*, **89**, 645 (1967).
86. I. Brown and O. E. Edwards, *Can. J. Chem.*, **43**, 1266 (1965).

CHAPTER 2

Electronic Structure and Spectra of NH and Nitrenes

R. STEPHEN BERRY

Department of Chemistry and The James Franck Institute,
The University of Chicago, Chicago, Illinois 60637

The electronic structures and spectra of nitrenes provide a range of theoretical and experimental problems fascinating for its breadth. At one extreme, one encounters some of the highest-resolution electronic spectroscopy and most elaborate and accurate computation from theory; at the other extreme, one deals with the most ad hoc sorts of empirical theory and with experiments designed to give only the grossest features of electronic spectra. It is our aim in this chapter to survey the theoretical and experimental information concerning electronic properties of nitrenes, over the range from the simplest molecules and most elaborate measurements and calculations, to the most elaborate molecules and crudest theoretical and experimental data. Hopefully, such a survey gives one enough insight to extend some of the detailed knowledge about small molecules to interpret the data we have concerning larger molecules.

I. THE ELECTRONIC STRUCTURE AND STATES OF NH AND ALKYLNITRENES

The simplest nitrene is NH, sometimes called imidogen; it is also the best known and understood. We shall therefore focus on it first, and then try to use NH as a prototype to study alkylnitrenes in much the same way that Mulliken first interpreted the properties of alkyl halides, starting with

13

properties of hydrogen halides. This discussion will provide us with
enough information about the behavior of monovalent nitrogen to be able
to go on to examine arylnitrenes, in which the monovalent nitrogen is not
necessarily the only functional or chromophoric group of interest. Finally,
we shall examine heteronuclear systems, the carbonyl nitrenes.

A. The NH Molecule: Electronic Structure

The NH molecule, chemically reactive as it is, nevertheless is a very
extensively characterized molecule. A list of its properties is collected in
Table I. The molecular ground state is $^3\Sigma^-$, a triplet as one would expect
from the isoelectronic oxygen atom, and a $^3\Sigma^-$ like CH_2 (1,2). The low
excited states are also analogous to those of the oxygen atom and of CH_2;
Figure 1 displays the lower-lying known states of all three. The lowest
states of NH and CH_2 are readily associated with the lowest electronic
configuration of the oxygen atom, $1s^2\,2s^2\,2p^4$. In an orbital model, the
NH molecule normally has occupied shells 1σ (like $1s$ of N), 2σ (mostly
like $2s$ of N), 3σ (becoming $2p\sigma$ of N when the internuclear separation is
large) and 1π (primarily the $2p\pi$'s of N). The ground state and first
excited states have the same molecular orbital configuration, namely
$1\sigma^2\,2\sigma^2\,3\sigma^2\,1\pi^2$; of course, in the ground state, one electron occupies each

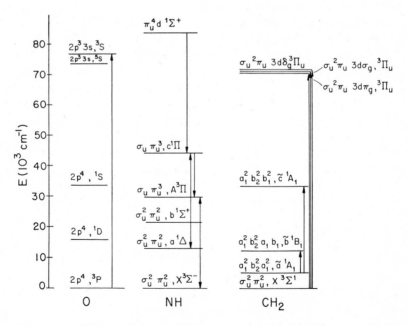

Fig. 1. Energies of low-lying states of the oxygen atom and of NH and CH_2.

TABLE I
Properties of NH [a]

Electronic State	Configuration	Excitation[b] Energy (theoret.)	Excitation Energy (expt.)	R_e	ω_e	B_e	Dissociation Energy	I.P.
$X^3\Sigma^-$	$1\sigma^2 2\sigma^2 3\sigma^2 1\pi^2$	0	0	1.0481 Å[c]	3125 cm^{-1}	16.3454 cm^{-1}[c]	3.21 eV[d,e]	13.1 ± 0.1 eV[f]
a^1	$..3\sigma^2 1\pi^2$	1.76 eV	a (ca 1.6 ev; see Sec. B)	1.0439	3186	16.453		
$b^1\Sigma^+$	$..3\sigma^2 1\pi^2$	3.04	$a + 1.05$ eV	1.0455	3480	16.401		
A^3	$..3\sigma 1\pi^3$	3.68	3.69	1.049[c]	3300	16.3221[c]		
c^1	$..3\sigma 1\pi^3$	5.76	$a + 3.94$	1.1252	2119	14.159		
$d^1\Sigma^+$	$...1\pi^4$			1.1282	2640	14.085		

[a] Except where otherwise noted, the values in this table are taken from G. Herzberg, *Molecular Spectra and Molecular Structure. I. Spectra of Diatomic Molecules*, D. Van Nostrand, New York, 1955, p. 556.

[b] See reference 6.

[c] The value of B_e is taken from reference 22, and R_e was calculated from the experimental B_e.

[d] K. E. Seal and A. G. Gaydon, *Proc. Phys. Soc. (London)* **89**, 459 (1966), based on the temperature dependence of the intensity of the $^3\Pi \leftarrow {}^3\Sigma$-transition.

[e] A. C. Hurley, *Proc. Roy. Soc. (London) Ser. A.* **248**, 119 (1958).

[f] S. Benson, *J. Chem. Ed.*, **42**, 507 (1965); a value of 3.8 eV is also frequently given (see reference 1). A. G. Gaydon and G. Pannetier, *J. Chim. Phys.* **48**, 221 (1951) give 3.9 ± 0.2 eV and D. deJager and L. Neven, *Mem. Soc. Roy. Sci. Liege*, **18**, 513 (1959) give 3.76 eV.

[g] See reference 1 ([a], above) and also R. I. Reed and W. Sneddon, *J. Chem. Soc.*, **1959**, 4132.

of the two degenerate π-orbitals, while in the $^1\Delta$, both electrons have the same direction for their 1 unit of λ, the angular momentum along the NH axis, and are thus described as both being in the same π orbital. (We choose our π orbitals most conveniently so that the individual orbitals have angular momentum $+1$ and -1 units of $\hbar$ angular momentum along the bond; thus, they vary as $\exp(\pm i\varphi)$ around the internuclear axis.) The $^1\Sigma^+$ is the singlet companion to the ground $^3\Sigma^-$ state. All states of higher energy require excitation of an electron from one orbital to another; for example, the spectroscopically known $^3\Pi$ and $^1\Pi$ states arise primarily from one-electron excitations $3\sigma \rightarrow 1\pi$.

The most accurate and most extensively analyzed calculations of the ground state wavefunctions of first-row hydrides, including NH, are almost certainly those of Cade and Huo (3); according to the authors, their functions are likely to be very good approximations to the true molecular Hartree-Fock functions. One can therefore use them with some confidence to infer properties of the ground states that are not associated directly with electron correlation. Theoretical work on excited states is more meager; Ransil (4) and Layton and Ruedenberg (5) studied the doubly excited $1\sigma^2\,2\sigma^2\,1\pi^4$, $^1\Sigma^+$ state; this state was chosen essentially for computational rather than physical reasons and does not correspond to any of the known low-lying states. Hurley (6) used a somewhat more empirical method in order to treat ground and excited states of the first-row hydrides in a way that would include the most important effects of electron correlation. The method, called the intraatomic correlation calculation, yields extremely good values for excitation energies, as Table I shows. For some time, this calculation was the only source for a value of the energy separation between the singlet and triplet manifolds of NH.

From the theoretical calculations for NH, we can make a number of inferences that can be generalized to organic nitrenes. First, the molecular orbitals in the ground state have energy values that vary somewhat with R, the N—H distance. When $R \leq 1.1$ Å, the highest occupied orbital is the 1π, and the 3σ (like $1s$ of H at such short distances) is just below it. At $R = R_e = 1.04$ Å, the 3σ orbital energy is 0.344 eV or approximately 8.0 kcal/mole below the 1π, indicating that if $R \cong R_e$, the ionization energy of a 3σ electron from NH is greater than the ionization energy of a 1π electron by about this amount. However, when $R \gtrsim 1.1$ Å, the energy of the 3σ orbital is higher than that of the 1π, indicative of its becoming the nitrogen $2p\sigma$. The orbital energies and ground state potential curve are shown in Figure 2. The classical amplitude of vibration for the $v = 0$ state, $(\hbar/2\pi m v)^{1/2}$, is approximately 0.1 Å for NH; consequently, those properties of NH (and similar species) which depend on orbital energies and ionization energies of individual electrons are likely to depend

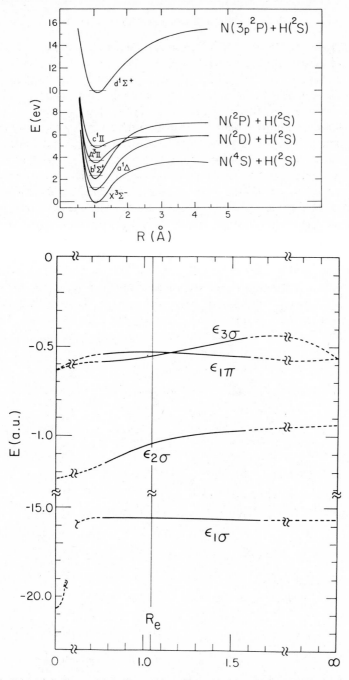

Fig. 2. Potential curves (*a*) and energies of the upper orbitals (*b*) of NH. Orbital
energies were taken from reference 3 (1 a.u. = 27.21 eV.).

17

sensitively on the N—H or N—R distance, and therefore on the vibrational state and temperature of the NH or NR.

To understand the electronic structure and bonding in NH and the higher nitrenes, it is useful to examine electron density contours. This has been done by Bader, Keaveny, and Cade (7) for many systems, including the first- and second-row diatomic hydrides; their analysis focuses on total and symmetrical orbital charge densities and on the changes in these densities associated with molecule formation. Layton and Ruedenberg have taken a complementary point of view (and used somewhat cruder wavefunctions), concentrating on the shapes of the *equivalent* orbitals, the most localized orbitals that give the same total wavefunction for all the electrons.

The total molecular charge density in NH is shown in Figure 3*a*. The distribution of charge is clearly asymmetric and even up to very small distances from the proton, the contour lines include both nuclei. The charge density difference (Figure 3*b*) shows $\rho_M - \rho_{SA}$, the charge density of the molecule less that of the separated atoms. It is clear that molecule formation not only involves removal of charge from "beyond" the proton, but also from a torus near the N atom, in the bonding region. Correspondingly, charge is increased in the bonding region near the proton, and

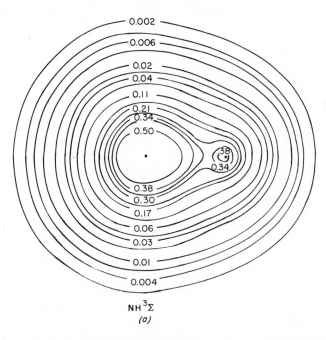

NH $^3\Sigma$

(a)

Fig. 3. Contour maps of charge density in NH: (*a*) total charge density ρ_m; ρ_{SA} at the same internuclear distance.

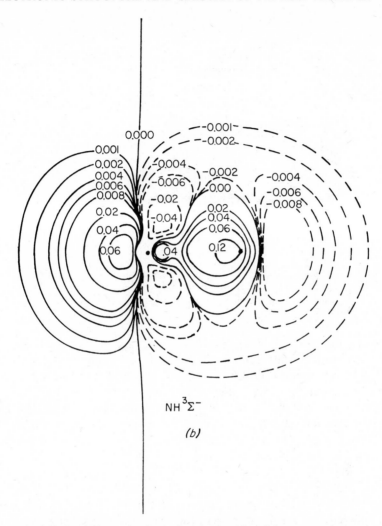

Fig. 3(*b*) Total charge density ρ_m less the densities of the separated atoms.

in the antibonding region "behind" the nitrogen atom. According to the analysis of Bader, Keaveny, and Cade (7), more charge moves into the antibonding region "beyond" the N atom (0.20*e*) than into the strongly bonding region between the nuclei (0.16*e*) when the N—H bond is formed from N + H. This separation was made with the contours of zero difference density (see below) as the dividing surfaces. It was not made on the basis of Berlin's definition of bonding and antibonding regions of space (8). However, the two definitions give similar pictures of how bond formation

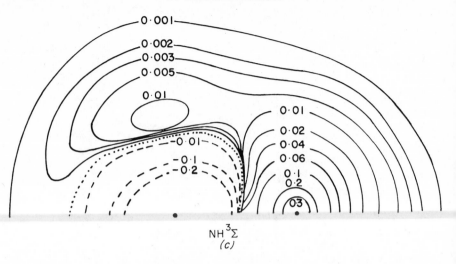

Fig. 3(c) ρ_m less the charge density of an oxygen atom located at the nitrogen nucleus. (Diagrams are all reproduced from reference 7 with permission of the authors and the *Journal of Chemical Physics*.)

produces the electronic charge distribution we have come to associate with the "lone pair" of nitrogen.

The difference map between the charge density of NH and that of the united oxygen atom was also computed by Bader, Keaveny, and Cade (7); this is reproduced in Figure 3c. The density change shown in this map corresponds to a depletion of valence shell charge around N with corresponding increase around the proton; the change is due largely to the electrons in the molecular 3σ orbital.

The orbital charge densities of the 2σ, 3σ, and 1π Hartree-Fock (symmetry) orbitals are shown in Figure 4 together with one of the equivalent trigonal hybrid orbitals determined by Layton and Ruedenberg (5) from the minimal basis set calculations of Ransil (4). Bader, Keaveny, and Cade (7) have analyzed the forces on each nucleus due to the electrons in each Hartree-Fock orbital. These forces in turn, have been broken down into an atomic part, an overlap part, and a screening part. For a molecule AH, the atomic part represents the force on nucleus A due to the charge distribution on A itself; the overlap part is the force due to the overlap charge density, essentially the charge density in the bonding region where the A and H atomic orbitals overlap most, and the screening force is the force on nucleus A due to electronic charge on H; when the internuclear distance is large this last must become equal to the negative of the nuclear–nuclear repulsive force, $-Z_A e^2/R$.

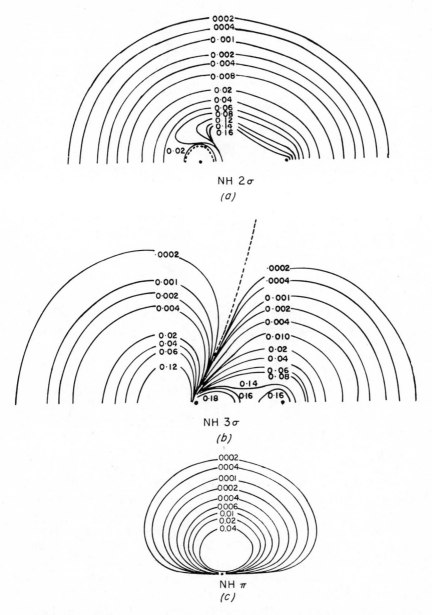

Fig. 4. Contour maps of orbital charge density in NH: (a) 2σ Hartree-Fock; (b) 3σ Hartree-Fock; (c) 1π Hartree-Fock.

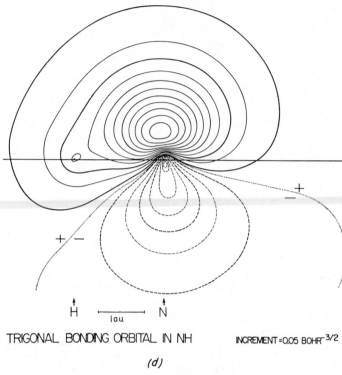

TRIGONAL BONDING ORBITAL IN NH   INCREMENT = 0.05 BOHR$^{-3/2}$

(d)

Fig. 4(*d*) trigonal-bonding orbital. (Hartree-Fock orbitals are based on the ground state function and are taken from reference 3; the trigonal orbital is based on a minimal basis calculation of the $1\sigma^2 2\sigma^2 1\pi^4 \, {}^1\Sigma^+$ state and is taken from reference 5; other diagrams are taken from reference 7, with permission of the authors and the *Journal of Chemical Physics*.)

The analysis of Bader, Keaveny, and Cade for NH shows that the 2σ, 3σ, and 1π orbitals contribute to bonding in the following ways. The 2σ orbital density exerts a binding force on both N and H; the force on the proton is principally due to screening of the N nucleus, while the strong force on the nitrogen is due mostly to its own atomic contribution and about half as much, to the force of the screening charge. The 3σ orbital's charge is quite different. It also exerts a binding force on the proton, but an antibinding force on the nitrogen. The binding force on H is again due mostly to screening and, about one-third as much, to overlap. The large antibinding force on N comes mostly from the antibinding contribution exerted by the "lone pair" charge density near the N nucleus, which tends to pull the N nucleus away from the proton. Binding, overlap, and screening contributions are just not enough to make this orbital exert a binding force on N. In other words, the lone pair does not act in a nonbinding manner, but is characteristically *antibinding*. Finally the 1π orbital is antibinding toward H because A screens the N nucleus, and is slightly binding with respect to nitrogen, because of its polarization toward the proton.

The composition of the orbitals, particularly the 2σ and 3σ orbitals, is worth some comment. Semiempirical calculations designed to take into account the intraatomic (N-atom) part of the electron–electron correlation were carried out by Moffitt (9) and later by Companion and Ellison (10). Both of these calculations gave results in accord with the popular conception that the 3σ orbital should contain very little nitrogen $2s$ character; in fact these calculations gave 6 and 5% nitrogen $2s$ character, respectively. The Hartree-Fock-Roothaan functions of Cade and Huo do not conform so simply to the popular notion. Their 3σ orbital, made from a basis set of 16 functions, does indeed contain more nitrogen $2p$ than anything else, but the sum of the nitrogen $2p$ contributions is certainly not an order of magnitude greater than that of the $2s$. Because of the somewhat ambiguous connection between the general basis set of Cade and Huo (3) and the atomic orbitals of an isolated nitrogen atom, it is not useful to interpret the Cade-Huo function in terms of a simple hybrid orbital picture in a quantitative manner. It suffices to say that for $R \cong R_e$, there is about 10–15% nitrogen $2s$ character in the molecular 3σ Hartree-Fock-Roothaan orbital, as well as some 5% hydrogen $1s$.

B. The NH Molecule: Spectral Properties

Six electronic states of NH are known spectroscopically. These are the ground state $X^3\Sigma^-$, the first excited triplet $A^3\Pi$, the lowest singlet $a^1\Delta$, and the three excited singlets $b^1\Sigma^+$, $c^1\Pi$, and $d^+\Sigma^+$. The origins and correlations of these states with united and separated atom limits are indicated in Figure 1. The best limitation on the energy a separating the $^1\Delta$ and ground $^3\Sigma^-$ states comes from the spectral studies of Okabe and Lenzi (11). They irradiated ammonia, and determined the minimum incident energy which leads to NH fluorescence in the $^1\Pi \rightarrow {}^1\Delta$ system. If their minimum, 9.35 ± 0.06 eV (1325 ± 7 Å), is the true threshold, then the energy separation would be

$$\mathbf{a} = 9.35 - \Delta H(NH_3 \rightarrow NH^3\Sigma^- + H_2) - h\nu_{0-0}(^1\Pi \rightarrow {}^1\Delta),$$

where the standard heat of reaction for forming NH and H_2 from NH_3 is 3.9 ± 0.1 eV (12) and the 0–0 transition of the $^1\Pi$–$^1\Delta$ system lies at 3.813 eV(1). This gives a value

$$\mathbf{a} \sim 1.6 \pm 0.1 \text{ eV}$$

However, the value 9.35 eV may still leave some relative kinetic energy in the NH–H$_2$ system, so that 1.6 eV should be taken only as an upper limit. A lower bound is not really obtainable from the nonappearance of fluorescence insofar as negative evidence can also mean that the transition probability for the process is low, due, for example, to a small Franck-Condon factor. Nevertheless, we may give some credence to the observation of Becker and Welge (13) that irradiation of NH$_3$ with the strong xenon line at 1470 Å (8.4 eV) gives no NH($^1\Pi$–$^1\Delta$) fluorescence, so that it is not likely that **a** is less than 0.6 eV. (Note that Becker and Welge set $1.05 \leq$ **a** ≤ 2.25 eV, but this was based on a ΔH of 3.5, rather than 3.9 eV.)

The known spectral bands are four in number. The sharp $A^3\Pi \leftrightarrow X^3\Sigma^-$ is readily observed in absorption and emission, at approximately 3360 Å. The broader $c^1\Pi \to a^1\Delta$ is well known in emission, but is peculiarly difficult to observe in absorption; Herzberg has reported it once in a comment (14) and one transient line at 30,466 cm^{-1} has been reported by McCarty and Robinson (15) in their observation of NH trapped in solid argon. Emission from the $c^1\Pi \to b^1\Sigma^+$ and $d^1\Sigma^+ \to c^1\Pi$ transitions has been observed at 4530 Å (5.6 eV) and 2540 Å (3.15 eV), respectively.

Oscillator strengths and radiative lifetimes have been determined by Bennett and Dalby (16) for the triplet–triplet transition and by Fink and Welge (17) for both the triplet–triplet and lowest singlet–singlet transitions. The two sets of experiments agree nicely on the values of oscillator strengths $[(8 \pm 1.1) \times 10^{-3}$ and $(7.3 \pm 1.3) \times 10^{-3}]$ and of lifetimes (0.425 ± 0.06 and 0.46 ± 0.08 μsec, respectively); the absorption oscillator strength for the $^1\Pi \leftarrow ^1\Delta$ transition is $(3.6 \pm 0.4) \times 10^{-3}$ and the corresponding radiative lifetime is 0.435 ± 0.04 μsec.

The singlet–triplet intersystem combination transitions have not been directly observed.

A broad variety of methods and materials may be used to produce NH, either for spectroscopic observation or for chemical reactions. A large variety of flames exhibit NH bands in their spectra (18). Other methods, including flash and continuous photolysis, electric discharges, thermal decomposition, shock waves, and electron bombardment, are presented in Table III. This table is intended to be representative rather than comprehensive. It is clear from its extent that NH has attracted intense attention and considerable fruitful study, particularly since about 1960.

The principal precursors of NH have been hydrazoic acid, hydrazine, and ammonia for photolytic generation, and virtually any mixture of nitrogen and hydrogen compounds if one uses an electric discharge. The hydrazoic acid photolysis proceeds directly through HN$_3$ + $h\nu \to$ NH + N$_2$. The photolytic decomposition of HNCO was assumed by Back and Mui to follow the same path; it certainly does produce NH (21, 22). However, it may also be that NHCO can split under photolysis to give

TABLE II

Partial Forces in NH [a]

| | Partial Forces on H | | | |
Orbital	$f_H(R_e)$	Atomic	Overlap	Screening	$f_H(\infty)$
2σ	2.366	0.046	0.533	1.769	2
3σ	1.370	0.068	0.363	0.939	1
1π	1.335	0.001	0.043	1.291	2

| | Partial Forces on N | | | |
Orbital	$f_N(R_e)$	Atomic	Overlap	Screening	$f_N(\infty)$
2σ	1.109	0.623	0.423	0.063	0
3σ	-0.485	-1.489	0.809	0.195	1
1π	0.158	0.116	0.041	0.001	0

[a] From reference 7, based on wavefunction of reference 3.

NCO + H, while HN_3 appears to give N_3 only through subsequent reactions (23).

Photolysis and thermal (shock) decomposition of N_2H_4 and of NH_3 seem to give NH only when the systems are hot. Presumably at low temperatures, decomposition proceeds through chains involving NH_2, producing explosive decomposition. There is one notable exception to this pattern, namely photolysis with very high energy light. Welge and his coworkers (13,24,25) have shown that the primary process from $NH_3 + h\nu$ ($\lambda < 1470$) can produce excited ($^1\Pi$)NH, but that with $\lambda \geq 1470$, they detect no NH. Similarly N_2H_4 requires a krypton lamp ($\lambda = 1235, 1165$ Å) to exhibit NH fluorescence, and shows none with a xenon lamp ($\lambda = 1470$, 1295 Å).

C. The NH Molecule: Reactions

The NH molecule undergoes a variety of reactions with itself and with many of its precursors; the reactions of NH with other species are relatively unstudied, by comparison.

Foremost among the reactions of NH are the two principal processes arising from collision of two NH molecules:

$$2NH \longrightarrow N_2 + H_2 \tag{1}$$

and

$$2NH \longrightarrow N_2H_2, \tag{2}$$

TABLE III

The Production and Observation of NH: A Survey

Method of production	Precursor	Method of observation	Observed properties	Ref.
Pyrolysis	HN_3	Product analysis with added CO	HNCO formed	33
Shock pyrolysis	NH_3; HN_3	Optical emission	$A^3\Pi \rightarrow X^3\Sigma^-$ from HN_3 only	a, b
	N_2H_4	Mass spectrum	Decay mechanism	c
	$N_2 + H_2$; NH_3	Optical emission	$A^3\Pi \rightarrow X^3\Sigma^-$; excitation mechanism	d
	N_2H_4	Optical emission	$A^3\Pi \rightarrow X^3\Sigma^-$ time dependence; excitation mechanism	29
Steady photolysis	NH_3	Optical emission	Decay mechanism	e
	HN_3	Composition of products	Decomposition as a primary process	34, 35
	HN_3	Optical emission and absorption	$c^1\Pi \rightarrow a^1\Delta$, $c^1\Pi \rightarrow b^1\Sigma^+$ and $A^3\Pi \rightarrow X^3\Sigma^-$; singlet as primary dissociation product	25b
	HN_3 (in solid matrix)	Infrared absorption	Inference of NH	f
	NH_3 (in solid matrix)	Optical absorption	$A^3\Pi \leftarrow X^3\Sigma^-$; wavelength dependence of NH_3 dissociation	g
	HN_3	Product composition	Reactions with hydrocarbons	41
	NH_3	H-atom scavenging by C_2D_4	Wavelength dependence of dissociation	h

	NH$_3$; N$_2$H$_4$	Fluorescence	$A^3\Pi \to X^3\Sigma^-$, $c^1\Pi \to a^1\Delta$; dissociation mechanism; threshold wavelength for fluorescence (<1470 Å).	13, 24, 25
	HNCO	Product composition	Reactions with hydrocarbons	19, 20
	HN$_3$	Infrared absorption	NH vibration frequency	i
	HN$_3$	Infrared absorption	NH, N$_2$H$_2$ vibration bands	28
Flash photolysis	NH$_3$	Optical absorption	Existence of NH; $A^3\Pi \leftarrow X^3\Sigma^-$	j
	NH$_3$	Optical absorption	Existence and molecular parameters of NH $A^3\Pi \leftarrow X^3\Sigma^-$	k
	HN$_3$	Optical absorption	$A^3\Pi \leftarrow X^3\Sigma^-$	23
	HNCO	High-resolution optical absorption	Rotational analysis of $A^3\Pi \leftarrow X^3\Sigma^-$	21, 22
	NH$_3$	Time-dependent optical absorption	$A^3\Pi \leftarrow X^3\Sigma^-$; mechanisms of dissociation and of NH vibrational relaxation	30
	NH$_3$	Optical absorption	$A^3\Pi \leftarrow X^3\Sigma^-$; mechanism of reaction	l
	HN$_3$	Optical absorption; product composition	Reactions of NH	42
Electric discharge	HN$_3$	Analysis of trapped products	Unstable intermediates	31, 32
	HN$_3$	Mass spectrum	N$_2$H$_2$ and other species (NH inferred indirectly only)	26

(continued)

TABLE III (continued)

Method of Production	Precursor	Method of Observation	Observed Properties	Ref.
	$N_2 + H_2$; N_2H_4; NH_3	Optical absorption spectra of cold-trapped species	$A^3\Pi \leftarrow X^3\Sigma^-$; $c^1\Pi \leftarrow a^1\Delta$	15
	N_2H_4	Infrared absorption of trapped products	N_2H_2 formation	m
	$N_2 + H_2$	Optical emission	$A^3\Pi \rightarrow X^3\Sigma^-$; $c^1\Pi \rightarrow a^1\Delta$; possibly $c^1\Pi \rightarrow b^1\Sigma^+$	n
	N_2H_4	Infrared absorption and mass spectra of gaseous species	N_2H_2 formation	o
Electron impact	NH_3	Optical emission	$A^3\Pi \rightarrow X^3\Sigma^-$ lifetime	16
	NH_3	Optical emission	$A^3\Pi \rightarrow X^3\Sigma^-$ and $c^1\Pi \rightarrow a^1\Delta$ lifetimes	17

a. H. Guenebat, G. Pannetier, and P. Goudmand, *Compt. Rend.*, **251**, 1166, 1480 (1960).

b. H. Guenebat, G. Pannetier, and P. Goudmand, *Bull. Soc. Chim.*, **1962**, 80.

c. R. Diesen, *J. Chem. Phys.*, **39**, 2121 (1963).

d. H. E. Avery, J. N. Bradley, and R. Tufnell, *Trans. Faraday Soc.*, **60**, 335 (1964).

e. H. E. Avery and J. N. Bradley, *Trans. Faraday Soc.*, **60**, 857 (1964).

f. E. D. Becker, G. C. Pimentel, and M. Van Thiel, *J. Chem. Phys.*, **26**, 145 (1957).

g. O. Schnepp and K. Dressler, *J. Chem. Phys.*, **32**, 1682 (1960).

h. J. R. McNesby, I. Tanaka, and H. Okabe, *J. Chem. Phys.*, **36**, 605 (1962).

i. D. E. Milligan and M. E. Jacox, *J. Chem. Phys.*, **41**, 2838 (1964).

j. G. Herzberg and D. A. Ramsay, *J. Chem. Phys.*, **20**, 347 (1952).

k. D. A. Ramsay, *J. Chem. Phys.*, **21**, 165 (1953).

l. D. Husain and R. G. W. Norrish, *Proc. Roy. Soc. (London) Ser. A*, **273**, 145 (1963).

m. E. J. Blau, B. F. Hochheimer, and H. J. Unger, *J. Chem. Phys.*, **34**, 1060 (1961).

n. G. G. Mannella, *J. Chem. Phys.*, **37**, 678 (1962).

o. E. J. Blau and B. F. Hochheimer, *J. Chem. Phys.*, **41**, 1174 (1964).

diimide. The presence of the volatile yellow diimide was first verified in the mass spectral analysis of products from an electric discharge in HN_3 by Foner and Hudson (26). The subject of diimide (or diimine) was reviewed by Hünig, Müller, and Thier (27). At almost the same time, Rosengren and Pimentel (28) presented the latest (and nearly definitive) demonstration that N_2H_2 exists in solid matrices, in both *cis* and *trans* forms. The energy of the N—N bond of diimide, presumably an average of the *cis* and *trans* values, is 104 ± 6 kcal or 4.5 eV, according to the analysis of Foner and Hudson.

The ultimate thermodynamic fate of NH in a system containing only N and H must in general be the net process

$$2NH \longrightarrow N_2 + H_2 + 158 \text{ kcal/mole} \quad \text{(ref. 29)} \quad \text{(or 153 kcal/mole)} \quad \text{(ref. 30)}$$

However, the lifetime of free NH seems to be relatively long, over 30 μsec in a gas of partially photolyzed NH_3 ($< 10\%$ in a partial ammonia pressure of up to ca. 6 torr) with added argon at pressures up to 400 torr (30). The lifetime of NH is reported to be as long as a millisecond in gas discharges, according to Rice and co-workers (31,32).

One reaction product of NH is the famous blue compound with composition $(NH)_x$ (29,31–33). The structure and formula of this compound are still not known. It is only recovered by cryogenic trapping of products from pyrolytic, discharge, or photolytic generators of NH, and reverts to ammonium azide on warming. Ammonium azide is a common product, often representing a large part of the yield from NH preparations.

The NH species reacts in the gas with at least one of its precursors, HN_3. This gives $H_2 + 2N_2$ according to the traditional scheme proposed first by Beckman and Dickinson (34,35) but may also give $NH_2 + N_3$, insofar as Thrush (23) was able to detect N_3 soon after flash photolysis of HN_3. Mui and Back (19) infer from their product analyses that the analogous reaction occurs between NH and HNCO, giving $NH_2 + NCO$, although it is possible that NCO could also be a primary photolysis product (36). Reactions with other precursors, e.g., NH_3, have been suggested but seem to be slow, at best; Stuhl and Welge infer that the reaction proposed by Willey (37), of $NH + NH_3$ to give excited N_2H_4, must require at least 10^4 collisions.

It has been suggested (30), as one alternative to explain the remarkable ease with which NH appears as triplet instead of singlet from photolysis of NH_3, that the reaction

$$NH(^1\Delta, {}^1\Sigma^+) + NH_3 \longrightarrow NH(^3\Sigma^-) + NH_3{}^*$$

is very rapid, requiring less than 10 collisions. The alternative mechanism would appear to be a direct violation of the Wigner spin selection rules, namely,

$$NH_3 + h\nu \longrightarrow NH(^3\Sigma^-) + H_2$$

This is not altogether implausible if an intermediate Rydberg state is involved in the initial absorption process. That is, if the process is actually

$$NH_3 + h\nu \longrightarrow NH_3^* \quad \text{(singlet Rydberg state)}$$

then the excited state may live long enough to suffer a few collisions. Rydberg electrons are only weakly coupled to the core electrons and can be expected to undergo spin flips very readily by comparison with ordinary valence electrons. Then, once the spin is flipped, the excited NH_3 (triplet) can dissociate directly to $NH(^3\Sigma^-) + H_2$.

With oxygen, in a matrix of solid N_2, NH (from HN_3 photolysis) reacts with molecular oxygen to give *cis* and *trans* HONO (38). In matrices of CO_2 at 20 or 53°K, HN_3 and DN_3 decompose under steady photolysis to give NH and DH which react with the host to give both HNO (or DNO) and a binary $NH–CO_2$ compound of undetermined structure (39).

Probably the most interesting reactions of NH, and the most pertinent to the present context, are those with simple hydrocarbons. The reaction of NH with ethylene in solid argon was studied by Jacox and Milligan (40) by following the infrared spectra of NH_3 photolysis products. The sole product they found was ethyleneimine,

$$\begin{array}{c} H_2C\text{———}CH_2 \\ \diagdown \quad \diagup \\ N \\ H \end{array}$$

They also studied the reaction of NH with acetylene and did indeed find a product, but were unable to identify it.

The reactions of NH with hydrocarbons in the gas phase has been studied by Miller (41) and by Cornell, Berry, and Lwowski (42). Miller surveyed a number of hydrocarbons: CH_4, C_2H_4, $(CH_3)_4C$ and, in less detail, C_3H_8, C_4H_{10}, and $(CH_3)_3CH$. His experiments, involving steady photolyses of HN_3 + hydrocarbon, were analyzed by studying products and mass balances. The principal products, other than NH_4N_3, included amines and amine salts (azides) (from alkanes) HCN (from C_2H_4), traces of saturated hydrocarbons and, from C_2H_4, a trace of CH_3CN. Ethylene proved decidedly more susceptible to attack by NH than methane, but less so than $(CH_3)_4C$. No ethyleneimine could be found in the reaction of $NH + C_2H_4$.

Cornell, Berry, and Lwowski (42) examined gas-phase reactions of NH and in some cases, DH, with several hydrocarbons—ethylene, ethane, methane, butene-1, heptene-3, and 2,3-dimethylbutene-2—in order to study the reaction of NH with simple olefins. Both steady and flash photolysis of HN_3 were used; in the flash experiments, the ultraviolet absorption spectrum showed that reaction occurred as the triplet NH disappeared, and no singlet NH was observed; it was therefore inferred that the reactions were all those of triplet NH, presumably in its ground state. Adiabatic reactions gave CN, while reactions with inert quenching gases led to HCN and alkyl nitriles. The mechanism for addition of NH to $H_2C{=}CH_2$ seems to be

$$NH + C_2H_4 \longrightarrow HN{-}\overset{\cdot}{C}H_2{-}\overset{\cdot}{C}H_2 \longrightarrow :N{-}CH_2{-}CH_3$$

followed by dissociation of any two of the three bonds to the α-carbon to give HCN or CH_3CN. The products from the flash-initiated reaction of NH + C_2H_4 were the same as those from photolysis of ethyl azide, $CH_3CH_2N_3$, which supports the mechanism involving the ethyl nitrene intermediate CH_3CH_2N. It appears that ethyl nitrene exists long enough to become vibrationally equilibrated, at least ca. 10^{-6} to 10^{-7} sec, under the conditions of the experiment. The rate constant for addition of NH to C_2H_4 was estimated as 10^7 liters/mole-sec.

No ethyleneimine could be found in the products obtained by Cornell, Berry, and Lwowski (42). The gas-phase reactions observed by these authors and by Miller (41) are clearly very different from that found by Jacox and Milligan (40). It remains to be determined whether the solid- and gas-phase reactions involve different electronic states, meaning that the solid reaction would be that of the singlet, or whether the solid reaction is that of the triplet but the matrix is such a good quencher that a second C—N bond forms before the hydrogen atom can migrate from N to C_β.

D. Alkylnitrenes

Data on the electronic states and electronic spectra of alkylnitrenes are almost nonexistent. No electronic spectra have been reported for these species. Wasserman, Smolinsky, and Yager (43) have reported electron spin resonance (ESR) spectra of triplet states of n-propyl-, 2-octyl-, cyclopentyl-, cyclohexyl-, t-butyl-, and α-carbethoxybenzylnitrenes, all taken at $4°K$. The existence of these ESR spectra demonstrates that in an alkylnitrene, the ground state or a state within a few cm^{-1} of the ground state is a triplet, and is strongly suggestive that the ground states of alkyl-nitrenes *are* triplets, in close analogy to NH. Chemical evidence for alkyl-nitrenes in the gas phase comes from the observations of rearrangement products in the materials recovered after pyrolyses of alkyl azides (44). (See Chapter 3.) The principal rearrangement reaction seems to be the production of imines by a 1,2-hydride shift. The second in importance (first when the α-carbon is tertiary) is the corresponding 1,2-alkyl shift.

Let us examine the alkylnitrenes by using the analogy between HN and RN in much the same way Mulliken used the HX–RX analogy to relate properties of hydrogen and alkyl halides (45,46). We have seen that the lowest state of NH is a triplet, $^3\Sigma^-$, with a fully occupied N—H σ-bonding orbital and a half-filled π-orbital largely on nitrogen. We also noted that the energy of this π orbital is rather close to that of the bonding σ orbital, and even falls below that of the σ when the N—H distance is large. The implication of the ESR experiments on alkylnitrenes is that, most probably, the ground state of RN has the electronic configuration $\ldots\sigma^2\pi^2$ and, in the threefold-symmetric prototype CH_3N, has the spin-and-symmetry designation $^3A_2$ (Chapter 13). This corresponds directly to the $^3\Sigma^-$ ground state designation of the axially symmetric NH.

The excited states of alkylnitrenes presumably correspond also to excited states of NH. The $^1\Delta$ of NH would become ^1E in CH_3N, as would the $^1\Pi$; $^3\Pi$, correspondingly would become ^3E. The $^1\Sigma^+$ states of NH correspond to totally symmetric $^1A_1$ states of CH_3N. The optical selection rules are the same for alkylnitrenes as for NH, at least to a good first approximation. The singlet–triplet transitions are quite forbidden; the triplet–triplet, $^3E \rightarrow {}^3A_2$ transition is allowed and polarized perpendicular to the C—N axis as is the transition from the higher-energy $^1\Sigma^+ \rightarrow {}^1\Delta$ or $^1A_1 \rightarrow {}^1E$. The $^1E(^1\Pi) \rightarrow {}^1E(^1\Delta)$ is allowed in both polarizations and the $^1A(^1\Sigma^+) \rightarrow {}^1E(^1\Delta)$ from the lower $^1A_1(^1\Sigma^+)$ is forbidden in first approximation because both states have the same electronic configuration.

From the general character of the orbitals and the electron-releasing nature of alkyl groups relative to hydrogen, we can make some qualitative statements about what can be expected in the electronic spectra of alkyl-nitrenes.

The most prominent transition, which we can call the $A \rightarrow X$ by analogy with the $A^3\Pi \rightarrow X^3\Sigma^-$ transition of NH, is presumably shifted about 2000 cm^{-1} toward longer wavelengths by substituting CH_3 for H. This inference is not based on a theoretical estimate of the effect of level shifting but on the comparison of the corresponding transitions in OH and CH_3O. Style and Ward (47) have attributed a series of bands in the region 3200–4200 Å to CH_3O, and if this assignment is correct, the transition is virtually certain to correspond to the $\sigma \rightarrow \pi$ occurring in the spectrum of OH in about the same wavelength region (1). The 0–0 transition of OH is approximately at 32,400 cm^{-1}, while that of CH_3O is tentatively assigned by Style and Ward (47) at 30,465 cm^{-1}. The differences between the oxygen compounds and their nitrogen homologs are (a) the additional electron in the oxygen compounds and (b) the greater nuclear charge and electronegativity and concomitant smaller size of oxygen.

The similarity of the $A \rightarrow X$ ($\sigma \rightarrow \pi$) transitions of NH and OH indicate that the *net* effects of the differences between NH and OH are small; the 0–0 separation between A and X states is only about 3000 cm^{-1} greater in OH than in NH. Furthermore, the $A \leftrightarrow X$ transition of OH$^+$, iso-electronic with NH, is also in the same energy range (1), ca. 28,937 cm^{-1}, reinforcing the inferences that the most important aspect of the $A \rightarrow X$ transitions in these species is the fact that they all involve a transition of an electron from the partly filled π-orbital in the upper electronic state down to the σ-bonding orbital, leaving it doubly occupied in the lower state.

Presumably the spectral bands of the singlet manifold, the $c \rightarrow a$ (at about 30,500 cm^{-1} in NH) and $c \rightarrow b$ (at about 22,000 cm^{-1} in NH) will show a red shift with alkylation much like the triplet $A \rightarrow X$ is expected to do. This qualitative inference follows because precisely the same orbitals are involved for these transitions in the singlet and triplet systems. Even the $d \rightarrow c$ transition will show about the same red shift for RN compared with NH, because it also is basically a one-electron transition from an upper π-orbital to the bonding σ-orbital at lower energy.

The intensities, in alkylnitrenes, of all the analogs of the observed NH transitions should be higher than in NH. This is a result of the greater localization of π-charge on the nitrogen in the alkylnitrenes; this leads to greater movement of charge in σ–π optical transitions in RN than in NH and in larger transition dipoles. Despite their intensity, the absorption spectra of alkylnitrenes unfortunately may be largely continuous and therefore hard to identify unambiguously. One cause for continuous rather than banded absorption is the possible situation that the optically accessible excited states of RN are not bound. In NH, the states with configuration $\ldots 3\sigma\, 1\pi^3$ are bound, indicating that the bonding force of the one σ electron, plus what little bonding force the π electrons contribute, is enough to compensate for the antibonding π charge. When an alkyl group replaces the hydrogen atom, electron charge density in the nitrogen π orbitals is increased, and, very likely, the center of charge of nitrogen π orbitals moves in the direction of the antibonding region "beyond" the nitrogen nucleus. If this does occur, than it is fairly likely that the states with one σ and three π-electrons will not be bound.

There is only one way to reconcile the ideas that (*1*) the optical transitions would be in about the same regions for NH and alkynitrenes and (*2*) the alkylnitrenes' $\ldots\sigma\pi^3$ states dissociate spontaneously or at least are less stable than the corresponding states of NH. This is to suppose that the potential curves for alkylnitrenes, or more specifically the curves for the C—N stretching motion, are all pushed up by about the same amount. If the $\ldots\sigma\pi^3\; {}^3A$ state, the lowest $\sigma\pi^3$ state, were to have no bound vibrational levels, then the potential curves at their minima would all have to be shifted upward by an amount equal to the dissociation energy of the A

state. In NH, the dissociation energy of this state is approximately 2.8 eV; it is surely less for CH_3N, but will still be about 2 eV. This is turn implies that the dissociation energy of the C—N bond of an alkylnitrene will be of order 1.5 eV, rather than 3.5 eV as it is in NH.

II. ARYLNITRENES

The state of knowledge of optical properties and electronic states of arylnitrenes is far more advanced than it is for alkylnitrenes, largely due to the electron paramagnetic resonance work of the Bell Laboratories group (48–50); (see Chapter 13) and to the optical spectroscopy of azide photolysis products, carried out by Reiser and co-workers (51–55).

The paramagnetic resonance spectra show that the arylnitrenes have triplet ground states or a triplet lying within a few cm^{-1} of the ground state, just as in the case of the alkylnitrenes. Moreover the dinitrene has a corresponding *quintet* state. The most reasonable presumption is that the ground states of arylnitrenes are in general the states of highest spin multiplicity. The electronic configurations are fairly closely analogous to the alkylnitrenes and NH except that one of the nitrogen $2p\pi$ orbitals can mix somewhat with the ring π orbitals. In NH, there are, for all intents and purposes, no $p\pi$ orbitals available on the hydrogen, while in CH_3N, the two $H_3 \equiv C$ quasi-π orbitals can mix with the nitrogen π's.

Theoretical consideration of arylnitrenes has consisted only of the qualitative interpretation used by Reiser, Bowes, and Horne (53) to assign the spectra they observed. The electronic situation in phenylnitrene, for example, is as follows. The twofold degeneracies of the e_{1g} and e_{2u} orbitals of the benzene ring are split both by the replacement of a hydrogen with a foreign substituent and by direct interaction with one of the nitrogen π orbitals. The orbital structure is qualitatively as shown in Figure 5; the orbital symmetry labels are appropriate for phenylnitrene, with C_{2v} symmetry, with axes and symmetry characters as shown in Figure 6. The splitting of the $3b_2$ and $1b_1$ orbitals, whose parents are the $2p\pi$ orbitals of nitrogen, must be smaller than the exchange interaction between these orbitals in order that the ground state (or a very low-lying state) of C_6H_5N be a triplet.

The low-lying allowed optical transitions are (in the one-electron model):

(1) $1b_1 \rightarrow 2a_2$, x-polarized, $n \rightarrow \pi^*$-like, with charge transfer character
(2) $3b_2 \rightarrow 2a_2$, y-polarized, $\pi \rightarrow \pi^*$ with charge transfer character
(3) $1a_2 \rightarrow 1b_1$, x-polarized, $\pi \rightarrow n$-like with reverse $(C \rightarrow N)$ charge transfer character
(4) $1a_2 \rightarrow 3b_2$, y-polarized, $\pi \rightarrow \pi^*$ with reverse charge transfer character
(5) $1a_2 \rightarrow 2a_2$, z-polarized, $\pi \rightarrow \pi^*$-like

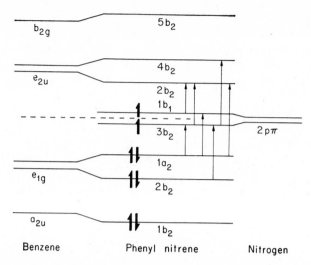

Fig. 5. Schematic diagram of the aromatic and nitrogen $2p$ orbitals in phenyl nitrene.

(6) $3b_2 \rightarrow 4b_2$, z-polarized, $\pi \rightarrow \pi^*$-like, with charge transfer character

(7) $2b_2 \rightarrow 3b_2$, z-polarized, $\pi \rightarrow \pi^*$-like, with reverse charge transfer character

The excited states arising from these seven transitions are not all of different symmetry: in fact, those transitions having the same polarization are associated with excited states of the same symmetry. Excited states from transitions (1) and (3) must mix, as will those from (2) and (4), and from (5), (6), and (7). The normal phenyl $\pi \rightarrow \pi^*$ transition (5) presumably

C_{2v}	E	C_2	$\sigma_v(yz)$	$\sigma_v(xz)$	
A_1	I	I	I	I	z
A_2	I	I	-I	-I	
B_1	I	-I	I	-I	y
B	I	-I	-I	I	x

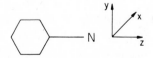

Fig. 6. Group characters and coordinates for phenyl nitrene.

mixes less with (6) and (7) than they mix with each other, and we shall neglect any mixing of (5) to simplify our discussion.

For convenience, let us now use the symbol [j] to represent the upper molecular state in the one-electron transition of type (j). Thus the mixing of excited one-electron states is described in terms of wavefunctions of the type [j] ± [k].

In studying the optical properties of the benzyl radical, several authors have examined transitions of the types (5), (6), and (7) and of (2) and (4) as well (56–58). The mixing of [6] with [7] and of [2] with [4] generates two new pairs of states. The analyses indicate that for both pairs, [6] ± [7] and [2] ± [4], the lower-energy member of each pair is responsible for by far the weaker optical transition. On this basis, the observed absorption (59) and emission spectra (60) at ca. 31,600 cm^{-1}, respectively, assigned to the benzyl radical have been interpreted as transitions between the ground state and the strong upper and weak lower components of the y-polarized [2] ± [4] transitions. The weaker $\pi \rightarrow \pi^*$ transition is presumably obscured or too weak to be seen in absorption and is not populated enough to be observed in emission.

The arylnitrene spectra reported by Reiser and co-workers contain three main regions of absorption. Typically, these have maxima in the 3700–3900 Å, 3100–3700 Å, and 2200–2600 Å regions, with maximum extinction coefficients in increasing order. Table IV, taken from reference 53, summarizes the results for the seven reported mononitrenes; Figure 6, based also on reference 53, shows a number of the spectra together with the parent azides from which they were prepared. Reiser and co-workers have interpreted these absorption spectra in terms of the same states that were invoked for the benzyl radical. The lowest-energy and weakest transitions were attributed to the weaker of the y-polarized transitions, the intermediate transition was assigned to the stronger y-polarized transition and the strongest and highest-energy transition was in essence assigned as the $\pi \rightarrow \pi^*$. This set of assignments is plausible not only on the basis of comparison with the theoretical calculations for benzyl but also with the frequency and intensity dependence found for the isoconjugate series of radicals benzyl, anilino and phenoxy. It appears that phenylnitrene may be much like the C_6H_5NH (anilino) radical in its optical properties.

One point of caution about the interpretation and some future experimental considerations are in order here. First, we must ask where the transitions fall that involve the [1] ± [3], x-polarized transitions, the mixtures of $\pi \rightarrow n$ and $n \rightarrow \pi^*$ transitions. These transitions are probably somewhat less intense than their y- and z-polarized counterparts, but it is fairly likely that the stronger x-polarized transition carries as much intensity as the weaker y-polarized transition. We are therefore tempted to

TABLE IV

Absorption Spectra of Aryl Mononitrenes: Wavelengths, λ_m, and Extinction Coefficients, ϵ_m, at Band Maxima [a] and Wavelengths λ_{00}, Assigned to the 0–0 Transitions

	[5] (z-polarized)		[2] − [4] (y-polarized)			[2] + [4] (y-polarized)		
	λ_m, Å	log ϵ_m	λ_m, Å	λ_{00}, Å	log ϵ_m	λ_m, Å	λ_{00}, Å	log ϵ_m
Nitreno benzene	2410	3.7	3030	3140	2.8	3680	4020	2.5
Nitreno mesitylene	2550	3.7		3200	3.4	3910	4250	2.4
3-Nitreno biphenyl	2420	4.8		3300	3.5	3930	4080	3.1
2-Nitreno biphenyl				3430	3.7	3730	3990	3.0
1-Nitreno naphthalene	2230	4.5	3210	3670	4.0		5360	3.2
3-Nitreno quinoline			3900	4270	3.4	5500		2.3
1-Nitreno anthracene	2400	5.3	4480	5280	4.1		6970	3.2
	3400	4.7						

point out the possibility that the lowest-energy transitions in Table IV and Figure 6 are due in part or entirely to the more intense x-polarized transitions involving the nitrogen "in-plane" $2p$ electron.

An analogous question was raised in connection with the absorption spectra of the diphenylcarbenes by Trozzolo and Gibbons. Although they recognized that one should expect to find at least one x-polarized transition, Trozzolo and Gibbons (61) felt that it was preferable to assign all the observed transitions as $\pi \rightarrow \pi^*$, analogous to the way Reiser and his co-workers assigned the nitrene transitions. This assignment was chosen because of the similarity between the spectra of the carbenes and the corresponding diphenylmethyl radicals, in which only $\pi \rightarrow \pi^*$ transitions are expected.

Asking about the location of the x-polarized transitions suggests some experimental possibilities. First, it would be very desirable to determine the polarizations of the various transitions, either by preparing the nitrenes as guests in single crystals of a suitable host, or by studying vapor spectra under high resolution. Trozzolo (62) has made the same point in connection with the carbenes. Second, it would be very desirable to study the fluorescence spectra of nitrenes, if conditions can be found under which the

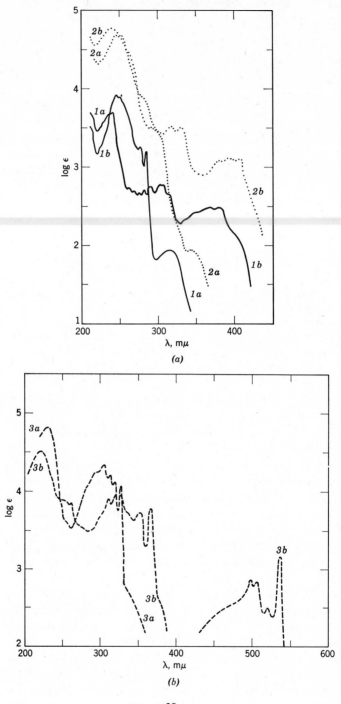

(a)

(b)

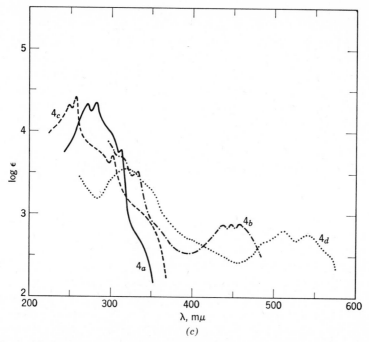

Fig. 7. Absorption spectra of several nitrenes and parent azides, all taken at 77°K. [Reproduced from Ref. 51, with permission of the Faraday Society]. 1a, phenyl azide; 1b, phenyl nitrene; 2a, 3-azidobiphenyl; 2b, 3-nitrenobiphenyl; 3a, 1-azido-naphthalene; 3b, 1-nitrenonaphthalene; 4a, 1,4-diazidobenzene; 4b, the corresponding mono-azide, mono-nitrene; 4c, the dinitrene, and 4d, the corresponding dinitrene cation.

nitrenes emit. Third, it is interesting to speculate about the energy of the lowest-energy transition, which is $1b_1 \rightarrow 2a_2$ in terms of the one-electron picture but is really a $[1] \pm [3]$ type. This transition might lie at an energy somewhat lower than those of the y- and z-polarized transitions, and might even be in the visible or near infrared region.

Having set the context for new experiments, let us review now the successful spectroscopic experiments that have been done. The principal problem, unambiguous preparation of the nitrenes, has been solved both by steady photolysis and flash photolysis of azides. Steady photolysis with a mercury arc of the corresponding azides in EPA (ethyl ether–pentane–ethanol) glasses at 77°K led to production of seven different mononitrenes: nitreno benzene, nitreno mesitylene, 3-nitreno biphenyl, 2-nitreno biphenyl, 1-nitreno naphthalene, 3-nitreno quinoline, and 1-nitreno anthracene. Examples of the spectra of aryl nitrenes and their precursors are shown in Figure 7. The spectra were identified on the bases that: The observed spectra were clearly distinguishable from the starting azides, that all the bands assigned to nitrenes appeared at the same rate; that the spectra exhibited isosbestic points with the parent azide spectra; that the spectra were essentially unaffected by solvent composition; that the absorbing species were formed in high quantum yield under the same conditions used by Smolinsky, Wasserman, and Yager to generate the EPR spectra assigned to nitrenes (see Chapter 13); that the intermediates disappeared at 90°K, when the glasses softened; that the spectra are similar to aryl carbene and aryl methyl spectra, and that in the case of 2-nitreno biphenyl, carbazole was identified as a product.

Only 1-azidoanthracene was studied by flash photolysis; the experiments were carried out at room temperature, in ethanol (5×10^{-4} M soln). The lifetime of the absorption at 3420 Å indicated that the half-life of the 1-nitrenoanthracene is between 3 and 10 sec under the conditions used by Reiser, Terry, and Willets (54).

Direct irradiation of matrix-isolated diazides at 77°K gives dinitrenes, in two distinguishable steps; further irradiation produced still other photo-products which were assigned as dinitrene cations. Reiser, Wagner, Marley, and Bowes (53) found dinitrenes this way starting with six diazides: $N_3C_6H_4N_3$, $(N_3C_6H_4)_2$, $N_3C_6H_4CH=CHC_6H_4N_3$, $N_3C_6H_4(CH=CH)$-$C_6H_4N_3$, $N_3C_6H_4N=NC_6H_4N_3$, and $(N_3C_6H_4CH=N)_2$. The compound $(N_3C_6H_4)_2CH_2$ also gave a dinitrene but whether there was an intermediate or not could not be determined because the spectrum of the intermediate in this case is expected to be approximately that of one azide and one nitrene absorber. As functions of irradiation time, the precursors decreased mono-tonically and apparently exponentially, the intermediate mononitrene absorption increased at a decreasing rate to a maximum in a time of order 5 min and then decayed, while the dinitrene rose, first slowly, then more rapidly, and finally at a decreasing rate to approach a maximum. The ratio of quantum efficiencies for production of dinitrene from mononitrene increased with separation of the azide groups, from 2.2 ± 0.7 for $N_3C_6H_4N_3$ to 3.0 ± 1.0 for $N_3C_6H_4(CH=CH)_3C_6H_4N_3$.

Dinitrene electronic spectra (53) are similar to diazide spectra; this is attributed to the closed-shell structure of both species. Similarly, the mononitrenes and the dinitrene cations have similar spectra because of their closely related open-shell configurations. This is readily seen by considering two species, each with orbital structure like that of Figure 5, and allowing the two to couple. The degree of coupling between orbitals of the two "monovalent" nitrogens of dinitrene is a function of molecular structure and of whether the orbitals are π- or n-like. In the *para* compound p-N—C_6H_4—N, the ground state is a triplet (48), indicating that one electron of each nitrogen remains unpaired. In terms of an orbital structure, we say that (presumably) the nitrogen π-orbitals couple strongly enough through the aromatic system to give one low-energy C—N π-bonding orbital which is doubly occupied, and one high-energy C—N antibonding orbital that remains unoccupied. The localized in-plane orbitals of the nitrogen atoms couple weakly in the para dinitrene. Hence, whether we speak of two localized n-orbitals or two delocalized n-orbitals, both remain singly occupied, and consequently the molecular ground state is a triplet. The meta isomer of NC_6H_4N stands in sharp contrast. Its gound state is a quintet (50), implying that both n and π orbitals are weakly coupled. Chemically this is not surprising, insofar as the para compound can take on a quinoid structure while the meta compound cannot.

The dinitrenes investigated by Reiser, Wagner, Marley, and Bowes are all para. In all cases except $(NC_6H_4)_2CH_2$, the ground states are presumably triplets, in analogy with the first member of the list. The exceptional compound appears to have the π–π coupling interrupted by the methylene insulator between the two phenyl rings. This indicates that $(NC_6H_4)_2CH_2$ probably has a quintet ground state and behaves like it contains two almost noninteracting nitrenes. Its spectrum certainly resembles that of a nitrene much more than it does the other aromatic dinitrenes, insofar as it has a strong bond in the 450–500 mμ region and no strong bond in the 330–420 mμ region. Furthermore it is the only compound in the series we listed for which the quinoid structure cannot be drawn.

III. CARBONYLNITRENES

Substantial chemical evidence indicates that pyrolytic or photolytic decomposition of azidoformates generates nitrene intermediates. This subject is discussed extensively in Chapter 6, particular in reference to the nitrenes in solution. Ethyl azidoformate vapor, flash photolyzed, appears to give the nitrene $C_2H_5O\overset{\displaystyle O}{\overset{\|}{C}}N$, insofar as the appearance of the chemical trapping reaction with cyclohexene gives the product shown (63):

$$C_2H_5O\overset{O}{\overset{\|}{C}}N_3 + h\nu \longrightarrow C_2H_5O\overset{O}{\overset{\|}{C}}-N + N_2$$

$$C_2H_5O\overset{O}{\overset{\|}{C}}-N + \text{(cyclohexene)} \longrightarrow C_2H_5O\overset{O}{\overset{\|}{C}}-N\text{(cyclohexane ring)}$$

From the trapping experiments, a crude range of $3 \times 10^{-7} < \tau < 10^{-5}$ sec was determined for the lifetime τ of the nitrene. However, further or

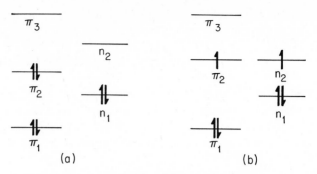

Fig. 8. Alternative orbital diagrams corresponding to

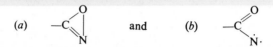

an alternative decomposition must be available for azidoformates; the principal spectroscopically observable transient is the NCO radical. In fact, no spectrum was found that could be attributed to the nitrene.

No quantum-mechanical calculations on the electronic structure of carbonyl nitrenes have been reported. We can nevertheless construct an orbital energy level diagram and argue qualitatively to ascertain what possible situations may actually exist. The five essential levels are three π-orbitals and two n orbitals; in terms of their atomic parentage, the π's come from oxygen, carbon, and nitrogen and the n's, from oxygen and nitrogen. The orbitals are shown in Figure 8. The nature of the states of

$$RO\overset{\overset{\textstyle O}{\|}}{C}-N$$

depends very sensitively on the composition and charge distribution of the five orbitals. In the framework of a simple nitrene model, one would expect π_1 to be the carbonyl-bonding π orbital, n_1 to be the oxygen lone pair orbital, π_3 to be the antibonding carbonyl π-orbital and π_2 and n_2 to be the π and n orbitals of the nitrogen. The ethyl compound has a triplet ground state, implying that π_2 and n_2 are close in energy, as in Figure 8b. If the splitting were large compared with the π_2–n_2 exchange energy, both electrons would go into the orbital of lower energy, to give the system a singlet ground state, as indicated in Figure 8a. The actual triplet nature of the ground state implies that the π_2 and n_2 orbitals must be physically similar, which in turn indicates that the interaction between carbonyl oxygen and the monovalent nitrogen is weak.

The mechanism for photolytic production of carbethoxynitrene is not understood. The Wigner spin conservation rule would imply that the nitrene should be formed as a singlet, under the conditions of experiments

$$\overset{\text{O}}{\underset{\|}{}}$$

so far reported. The ground state of $C_2H_5O\overset{\overset{\text{O}}{\|}}{C}$—N is a triplet, as shown by the stable low-temperature electron paramagnetic resonance it exhibits (64). Photolysis of ethyl azidoformate in solution indicates that both singlet and triplet nitrenes are formed (65). Moreover, the evidence is consistent with *both* singlet and triplet being primary photolysis products, in a ratio of about 2:1 at 38°C and about 3:1 at 12°C. Specifically, the experiments, based on the ratio of stereospecific to nonstereospecific addition of nitrene to the olefin *cis*-4-methylpent-2-ene, showed that the ratio of singlet to triplet nitrene is determined more rapidly than the reaction of either species with the olefin substrate. Interestingly, the thermal decomposition of ethyl azidoformate gives 100% singlet in the primary step. In either thermolysis or photolysis, the singlet may react with substrate or relax to the lower triplet state. The ratio of rate constants $k_{s \to t}/k_{react}$ is about 10^{-2} mole % with a substrate of 4 methylpentene-2, and decreases with temperature. The experiments of McConaghy and Lwowski spanned the olefin concentration range for which the ratio of reaction rates varied between about 0.1 and 2.5. An estimate of an upper limit for $k_{s \to t}$ can be obtained by assuming that the bimolecular nitrene–olefin reaction is diffusion controlled, so as to have a rate constant of about 10^9 liters/mole-sec. Under the conditions of the experiment, this corresponds to $k_{s \to t}/k_{react} \cong 10^{-2}$ mole % or 8×10^{-2} moles/liter. This implies that $k_{s \to t}$ has an *upper limit* of about 8×10^5 sec^{-1}, much slower than the vibrational relaxation time for the excited singlet. This is consistent with the rate expected for intersystem crossing in a moderately small molecule (66).

We may conjecture that the singlet–triplet branching ratio is determined by the vibrational relaxation or after relaxation has occurred. A more specific possibility follows from one consideration of the electronic energy levels. We have seen that the π_2 and n_2 orbitals must be quite similar; that is, they must be quite similar when the molecule is in or near the equilibrium geometry of the triplet ground state. It must be in a geometry similar to this when it undergoes intersystem crossing from the excited singlet. However, this crossing occurs when the molecule is vibrationally relaxed, so that the equilibrium configuration of the triplet must be similar to that of a minimum, at least a local minimum, of the singlet potential surface. Yet about $\frac{2}{3}$ of the vibrationally relaxed singlets remain as such. This implies that there may be a second local minimum on the potential surface somewhat removed from that of the triplet, and that nitrene molecules produced by photolysis can find their way into one or the other of the singlet minima as they relax. Thus, according to our model, one minimum serves to trap singlets while the other offers a channel into the triplet. The singlet–triplet branching ratio is probably not thermodynamically determined; this follows from the fact that it is 2:1 at $311°K$ and 3:1 at $285°K$, inconsistent with the Boltzmann distribution function unless the experimental uncertainty is very large. We can push the model one step further by asking what physical representation one could find for the singlet trap minimum. One possibility suggests itself, namely the $\text{ROC}\begin{smallmatrix} \diagup O \\ \Big| \\ \diagdown N \end{smallmatrix}$ structure or something near it, which would be an appropriate description if the n orbitals of nitrogen and oxygen were to mix.

REFERENCES

1. G. Herzberg, *Molecular Spectra and Molecular Structure. I. Spectra of Diatomic Molecules*, D. Van Nostrand, New York, 1955.
2. G. Herzberg, *Molecular Spectra and Molecular Structure. III. Electronic Spectra and Electronic Structure of Polyatomic Molecules*, D. Van Nostrand, New York, 1966, esp. Chap. III.
3. P. E. Cade and W. Huo, *J. Chem. Phys.*, **47**, 614 (1967).
4. B. J. Ransil, *Rev. Mod. Phys.*, **32**, 235, 239, 245 (1960).
5. E. M. Layton and K. Ruedenberg, *J. Phys. Chem.*, **68**, 1654 (1964).
6. A. C. Hurley, *Proc. Roy. Soc. (London) Ser. A*, **249**, 402 (1959).
7. R. F. W. Bader, I. Keaveny, and P. E. Cade, *J. Chem. Phys.*, **47**, 3381 (1967).
8. T. Berlin, *J. Chem. Phys.*, **19**, 208 (1951).
9. W. Moffitt, *Proc. Roy. Soc. (London) Ser. A*, **202**, 548 (1950).
10. A. L. Companion and F. O. Ellison, *J. Chem. Phys.*, **32**, 1132 (1960).
11. H. Okabe and M. Lenzi, *J. Chem. Phys.*, **47**, 5241 (1967).
12. JANAF Thermochemical Tables, D. R. Stull, Ed., Dow Chemical Co. Midland, Michigan, 1966. This value is used in reference 11, but 3.5 is taken in reference 13, which would raise *a* to 2.0 eV.
13. K. H. Becker and K. H. Welge, *Z. Naturforsch.*, **18a**, 600 (1963).
14. G. Herzberg, in *Energy Transfer in Gases*, R. Stoops, Ed., Interscience, New York, 1963, p. 167.
15. M. McCarty and G. W. Robinson, *J. Am. Chem. Soc.*, **81**, 4472 (1959).
16. R. G, Bennett and F. W. Dalby, *J. Chem. Phys.*, **31**, 434 (1959).
17. E. Fink and K. H. Welge, *Z. Naturforsch.*, **19a**, 1193 (1964).
18. A. G. Gaydon, *Spectroscopy of Flames*, Chapman and Hall, London, 1957.
19. J. Y. P. Mui and R. A. Back, *Can. J. Chem.*, **41**, 826 (1963).
20. R. A. Back, *J. Chem. Phys.*, **40**, 3493 (1964).
21. R. Holland, D. W. G. Style, R. N. Dixon, and D. A. Ramsay, *Nature*, **182**, 336 (1958).
22. R. N. Dixon, *Can. J. Phys.*, **37**, 1171 (1959).
23. B. A. Thrush, *Proc. Roy. Soc. (London) Ser. A*, **235**, 143 (1956).
24. K. D. Bayes, K. H. Becker, and K. H. Welge, *Z. Naturforsch.*, **17a**, 676 (1962).
25. (a) K. H. Becker and K. H. Welge, *Z. Naturforsch.*, **19a**, 1006 (1964); (b) K. H. Welge, *J. Chem. Phys.*, **45**, 4373 (1966).

26. S. N. Foner and R. L. Hudson, *J. Chem. Phys.*, **28**, 719 (1958); *Ibid.*, **29**, 442 (1958).
27. S. Hünig, H. R. Müller, and W. Thier, *Angew. Chem.*, *Int'l. Ed.*, **4**, 271 (1965).
28. K. Rosengren and G. C. Pimentel, *J. Chem. Phys.*, **43**, 507 (1965).
29. H. E. Avery and J. N. Bradley, *Trans. Faraday Soc.*, **60**, 851 (1964).
30. F. Stuhl and K. H. Welge, *Z. Naturforsch.*, **18a**, 900 (1963).
31. F. O. Rice and M. Freamo, *J. Am. Chem. Soc.*, **73**, 5529 (1951).
32. F. O. Rice and C. Grelecki, *J. Am. Chem. Soc.*, **79**, 1880 (1957).
33. F. O. Rice and M. Freamo, *J. Am. Chem. Soc.*, **75**, 548 (1953), and reference 31.
34. A. O. Beckman and R. G. Dickinson, *J. Am. Chem. Soc.*, **50**, 1870 (1928).
35. A. O. Beckman and R. G. Dickinson, *J. Am. Chem. Soc.*, **52**, 124 (1930).
36. D. W. Cornell, R. S. Berry, and W. Lwowski, *J. Am. Chem. Soc.*, **87**, 3626 (1965).
37. E. J. B. Willey, *Trans. Faraday Soc.*, **30**, 238 (1934).
38. J. D. Baldeschwieler and G. C. Pimentel, *J. Chem. Phys.*, **33**, 1008 (1960).
39. D. E. Milligan, M. E. Jacox, S. W. Charles, and G. C. Pimentel, *J. Chem. Phys.*, **37**, 2302 (1962).
40. M. E. Jacox and D. E. Milligan, *J. Am. Chem. Soc.*, **85**, 278 (1963).
41. E. D. Miller, Ph.D. Dissertation, Catholic University of America, Catholic University of America Press, Washington, D.C., 1961.
42. D. W. Cornell, R. S. Berry, and W. Lwowski, *J. Am. Chem. Soc.*, **88**, 544 (1966).
43. E. Wasserman, G. Smolinsky, and W. A. Yager, *J. Am. Chem. Soc.*, **86**, 3166 (1964).
44. W. Pritzkow and D. Timm, *J. Prakt. Chem.*, **32**, 178 (1966).
45. R. S. Mulliken, *J. Chem. Phys.*, **8**, 382 (1940).
46. R. S. Mulliken, *Phys. Rev.*, **61**, 277 (1942).
47. D. W. G. Style and J. C. Ward, *Trans. Faraday Soc.*, **49**, 999 (1953).
48. G. Smolinsky, E. Wasserman, and W. A. Yager, *J. Am. Chem. Soc.*, **84**, 3220 (1962).
49. G. Smolinsky, L. C. Snyder, and E. Wasserman, *Rev. Mod. Phys.*, **35**, 576 (1963).
50. E. Wasserman, R. W. Murray, W. A. Yager, A. M. Trozzolo, and G. Smolinsky, *J. Am. Chem. Soc.*, **89**, 5076 (1967).
51. A. Reiser and V. Frazer, *Nature*, **208**, 682 (1965).
52. A. Reiser, H. Wagner, and G. Bowes, *Tetrahedron Letters*, **23**, 2635 (1966).
53. A. Reiser, G. Bowes, and R. J. Horne, *Trans. Faraday Soc.*, **62**, 3162 (1966).
54. A. Reiser, G. C. Terry, and F. W. Willets, *Nature*, **211**, 410 (1966).
55. A. Reiser, H. M. Wagner, R. Marley, and G. Bowes, *Trans. Faraday Soc.*, **63**, 2403 (1967).
56. W. Bingel, *Z. Naturforsch.*, **10a**, 459 (1955).
57. M. J. S. Dewar and H. C. Longuet-Higgins, *Proc. Phys. Soc.* (*London*), *A*, **67**, 795 (1954).
58. H. C. Longuet-Higgins and J. C. Pople, *Proc. Phys. Soc.* (*London*), *A*, **68**, 591 (1955).
59. G. Porter and J. Norman, *Nature*, **174**, 508 (1954).
60. H. Schüler and A. Michel, *Z. Naturforsch.*, **10a**, 459 (1955).
61. A. M. Trozzolo and W. A. Gibbons, *J. Am. Chem. Soc.*, **89**, 239 (1967).
62. A. M. Trozzolo, private communication.
63. R. S. Berry, D. W. Cornell, and W. Lwowski, *J. Am. Chem. Soc.*, **85**, 1199 (1967).
64. E. Wasserman, private communication.
65. J. S. McConaghy and W. Lwowski, *J. Am. Chem. Soc.*, **89**, 2357 (1967); **89**, 4450 (1967).
66. J. Jortner and R. S. Berry, *J. Chem. Phys.*, **48**, 2757 (1968).

CHAPTER 3

Alkylnitrenes

FREDERICK D. LEWIS and WILLIAM H. SAUNDERS, JR.
*Department of Chemistry, University of Rochester,
Rochester, New York* 14627

I. INTRODUCTION

Alkylnitrenes have long been proposed as intermediates in a variety of rearrangement reactions. As early as 1896 Stieglitz (1) proposed a univalent nitrogen species as the intermediate in the Beckmann and related rearrangements. Nitrene formation in the decomposition of an alkyl azide was first postulated to explain the rearrangement of trityl azide to benzophenone phenylimine (eq. 1) (2). This and the related rearrangements of trityl hydroxylamines (3) and N-halo amines (4) (eqs. 2 and 3), which were postulated to give a nitrene by α-elimination, are usually included in the term "Stieglitz rearrangement." The establishment of a concerted

$$(C_6H_5)_3CN_3 \xrightarrow{-N_2} (C_6H_5)_2C=NC_6H_5 \qquad (1)$$

$$(C_6H_5)_3CNHOH \xrightarrow{-H_2O} (C_6H_5)_2C=NC_6H_5 \qquad (2)$$

$$(C_6H_5)_3CNHX \xrightarrow{-HX} (C_6H_5)_2C=NC_6H_5 \qquad (3)$$

mechanism for the Beckmann rearrangement (5–7) led to a general decline in interest in nitrene intermediates. As recently as 1954 a review on azide chemistry (8) contained little more than some extensions of Stieglitz's work and a few kinetic studies of simple alkyl azide pyrolyses.

The rapid development of alkylnitrene chemistry in the past decade is to a large extent due to the technological developments which have made the studies of rapid reactions and short-lived intermediates possible. Of particular importance has been the application of electron spin resonance spectroscopy and flash spectroscopy which allow direct observation of nitrenes. These topics are of sufficient interest to warrant chapters elsewhere in this volume. While physical studies have provided a firm foundation upon which to base belief in discrete nitrene intermediates, the volume of literature in which alkylnitrenes have been postulated as intermediates has expanded rapidly. In many of these reports the existence of a discrete nitrene is far from being clearly established. It is the objective of this chapter to consider critically all reactions where alkylnitrene intermediates have been proposed and to offer alternative mechanisms where current evidence does not fully support a nitrene mechanism.

The majority of studies of alkylnitrenes have involved decomposition of alkyl azides. It is found that the reactivity of the nitrene is often dependent upon the method by which the azide is decomposed. Furthermore, other functional groups in the molecule can exert a marked effect upon the fate of the nitrene. Vinylnitrenes are included in this discussion and are of particular interest as a route to azirines and as postulated intermediates in the Neber rearrangement (eq. 4). Nitrenes and dehydronitrenes have also

$$\underset{\substack{\| \\ N-X}}{RCH_2C-R} \xrightarrow{OH^-} \underset{\substack{| \quad \| \\ NH_2 \ O}}{R-CH-C-R} \qquad (4)$$

been suggested as intermediates in the reactions of active nitrogen and nitrene ($:NH$) itself with organic molecules. These and other miscellaneous methods of preparing nitrenes usually depend upon the similarity of their products to those obtained from alkyl azides for verification.

II. NITRENE FORMATION FROM SATURATED
ALKYL AZIDES

Alkyl azides can be decomposed by several methods. The method most frequently encountered in the early literature is acid hydrolysis (8,9). Acid-catalyzed hydrolyses of azides display second-order kinetics and thus must proceed with protonation prior to or concerted with loss of nitrogen. Hence a discrete nitrene intermediate is not formed, but rather a protonated nitrene or nitrenium ion, whose reactions are discussed elsewhere in this volume. It should be noted that the acid-catalyzed decomposition products are often quite different from those obtained by thermal decomposition. Alkyl azides are thermally stable at room temperature, but at temperatures generally in excess of 100° nitrogen is evolved in a first-order homogeneous process (10). Low molecular weight alkyl azides explode violently at elevated temperatures with the evolution of nitrogen and other products observed in the thermal decomposition (11). Alkyl azides can also be decomposed photolytically either by direct photolysis or in the presence of triplet or singlet sensitizers. The photolysis is a first-order, temperature-independent process under most conditions. Alkyl azides can be photolyzed by excitation at either of the two broad absorption maxima which occur around 2160 Å ($\epsilon \simeq 500$) and 2870 Å ($\epsilon \simeq 25$). The 2870 Å band has been assigned to a perpendicular $\pi \rightarrow \pi^*$ transition on the basis of its insensitivity to solvent polarity and low intensity (12). The excited azide could lose nitrogen directly from the singlet excited state or from a triplet azide formed from intersystem crossing. Neither fluorescent nor phosphorescent emission can be detected from low-temperature glasses of alkyl azides (13,14). Hence radiative return to the ground state is not an important process. This is in accord with the high-quantum yields observed for the photolysis of alkyl azides (15,16b).

Once formed, the alkylnitrene is a highly reactive species. As it has two unshared sets of electrons, it is capable of existing as either a singlet or triplet which might be expected to show different reactivities. Electron spin resonance studies of several alkyl azides in low-temperature glasses show the triplet to be the ground state (17). An alkylnitrene would be expected to be more reactive than the isoelectronic carbene due to the higher electron affinity of nitrogen. Unlike aryl- and carbonylnitrenes, delocalization of the radical character cannot occur for alkylnitrenes. This results in a more highly reactive nitrene species. Arylnitrenes are stable in organic matrices at 77°K as shown by electron spin resonance (18) and ultraviolet spectroscopy (19). Alkylnitrenes, on the other hand, cannot be observed at 77°K (18) but are stable at 4°K (17). Phenylnitrene dimerizes to azobenzene (20) whereas formation of azo compounds is not observed with alkylnitrenes, with the exception of perhaloalkylnitrenes. Carbethoxynitrene (21) and cyanonitrene (22) add to olefins in moderate yields whereas alkylnitrene additions to olefins have not been observed. The failure of alkylnitrenes to undergo intermolecular reactions is due not only to their lack of stabilization, but also to the facility of 1,2-hydrogen or alkyl migration, processes which cannot occur for aryl-, carbonyl-, or cyanonitrenes in which no α-hydrogens are present. The propensity of alkylnitrenes for imine formation makes studies of the intermediate nitrene and assignment of a detailed mechanism very difficult. Trapping techniques developed in carbene chemistry have not been successfully applied to alkylnitrenes as they have been to carbonylnitrenes (21) and cyanonitrenes (22). Hence the assignment of a nitrene mechanism to the decomposition of alkyl azides rests principally on product studies which often do not allow differentiation between a concerted rearrangement and a discrete nitrene intermediate. The experimental basis for belief in alkylnitrenes is perhaps the most tenuous of any of the classes of nitrene intermediates discussed in this volume.

A. Methyl Azide

Methyl azide is the simplest and most extensively studied of the alkyl azides. However, its reactions are far from simple due to the instability of methyleneimine formed as the primary product from methylnitrene. Bond energies are not known for methyl azide but they can be estimated from the values for hydrazoic acid. The bond dissociation energy for $HN—N_2$ was found by electron impact studies to be 37 kcal/mole (23). An often quoted estimate for the $CH_3N—N_2$ bond is 40 kcal/mole (24). It is clear from appearance potentials and relative abundances from electron impact studies that rupture of the $CH_3N—N_2$ bond is the most favorable mode of decomposition for the methyl azide molecular ion (23). This is in accord with the results of Evans, Yoffe, and Gray which show that on the basis of correlation rules and thermochemical data hydrazoic acid should decompose to nitrene and nitrogen rather than hydrogen and azide radical (25).

The gas-phase pyrolysis of methyl azide was first studied by Ramsperger in 1929 (26). The reaction was found to be homogeneous and unimolecular with the rate falling off at low pressures. Although the products were not studied it was incorrectly postulated that hydrazoic acid and ethylene were the only major reaction products. Leermakers (10) studied the products of the thermolysis at 200–240° and found a complex array of products including ethylene, hydrazoic acid, hexamethylenetetramine, ammonia, and nitrogen. As the reaction was carried to completion it is likely that the initially formed products were degraded by further pyrolysis. Leermakers postulated the existence of two competing primary processes (eqs. 5 and 6).

$$0.25CH_3N_3 \longrightarrow 0.25CH_2\!: + 0.25HN_3 \qquad (5)$$

$$0.75CH_3N_3 \longrightarrow 0.75CH_3\overset{..}{N}\!: + 0.75N_2 \qquad (6)$$

Such large amounts of hydrazoic acid seem unreasonable in view of the large carbon–nitrogen bond dissociation energy. Subsequent workers have in fact observed only trace amounts of hydrazoic acid formation from the decomposition of methyl azide.

More recently, the thermolysis of methyl azide has been studied in a high-speed flow system at 900°C (27). The condensable products were trapped on a liquid nitrogen cold finger and analyzed by mass spectroscopy. As nitrogen formation was essentially quantitative, nitrene formation (eq. 7) is the major primary process. Hydrogen cyanide formation is observed at 900° but not at 200–240° and hence is probably due to loss of hydrogen from a vibrationally excited nitrene (eq. 8). The polymeric material

$$CH_3N_3 \xrightarrow{900°} CH_3\ddot{N}: + N_2 \qquad (7)$$

$$0.25CH_3\ddot{N}: \xrightarrow{900°} 0.25HCN + 0.25H_2 \qquad (8)$$

obtained is probably formed on warming the trapped nitrene or methyleneimine (eq. 9). Rapid gas-phase pyrolysis of methyl azide at 350–410° has

$$0.75CH_3\ddot{N}: \longrightarrow 0.3NH_3 + 0.15C_5H_9N_3 \qquad (9)$$

also been shown to give mainly polymer (28). A small amount (1%) of methylamine was also formed indicating that the nitrene intermediate existed long enough for hydrogen abstraction to occur. When the pyrolysis was carried out in the presence of a 7:1 excess of cyclohexane, a small amount (0.4%) of N-methylcyclohexylamine was obtained (eq. 10). Even

$$CH_3\ddot{N}: + \bighexagon_{S} \longrightarrow \bighexagon_{S}^{NHCH_3} \qquad (10)$$

less N-methylaniline was obtained when benzene was used. Apparently methylnitrene rearranges to the imine too rapidly to allow intermolecular insertion reactions to compete to an appreciable extent.

Methylnitrene appears to be the only major primary product in the photolysis of methyl azide in the gas phase or in low-temperature matrices. The photolysis of methyl azide and methyl azide-d_3 in an argon matrix at 4°K and in carbon dioxide at 50°K have been studied by infrared spectroscopy (29). Only methyleneimine and methyleneimine-d_3 could be detected, indicating that methyl nitrene formation was the only primary process. On heating, methyleneimine polymerizes with the evolution of some ammonia. Prolonged photolysis in the argon matrix produced $C\equiv N-H$ by loss of hydrogen from methyleneimine (30). On warming, this isomerizes to hydrogen cyanide.

$$CH_2{=}NH \xrightarrow[-H_2]{h\nu} C\equiv N{-}H \longrightarrow H{-}C\equiv N \qquad (11)$$

Hence, hydrogen cyanide formation is a secondary process in the photolysis as well as the thermolysis of methyl azide. The gas-phase photolysis of methyl azide has been extensively studied by Currie and Darwent (24). The effects of temperature, pressure, light intensity, and wavelength were studied. It was found that at low conversions (ca. 1%) the products under all conditions were nitrogen, small amounts of hydrogen and hydrogen cyanide (5–11%), traces of methane, ethane, and ethylene, and a condensate of empirical formula $(CH_3N)_x$. The condensate also contained a trace of hexamethylenetetramine. The quantum yield was found to depend upon the wavelength of irradiation, increasing from 1.7 at 2750–3130 Å to 2.3 at 2537 Å. Addition of the radical inhibitors azomethane or ethylene reduced the rate of nitrogen production by up to 45%. This is taken as evidence for a short-chain mechanism carried by the nitrene (eqs. 12–14).

$$CH_3N_3 \xrightarrow{h\nu} CH_3\ddot{N}: + N_2 \qquad (12)$$

$$CH_3\ddot{N}: + CH_3N_3 \longrightarrow (CH_3N)_2 + N_2 \qquad (13)$$

$$CH_3\ddot{N}: + (CH_3N)_2 \longrightarrow (CH_3N)_3 \qquad (14)$$

Addition of an inert gas decreased the rate of nitrogen evolution as much as 20% indicating that in addition to radical reactions, excited species (CH_3N^* or N_2^*) are involved in the reaction (eqs. 15 and 16).

$$CH_3\ddot{N}:^* + CH_3N_3 \longrightarrow 2CH_3\ddot{N}: + N_2 \qquad (15)$$

$$N_2^* + CH_3N_3 \longrightarrow CH_3\ddot{N}: + 2N_2 \qquad (16)$$

No noticeable effect on the quantum yield was observed upon changing the temperature or pressure of the azide or light intensity.

Koch (15) has studied the photolysis of methyl azide in a variety of solvents over a wide temperature range. Above $-60°$ the rate of nitrogen evolution from methanol and ethanol solutions is independent of temperature. One mole of nitrogen is formed for each mole of azide consumed. Contrary to the results obtained for pyrolysis and gas-phase photolysis, hexamethylenetetramine is the major polymeric product, constituting approximately 30% of the reacted azide. In reactive solvents such as cyclohexene or tetramethylethylene, considerable reaction with solvent occurred; however, the products were not identified. The quantum yield was found to be highly dependent upon concentration reaching a maximum near 2.0 for 0.5 M methyl azide. Quantum yields were determined using polychromatic light and a uranyl oxalate actinometer and hence are probably not very accurate. However, a quantum yield of 2.0 is in reasonable accord with the value obtained by Currie and Darwent (24) when one considers that vibrationally excited species would not be important in solution. Although these results seem in agreement with a short-chain mechanism carried by nitrene, Koch (15) argues that monomolecular decay of the excited azide to a nitrene is an insignificant process and that the results are better explained by attack of an excited azide upon a second molecule of azide or solvent. Although such a mechanism cannot be ruled out it seems rather unlikely. As previously mentioned, attempts to observe an excited azide by emission spectroscopy have been unsuccessful (13,14). Furthermore, attempts to quench an excited azide with singlet or triplet quenchers have been unsuccessful (see below). If the excited azide persisted in solution long enough to undergo bimolecular reactions with other azide molecules it certainly would be expected to undergo quenching at a diffusion controlled rate.

From these results a reasonably coherent picture of the formation and reactions of methylnitrene can be drawn. Under all reaction conditions, the primary process in the decomposition of methyl azide is loss of nitrogen and nitrene formation. The reactions of the nitrene depend to a large extent on the energy with which it is formed and its environment. In low-temperature matrices where further reaction via bimolecular collision are limited, methyleneimine formation predominates. In the gas phase a short-chain mechanism carried by the nitrene is possible. Prolonged photolysis or elevated temperatures result in loss of hydrogen and hydrogen cyanide formation. The presence of other reactive molecules can lead to small amounts of hydrogen abstraction and carbon–hydrogen insertion reactions either in the gas phase or in solution. In all cases a polymer is formed indicating the instability of the primary products. The exact nature of this polymer is unclear at this time.

B. Other Alkyl Azides: Alkyl Nitrene Rearrangements

The pyrolyses and photolyses of higher alkyl azides are in many ways analogous to those of methyl azide. One notable difference is that the imines formed by 1,2-hydrogen migration are more stable than methylene-imine and can often be isolated as reaction products. As a result polymers are not generally major products except in the case of ethyl azide. The gas-phase pyrolysis of ethyl azide was found to be a first-order homogeneous reaction by Leermakers (10). The kinetics of the two isomeric propyl azides have been determined by Geiseler and König (11). The Arrhenius parameters are given in Table I. The energy of activation is found to be approximately 40 kcal/mole for all three azides. These values are in good agreement with the results of Franklin et al. (23) for the dissociation energy of hydrazoic acid. The major product from the decomposition of ethyl azide was aziridine with lesser amounts of piperazine and higher polymers also present (eq. 17) (11). Trace amounts of hydrazoic acid, ethylamine,

$$C_2H_5N_3 \xrightarrow{\ 200^\circ\ } C_2H_5\ddot{N}: \longrightarrow \underset{NH}{H_2C\!\!-\!\!CH_2} + \underset{H_2C\diagdown_{N}\diagup CH_2}{\overset{H_2C\diagup^{N}\diagdown CH_2}{}} \quad (17)$$

TABLE I

Arrhenius Parameters for Simple Alkyl Azides (11)

	E_{act}, kcal/mole	log A	$k_{200°} \times 10^5$	$\Delta S_{200°}^{\ddagger}$, cal/mole-deg
Ethyl azide	40.1	14.5	9.58	7.0
n-Propyl	39.4	14.2	10.2	7.4
Isopropyl azide	38.5	13.9	16.6	4.0

acetonitrile, ethylene, butane, and ammonia were detected by infrared analysis. Ethylamine and acetonitrile are postulated to arise from disproportionation of ethylnitrene. This seems unlikely in view of the short lifetime of the nitrene intermediates. Acetonitrile more likely is formed by loss of hydrogen from an excited nitrene while ethylamine can result from hydrogen abstraction from any of the species present. Pritzkow and Timm (28) have also studied the vapor-phase pyrolysis of ethyl azide at 380–400°. The products were analyzed after acid hydrolysis. In addition to aziridine and polymeric imine, a small amount of N-methyl methyleneimine was formed as a result of 1,2-alkyl migration (eq. 18).

$$C_2H_5N_3 \xrightarrow{400°} C_2H_5\ddot{N}: \longrightarrow (CH_3CH{=}NH)_n + H_2C\text{—}CH_2 + H_2C{=}N\text{—}CH_3$$

$$55\% \qquad \underset{NH}{\diagdown\diagup} \qquad 4\%$$

$$35\% \qquad (18)$$

The solution phase photolysis of ethyl azide has been studied by Koch (15). The rate of nitrogen evolution was virtually independent of temperature from $-60°$ to $+40°$ over a concentration range of 0.02–1.0 M. A concentration independent quantum yield of approximately unity was recorded. Polymeric acetaldimine accounted for about 70% of the reaction products. The nature of the polymer is not known. Lesser amounts of side products were formed than in the gas-phase pyrolysis indicating a greater selectivity for the photo reaction. Quite different results were obtained by Cornell, Barry, and Lwowski (31) for the gas-phase isothermal and flash photolysis of ethyl azide. In the flash photolysis, hydrogen cyanide and acetonitrile are the major nitrogen-containing products. Apparently the initially formed nitrene is sufficiently energetic to cause rupture of one or more C_α bonds. Isothermal gas-phase photolysis led to much less fragmentation due to the lower energy of the nitrene.

Quantum yields of nitrogen evolution for a series of alkyl azides in various solvents have been determined (Table II) (16b). The quantum yield is found to be independent of solvent or azide concentration during

TABLE II
Alkyl Azide Quantum Yields (16)

Azide	Wavelength, Å	Solvent	Quantum yield
Ethyl	2537	Methanol	0.88
	3130	Methanol	0.88
	3130	Ethyl ether	0.89
n-Propyl	3130	Methanol	0.83
	3130	Heptane	0.79
n-Butyl	3130	Hexane	0.78
Isobutyl	3130	Heptane	0.79
n-Hexyl	2537	Methanol	0.86
	3130	Methanol	0.71
	3130	Ethyl ether	0.71
	3130	Heptane	0.69
Cyclohexyl	3130	Heptane	0.68

the initial stages of the photolysis. Over the first 20–30% reaction a first-order rate constant applies, but falls off gradually as light-absorbing secondary photoproducts are formed. The quantum yields for azide disappearance have not been determined, but must be at least as large as those for nitrogen evolution. High quantum yields are not surprising in view of the large amount of energy supplied by irradiation. Light of 3130 Å is equivalent to 91 kcal/mole, far in excess of the 40 kcal/mole activation energy for rupture of the nitrogen–nitrogen bond. The high-quantum yields for nitrogen evolution mean also that nitrene formation must be the major primary photoprocess. The slight decrease in quantum yield with increasing size of the alkyl group may be due to the increased possibility of vibrational relaxation of the excited azide to the ground state.

Attempts to detect triplet azides in the direct photolysis by using piperylene or oxygen as quenchers failed to lower the quantum yield (16b). Hence loss of nitrogen probably takes place from the singlet azide without intersystem crossing to the triplet azide. This is in accord with the observation that carbethoxynitrene (21) and cyanonitrene (22) are formed initially in the singlet state and relax to the triplet. Alkyl azides can be decomposed with a variety of triplet sensitizers. In this case the triplet azide must be formed. Sensitized quantum yields near unity were obtained for n-hexyl isobutyl, and cyclohexyl azides using triplet sensitizers with high triplet energies such as acetophenone (73.6 kcal/mole) and cyclopropyl phenyl ketone (74.4 kcal/mole) (16c). The sensitized quantum yields fall off as the sensitizer energy is decreased indicating that the triplet energy of the alkyl azides is approximately 75–80 kcal/mole. In view of the relatively large singlet–triplet splitting (> 19 kcal/mole) it is unlikely that intersystem crossing would be efficient enough to explain the high-quantum yields observed in the direct photolysis of alkyl azides. Thus it seems probable that direct photolysis of alkyl azides proceeds by loss of nitrogen from a singlet azide, in accord with the failure to observe triplet quenching. Alkyl azides are also efficient singlet quenchers (31a). The fluorescence intensities of several aromatic hydrocarbons are quenched by hexyl azide, making them unsuited for use as triplet sensitizers. Singlet sensitization by phenanthrene leads to decomposition with an efficiency similar to that for the direct photolysis.

In all cases of primary and secondary alkyl azide pyrolyses and photolyses the predominant product is the aldimine formed by hydrogen migration. That this is the case is now well established; however, until recently there has been much confusion on this point in the literature due to the report of pyrrolidine formation from n-butyl azide and other long-chain alkyl azides. This reaction was described by Barton and Morgan (32) in 1961 and drew much attention as it seemed a convenient route to substituted pyrrolidines and natural products such as conessine. It has been widely cited in review articles on nitrenes (33,34) as well as texts on photochemistry (35,36). Several attempts have been made to repeat the pyrrolidine synthesis, but none has been successful. Barton and Starratt (37) found that photolysis of n-butyl and n-octyl azides gave only the imines in 60–75% yield. They also cautioned that the results of Barton and Morgan (32) should be regarded with reserve.

Further studies of alkyl azides have shown that intramolecular cyclization to pyrrolidines does not occur even under what would be considered favorable conditions. Intramolecular nitrene insertions are known to occur with arylnitrenes and carbonylnitrenes. For example, o-azidocumene forms 3-methyl-2,3-dihydroindole in 15% yield upon thermolysis (38) (eq. 19). However, the analogous formation of 2,3-dihydroindole from β-phenylethyl azide does not occur (eq. 20) (39). Similarly, photolysis of

(19)

15%   37%

(20)

t-butylazidoformate gives 5,5-dimethyl-1,3-oxazolidin-2-one (**1**) in good yield (eq. 21) (40). The corresponding alkyl azide, 3,3-dimethylbutyl azide, as well as 4-methylpentyl azide, fail to give any pyrrolidine (eq. 22) (28). Similarly, no 4-phenylpyrrolidine is formed in the photolysis of 4-phenyl-1-butyl azide (13), a case which would seem particularly favorable as a benzyl hydrogen would be abstracted by the nitrene (eq. 23). Photolysis

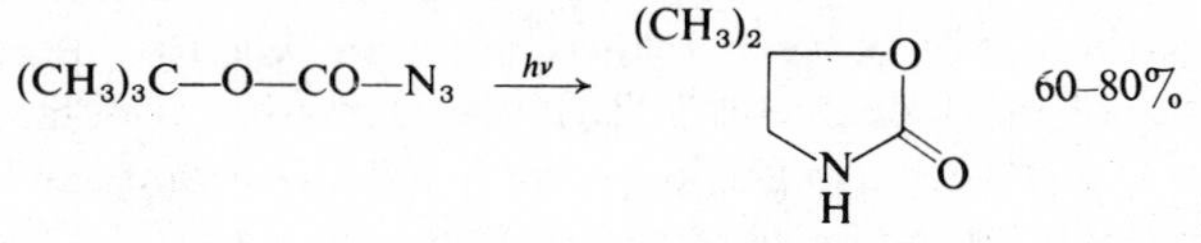
(CH3)3C—O—CO—N3
hν
(CH3)2
O
N
H
O
60–80%

(21)

(**1**)

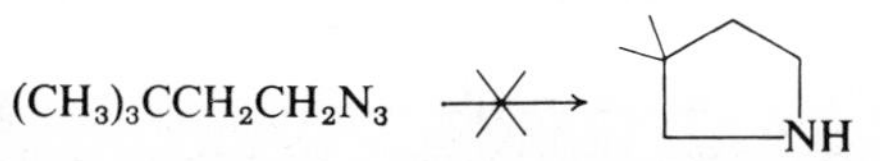
(CH3)3CCH2CH2N3
NH

(22)

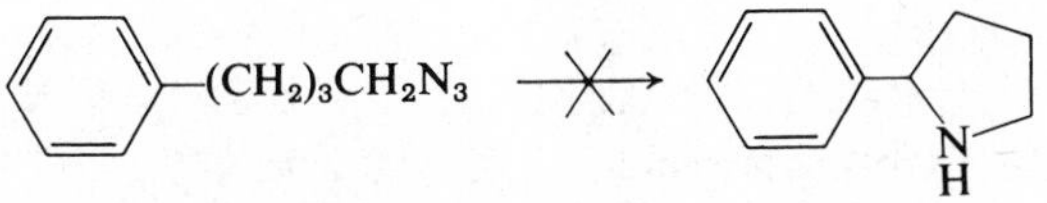
(CH2)3CH2N3
N
H

(23)

or pyrolysis of 6-β-azidocholestane (**2**) gives 6-iminocholestane and other unidentified products (eq. 24) but none of the desired heterocyclic steroidal pyrrolidine (41). Recently, a small amount of intramolecular nitrene insertion has been reported for the thermolysis of 1-biphenyl-2-yl-1-methyl azide (eq. 24a) (41a). Small amounts of methyl and aryl migration products are also formed; however, the major product is 2-isopropenylbiphenyl (62.0%) which formally occurs by loss of N_3H. Photolysis does

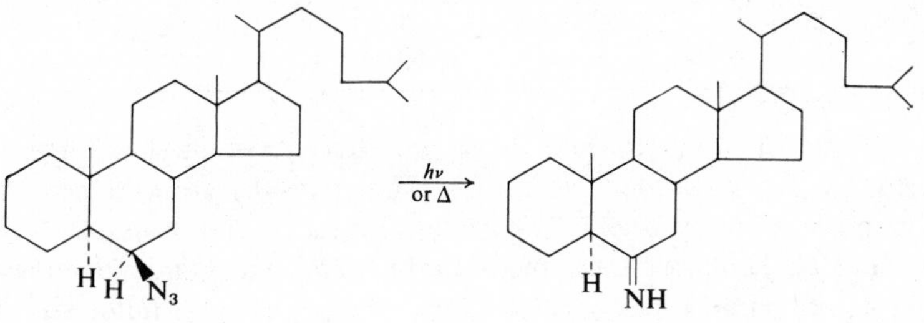
H
H
N3
hν
or Δ
H
NH

(24)

(**2**)

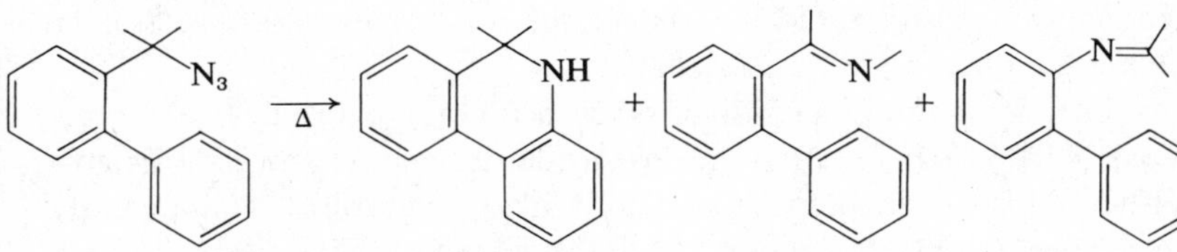
N3
Δ
NH
+
N
+
N

(24a)

6.5% 7.6% 7.4%

not lead to the insertion product. Methyl (62.2%) and aryl (8.0%) migration are the major nitrene reactions.

The failure of alkylnitrenes to form pyrrolidines is probably due to the ease of 1,2-hydrogen migration. Azidoformates and aryl azides obviously have no α-hydrogens and hence cannot form imines. Another indication of the short lifetime of the alkylnitrenes is the small amount of amine formed by intermolecular hydrogen abstraction. At most, a few percent amine is observed even in the presence of hydrogen donors whereas in the above-cited example of *o*-azidocumene 37% amine formation is obtained (39). Aside from 1,2-hydrogen migration several less important pathways of alkylnitrene stabilization are observed. Among these are 1,2-alkyl migration, aziridine formation, hydrogen abstraction from solvent, and insertion in solvent. All of these pathways have been observed in the pyrolysis of a series of alkyl azides (28). The results for gas-phase pyrolysis in the absence of other hydrocarbons are summarized in Table III. In the presence of

TABLE III
Pyrolysis Products of Alkyl Azides (28)

Azide	1,2-Hydrogen migration, %	1,2-Alkyl migration, %	Amine, %	Aziridine, %
Ethyl	55	4	—	35
n-Butyl	82	7	1	—
n-Octyl	75	15	1	—
Isobutyl	85	5	3	0.02
4-Methylpentyl	80	10	1	—
3,3-Dimethylbutyl	80	16	—	—
t-Butyl	—	60	—	1

excess cyclohexane *n*-butylnitrene forms only traces *N-n*-butylcyclohexyl-amine (0.6%) and *n*-butylamine (1.1%). Similar results are obtained by solution phase photolysis of alkyl azides. Moriarty and Rahman (13) found about 70% butyraldimine and 5% *n*-butylamine and an unidentified polymer upon photolyzing *n*-butyl azide in ethyl ether. Photolysis in cyclohexane gave only 1% *N-n*-butylcyclohexylamine. Use of benzophenone as a sensitizer did not change the product distribution. Similar results were obtained with *n*-octyl azide.

An interesting result is obtained when perfluoroalkyl azides are thermolyzed (42,43). 2-H-hexafluoropropyl azide decomposes at 200° to give 3-H-hexafluoro-2-azabutene (eq. 25). Alkyl migration is apparently favored over fluorine migration due to the strength of the carbon–fluorine bond. It has been found that cyclohexane can be used to trap the nitrene formed in this reaction (43).

$$CF_3CFHCF_2N_3 \xrightarrow{200°} CF_3CFH-N=CF_2 \qquad (25)$$

III. NITRENES FROM FUNCTIONALIZED ALKYL AZIDES

Incorporation of other functional groups into alkyl azides often results in interesting changes in the mechanism and products of decomposition. In some cases the relative ease with which the azides decompose makes a concerted reaction with participation of the functional group seem more likely than a discrete nitrene mechanism; however, it is usually difficult to distinguish between these two possibilities. Thermolysis and photolysis of functionalized alkyl azides often lead to markedly different products. The photolysis products are most often more consistent with a discrete nitrene mechanism than are the thermolysis products.

A. Aralkyl Azides: Trityl and Benzyl Azides

Perhaps the most extensively studied functionalized azides are trityl azide and its derivatives. Recent investigations by Saunders and co-workers (14,44,45) have largely elucidated the mechanism of this variation of the Stieglitz rearrangement. The thermolysis of triarylmethyl (trityl) azides was first studied by Senior (2), a student of Stieglitz. Triphenyl-methyl azide was found to be remarkably stable toward acid but to give benzophenone phenylimine upon thermolysis at 225°. p-Chlorophenyl-(diphenyl)-methyl azide was claimed to give a statistical product ratio: 2 parts p-chlorobenzophenone phenylimine and 1 part benzophenone p-chlorophenylimine. A nitrene intermediate was postulated to explain these results. Saunders and Ware (44) examined the pyrolysis of a series of triarylmethyl azides of the type p-X-$C_6H_4(C_6H_5)_2CN_3$ where X = H, CH_3O, Cl, NO_2, and $N(CH_3)_2$. The resulting migration aptitudes (corrected for a 2:1 statistical preference for unsubstituted phenyl) and activation parameters are given in Table IV. The enthalpies of activation are distinctly lower than those observed for simple alkyl azides (11) or cyclohexyl azide (47.5 kcal/mole) (46). The effects of substituents on activation enthalpies follow an inductive order and thus are compatible with the formation of a discrete nitrene intermediate. However, the enhanced rates displayed by all substituted trityl azides and the low activation enthalpies argue in favor of a small amount of aryl participation in the transition state (eq. 26). A pseudo three-membered ring transition state (3) in which the migrating group is relatively positive and nitrogen relatively negative is consistent with the kinetic data. The more electron donating the p-substituent, the greater the contribution of 3 and hence the

TABLE IV

Migration Aptitudes and Activation Parameters for Pyrolysis of
Triarylmethyl Azides (44)

X	% Yield imines	Mig. apt. p-X-C$_6$H$_4$ vs. C$_6$H$_5$	k_{rel}	$\Delta H^{\ddagger}$, kcal/mole	$\Delta S^{\ddagger}$, kcal/mole
(CH$_3$)$_2$N	70	6.7	2.50	25.4	−22.5
CH$_3$O	70	2.5	1.53	28.9	−16.0
CH$_3$	66	1.8	1.08	29.0	−16.1
H	75	1.0	1.0	32.0	−9.8
Cl	43	0.39	1.17	33.8	−5.4
NO$_2$	75	0.20	1.07	34.3	−4.6

$$\text{(3)} \qquad (26)$$

lower the activation energy. Similarly, increased charge separation in the transition state and the resulting increase in the orientation of solvent molecules accounts for the effect of p-substituent on the entropy of activation. The rates of nitrogen evolution were slightly faster in more polar solvents.

The direct (45) and sensitized (14) photolyses of triarylmethyl azides have been investigated. Direct irradiation of several p-substituted triarylmethyl azides in hexane solution with a low-pressure mercury lamp gives both possible imines in statistical amounts (45). That is, the phenyl substituents neither hinder nor aid migration as they do in the thermal rearrangement. Also studied were the thermal and photolytic rearrangements of 1,1-diphenylethyl azide and 2-phenyl-2-propyl azide. The results are shown in Table V. The failure of 1,1-diphenylethyl azide to give a statistical photolytic migration aptitude seems incongruous with the results observed for other tertiary azides. This reaction has been recently examined (16b) and found to give a nearly statistical (0.94) migration aptitude for the first few percent reaction. On prolonged photolysis the migration aptitude increased, apparently due to photodecomposition of the benzophenone methylimine formed by methyl migration.

TABLE V

Migration Aptitudes of 1,1-Diphenylethyl Azide and
2-Phenyl-2-propyl Azide (45)

Azide	Reaction	% Imine	Mig. apt. C_6H_5/CH_3
1,1-Diphenylethyl	Photolysis	20–32	2.18
	Thermolysis	100	2.36
2-Phenyl-2-propyl	Photolysis	26–51	0.96
	Thermolysis	83	4.05

It can be concluded that for all the tertiary azides studied no preference for methyl versus phenyl or phenyl versus substituted phenyl is shown in the unsensitized photolyses. These results are best explained in terms of formation of a highly reactive nitrene intermediate without any alkyl or aryl participation. The migration of alkyl or aryl to nitrogen could involve either a singlet or triplet nitrene.

In an attempt to determine whether the nitrene intermediate is singlet or triplet, the effect of known triplet sensitizers upon the rearrangement of trityl azides was studied (14). It was found that a variety of triplet sensitizers with triplet energies ranging from 67.6 kcal/mole (fluorene) to 48.7 kcal/mole (pyrene) could effect the rearrangement of trityl azide in low quantum yield. Furthermore, the migration aptitudes for five triarylmethyl azides were completely statistical as was the case for the direct photolysis (45). The mechanism of the sensitized reaction can be represented as shown in equations 27–31. The similarity of migration aptitudes for the direct

$$\text{Sens} \longrightarrow \text{Sens*}^1 \xrightarrow[\text{crossing}]{\text{intersystem}} \text{Sens*}^3 \qquad (27)$$

$$\text{Sens*}^3 \longrightarrow \text{Sens} \qquad (28)$$

$$\text{Sens*}^3 + \text{Azide} \longrightarrow \text{Sens} + \text{Azide*}^3 \qquad (29)$$

$$\text{Azide*}^3 \longrightarrow \text{Nitrene*}^3 + \text{N}_2 \qquad (30)$$

$$\text{Nitrene*}^3 \xrightarrow[\text{2. spin inversion}]{\text{1. migration}} \text{Imine} \qquad (31)$$

and sensitized photolyses is suggestive evidence that both proceed through a triplet nitrene and perhaps also a triplet azide. Attempts to observe a triplet azide in the direct photolysis by phosphorescent emission (14) or triplet quenching (16b) have been unsuccessful. Thus, if a triplet azide is formed in the direct photolysis it must lose nitrogen much faster than it undergoes quenching. When the direct photolysis is carried out in the presence of hydrogen donors such as sec-butyl mercaptan or tri-n-butyltin hydride several percent triphenylmethylamine is obtained (16b) thus providing further evidence for a discrete nitrene intermediate. Unfortunately, the facility with which this nitrene undergoes phenyl migration makes it impossible to distinguish between a singlet and triplet nitrene by trapping experiments.

The quantum yield for trityl azide photolysis is approximately 0.8 (16b) in good agreement with the values obtained for simple alkyl azides. However, after a few percent reaction the solution becomes yellow and the rate of nitrogen evolution falls off rapidly. It was thought that the highly absorbing trityl radical could be formed from trityl azide by carbon–nitrogen cleavage (eq. 32) and lower the quantum yield by internal

$$(C_6H_5)_3CN_3 \xrightarrow[?]{h\nu} (C_6H_5)_3C\cdot + N_3\cdot(?) \tag{32}$$

filtering. Electron spin resonance studies of the photolysis of trityl azide established the presence of a small amount of trityl radical. Hence carbon–nitrogen cleavage is shown to be a minor reaction competing with nitrene formation. In order to assess the importance of this route in both the photolysis and thermolysis, nitrogen-15 labeling studies have been carried out (16a). Trityl azide terminally labeled with nitrogen-15 was synthesized by diazotization of tritylhydrazine with labeled sodium nitrite. Mass spectral analysis of the nitrogen evolved in the photolysis revealed that 91% of the label was retained in the terminal position. Scrambling could occur either by a concerted cyclic rearrangement, an ion pair, or by a radical recombination pathway. The latter possibility (eq. 33) seems more likely in view of the detection of trityl radicals in the photolysis. Unimolecular decomposition of the azide radical could account for the observed loss of terminal label, but this process is forbidden by correlation rules (25). Bimolecular decomposition of the azide radical (eq. 34) is

$$(C_6H_5)_3C\cdot + \cdot N\!-\!N\!-\!N^{15} \longrightarrow$$
$$(C_6H_5)_3C\!-\!N\!-\!N\!-\!N^{15} + (C_6H_5)_3C\!-\!{}^{15}N\!-\!N\!-\!N \tag{33}$$

$$2N\!-\!N\!-\!N\cdot \longrightarrow 3N_2 \tag{34}$$

allowed and highly exothermic. However, some nitrogen-30 would be formed by this reaction and no excess nitrogen-30 is observed mass spectroscopically. Hence, recombination of azide radical with trityl radical (eq. 33) is probably the major reaction of these radicals. Since the direct photolysis quantum yield is approximately 0.8 the nitrogen evolved would be expected to retain at least 80% of the nitrogen-15 label. If radical recombination does occur, at least 50% of the label will return to the terminal position leading to a predicted 90% retention of terminal label, in excellent agreement with the experimental results. Thus, it appears that in the photolysis of trityl azide every photon absorbed leads to a bond cleavage reaction.

Extensive scrambling is observed in the thermolysis of labeled trityl azide. The rate of scrambling is several times that for nitrogen evolution resulting in only $55 \pm 1\%$ retention of label when the thermolysis is carried to completion. It is unlikely that sufficient energy is supplied in the thermolysis to cause carbon–nitrogen cleavage. Thus scrambling probably occurs via a tight ion pair or a concerted cyclic rearrangement. This would explain the failure to observe trityl radicals in the thermolysis of trityl azide by ESR spectroscopy.

Triarylmethyl azides in which two of the aryl groups are fused are capable of undergoing ring expansion reactions, in a variation of the Stieglitz rearrangement. Pinck and Hilbert (47) found 9-(α-naphthyl)-9-fluorenyl azide (4) to give only 9-(α-naphthyl)phenanthridine (5) upon pyrolysis (eq. 35). The 9-methyl and 9-phenyl compounds gave a similar

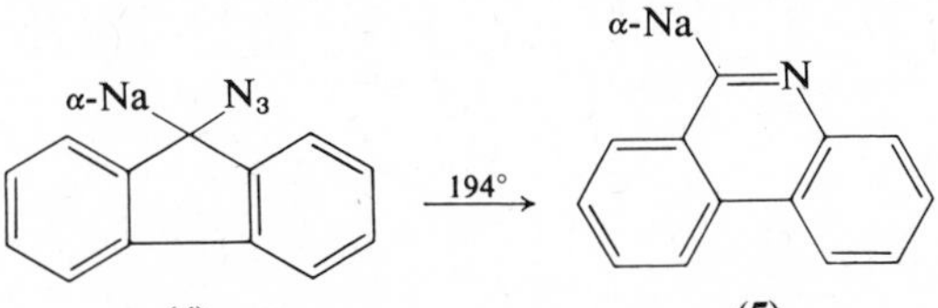

$$(35)$$

(4)   **(5)**

result. The preference for ring expansion to fluoreneimine formation is explained as being due to relief of strain in going from a five- to six-membered ring. When the 9-position bears a hydrogen, ring expansion does not occur and the fluorenone imines are formed exclusively (48). This marked preference for hydrogen versus aryl migration is also observed for the rearrangements of α-azido ketones (49). Six-membered rings do not undergo ring expansion as readily as the five-membered analogs. 9-Phenylthiaxanthyl azide (6) gives a small amount of thiazepine (7) by ring expansion along with larger amounts of the phenyl imine (eq. 36) (50). 9-Phenyl-9-xanthyl azide behaves similarly.

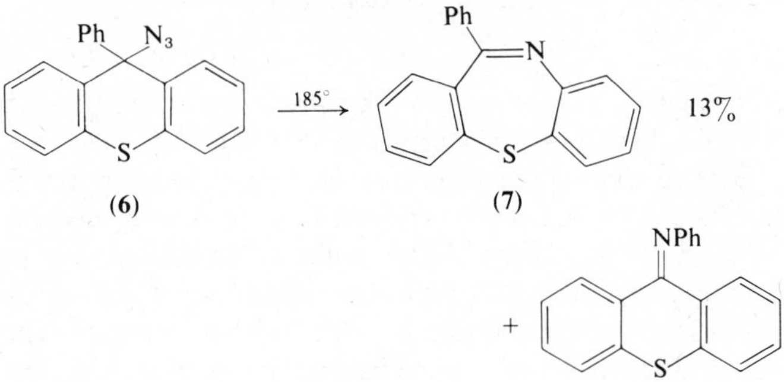

$$(36)$$

(6)   **(7)**

The pyrolysis of benzyl azide was studied by Curtius and co-workers (51,52). Pyrolysis in xylene or glacial acetic acid gave complex products assumed to arise from benzylnitrene and phenylcarbene. No evidence for carbene formation is given except as a rationalization of the trimeric imine (8) formed (eq. 37). Recently the decomposition of some azido-

$$3(C_6H_5)CH_2N_3 \xrightarrow{138°} (C_6H_5)CH_2N{=}\underset{\underset{(C_6H_5)}{|}}{C}{-}NHCH_2(C_6H_5) \qquad (37)$$

(8)

methylbiphenyls have been studied by Coffin and Robbins (53). Photolysis of 2-azidomethylbiphenyl gave only 2-iminomethylbiphenyl (9) in 8%

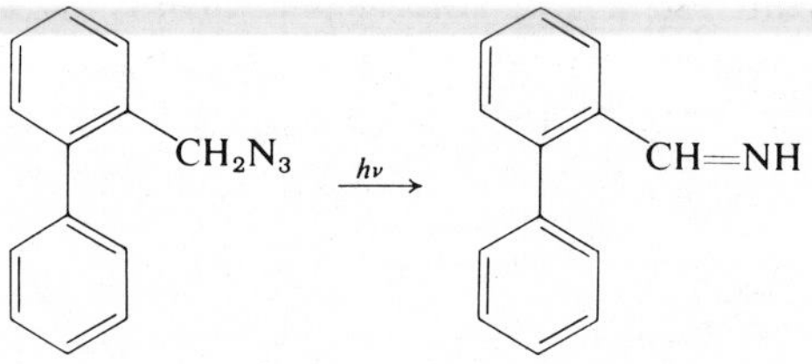

$$8\% \qquad (38)$$

(9)

yield (eq. 38). Thermolysis in refluxing diphenyl ether gave N-2-biphenyl-methyl-2-biphenylmethyleneamine (10) and lesser amounts of 2-amino-biphenyl (eq. 39), but no isolable 2-iminomethylbiphenyl. The authors

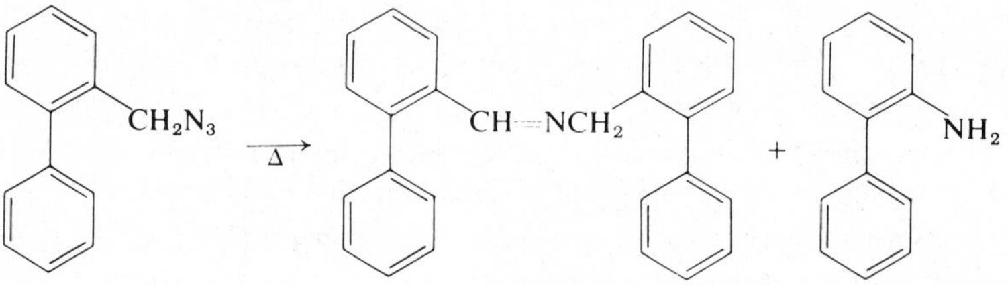

$$65\% \qquad 19\% \qquad (39)$$

(10)

postulated anil formation via coupling of a nitrene and a carbene in analogy to the earlier hypothesis of Curtius (51,52). Such a mechanism must be viewed with reservations since no symmetrical carbene or nitrene coupling products are observed. Furthermore, coupling of two unstable intermediates in the solution phase is highly unlikely. The mechanism of this reaction may involve attack of a nitrene or azide upon a previously formed imine followed by decomposition with loss of nitrogen. The reaction is doubtless a complex one. Amine formation can be explained as arising from aryl migration to nitrogen to give the formaldimine (**11**) followed by hydrolytic cleavage during workup (eq. 40). 2,2'-Di(azidomethyl)biphenyl gives phenanthridine and dibenzapepine (eq. 41) (53). A dual carbene–nitrene mechanism is also suggested to explain these products. Photolysis or thermolysis of **11a** in inert solvents leads to loss of nitrogen and exclusive phenyl migration (eq. 41a) (53a).

$$CH_2N_3 \longrightarrow N{=}CH_2 \xrightarrow{H_2O} NH_2 \tag{40}$$

(**11**)

$$N_3H_2C \quad CH_2N_3 \xrightarrow{\Delta} \quad + \quad \tag{41}$$

$$29\% \qquad 21\%$$

$$Ph_2C \overset{N{=}CPh_2}{\underset{N_3}{\diagdown}} \xrightarrow[\text{or } \Delta]{hv} Ph{-}C \overset{N{=}CPh_2}{\underset{N{-}Ph}{\diagdown}} \tag{41a}$$

B. Unsaturated Azides: Olefinic and Vinyl Azides

The presence of an unsaturated center in an alkyl azide gives rise to the possibility of interactions between the double bond and the azido group. Since azides and olefins are known to form triazolines by 1,3-dipolar additions (34) it would be expected that a properly constructed olefinic azide might undergo internal 1,3-dipolar addition to form a bicyclic triazoline. Logothetis (54) recently found that a triazoline (**12**) is indeed formed when 4-azido-5-methyl-1-hexene or 4-azido-1-hexene are heated at 50°. More vigorous conditions (80–160°) or further heating of the triazolines leads to loss of nitrogen and formation of cyclic imines (**13**) and lesser amounts of 1-azabicyclo[3.1.0]-hexanes (**14**) in various ratios (eq. 42). Isolation of the triazoline and absence of products arising from 1,2-hydrogen or alkyl migration to nitrogen rule out a nitrene mechanism.

$$(42)$$

Photolysis, on the other hand, probably involves a nitrene intermediate as only traces of the thermal products were observed from 5-azido-1-hexene and 5-azido-2-methyl-1-hexene. The major products were high boiling amines which were not identified. It is not surprising that no triazoline is formed in the photolysis as sufficient energy is supplied to decompose the azido group without the necessity of anchimeric assistance. In fact, loss of nitrogen and formation of imine or amine are probably much too rapid to allow for any interaction with the olefinic portion of the molecule.

Gas-phase pyrolysis of allyl azide at 500° leads to a number of fragmentation products (54). Allyl azide reacts slowly at 75° in solution to give a dimer of unknown structure (54,55). Substituted allylic azides undergo rapid thermal rearrangement (eq. 43) (56,57). A cyclic transition state (15) is indicated by the negative entropy and the small solvent effect for the rearrangement (57). Photolysis of 3-azido-2-phenyl-1-propene yields 3-phenyl-1-azabicyclobutane and an imine (eq. 42a) (57a).

$$(42a)$$

$$(43)$$

Vinylnitrenes, while not strictly speaking alkylnitrenes are included in this discussion of unsaturated alkylnitrenes. In addition to being the probable intermediates in the decomposition of vinyl azides, vinylnitrenes have been postulated as intermediates in the Neber rearrangement, which will be discussed later. A review of vinyl azide reactions has recently appeared (57b). Smolinsky (58,59) showed that nonterminal azides form azirines in good yield upon vapor-phase pyrolysis. Thus α-azidostyrene forms 2-phenylazirine along with some N-phenyl-ketenimine (eq. 44). Upon standing for one month at room temperature α-azidostyrene forms 2-phenylazirine (10%) along with some dimer (57c). Several possible intermediates have been suggested for azirine formation. The intermediacy of a triazoline seems unlikely as the azido group is orthogonal to the vinyl

$$
\underset{R}{\overset{CH_2}{\underset{\overset{|}{C}}{\parallel}}}_{N_3} \quad \xrightarrow{\Delta} \quad \underset{R}{\overset{CH_2}{\underset{}{\diagup\backslash}}}C\!\!=\!\!N \quad + \quad R\!\!-\!\!N\!\!=\!\!C\!\!=\!\!CH_2 \qquad (44)
$$

$$50\text{--}60\% \qquad 5\text{--}6\%$$

$$R = C_6H_5, \ o\text{-}CH_3C_6H_4, \ n\text{-}C_4H_9$$

π-electrons making ring closure impossible from the ground state (60). Furthermore, triazolines have not been isolated from vinyl azides except in one instance in which a vinyl proton was first removed (61). α-Azido-o-methylstyrene does not form products from intramolecular carbon–hydrogen insertions (59), nor have attempts to trap a vinylnitrene with 2-methyl-2-butene been successful (60). However, this does not constitute serious evidence against a nitrene mechanism as alkylnitrenes also fail to give these reactions. A vinylnitrene would be expected to be even more reactive (electron deficient) than an alkylnitrene. Azirine formation from a vinylnitrene certainly would be the most facile reaction pathway and would be expected by analogy to cyclopropene formation from alkenyl-carbenes (62). N-Substituted ketenimines could be formed by a carbon to nitrogen migration similar to that observed in the Curtius rearrangement. Ketenimine formation is the predominant reaction for thermolysis of 2-azidoethylene-1,1-dicarbonitriles (62a). In inert solvents only polymers are formed, but in active solvents addition to the ketenimine occurs with 35–86% yields (eq. 44a).

$$
\underset{NC}{\overset{NC}{\diagdown}}C\!\!=\!\!C\underset{N_3}{\overset{R}{\diagup}} \quad \xrightarrow{60\text{--}70^\circ} \quad \left[\underset{NC}{\overset{NC}{\diagdown}}C\!\!=\!\!C\!\!=\!\!N\!\!-\!\!R\right] \quad \xrightarrow{HX} \quad \underset{NC}{\overset{NC}{\diagdown}}C\!\!=\!\!C\underset{X}{\overset{NHR}{\diagup}} \qquad (44a)
$$

Further evidence for a discrete vinylnitrene is provided by the similarity of products obtained in the pyrolysis (58,59) and photolysis (63,64) of α-azidostyrene. 2-Phenylazirine is isolated in 58% yield from the photolysis of α-azidostyrene in benzene solution (63). Further photolysis leads to formation of 2,5-diphenyl-3,6-dihydropyrazine (eq. 45). The dihydro-pyrazine is also formed by treatment of 2-phenylazirine with dilute acid (59). Recently (64) a second product from the photolysis of α-azidostyrene

$$
\underset{H_5C_6}{\overset{CH_2}{\underset{\overset{\|}{C}}{}}}\text{---}N_3 \quad \xrightarrow{h\nu} \quad \underset{H_5C_6}{\overset{N}{\underset{}{}}}\overset{}{C}\text{---}CH_2 \quad \xrightarrow{h\nu} \quad \underset{H_5C_6}{\overset{N}{\underset{N}{}}}\overset{C_6H_5}{\underset{}{}} \tag{45}
$$

58%

has been isolated in 10% yield and identified as 4-phenyl-3-phenylimino-1-azabicyclo-[2.1.0]pentane (16). This product is postulated to arise from 1,2-photocycloaddition of phenylazirine and N-phenylketenimine (eq. 46). Both cis- and trans-2-azido-1-phenylpropane form 2-methylindole in 86% yield upon thermolysis at 287° (64a). Photolysis or thermolysis at 100° gives a high yield of azirine which is converted to 2-methylindole upon heating. Thus the azirine is probably an intermediate in indole formation.

$$
\underset{H_5C_6}{\overset{N}{\underset{}{}}}\overset{}{C}\text{---}CH_2 \; + \; C_6H_5N{=}C{=}CH_2 \quad \xrightarrow{h\nu} \quad \underset{H_5C_6}{\overset{H_2C}{\underset{}{}}}\overset{N\text{---}CH_2}{\underset{C\text{---}C}{}}\overset{}{\underset{NC_6H_5}{}} \tag{46}
$$

(16)

β-Azidocrotonates rearrange upon photolysis to give a 93% yield of a 4:1 mixture of azirine and ketenimine (eq. 47) (60). Hassner and co-workers (65–67) have investigated the syntheses and photolyses of several vinyl

$$
\underset{N_3}{\overset{R}{\underset{|}{}}}\overset{|}{CH_3C{=}CCO_2C_2H_5} \quad \xrightarrow{h\nu} \quad \underset{H_3C}{\overset{N}{\underset{}{}}}\overset{R}{\underset{C\text{---}C}{}}\overset{}{\underset{CO_2C_2H_5}{}} \; + \; CH_3N{=}C{=}\overset{R}{\underset{|}{}}CCO_2C_2H_5 \tag{47}
$$

R = H, CH₃

azides. Photolysis of α-azido-trans-stilbene in pentane solution gives 2,3,5,6-tetraphenylpyrazine, presumably via dimerization of an azirine (eq. 48) (65). The first examples of fused azirines were obtained by photolysis of

$$
\underset{}{\overset{H \; N_3}{\underset{|\;\;|}{}}}C_6H_5C{=}CC_6H_5 \quad \xrightarrow{h\nu} \quad \left[\underset{H_5C_6}{\overset{N}{\underset{}{}}}\overset{}{\underset{C\text{---}C}{}}\overset{H}{\underset{C_6H_5}{}} \right] \longrightarrow \underset{H_5C_6}{\overset{H_5C_6 \; N \; C_6H_5}{\underset{N \;\; C_6H_5}{}}} \tag{48}
$$

40%

1-azidocyclooctene (eq. 49) and 2-azido-1,3-cyclooctadiene (67). Photolysis of a series of vinyl azides in hydrocarbon solvents results in high yields (81–100%) of azirines (67a). No ketenimines are observed. Anchimeric assistance from the double bond would not be expected in the photolyses of vinyl azides for reasons previously mentioned. Hence, it seems probable that a vinylnitrene is formed in these reactions.

$$\text{(cyclooctene with } N_3) \xrightarrow{h\nu} \text{(bicyclic azirine with } N) \qquad (49)$$

Terminal vinyl azides do not generally form stable azirines upon decomposition. Vinyl azide itself is unstable at room temperature (67) and polymerizes in the presence of radical initiators (68). β-Styryl azide gives phenylacetonitrile under both photolytic and pyrolytic conditions (69). The same product is also obtained from photolysis of β-styryl isocyanate and deoxygenation of β-nitrosostyrene (eq. 50). These results suggest

$$
\begin{array}{l}
C_6H_5CH{=}CHN_3 \\
\qquad\qquad\qquad \searrow{\underset{-N_2}{\Delta \text{ or } h\nu}} \\
C_6H_5CH{=}CHNCO \xrightarrow[-CO]{h\nu} [C_6H_5CH{=}CH\ddot{N}] \\
\qquad\qquad\qquad \nearrow{R_3P} \\
C_6H_5CH{=}CHNO_2
\end{array}
\quad
\begin{array}{l}
\nearrow C_6H_5CH{=}C{=}NH \\
\qquad\downarrow \\
\searrow C_6H_5CH_2C{\equiv}N
\end{array}
\qquad (50)
$$

strongly that a vinylnitrene is a common intermediate for all these reactions. Photolysis of both the azide and the isocyanate also gave unidentified, apparently polymeric, material. Azirines were not observed, nor were indoles. Nitrile formation could proceed either by migration of hydrogen from the β- to the α-carbon or by intermediate formation of a ketenimine followed by tautomerization. Ketenimine formation is analogous to alkyl imine formation for primary and secondary alkylnitrenes. Recently, photolysis of β-styrylazide at $-30°$ has been reported to give 3-phenylazirine as the major product (64a). This azirine was unstable at room temperature. Pyrolysis at $287°$ of either the azirine or the β-azidostyrene gave equal amounts of phenylacetonitrile and indole. An azirine is also the probable primary product in the photolysis of trans-β-azidovinyl-p-tolylsulfone (eq. 50a), as is shown by the formation of the ditoluenesulfonylaziridine (69a). Thermolysis results in only p-toluenesulfonylacetonitrile.

$$
\underset{N_3}{\overset{Ts}{\diagdown}}\!\!=\!\! \xrightarrow[H_2O,\ C_2H_5OH]{h\nu} \left[\underset{Ts}{\overset{N}{\triangle}} \right] \xrightarrow{C_7H_7SO_2H} \underset{Ts \qquad Ts}{\overset{\overset{H}{N}}{\triangle}} \qquad (50a)
$$

One case in which a terminal vinyl azide forms a stable azirine has been reported (70,71). In 1960 Knunyants and Bykhovskaya (72) reported the formation of perfluoro-2-azetine from perfluoropropenyl azide. More recent investigations (70,71) have shown this product assignment to be incorrect. The product obtained is perfluoro-2-methylazirine (eq. 51). In view of the rapid decomposition of this vinyl azide at room temperature, Cleaver and Krespan (70) argued that participation of the double bond in

$$CF_3CF{=}CFN_3 \xrightarrow[\text{temp.}]{\text{room}} \underset{CF_3}{\overset{N}{\diagdown}}CF{-}CF \qquad (51)$$

the transition state must occur and that a free nitrene is not formed. Banks and Moore (71) were unable to trap a vinylnitrene with cyclohexane, but this evidence is inconclusive in view of other failures to trap vinylnitrenes with more reactive trapping agents.

β-Azidovinyl ketones lose nitrogen to form both nitriles and substituted isoxazoles. 1-Benzoyl-2-azidoethylene on treatment with acid gives 5-phenylisoxazole (17) and benzoylacetonitrile (eq. 52) (73). The presence of acid may give rise to a nitrenium rather than a free nitrene intermediate.

$$\underset{O}{\overset{O}{\underset{\|}{C_6H_5CCH}}}{=}CHN_3 \xrightarrow[\text{HOAc}]{\text{HCl}} \qquad \underset{7\%}{\underset{(17)}{\text{[isoxazole]}}} + \underset{82\%}{C_6H_5\overset{O}{\overset{\|}{C}}CH_2C{\equiv}N} \qquad (52)$$

Thermolysis of 1-benzoyl-2-azidoethylene at 50° for eight hours similarly gave 50% benzoylacetonitrile and 3% 5-phenylisoxazole (74). Photolysis in benzene at 20° gave 40% isoxazole and 10% nitrile. Higher yields of isoxazoles were also obtained with α-substituted β-azidovinyl ketones (eqs. 53–55) (74). Treatment of the iodo azide adduct of chalcone with diazabicyclooctane at room temperature gives 3,5-diphenylisoxazole in 53%

$$C_6H_5\overset{O}{\overset{\|}{C}}\underset{C_6H_5}{\overset{}{C}}{=}CHN_3 \xrightarrow[\text{DMF}]{90°} \underset{65\%}{\text{[diphenylisoxazole]}} \qquad (53)$$

$$\text{[cyclopentene azide]} \xrightarrow[\text{DMF}]{50°} \text{[fused isoxazole]} \qquad (54)$$

$$C_6H_5\overset{O}{\overset{\|}{C}}\underset{CH_3}{\overset{}{C}}{=}CHN_3 \xrightarrow{110°} \underset{43\%}{\text{[isoxazole]}} + \underset{16\%}{C_6H_5\overset{OH}{\overset{\|}{C}}C{\equiv}N} \qquad (55)$$

$$\xrightarrow{h\nu} \qquad 43\% \qquad 9\%$$

$$\text{C}_6\text{H}_5\overset{\overset{\displaystyle \text{N}_3}{|}}{\text{C}}\text{H}\overset{\overset{\displaystyle \text{O}}{\|}}{\text{CH}}\text{C}\text{C}_6\text{H}_5 \quad \xrightarrow[-\text{HI}]{} \quad \left[\text{C}_6\text{H}_5\overset{\overset{\displaystyle \text{N}_3}{|}}{\text{C}}{=}\text{CH}\overset{\overset{\displaystyle \text{O}}{\|}}{\text{C}}\text{C}_6\text{H}_5 \right] \quad \xrightarrow{-\text{N}_2}$$

$$\left[\begin{array}{c} \end{array} \right] \longrightarrow \quad (56)$$

53%

yield (67) (eq. 56). Presumably dehydrohalogenation and nitrene formation precede cyclization. Isoxazole formation most likely occurs by attack of the initially formed nitrene on the nonbonding orbitals of oxygen. Although these reactions can be nicely explained in terms of nitrene intermediates, a concerted mechanism cannot be ruled out at this time. The synthetic promise of vinyl azides as routes to azirines and isoxazoles will doubtless give rise to further investigations into the mechanisms of these reactions.

C. α-Azido Ketones, Acids, and Hydroxyl Compounds

The pyrolysis of α-azido ketones has been studied by Boyer and Straw (49,75,76). Phenacyl azides form imidazoles when heated at 180–240° in diphenyl ether (eq. 57) (49,76). Presumably loss of nitrogen gives a nitrene which undergoes 1,2-hydrogen migration to the imine. Dimerization and dehydration account for imidazole formation. o-Aminophenacyl azide gave only indigo in 85% yield (eq. 58) (76). The originally formed imine is

$$\text{R}\overset{\overset{\displaystyle \text{O}}{\|}}{\text{C}}\text{CH}_2\text{N}_3 \quad \xrightarrow[180\text{--}240°]{-\text{N}_2} \quad \left[\text{R}\overset{\overset{\displaystyle \text{O}}{\|}}{\text{C}}\text{CH}_2\overset{..}{\text{N}}{:} \right] \quad \xrightarrow{\sim\text{H}} \quad \left[\text{R}\overset{\overset{\displaystyle \text{O}}{\|}}{\text{C}}\text{CH}{=}\text{NH} \right] \quad \xrightarrow{-\text{H}_2\text{O}} \quad (57)$$

$$R = \text{C}_6\text{H}_5,\ p\text{-Br}{-}\text{C}_6\text{H}_4,\ p\text{-C}_6\text{H}_5{-}\text{C}_6\text{H}_4,\ \beta\text{-C}_{10}\text{H}_7,\ o\text{-HO}{-}\text{C}_6\text{H}_4$$

$$\xrightarrow[240°]{-\text{N}_2} \quad \xrightarrow{\sim\text{H}} \quad \longrightarrow \quad (58)$$

postulated to react with the *o*-amino group to give 2-amino-3-hydroxy-indole which then dimerizes under the reaction conditions. When the α-carbon contains hydrogen, alkyl, or aryl substituents, marked differences in migration aptitudes are shown (eq. 59) (49). Hydrogen migrates ex-

$$R-\overset{\overset{\displaystyle O}{\|}}{C}-\overset{\overset{\displaystyle R'}{|}}{\underset{\underset{\displaystyle R''}{|}}{C}}-N_3 \xrightarrow[200° \pm 20°]{-N_2} \left[R-\overset{\overset{\displaystyle O}{\|}}{C}-\overset{\overset{\displaystyle R'}{|}}{\underset{\underset{\displaystyle R''}{|}}{C}}-\ddot{N}: \right] \longrightarrow R-\overset{\overset{\displaystyle O}{\|}}{C}-\overset{\overset{\displaystyle R'}{|}}{C}=N-R'' \quad (59)$$

clusively when R″ = H and R′ = CH$_3$ or C$_6$H$_5$. If R″ = C$_6$H$_5$ and R′ = CH$_3$ then phenyl migration predominates. Acyl migration is never observed. Azidoacetone is observed to undergo aluminum chloride-catalyzed electrophilic aromatic substitution (eq. 60) (77). Clearly a nitrene

$$CH_3-\overset{\overset{\displaystyle O}{\|}}{C}-CH_2-N_3 \xrightarrow{AlCl_3} \left[CH_3-\overset{\overset{\displaystyle O}{\|}}{C}-CH_2-\overset{\overset{\displaystyle AlCl_3^{\theta}}{\diagup}}{\underset{\underset{\displaystyle N\equiv N}{\diagdown \oplus}}{N}} \right] \xrightarrow[-N_2]{C_6H_6}$$

$$CH_3-\overset{\overset{\displaystyle O}{\|}}{C}-CH_2-NHC_6H_5 \quad (60)$$
$$35\%$$

is not involved. A complex *N*-diazonium salt has been suggested as the intermediate in this reaction. Several 2-azido-2-substituted-1,3-indane-diones have recently been synthesized (78). They are highly unstable and evolve nitrogen, but the decomposition products have not been identified. No studies of photolytic decomposition of α-azido ketones have been reported. Such studies might be quite interesting as both the azido and carbonyl groups could be excited either simultaneously, or possibly independently. Sensitized photolyses could be particularly revealing as to the interactions of the two chromophores.

α-Azido carboxylic acids and esters undergo acid catalyzed rearrangement to form the corresponding imino esters (79). Acid catalysis probably rules out a nitrene intermediate. Recently, the photolysis of α-azido acids and esters have been studied by Moriarty and Rahman (13,80). Ester photolysis gives the α-imino ester (18) plus a small amount of the α-amino ester (eq. 61). A nitrene intermediate is clearly indicated by the presence of the amino ester. α-Azido acids undergo decarboxylation as well as loss of nitrogen upon photolysis to give the lower homologous nitrene. For example, α-azidobutyric acid gives propionaldimine and carbon dioxide when photolyzed in methanol solution (eq. 62). Such imines decompose

$$CH_3CH_2\overset{N_3}{\underset{|}{CH}}-C\overset{\displaystyle O}{\underset{\displaystyle OC_2H_5}{\diagdown}} \xrightarrow{h\nu} \left[CH_3-CH_2\overset{\overset{..}{\overset{..}{N:}}}{\underset{|}{CH}}-C\overset{\displaystyle O}{\underset{\displaystyle OC_2H_5}{\diagdown}} \right] \longrightarrow$$

$$CH_3CH_2\overset{NH}{\underset{\parallel}{C}}-C\overset{\displaystyle O}{\underset{\displaystyle OC_2H_5}{\diagdown}} + CH_3CH_2\overset{NH_2}{\underset{|}{CH}}-C\overset{\displaystyle O}{\underset{\displaystyle OC_2H_5}{\diagdown}} \qquad (61)$$

(18) 25%   5%

$$CH_3CH_2\overset{N_3}{\underset{|}{CH}}C\overset{\displaystyle O}{\underset{\displaystyle OH}{\diagdown}} \xrightarrow[-N_2]{h\nu} CH_3CH_2CH=NH + CO_2 \qquad (62)$$

                                                 25%   60%

under photolytic conditions, so the yields are relatively poor. When azido-diphenylacetic acid was photolyzed, two imines were observed (eq. 63)

$$(C_6H_5)_2\overset{N_3}{\underset{|}{C}}-C\overset{\displaystyle O}{\underset{\displaystyle OH}{\diagdown}} \xrightarrow[-N_2]{h\nu} \left[(C_6H_5)_2\overset{:N:}{\underset{|}{C}}-C\overset{\displaystyle O}{\underset{\displaystyle OH}{\diagdown}} \right] \xrightarrow[\text{path } a]{-CO_2} (C_6H_5)_2C=NH$$

path b

$$\left[C_6H_5-\overset{NC_6H_5}{\underset{\parallel}{C}}-C\overset{\displaystyle O}{\underset{\displaystyle OH}{\diagdown}} \right] \xrightarrow[-CO_2]{h\nu} C_6H_5-CH=NC_6H_5$$

                                                                      (63)

which can be explained by phenyl migration either prior to decarboxylation (path a) or following it (path b). Whether decarboxylation of α-azido acids bearing an α-hydrogen proceeds via concerted loss of nitrogen and carbon dioxide or by photolysis of an intermediate α-imino acid remains uncertain. Pyruvic acid N-methylimine and benzoylformic acid N-phenyl-imine both undergo photodecarboxylation.

The only reported study of the decomposition of an α-hydroxyalkyl azide is the photolysis of 5α-hydroxy-6β-azido steroids (81). The α-hydroxy azide (19) upon photolysis gives an α-hydroxy ketone (20) which most likely arises via 1,2-hydrogen migration to nitrene followed by hydrolysis of the imine by water (eq. 64). Photolysis of the α-hydroxy azide (21) gave more than 10 products. Compounds (22) and (23) were isolated (eq. 65). Compound (23) apparently is formed by a retroaldol-type reaction to give the intermediate keto imine (24). Such a cleavage reaction has not been previously reported, but may be analogous to the decarboxylation of

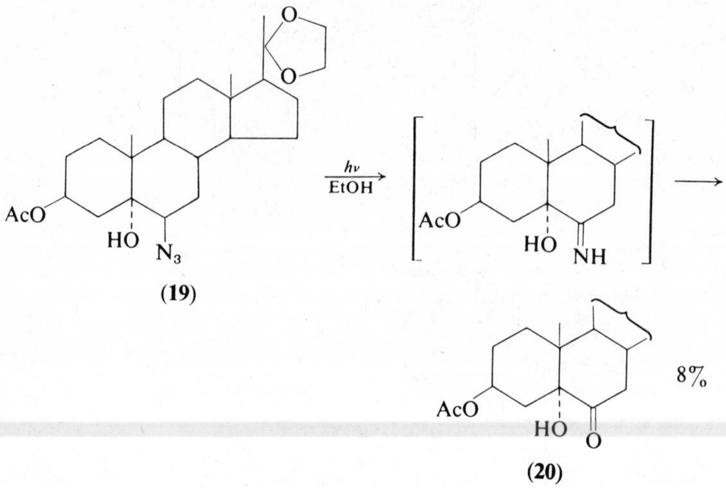

(19)

$\xrightarrow[\text{EtOH}]{h\nu}$

(20) 8% (64)

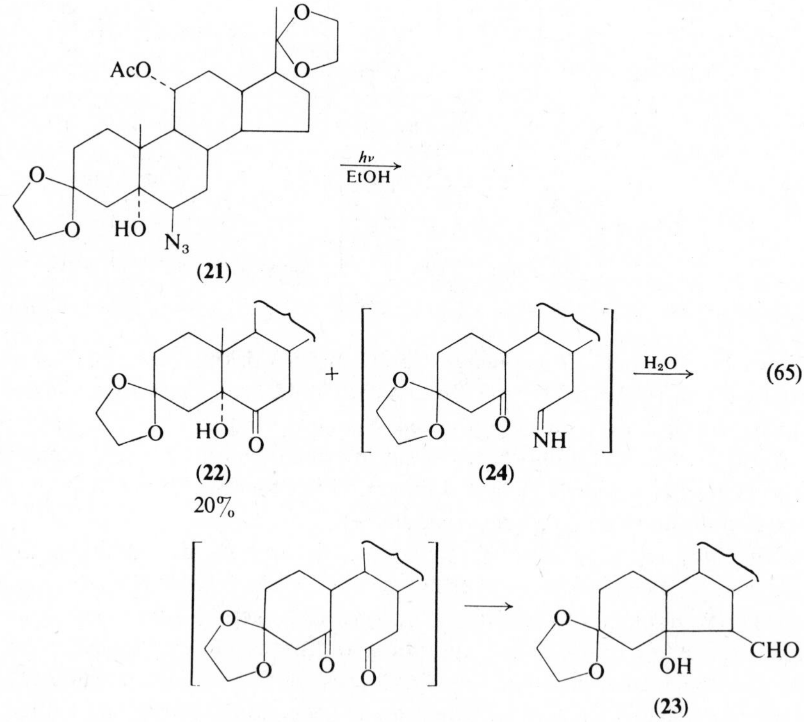

(21)

$\xrightarrow[\text{EtOH}]{h\nu}$

(22) 20%

+

(24)

$\xrightarrow{H_2O}$ (65)

(23) 8%

α-azido acids. Photolysis of compound (21) in benzene solution led to olefin formation with loss of azide and hydroxyl. No olefin was observed when the photolysis was carried out in 95% ethanol. The mechanism of this elimination reaction is too obscure to warrant conjecture. Further investigation of simple α-hydroxy azides will be necessary to establish the generality of the retroaldol-type pathway.

D. Geminal Diazides

The synthesis and thermal decomposition of geminal diazides are discussed in several early papers in the German literature (82–85). Benzophenone diazide was found to give 1,5-diphenyltetrazole (25) upon heating in various solvents (82). Evidence for the tetrazole structure was provided by the formation of the identical compound by addition of sodium azide to benzyl phenylimide chloride (eq. 66) (83). p-Nitrobenzophenone diazide

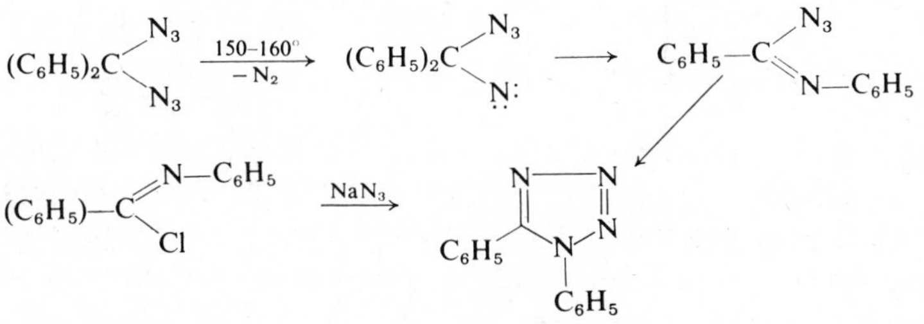

(66)

(25)

also formed a tetrazole but the identity of the migrating group was not established (83). Pivalophenone diazide formed solely 5-phenyl-1-t-butyl-tetrazole (eq. 67) (84). Migration of t-butyl in preference to phenyl is

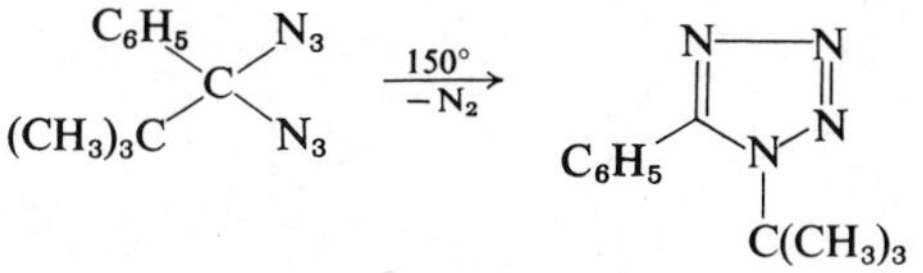

(67)

difficult to explain in view of the migratory aptitudes observed by Boyer and Straw (49) and by Saunders and Caress (45) (see above).

Lindemann and Mühlhaus (86) studied several ortho- and para-hydroxy-benzyl diazides. Pyrolysis in benzene or nitrobenzene gave the corresponding nitrile (eq. 68). The mechanism proposed by the authors involves loss of hydrazoic acid to give a vinyl azide intermediate, loss of nitrogen to form a vinylnitrene, and rearrangement to a nitrile. This mechanism involves energetically unfavorable dearomatization of a benzene nucleus and thus seems unlikely. A possible alternative mechanism could proceed

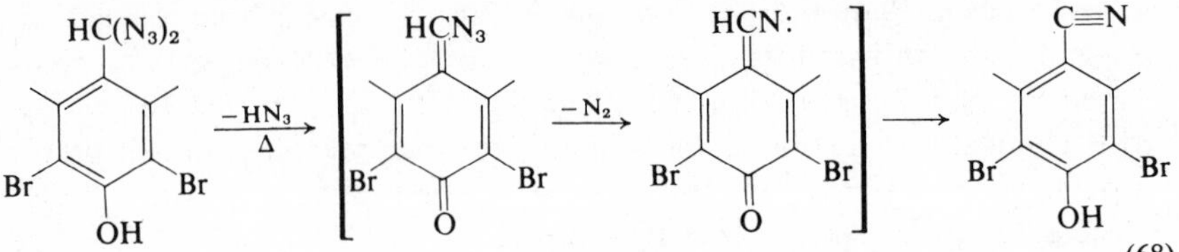

(68)

by imine formation followed by loss of hydrazoic acid (eq. 69). The formation of a nitrile from unsubstituted benzyl diazide would provide evidence for this alternative mechanism.

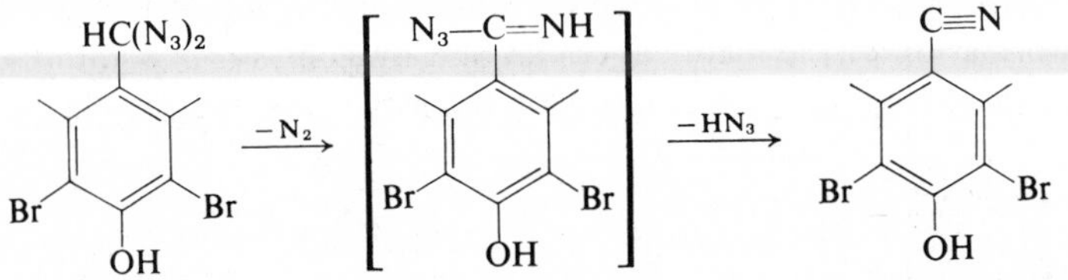

(69)

The photolysis of benzophenone diazide has recently been investigated both in low-temperature matrices (87) and in dilute fluid solution (88). Barash, Wasserman, and Yager (87) have observed that sensitized photolyses of benzophenone diazide and similar diazides in rigid matrices at 77°K yield the corresponding methylenes (eq. 70). Use of ESR spectroscopy

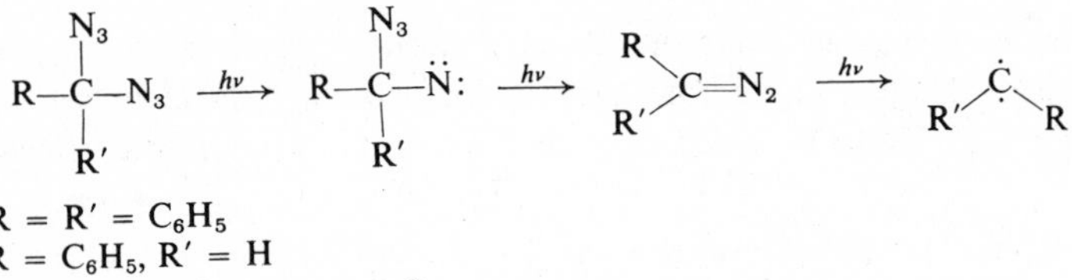

a. $R = R' = C_6H_5$
b. $R = C_6H_5, R' = H$

c. $R + R' =$

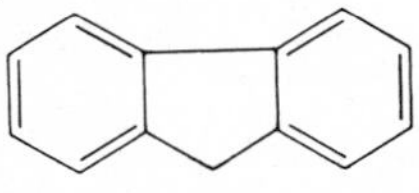

allows observation of the initially formed triplet α-azidonitrene which is further photolyzed to the methylene. A markedly different result is reported by Moriarty and Kliegman (88) for photolysis in benzene solution (eq. 71). Three products are obtained, one of which is the 1,5-diphenyltetrazole observed by Schroeter for the thermal decomposition. As further photolysis of 1,5-diphenyltetrazole gives 2-phenylbenzimidazole (26) as the only isolable product, the former is probably the major primary product of the diazide photolysis. The origin of carbodiimide trimer is not clear as it is not obtained by prolonged photolysis of 1,5-diphenyltetrazole.

$$H_5C_6-\underset{\underset{C_6H_5}{|}}{\overset{\overset{N_3}{|}}{C}}-N_3 \xrightarrow{hv} \left[H_5C_6-\underset{\underset{C_6H_5}{|}}{\overset{\overset{N_3}{|}}{C}}-\ddot{N}: \longrightarrow \underset{H_5C_6}{\overset{N_3}{\diagdown}}C=N_{\diagdown C_6H_5} \longrightarrow \underset{H_5C_6}{\overset{:\ddot{N}}{\diagdown}}C=N_{\diagdown C_6H_5} \right]$$

(71)

$$\underset{\underset{\underset{C_6H_5}{|}}{H_5C_6}}{\overset{N=N}{\diagup}}\ \xrightarrow{hv}\ \text{(benzimidazole)}-C_6H_5 \qquad (C_6H_5N{=}C{=}NC_6H_5)_3$$

14%   52%   10%

(25)   (26)

It is interesting to note that thermolysis of 1,5-diphenyltetrazole gives both 2-phenylbenzimidazole and the carbodiimide trimer (89). Clearly, solution-phase photolysis and pyrolysis of benzophenone diazide give highly similar results, but these results are quite different from those obtained at low temperatures in rigid matrices. Several explanations for these differences have been advanced (88). Most likely, phenyl migration is unfavorable in the low-temperature matrix due to insufficient vibrational energy or unfavorable orientation with respect to the electron-deficient nitrogen. Stabilization thus occurs by loss of two more molecules of nitrogen rather than by phenyl migration.

Irradiation of a benzene solution of dimethyl diazidomalonate gives dimethyl tetrazole-1,5-dicarboxylate (27) in 48% yield (eq. 72) (90). Further irradiation causes loss of a second mole of nitrogen and the formation of an oxadiazole (28). Tetrazole formation requires migration of a carbomethoxy group, a transformation not previously observed. Oxadiazole

$$H_3CO_2C-\underset{\underset{CO_2CH_3}{|}}{\overset{\overset{N_3}{|}}{C}}-N_3 \xrightarrow[-N_2]{hv} \underset{\underset{CO_2CH_3}{|}}{\overset{N-N}{H_3CO_2C\diagup\diagdown N}}\ \xrightarrow[-N_2]{hv}\ \underset{H_3CO_2C}{\overset{N-O}{\diagup\diagdown OCH_3}}$$

(27)   (28)   (72)
48%

$$H_3CO_2C-\overset{N-N}{\underset{\underset{CO_2CH_3}{|}}{\diagup\diagdown N}} \longrightarrow \left[H_3CO_2C-\overset{:\ddot{N}}{\underset{N}{\diagup\diagdown}} \overset{:\ddot{O}}{OCH_3} \right] \longrightarrow \underset{N}{\overset{N-O}{H_3CO_2C-\diagup\diagdown OCH_3}}$$

(73)

formation is probably analogous to isoxazole formation from β-azidovinyl ketones (eq. 73).

IV. NITRENES FROM OTHER SOURCES

Alkylnitrenes have been postulated as intermediates in many reactions other than those of alkyl azides. The same problems that beset mechanistic investigations of alkyl azide decompositions are present in these other methods of preparing alkylnitrenes. In none of these reactions have ESR or spectroscopic investigations been used to establish the presence or absence of an alkylnitrene intermediate. In most cases trapping experiments have either not been carried out or have been unsuccessful. Hence, evidence for nitrene intermediates often rests solely on the products of the reaction and analogy to the reactions of alkylnitrenes formed from alkyl azides. Of course, in cases where the same species is formed from both an alkyl azide and a precursor other than an alkyl azide, strong evidence for a discrete nitrene intermediate is provided.

A. The Neber Rearrangement and Related Reactions

The various reactions which involve migration of carbon to nitrogen form an enormous part of the chemical literature. As early as 1896 Stieglitz (1) sought to generalize all rearrangements such as the Beckmann, Hofmann, Curtius, and related reactions as all proceeding through a univalent nitrogen or nitrene intermediate. Since the Beckmann rearrangement often gives different geometric isomers depending upon which isomeric oxime is used (5–7), it became clear that a univalent nitrogen intermediate could not be involved. The synchronous nature of the Beckmann rearrangement was unequivocally established by the work of Campbell, Kenyon, and Young (91,92) which showed that the migration occurred with retention of configuration about the migrating carbon atom. While it was thus established that the Beckmann rearrangement is a concerted reaction, no such conclusive evidence exists for several of the other carbon to nitrogen rearrangements. The Hofmann, Curtius, and Lossen rearrangements are often postulated to proceed via an acylnitrene intermediate (33,93); however, there is still much debate on this point. These reactions are discussed with other carbonylnitrene reactions elsewhere in this volume. The Schmidt reaction, like the acid-catalyzed decomposition of azides, could either be concerted or involve a protonated nitrene or nitrenium ion. Such intermediates are also discussed elsewhere in this volume.

Closely related to the Beckmann rearrangement is the rearrangement of ketoxime derivatives to α-amino ketones, known as the Neber rearrangement (94,95). A review of the Neber rearrangement has recently appeared

(96). Thus, only material pertinent to the mechanism of the reaction will be included in this discussion. The Neber rearrangement was discovered over forty years ago by Neber and Friedolsheim (94). They found that, on treatment with base followed by hydrolytic workup, oxime tosylates having an α-hydrogen form an α-amino ketone (eq. 74) rather than the

$$\underset{\text{NOTs}}{\overset{\parallel}{C_6H_5CH_2C}}\text{—CH}_3 \xrightarrow[\substack{\textit{3. HCl}}]{\substack{\textit{1. KOEt} \\ \textit{2. HOAc}}} \underset{\underset{NH_2}{\mid}}{C_6H_5CH}\text{—}\underset{\underset{O}{\parallel}}{C}\text{—CH}_3 \qquad (74)$$

expected Beckmann product. Subsequent investigations using mild reaction conditions led to the isolation of a cyclic intermediate complex (29) assigned an azirine structure (eq. 75) (95). The mechanism of the Neber

$$\text{(75)}$$

(29)

rearrangement received little further attention until the work of Cram and Hatch (97,98) which confirmed the azirine structure for the isolated intermediate. They also found that the product is not determined by the configuration of the oxime. A base-induced γ-elimination mechanism was proposed to explain the lack of configurational discrimination in the reaction. The resulting intermediate is formally a vinylnitrene which can be represented by several resonance forms (eq. 76) (98). Most subsequent workers have also favored a vinylnitrene intermediate.

$$\text{(76)}$$

House and Berkowitz (99,100) discussed but ruled out a possible mechanism initiated by attack of alkoxide on the carbon–nitrogen double bond followed by loss of tosylate to give the alkylnitrene (eq. 77). Such a

$$RCH_2C-CH_2R' \xrightarrow{R''O^-} RCH_2-\underset{\underset{\ominus N-OTs}{|}}{\overset{\overset{OR''}{|}}{C}}-CH_2R' \longrightarrow RCH_2-\underset{\underset{:N:}{|}}{\overset{\overset{OR''}{|}}{C}}-CH_2R' \quad (77)$$

with $NOTs$ under the first carbon.

nitrene should give both Neber products if the azirine were formed by nitrene carbon–hydrogen insertion. It was found that when

$$R = p\text{-}NO_2C_6H_4 \quad \text{and} \quad R' = p\text{-}CH_3OC_6H_4$$

only one product was formed in which the amino group was adjacent to the nitrophenyl group. This is in accord with Cram's mechanism, as base should remove the most acidic proton.

Smith and Most (101) found that dimethylhydrazone methiodides also undergo Neber rearrangement. Parcell (102) has studied the mechanism of this reaction. Under mildly basic conditions compound **30** forms an azirine (**31**) in 85% yield (eq. 78). With excess base an alkoxy aziridine (**32**) is isolated in 82% yield. Thus the azirine must be the first isolable inter-

mediate in the Neber rearrangement and reacts in turn to form an alkoxy aziridine which is hydrolyzed to an α-amino ketone. Isolable azirine intermediates have also been found in the Neber Rearrangement of steroidal dimethylhydrazone methiodides (eq. 79) (103). The formation of the azirine **33** seems anomalous since the least acidic α-hydrogen must be abstracted to give the observed product. It is not clear why this should be the case.

Azirine intermediates are thus well established in the Neber rearrangement. The question remains, however, whether these azirines arise from vinylnitrene (path a or b) or by a concerted displacement of the substituent on nitrogen by an initially formed carbanion (path c) (eq. 80). Further, a

$$R_2CH-\overset{\overset{\displaystyle N-X}{\|}}{C}-R \xrightarrow{a} R_2C=\overset{\overset{\displaystyle :N:}{|}}{C}-R$$

$$\xrightarrow{b} R_2\overset{\ominus}{C}-\overset{\overset{\displaystyle N-Y}{\|}}{C}-R \xrightarrow{-X} R_2C=\overset{\overset{\displaystyle :N:}{|}}{C}-R \longrightarrow R_2C\overset{\displaystyle N}{\diagup\diagdown}C-R \quad (80)$$

$$\xrightarrow{c} R_2\overset{\ominus}{C}-\overset{\overset{\displaystyle N-X}{\|}}{C}-R$$

vinylnitrene could arise from either a γ-elimination (path a) or carbanion formation followed by loss of the oxime substituent (path b). Clearly the vinylnitrene mechanism is attractive in view of azirine formation from nonterminal vinyl azides. One must keep in mind, however, that no definitive evidence for vinylnitrenes exists, either for vinyl azide decompositions or for the Neber rearrangement. In both cases the results are compatible with concerted azirine formation. One objection to a concerted mechanism (path c) is that it would require a front side displacement of a syn-oxime tosylate, since it is known that the oxime tosylates do not equilibrate as rapidly as they undergo Neber rearrangement (100). Recent evidence favoring a vinylnitrene intermediate has been obtained for the anomalous Neber rearrangement of dimethylhydrazone methiodide acetals to give α-imino orthoesters (eq. 81) (104). The reaction is postulated

$$R-\overset{\overset{\displaystyle N-\overset{+}{N}(CH_3)_3 I^-}{\|}}{C}-CH(OR')_2 \xrightarrow[-H^+]{i\text{PrO}^-} R-\overset{N}{C}\overset{\overset{\displaystyle \overset{+}{N}(CH_3)_3}{\diagdown}}{\underset{\overset{\displaystyle \ominus}{C}(OR')_2}{}} \longrightarrow R-\overset{\overset{\displaystyle :N:}{\|}}{C}=C\overset{OR'}{\diagdown OR'}$$

$$R-\overset{\overset{\displaystyle NH}{\|}}{C}-\overset{OR'}{\underset{\displaystyle OCH(CH_3)_2}{C-OR'}} \longleftarrow R-\overset{\overset{\displaystyle :\ddot{N}{}^{\ominus}}{\|}}{C}-\overset{OR'}{\underset{\displaystyle OCH(CH_3)_2}{C-OR'}} \xleftarrow{i\text{PrO}^-} R-\overset{\overset{\displaystyle :\ddot{N}{}^{\ominus}}{\|}}{C}-\overset{OR'}{\underset{\displaystyle OR'}{C+}} \quad (81)$$

to involve trapping of an intermediate vinylnitrene by isopropoxide. A charge-separated resonance hybrid might be favored in this case due to the charge-delocalizing effect of the alkoxy groups.

Further evidence on the mechanism of the Neber rearrangement could be provided by appropriate kinetic studies. First it should be established that the rate of the reaction as well as the product is not dependent upon the configuration of the oxime tosylate. The effect of deuterium substitution for the abstracted hydrogen upon the reaction rate would establish that this is the rate-determining step and give a measure of the extent of carbon–hydrogen bond breaking in the transition state. Use of sulfonates other than tosylate in determining the reaction rate should establish whether the leaving group is involved in the rate-determining step or is lost subsequent to the rate-determining step.

Several other rearrangements have been reported to possess mechanisms similar to that of the Neber rearrangement. Baumgarten and Bower (105) found that treatment of N,N-dichloro-*sec*-alkyl amines with sodium methoxide gave the corresponding α-amino ketones in yields comparable to those for the Neber rearrangement (eq. 82). This reaction has been

shown to proceed by dehydrochlorination to the N-chloroketimine (106, 107). Evidence for a subsequent azirine intermediate was obtained by reduction with lithium aluminum hydride to give an aziridine (108). Further evidence for the similarity of N,N-dichloroamine reactions and the Neber rearrangement is provided by the identical nature of the products obtained for several pairs of compounds (e.g., eq. 83) (109). Pyrolysis (110) and reduction (111,112) of ketoximes have also been postulated to be mechanistically similar to the Neber rearrangement. Wenkert and Barnett (110) found that pyrolysis of α-phenylcyclohexanone oxime gave tetrahydrocarbazole (34). They suggested a vinylnitrene as a possible intermediate (eq. 84). Smolinsky (59) sought to test this mechanism by decomposing 1-azido-2-phenylcyclohexene, but was unable to prepare the starting material.

Benzylketoximes and benzobicyclo[2.2.2]octenone oximes are found to form aziridines on reduction with lithium aluminum hydride (111,112). A Neber-like mechanism was originally proposed for this reaction (eq. 85)

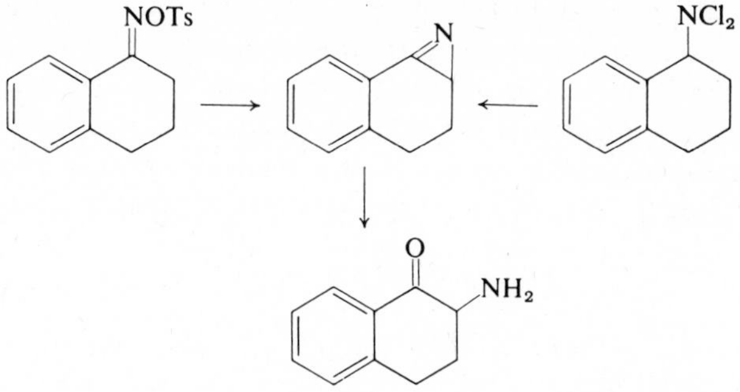

(83)

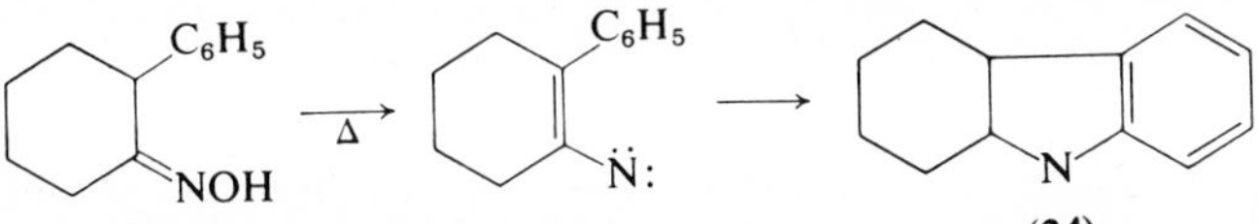

(84)

(34)

(111). Further investigation showed that the reaction products are influenced by the stereochemistry of the oximes used, thus ruling out a vinylnitrene intermediate (112).

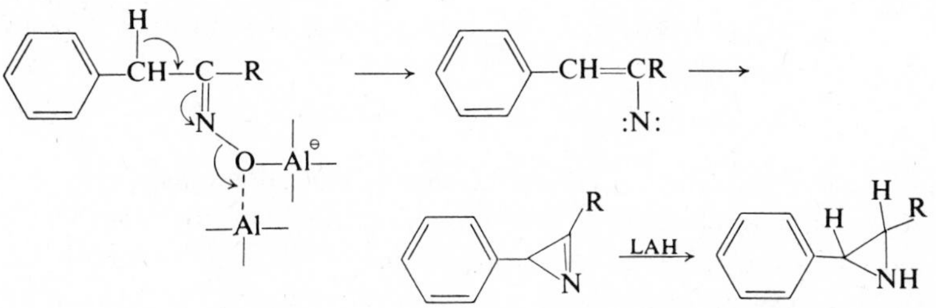

(85)

An imidylnitrene has been postulated as an intermediate in the halogenation of aryl or alkyl amidines to give substituted diazirines (eq. 86) (113). The mechanism of this reaction is complex but may be related to the Neber rearrangement.

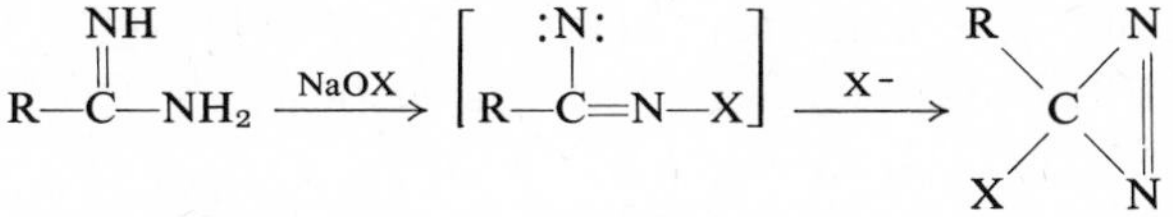

(86)

B. The Stieglitz Rearrangement: *N*-Halo and Hydroxyl Amines

The rearrangements of alkyl haloamines in the presence of strong base have been postulated to involve a nitrene intermediate. Such a mechanism would require an α-elimination analogous to the known carbene-forming α-elimination reactions. α-Eliminations of *N*-haloamines to give acyl-nitrenes in the Hofmann rearrangement have also been postulated and are discussed elsewhere in this volume. Stieglitz and co-workers (4,114,115) studied the rearrangements of several trityl haloamines. *N*-Bromotrityl-amine was found to give benzophenone phenylimine on treatment with sodium hydroxide (eq. 87) (4,115), *p*-Halotrityl-*N*-haloamines were re-

$$(C_6H_5)_3CNHBr \xrightarrow{\text{NaOH}} (C_6H_5)_2C{=}N(C_6H_5) \qquad (87)$$

ported to give a statistical distribution of products derived from phenyl and *p*-halophenyl migration (114,115). In view of the results of Saunders and Ware (44) for trityl azide rearrangements, it is likely that the methods of analysis used by Stieglitz and co-workers were not sufficiently accurate to detect small differences in migratory aptitudes. Under more vigorous conditions *N,N*-dichlorotritylamine (114) and *N,N*-dichloro-1,1-diphenyl-ethylamine (115) undergo rearrangement (eqs. 88 and 89). On the other hand, *N*-methyl-*N*-chlorotritylamine does not undergo base-induced re-arrangement (eq. 90) (116). Stieglitz proposed a tritylnitrene intermediate

$$(C_6H_5)_3CNCl_2 \xrightarrow[120-130°]{\text{soda lime}} (C_6H_5)_2C{=}NC_6H_5 \qquad (88)$$

$$(C_6H_5)_2(CH_3)CNCl_2 \xrightarrow[\text{soda lime}]{130°} C_6H_5C(CH_3){=}NC_6H_5 \qquad (89)$$

$$(C_6H_5)_3CNClCH_3 \xrightarrow{\text{NaOH}} \text{No reaction} \qquad (90)$$

for all these rearrangements. The initial step in the haloaminere actions is doubtless abstraction of the *N*-hydrogen. For the dichloroamines Cl⁺ is removed by base to give the monochloro anion. Whether halogen is lost prior to, or concerted with phenyl migration is not clear. A reexamination of the migratory aptitudes for phenyl-substituted trityl haloamines might establish whether the reaction is concerted or not. Other reactions of haloamines make an α-elimination, free nitrene mechanism seem unlikely (see below).

A further example of the Stieglitz rearrangement of haloamines is the (formation of 9-substituted phenanthridines from 9-fluorenyl haloamines eq. 91) (47). This rearrangement is analogous to that obtained from the corresponding 9-fluorenyl azides. Again either a nitrene or an internal S_N2 mechanism may be involved.

As previously discussed, secondary alkyl dichloroamines undergo base-catalyzed Neber rearrangement. Reactions of primary haloamines and chloramine itself with base have been postulated to involve nitrene intermediates; however, the bulk of the evidence favors an S_N2 mechanism for these reactions (117). Raschig (118) originally suggested that the synthesis of hydrazine from hypochlorite and ammonia involved attack of NH upon ammonia. This view was commonly held until recently. Yagil and Anbar (119) have shown that the hydrazine synthesis is first order in both chloroamine and ammonia thus establishing the S_N2 mechanism first suggested by Bodenstein (eqs. 92 and 93) (120). 1-N-Chloro-1,3-diaminopropane was

$$NH_3 + NH_2Cl \xrightarrow{S_N2} N_2H_5^+ + Cl^- \tag{92}$$

$$N_2H_5^+ + OH^- \longrightarrow N_2H_4 + H_2O \tag{93}$$

reported to form pyrazolidine via a nitrene intermediate (eq. 94) (121). Horner, Christmann, and Gross (63) showed that a nitrene is probably

$$H_2N-(CH_2)_3-NHCl \xrightarrow{NaOH} [H_2N-(CH_2)_3-\ddot{N}] \longrightarrow \underset{HN-NH}{\square} \tag{94}$$

not involved in this reaction by photolyzing 3-azidopropylamine. No ring formation was observed. The only products were polymers arising from the initially formed imine (eq. 95). Pyrazolidine formation from 1-N-

$$H_2N-(CH_2)_3-N_3 \xrightarrow{h\nu} [H_2N-(CH_2)_3-\ddot{N}:] \longrightarrow$$
$$[H_2N(CH_2)_2CH=NH] \longrightarrow Polymer \tag{95}$$

chloro-1,3-diaminopropane most likely occurs via an intramolecular S_N2 reaction. Several other interesting applications of N-haloamines to heterocyclic syntheses have recently been developed (122). A few representative examples from the work of Schmitz appear in equations 96–98 (123–125). The rate of product formation in these reactions depends on the nature of

$$\xrightarrow{H_3CNHCl}{OH^-} \tag{96}$$

$$\xrightarrow{H_3CNHCl}{OH^-} \tag{97}$$

$$\text{(98)}$$

the substrate, which would probably not be the case if nitrene formation were rate determining (126). Hence, these haloamine reactions are best regarded as nucleophilic substitution or addition reactions. The Hofmann-Löffler reaction of N-haloamines (eq. 99) does not involve a nitrene intermediate but rather a free radical mechanism (127).

$$RCH_2(CH_2)_3NXR' \xrightarrow[\text{heat}]{H^+} \xrightarrow{OH^-} \qquad \text{(99)}$$

Recently, the thermolysis (128) and photolysis (129) of N,N-dihaloperfluoroalkyl amines have been found to give symmetrical azo compounds, presumably by the coupling of a nitrene intermediate (eqs. 100–103).

$$2CClF_2CF_2NCl_2 \xrightarrow{200°} CClF_2CF_2N{=}NCF_2CClF_2 \qquad \text{(100)}$$

$$NCl_2CF_2CF_2CF_2NCl_2 \xrightarrow{200°} \qquad \text{(101)}$$

$$CF_3NClF \xrightarrow{h\nu} CF_3N{=}NCF_3 \qquad \text{(102)}$$

$$CF_3NCl_2 \xrightarrow{h\nu} CF_3N{=}NCF_3 \qquad \text{(103)}$$

It has been suggested that coupling of phenylnitrene is essentially a reaction of triplet nitrene, whereas insertion reactions are due to singlet nitrene (130). Since thermolysis or photolysis of N,N-dihaloperfluoroalkyl amines involves cleavage of two bonds to nitrogen it seems likely that a triplet nitrene is formed. Thus, by analogy to the coupling of phenyl-nitrene, the coupling observed for perfluoro nitrenes may be due to a triplet nitrene. In these gas phase reactions it appears that the C—F bond is sufficiently strong to prevent rapid isomerization to the fluoro imine. Thus the most favorable reaction is a dimerization similar to that observed for phenylnitrene.

Trityl hydroxylamines undergo Stieglitz rearrangement on warming with phosphorus pentachloride (3,131). p-Chlorotrityl hydroxylamine was reported to give statistical amounts of the two possible imines (132). o-Methyl- and o-benzyltrityl hydroxylamines are much more resistant to rearrangement and give only partial reaction even with strong heating in the presence of phosphorous pentachloride (133). Newman and Hay (134) reexamined the effect of substituents on the migration aptitudes of trityl hydroxylamines. *p*-Methoxyphenyl was found to migrate nine times more readily than phenyl. *p*-Chlorophenyl and *p*-nitrophenyl migrated less readily than phenyl. Though of only qualitative significance, these results show the same trends exhibited in the Stieglitz rearrangement of trityl azides (44).

Stieglitz proposed a nitrene mechanism for these rearrangements. In view of the migratory aptitudes observed by Newman and Hay (134) a concerted reaction with some phenyl participation in the transition state seems more likely. Smith (6) has suggested that a chlorophosphonate ester is the species which undergoes rearrangement with loss of phosphoryl chloride (eq. 104). Such intermediates have not been isolated, and in the absence of more definitive experimental evidence the mechanism of these reactions remains uncertain.

$$(C_6H_5)_3CNHOH \xrightarrow{PCl_5} [(C_6H_5)_3C\text{—}NH\text{—}O\overset{+}{P}Cl_3] \longrightarrow$$
$$(C_6H_5)_2C\text{=}NC_6H_5 + POCl_3 \quad (104)$$

C. Fluorination of Perhalonitriles

Fluorination of perfluoronitriles under mild fluorinating conditions is observed to give perfluoroazoalkanes. This reaction is postulated to involve formation of a perfluoro nitrene which dimerizes to form the azo compound. For example, gas-phase fluorination of pentafluoroethyl-nitrile at 54–65° gives an azo compound (eq. 105) whereas at higher tem-

$$C_2F_5CN \xrightarrow[54-65°]{F_2} [C_3F_7\ddot{N}:] \longrightarrow C_3F_7N\text{=}NC_3F_7 \quad (105)$$
$$13\%$$

peratures only the perfluoroamine is observed (135). Both hydrogen cyanide and cyanogen give the same products upon fluorination, among them being CF_3NF_2, $(CF_3)_2NF$, $(CF_3)_3N$, and $CF_3N\text{=}NCF_3$ (eqs. 106–109) (136). Fluorination of $CClF_2CN$ gives the perhaloamine, the un-

$$HCN \text{ or } NCCN \xrightarrow{F_2} FCN \quad (106)$$

$$FCN \xrightarrow{F_2} CF_3\ddot{N}: \quad (107)$$

$$CF_3\ddot{N}: \xrightarrow{F_2} CF_3NF_2 \quad 20\text{–}89\% \quad (108)$$

$$2CF_3\ddot{N}: \longrightarrow CF_3N\text{=}NCF_3 \quad 2\text{–}8\% \quad (109)$$

saturated fluoroimine, and an unsymmetrical azo compound (eqs. 110 and 111) (137). Direct jet fluorination of fluorinated aliphatic dinitriles gives

$$CClF_2CN \xrightarrow{F_2} CClF_2CF_2NF_2 + CClF_2CF=NF + [CClF_2CF_2\ddot{N}:]$$
$$\quad\quad\quad\quad\quad\quad 12\% \quad\quad\quad\quad 15\%$$
$$+ CClF_3 + [CF_2\ddot{N}:] \quad (110)$$

$$CClF_2CF_2\ddot{N}: + CF_3\ddot{N}: \longrightarrow CClF_2CF_2N=NCF_3 \quad (111)$$
$$\quad\quad\quad\quad\quad\quad\quad\quad\quad\quad 10\%$$

cyclization products (138,139). Tetrafluorosuccinonitrile forms perfluoro-pyrrolidine in 20% yield upon fluorination at 145° (eq. 112) (138). A possible mechanism for this reaction involves attack of a divalent nitrogen

$$N\equiv CCF_2CF_2C\equiv N \xrightarrow{F_2} \quad \begin{array}{c} F_2C\text{---}CF_2 \\ F_2C \quad CF \\ F_2N \quad N \end{array} \longrightarrow \begin{array}{c} F_2C\text{---}CF_2 \\ F_2C \quad CF_2 \\ N \end{array} \xrightarrow{F_2} \begin{array}{c} F_2C\text{---}CF_2 \\ F_2C \quad CF_2 \\ N \\ F \end{array} \quad (112)$$
$$\quad\quad\quad\quad\quad\quad\quad\quad\quad\quad\quad\quad\quad\quad\quad\quad\quad\quad\quad\quad\quad\quad\quad\quad\quad 20\%$$

radical with displacement of an NF_2 radical (137,139). Similarly, hexa-fluoroglutaronitrile gives nonafluoro-1-piperidine and perfluoropiperidine among other products (eq. 113) (139).

$$N\equiv C(CF_2)_3C\equiv N \xrightarrow{F_2} \quad \begin{array}{c} F_2 \\ C \\ F_2C \quad CF_2 \\ F_2C \quad CF \\ N \end{array} + \begin{array}{c} F_2 \\ C \\ F_2C \quad CF_2 \\ F_2C \quad CF_2 \\ N \\ F \end{array} \quad (113)$$

Fluorinated cyclic azo compounds are formed by indirect fluorination of dinitriles with argentic fluoride (eqs. 114–116) (140,141). The cyclic

$$HCN \xrightarrow[105-115°]{AgF_2} \quad \begin{array}{c} F_2C\text{---}CF_2 \\ N=N \end{array} \quad 90\% \quad (114)$$

$$NCCCl_2CN \xrightarrow[100°]{AgF_2} \quad \begin{array}{c} CCl_2 \\ F_2C \quad CF_2 \\ N=N \end{array} \quad 25\% \quad (115)$$

$$NCCF_2CN \xrightarrow[100°]{AgF_2} \quad \begin{array}{c} F_2 \\ C \\ F_2C \quad CF_2 \\ N=N \end{array} \quad 15\% \quad (116)$$

azo compounds are postulated to arise from intramolecular coupling of two nitrenes (141). Azoalkanes are also formed in high yield by treatment of perhalonitriles with argentic fluoride (eqs. 117 and 118) (141,142). Use of a more vigorous fluorinating agent, cesium fluoride, gave no unsaturated products (143). Argentic fluoride has also been used to convert an iso-cyanate to an azo compound (eq. 119) (144).

$$CClF_2CN \xrightarrow{\text{AgF}_2} \underset{45\%}{CClF_2CF_2N{=}NCF_2CClF_2} \qquad (117)$$

$$CF_3CN \xrightarrow{\text{AgF}_2} \underset{90\%}{C_2F_5N{=}NC_2F_5} \qquad (118)$$

$$C_2F_5{-}N{=}C{=}O \xrightarrow{\text{AgF}_2} \left[C_2F_5{-}\overset{..}{N}\underset{F}{\overset{F\cdot}{C}}{=}O \right] \longrightarrow$$

$$[C_2F_5\overset{..}{N}{:}] \longrightarrow C_2F_5N{=}NC_2F_5 \qquad 82\% \qquad (119)$$

The high yields of azo compounds obtained with argentic fluoride suggests that these reactions may be more complex than a simple coupling of two free nitrenes. A silver–nitrene complex may be formed which could enhance nitrene coupling both by increasing the lifetime of the nitrene and providing a catalytic surface on which the coupling could occur.

Other fluorinating agents have also been used to form azo compounds from perhalo nitriles. Cyanogen iodide and iodine pentafluoride have long been known to give hexafluoroazomethane (145). Mixed reagents such as argentous fluoride and chlorine or sodium fluoride and chlorine have also been successfully employed (146). The formation of trifluoromethyl isocyanate by reaction of hexafluoroazomethane with carbon monoxide at 325° and 650 atm indicates that the formation of perfluoro azo compounds may be reversible (eq. 120) (139). The formation of perfluoro azo

$$CF_3N{=}NCF_3 \longrightarrow 2[CF_3\overset{..}{N}{:}] \xrightarrow{\text{CO}} \underset{88\%}{CF_3N{=}C{=}O} \qquad (120)$$

compounds from the fluorination of perfluoro nitriles and isocyanates as well as from the photolysis and thermolysis of N,N-dihaloperfluoroalkyl amines (eqs. 100–103) suggests that both reactions proceed through the same intermediate. Stepwise addition of two fluorine atoms to a nitrile or iso-cyanate would be expected to give a triplet nitrene. Thus, it seems prob-able that the coupling of perfluoro nitrenes is due to a triplet intermediate.

D. Active Nitrogen and Nitrene Reactions with Organic Molecules

Nitrene intermediates have been postulated in the reactions of olefins and other organic molecules with active nitrogen. The literature of active nitrogen reactions with organic molecules is large and no attempt at a thorough review will be made. Only those results which bear upon the question of nitrene intermediates will be discussed. Most active nitrogen reactions have been carried out in flow reactors. Active nitrogen is normally generated by microwave or electric discharge. In many of the studies, kinetics of product formation has been followed spectroscopically or mass spectrometrically. The most extensively studied reaction is that of active nitrogen in its quartet ground state with simple olefins. The major nitrogen-containing product in the reaction of active nitrogen with ethylene and propylene is hydrogen cyanide (147–150) formed in 97 and 78% yield, respectively (148). The mechanism suggested to explain the formation of hydrogen cyanide and other reaction products involves attack of active nitrogen upon the olefin to form a triradical or dehydronitrene (147). Spin inversion, hydrogen migration, and carbon–carbon homolytic cleavage gives a net reaction (eq. 121) which is energetically favorable.

$$H_2C{=}CH_2 + N \longrightarrow \underset{\underset{\cdot}{\overset{|}{:N\cdot}}}{H_2C{-}CH_2} \longrightarrow \underset{\overset{|}{:N\cdot}}{HC{-}CH_3} \longrightarrow HCN + H_3C\cdot \quad (121)$$

Methyl radical can react further with active nitrogen to give hydrogen cyanide and two hydrogen atoms. The latter react with olefin to give the saturated hydrocarbons observed in these reactions. Kinetic studies support the formation of hydrogen cyanide by direct attack of nitrogen on ethylene rather than by some secondary reaction (151).

Active nitrogen reacts with butadiene to give pyrrole and *cis*- and *trans*-crotonitrile along with degradation products (152–154). The mechanism of this reaction probably involves nitrene formation followed by loss of hydrogen (eq. 122). Herron (150) suggests that pyrrole formation proceeds

$$H_2C{=}CH{-}CH{=}CH_2 + N \longrightarrow \cdot NCH_2{-}\overset{\overbrace{\hspace{3em}}}{CH{-}CH}{-}CH_2 \xrightarrow{-H}$$

$$\underset{15\%}{\underset{H}{\overset{H_3C}{>}}C{=}C\underset{CN}{\overset{H}{<}}} \quad + \quad \underset{9\%}{\underset{H}{\overset{H_3C}{>}}C{=}C\underset{H}{\overset{CN}{<}}} \quad + \quad \underset{28\%}{\boxed{}} \qquad (122)$$

through a five-membered cyclic intermediate which yields pyrrole by loss of a hydrogen atom. Lichtin and co-workers (155,156) have also studied the reaction of active nitrogen with ^{14}C-labeled propylene. They propose that bridged intermediates rather than nitrenes are involved. Insufficient evidence is available to decide on the validity of either mechanism.

Acetylene undergoes reaction with active nitrogen to give hydrogen cyanide and a polymer (157,158). The polymer is probably formed by a chain mechanism initiated by a nitrene triradical (eq. 123) (158). Alternatively, the nitrene could lose hydrogen to form cyanomethylene which would then polymerize with acetylene.

$$\text{H---C}\equiv\text{C---H} + \text{N} \longrightarrow \overset{\text{H}}{\underset{:\overset{\cdot}{\underset{\cdot}{\text{N}}}\cdot}{\diagdown}}\text{C}=\overset{\cdot}{\text{C}}\text{---H} \xrightarrow{\text{HC}\equiv\text{CH}} \text{Polymer} \qquad (123)$$

Alkyl chlorides react with active nitrogen to form hydrogen cyanide and hydrogen chloride as the principal products (159). The similarity of reaction rates and products to the analogous reactions of olefins with active nitrogen suggests that nitrogen attacks saturated carbon with expulsion of hydrogen chloride to give the dehydronitrene observed in the olefin reactions (eq. 124). This mechanism is supported by the absence of any

$$\text{C}_2\text{H}_5\text{Cl} + \text{N} \longrightarrow \underset{\cdot\overset{\cdot\cdot}{\text{N}}:}{\text{H}_2\text{C---CH}_2} + \text{HCl} \qquad (124)$$

molecular chlorine, which would be expected from a simple displacement of atomic chlorine.

Methanol (160) and ethanol (161) also react with active nitrogen at high temperatures to form hydrogen cyanide and other products. In this case loss of hydroxyl radical is postulated to give a saturated alkyl nitrene (eq. 125) (160). The formation of hydrogen and hydrogen cyanide is also

$$\text{CH}_3\text{OH} + \text{N} \xrightarrow{300\text{--}750°} \text{CH}_3\text{---}\overset{\cdot\cdot}{\text{N}}: + \overset{\cdot}{\text{O}}\text{H}$$

$$\text{CH}_3\overset{\cdot\cdot}{\text{N}}: \longrightarrow \text{HCN} + 2\text{H}\cdot \qquad (125)$$

observed in the high temperature pyrolysis of methyl azide (27) and the flash photolysis of ethyl azide (31). Hence it seems likely that an alkylnitrene is formed in the reactions of active nitrogen with alcohols.

Photolysis of gaseous hydrazoic acid in the presence of olefins (31) gives products similar to those obtained with active nitrogen and olefins. However, in this case nitrene is concluded to be the attacking species on the basis of spectroscopic evidence. The proposed reaction sequence involves attack of nitrene on the double bond followed by a 1,3-hydrogen shift to form an alkylnitrene (eq. 126). The fate of the alkylnitrene is

$$\text{H}_2\text{C}=\text{CH}_2 + \text{H}\overset{\cdot}{\underset{\cdot\cdot}{\text{N}}}\cdot \longrightarrow \underset{\text{H}\overset{\cdot}{\text{N}}\cdot}{\text{H}_2\text{C---CH}_2} \longrightarrow \underset{:\overset{\cdot}{\text{N}}\cdot}{\text{H}_2\text{C---CH}_3} \qquad (126)$$

determined by the energy of the nitrene from which it was formed. In the flash photolysis a highly energetic or "hot" intermediate is formed which breaks all three C_α bonds to give CN radicals. The continuous irradiation of hydrazoic acid with olefins gives both hydrogen cyanide and nitriles. It is clear from these results and those for gas-phase flash and continuous photolysis of ethyl azide (see above) that the products obtained from alkylnitrenes are more dependent upon the energy of the nitrene than the method by which it is formed. This result would only be expected to hold for gas-phase reactions in which nitrene rearrangements can occur more rapidly than loss of vibrational energy through collisions.

E. Photolysis of Isocyanates and Phosphineimines

Several interesting methods of preparing alkylnitrenes have been reported recently. As mentioned previously β-styryl isocyanate forms the same product as β-styryl azide upon photolysis (eq. 127) (69). Presumably

$$(C_6H_5)CH\!\!=\!\!CHNCO \xrightarrow[-CO]{h\nu} (C_6H_5)CH\!\!=\!\!CH\ddot{N}: \longrightarrow (C_6H_5)CH_2C\!\!\equiv\!\!N \quad (127)$$

a nitrene is formed by loss of carbon monoxide. The only previous example of probable nitrene formation by isocyanate photolysis was the photolysis of methyl isocyanate (eq. 128) (162). Although no products were identified,

$$CH_3NCO \xrightarrow[-CO]{h\nu} CH_3\ddot{N}: \longrightarrow ? \quad (128)$$

a white film such as is usually observed in methyl azide photolyses was present. Recently, o-biphenylyl isocyanate has been photolyzed to carbazole and phenanthridone (eq. 129) (163). Further examples are necessary

to establish the generality of isocyanate photolyses as a route to nitrenes.

The photolysis of triphenylphosphine-t-butylimine has been reported to give products arising from t-butylnitrene and t-butyl radical (eqs. 130–132)

$$(C_6H_5)_3P\!\!=\!\!NC(CH_3)_3 \xrightarrow{h\nu} CH_3C\ddot{N}: + (C_6H_5)_3P:$$

$$+ (CH_3)_3C\cdot + (C_6H_5)_3P\!\!=\!\!\ddot{N}: \quad (130)$$

$$(CH_3)_3C\ddot{N}: + (CH_3)_3C\cdot \longrightarrow ((CH_3)_3C)_2NH \quad (132)$$
$$25\%$$

(164). The products seem in reasonable accord with a nitrene mechanism except for the unusually high amount of cyclohexene insertion product. Alkylnitrenes from photolyses and thermolyses of alkyl azides give but small yields of intermolecular insertion products. It is possible that the *t*-butylnitrene formed in this reaction differs in spin multiplicity from the nitrene formed in the photolyses and thermolyses of alkyl azides. Further study of this novel reaction will be necessary to establish a nitrene mechanism.

REFERENCES

1. J. Stieglitz, *Am. Chem. J.*, **18**, 751 (1896).
2. J. K. Senior, *J. Am. Chem. Soc.*, **38**, 2718 (1916).
3. J. Stieglitz and P. N. Leech, *Chem. Ber.*, **46**, 2147 (1913).
4. J. Stieglitz and I. Vosburgh, *Chem. Ber.*, **46**, 2151 (1913).
5. L. G. Donaruma and W. Z. Heldt, *Org. Reactions*, **11**, Chap. 1 (1960).
6. P. A. S. Smith, *Molecular Rearrangements*, Vol. 1, P. de Mayo, Ed., Interscience, New York, 1963, Chap. 8.
7. P. A. S. Smith, *Open-Chain Nitrogen Compounds*, Vol. 2, Benjamin, New York, 1966, Chap. 8.
8. J. H. Boyer and F. C. Canter, *Chem. Rev.*, **54**, 1 (1954).
9. L. I. Smith, *Chem. Rev.*, **23**, 193 (1938).
10. J. A. Leermakers, *J. Am. Chem. Soc.*, **55**, 2719, 3098 (1933).
11. G. Geiseler and W. König, *Z. Phys. Chem.*, **227**, 81 (1964).
12. W. D. Closson and H. B. Gray, *J. Am. Chem. Soc.*, **85**, 290 (1963).
13. R. M. Moriarty and M. Rahman, *Tetrahedron*, **21**, 2877 (1965).
14. F. D. Lewis and W. H. Saunders, Jr., *J. Am. Chem. Soc.*, **89**, 645 (1967).
15. E. Koch, *Tetrahedron*, **23**, 1747 (1967).
16a. F. D. Lewis and W. H. Saunders, *J. Am. Chem. Soc.*, **90**, 3828 (1968).
16b. F. D. Lewis and W. H. Saunders, *J. Am. Chem. Soc.*, **90**, 7031 (1968).
16c. F. D. Lewis and W. H. Saunders, *J. Am. Chem. Soc.*, **90**, 7033 (1968).
17. E. Wasserman, G. Smolinsky, and W. A. Yager, *J. Am. Chem. Soc.*, **86**, 3166 (1964).
18. G. Smolinsky, E. Wasserman, and W. A. Yager, *J. Am. Chem. Soc.*, **84**, 3220 (1962).
19. A. Reiser, G. Bowes, and R. J. Horne, *Trans. Faraday Soc.*, **62**, 3162 (1966).
20. A. Bertho, *Chem. Ber.*, **57**, 1138 (1924).
21. J. S. McConaghy, Jr., and W. Lwowski, *J. Am. Chem. Soc.*, **89**, 2357 (1967).
22. A. G. Anastassiou, *J. Am. Chem. Soc.*, **89**, 3184 (1967).
23. J. L. Franklin, V. H. Dibeler, R. M. Reese, and M. Krauss, *J. Am. Chem. Soc.*, **80**, 298 (1958).
24. C. L. Currie and B. D. Darwent, *Can. J. Chem.*, **41**, 1552 (1963).
25. B. L. Evans, A. D. Yoffe, and P. Gray, *Chem. Rev.*, **59**, 515 (1959).
26. H. C. Ramsperger, *J. Am. Chem. Soc.*, **51**, 2134 (1929).
27. F. O. Rice and C. J. Grelecki, *J. Phys. Chem.*, **61**, 830 (1957).
28. W. Pritzkow and D. Timm, *J. Prakt. Chem.*, **32**, 178 (1966).
29. D. W. Milligan, *J. Chem. Phys.*, **35**, 1491 (1961).
30. D. E. Milligan and M. E. Jacox, *J. Chem. Phys.*, **39**, 712 (1963).
31. D. W. Cornell, R. S. Berry, and W. Lwowski, *J. Am. Chem. Soc.*, **88**, 544 (1966).

31a. T. D. Lewis and J. C. Dalton, unpublished results..
32. D. H. R. Barton and L. R. Morgan, Jr., *Proc. Chem. Soc.*, 206 (1961); *J. Chem. Soc.*, 622 (1962).
33. R. A. Abramovitch and B. A. Davis, *Chem. Rev.*, **64**, 149 (1964).
34. E. Lieber, J. S. Curtice, and C. N. R. Rao, *Chem. Ind.*, **1966**, 586.
35. J. C. Calvert and J. N. Pitts, *Photochemistry*, Wiley, New York, 1966, p. 474.
36. R. O. Kan, *Organic Photochemistry*, McGraw-Hill, New York, 1966, p. 248.
37. D. H. R. Barton and A. N. Starratt, *J. Chem. Soc.*, **1965**, 2444.
38. G. Smolinsky, *J. Am. Chem. Soc.*, **83**, 2491 (1961).
39. P. A. S. Smith and B. B. Brown, *J. Am. Chem. Soc.*, **73**, 2435 (1957).
40. R. Kreher and G. H. Bockhorn, *Angew. Chem.*, **76**, 681 (1964).
41. R. Ledger and J. McKenna, *Chem. Ind.*, **1963**, 1662.
41a. R. A. Abramovitch and E. P. Kyba, *Chem. Commun.*, 265 (1969).
42. I. L. Knunyants, E. G. Bykhovskaya, and V. N. Frosin, *Dokl. Akad. Nauk SSSR*, **132**, 513 (1960).
43. R. E. Banks and G. J. Moore, unpublished results.
44. W. H. Saunders, Jr. and J. C. Ware, *J. Am. Chem. Soc.*, **80**, 3328 (1958).
45. W. H. Saunders, Jr. and E. A. Caress, *J. Am. Chem. Soc.*, **86**, 861 (1964).
46. P. Walker and W. A. Waters, *J. Chem. Soc.*, **1962**, 1632.
47. L. A. Pinck and G. E. Hilbert, *J. Am. Chem. Soc.*, **59**, 8 (1937).
48. C. L. Arcus, R. E. Marks, and M. M. Coombs, *J. Chem. Soc.*, **1957**, 4064.
49. J. H. Boyer and D. S. Straw, *J. Am. Chem. Soc.*, **75**, 1642 (1953).
50. R. H. B. Galt, J. D. Loudon, and A. B. D. Sloan, *J. Chem. Soc.*, **1958**, 1588.
51. T. Curtius and G. Ehrhart, *Chem. Ber.*, **55**, 1559 (1922).
52. T. Curtius and K. Raschig, *J. Prakt. Chem.*, **125**, 466 (1930).
53. B. Coffin and R. F. Robbins, *J. Chem. Soc.*, **1965**, 1252.
53a. N. Koga, G. Koga and J. P. Anselme, *Can. J. Chem.*, **46**, 1143 (1969).
54. A. L. Logothetis, *J. Am. Chem. Soc.*, **87**, 749 (1965).
55. M. O. Foster and H. E. Fierz, *J. Chem. Soc.*, **1908**, 1174.
56. A. Gagneux, S. Winstein, and W. G. Young, *J. Am. Chem. Soc.*, **82**, 5956 (1960).
57. C. A. VanderWerf and V. L. Heasley, *J. Org. Chem.*, **31**, 3534 (1966).
57a. A. H. Hortman and J. E. Martinelli, *Tetrahedron Letters*, 6205 (1968).
57b. G. G. Smolinsky, *Trans. N.Y. Acad. Sci.*, **30**, 511 (1968).
57c. J. H. Boyer, W. E. Krueger, and R. Modler, *Tetrahedron Letters*, 5979 (1968).
58. G. Smolinsky, *J. Am. Chem. Soc.*, **83**, 4483 (1961).
59. G. Smolinsky, *J. Org. Chem.*, **27**, 3557 (1962).
60. G. R. Harvey and K. W. Ratts, *J. Org. Chem.*, **31**, 3907 (1966).
61. J. S. Meek and J. S. Fowler, *J. Am. Chem. Soc.*, **89**, 1967 (1967).
62. G. L. Closs and L. E. Closs, *J. Am. Chem. Soc.*, **83**, 2015 (1961).
62a. K. Friedrich, *Angew. Chem. Int. Ed. Engl.*, **6**, 959 (1967).
63. L. Horner, A. Christmann, and A. Gross, *Chem. Ber.*, **96**, 399 (1963).
64. H. Reimlinger, F. P. Woerner, and D. R. Arnold, *Angew. Chem. Int. Ed. Engl.*, **7**, 130 (1968).
64a. K. Isomura, S. Kobayashi, and H. Taniguchi, *Tetrahedron Letters*, 3499 (1968).
65. A. Hassner and L. A. Levy, *J. Am. Chem. Soc.*, **87**, 4203 (1965).
66. A. Hassner and F. W. Fowler, *Tetrahedron Letters*, **1967**, 1545.
67. F. W. Fowler, A. Hassner, and L. A. Levy, *J. Am. Chem. Soc.*, **89**, 2077 (1967).
67a. A. Hassner and F. W. Fowler, *J. Am. Chem. Soc.*, **90**, 2869 (1968).
68. R. H. Wiley and J. Moffat, *J. Org. Chem.*, **22**, 995 (1957).
69. J. H. Boyer, W. E. Krueger, and G. J. Mikol, *J. Am. Chem. Soc.*, **89**, 5504 (1967).

69a. J. S. Meek and J. S. Fowler, *J. Org. Chem.*, **33**, 3418 (1968).

70. C. S. Cleaver and C. G. Krespan, *J. Am. Chem. Soc.*, **87**, 3716 (1965).

71. R. E. Banks and G. J. Moore, *J. Chem. Soc.*, **1966**, 2304.

72. I. L. Knunyants and E. G. Bykhovskaya, *Dokl. Akad. Nauk SSSR*, **131**, 1338 (1960).

73. A. N. Nesmeyanov and M. I. Rybinskaya, *Izv. Akad. Nauk SSSR, Otd. Khim. Nauk*, **1962**, 816.

74. S. Maiorana, *Ann. Chim. (Rome)*, **56**, 1531 (1966).

75. J. H. Boyer and D. Straw, *J. Am. Chem. Soc.*, **74**, 4506 (1952).

76. J. H. Boyer and D. Straw, *J. Am. Chem. Soc.*, **75**, 2683 (1953).

77. R. Kreher and G. Jager, *Angew. Chem.*, **77**, 963 (1965).

78. E. Gudriniece, N. R. Burvele and A. Ievins, *Dokl. Akad. Nauk SSSR*, **171**, 869 (1966).

79. J. H. Boyer and J. Stocker, *J. Org. Chem.*, **21**, 1030 (1956).

80. R. M. Moriarty and M. Rahman, *J. Am. Chem. Soc.*, **87**, 2519 (1965).

81. W. J. Wechter, *J. Org. Chem.*, **31**, 2136 (1966).

82. G. Schroeter, *Chem. Ber.*, **42**, 2336 (1909).

83. G. Schroeter, *Chem. Ber.*, **42**, 3356 (1909).

84. G. Schroeter, *Chem. Ber.*, **44**, 1201 (1911).

85. S. Gotzky, *Chem. Ber.*, **64**, 1555 (1931).

86. H. Lindemann and A. Mühlhaus, *Ann.*, **446**, 1 (1925).

87. L. Barash, E. Wasserman, and W. A. Yager, *J. Am. Chem. Soc.*, **89**, 3931 (1967).

88. R. M. Moriarty and J. M. Kliegman, *J. Am. Chem. Soc.*, **89**, 5959 (1967).

89. P. A. S. Smith and E. Leon, *J. Am. Chem. Soc.*, **80**, 4647 (1958).

90. R. M. Moriarty, J. M. Kliegman, and C. Shovlin, *J. Am. Chem. Soc.*, **89**, 5958 (1967).

91. J. Kenyon and D. P. Young, *J. Chem. Soc.*, **1941**, 263.

92. A. Campbell and J. Kenyon, *J. Chem. Soc.*, 25 (1946).

93. C. K. Ingold, *Structure and Mechanism in Organic Chemistry*, G. Bell, London, 1953, pp. 497–500.

94. P. W. Neber and A. Friedolsheim, *Ann.*, **449**, 109 (1926).

95. P. W. Neber and A. Burgard, *Ann.*, **493**, 281 (1932).

96. C. O'Brien, *Chem. Rev.*, **64**, 81 (1964).

97. D. J. Cram and M. J. Hatch, *J. Am. Chem. Soc.*, **75**, 33 (1953).

98. M. J. Hatch and D. J. Cram, *J. Am. Chem. Soc.*, **75**, 38 (1953).

99. H. O. House and W. F. Berkowitz, *J. Org. Chem.*, **28**, 307 (1963).

100. H. O. House and W. F. Berkowitz, *J. Org. Chem.*, **28**, 2271 (1963).

101. P. A. S. Smith and E. E. Most, *J. Org. Chem.*, **22**, 358 (1957).

102. R. F. Parcell, *Chem. Ind.*, **1963**, 1396.

103. D. F. Morrow, M. E. Butler, and E. C. Y. Huang, *J. Org. Chem.*, **30**, 579 (1965).

104. K. R. Henery-Logan and T. L. Fridinger, *J. Am. Chem. Soc.*, **89**, 5724 (1967).

105. H. E. Baumgarten and F. A. Bower, *J. Am. Chem. Soc.*, **76**, 4561 (1954).

106. H. E. Baumgarten, J. E. Dirks, J. M. Petersen, and D. C. Wolf, *J. Am. Chem. Soc.*, **82**, 4422 (1960).

107. G. H. Alt and W. S. Knowles, *J. Org. Chem.*, **25**, 2047 (1960).

108. H. E. Baumgarten and J. M. Petersen, *J. Am. Chem. Soc.*, **82**, 459 (1960).

109. U. K. Pundit and H. O. Huisman, *Rec. Trav. Chim. Pays-Bas*, **85**, 311 (1966).

110. E. Wenkert and B. F. Barnett, *J. Am. Chem. Soc.*, **82**, 4671 (1960).

111. K. Kitahonoki, et al., *Tetrahedron Letters*, **1965**, 1059.

112. K. Kotera, T. Okada, and S. Miyazaki, *Tetrahedron Letters*, **1967**, 841.

113. W. H. Graham, *J. Am. Chem. Soc.*, **87**, 4396 (1965).

114. I. Vosburgh, *J. Am. Chem. Soc.*, **38**, 2081 (1916).

115. A. F. Morgan, *J. Am. Chem. Soc.*, **38**, 2095 (1916).

116. A. Neff, Ph.D. Thesis, University of Chicago, Chicago, Ill., 1927.

117. L. Horner and A. Christmann, *Angew. Chem.*, **75**, 707 (1963).

118. F. Raschig, *Schwefel- und Stickstoffstudien*, Verlag Chemie, Berlin, 1924.

119. G. Yagil and M. Anbar, *J. Am. Chem. Soc.*, **84**, 1797 (1962).

120. M. Bodenstein, *Z. Physik. Chem.*, **A139**, 397 (1928).

121. A. Lüttringhaus, J. Jander, and R. Schneider, *Chem. Ber.*, **92**, 1756 (1959).

122. E. Schmitz, *Angew. Chem.*, **73**, 23 (1961).

123. E. Schmitz, *Chem. Ber.*, **95**, 676 (1962).

124. E. Schmitz, *Angew. Chem.*, **72**, 579 (1960).

125. E. Schmitz and R. Ohme, *Chem. Ber.*, **94**, 2166 (1961).

126. E. Schmitz, *Angew. Chem.*, **76**, 197 (1964).

127. M. E. Wolff, *Chem. Rev.*, **63**, 55 (1963).

128. J. B. Hynes and T. E. Austin, *Inorg. Chem.*, **5**, 488 (1966).

129. J. B. Hynes, B. C. Bishop, and L. A. Bigelow, *Inorg. Chem.*, **6**, 417 (1967).

130. W. D. Crow and C. Wentrup, *Tetrahedron Letters*, **1967**, 4379.

131. J. Stieglitz and P. N. Leech, *J. Am. Chem. Soc.*, **36**, 272 (1914).

132. B. A. Stagner, *J. Am. Chem. Soc.*, **38**, 2069 (1916).

133. W. S. Guthmann and J. Stieglitz, *J. Org. Chem.*, **1**, 31 (1936).

134. M. S. Newman and P. M. Hay, *J. Am. Chem. Soc.*, **75**, 2322 (1953).

135. J. A. Attaway, R. H. Groth, and L. A. Bigelow, *J. Am. Chem. Soc.*, **81**, 3599 (1959).

136. P. Robson, V. C. R. McLoughlin, J. B. Hynes, and L. A. Bigelow, *J. Am. Chem. Soc.*, **83**, 5010 (1961).

137. B. C. Bishop, J. B. Hynes, and L. A. Bigelow, *J. Am. Chem. Soc.*, **86**, 1827 (1964).

138. B. C. Bishop, J. B. Hynes, and L. A. Bigelow, *J. Am. Chem. Soc.*, **84**, 3409 (1962).

139. B. C. Bishop, J. B. Hynes, and L. A. Bigelow, *J. Am. Chem. Soc.*, **85**, 1606 (1963).

140. H. J. Emeléus and G. L. Hurst, *J. Chem. Soc.*, **1962**, 3276.

141. J. B. Hynes, B. C. Bishop, and L. A. Bigelow, *J. Org. Chem.*, **28**, 2811 (1963).

142. O. Glenser, H. Schroder, and H. Haesler, *Z. Anorg. Allgem. Chem.*, **282**, 80 (1955).

143. J. K. Ruff, *J. Org. Chem.*, **32**, 1675 (1967).

144. J. A. Young, W. S. Durrell, and R. D. Dresdner, *J. Am. Chem. Soc.*, **82**, 4553 (1960).

145. O. Ruff and W. Willenberg, *Chem. Ber.*, **73**, 724 (1940).

146. W. J. Chambers, C. W. Tullock, and D. D. Coffman, *J. Am. Chem. Soc.*, **84**, 2337 (1962).

147. H. G. V. Evans, G. R. Freeman, and C. A. Winkler, *Can. J. Chem.*, **34**, 1271 (1956).

148. J. T. Herron, J. L. Franklin, and P. Bradt, *Can. J. Chem.*, **37**, 579 (1959).

149. J. T. Herron, *J. Phys. Chem.*, **69**, 2736 (1965).

150. J. T. Herron, *J. Phys. Chem.*, **70**, 2803 (1966).

151. L. I. Avramenko and V. M. Krasnen'kov, *Izv. Akad. Nauk SSSR, Ser. Khim.*, **4**, 600 (1964).

152. A. Tsukamoto and N. N. Lichtin, *J. Am. Chem. Soc.*, **82**, 3798 (1960).
153. A. Tsukamoto and N. N. Lichtin, *J. Am. Chem. Soc.*, **84**, 1601 (1962).
154. A. Fujino, S. Lunsted, and N. N. Lichtin, *J. Am. Chem. Soc.*, **88**, 775 (1966).
155. Y. Shinozaki, R. Shaw, and N. N. Lichtin, *J. Am. Chem. Soc.*, **86**, 341 (1964).
156. P. T. Hinde, Y. Titani, and N. N. Lichtin, *J. Am. Chem. Soc.*, **89**, 1411 (1967).
157. J. Versteeg and C. A. Winkler, *Can. J. Chem.*, **31**, 129 (1953).
158. G. McTurk and J. G. Waller, *J. Chem. Soc.*, 262 (1963).
159. B. Dunford, H. G. V. Evans, and C. A. Winkler, *Can. J. Chem.*, **34**, 1074 (1956).
160. M. J. Sole and P. A. Gartaganis, *Can. J. Chem.*, **41**, 1097 (1963).
161. P. A. Gartaganis, *Can. J. Chem.*, **43**, 935 (1965).
162. D. A. Bamford and C. H. Bamford, *J. Chem. Soc.*, **1941**, 30.
163. J. S. Swenton, *Tetrahedron Letters*, **1967**, 2855.
164. H. Zimmer and M. Jayawant, *Tetrahedron Letters*, **1966**, 5061.

CHAPTER 4

Arylnitrenes and Formation of Nitrenes by Rupture of Heterocyclic Rings

PETER A. S. SMITH

University of Michigan, Ann Arbor, Michigan 48104

The subject of this chapter overlaps to some extent the material in Chapter 2, "Electronic Structures and Spectra of NH and Nitrenes," Chapter 5, "Deoxygenation of Nitro and Nitroso Groups," and Chapter 13, "ESR Spectroscopy of Nitrenes." The emphasis in this chapter will be on the chemical behavior of nitrenes, with some attention also given to methods of formation to the extent that it is not to be found in Chapter 5. The overwhelming bulk of the work covered is thus concerned with thermolysis and photolysis of aryl azides. Homocyclic arylnitrenes will be treated separately from heterocyclic nitrenes, and a small third division will deal with ring fragmentation as a source of nitrenes.

Aryl nitrenium ions, $ArNH^+$, which may be important in some of the acid-catalyzed reactions of aryl azides, are not taken up in this chapter. Their chemistry differs markedly from that of nitrenes in the same way that carbonium ion chemistry differs from that of carbenes, and commonly involves so-called "aromatic rearrangements." These have recently been reviewed (1).

I. HOMOCYCLIC ARYLNITRENES

A. Methods of Formation

Arylnitrenes may be formed most simply by removal of a doubly bonded group from trivalent nitrogen or by 1,1-elimination of two singly

99

bonded groups. The first type includes removal of oxygen from nitroso compounds (used as such, or generated *in situ* from nitro compounds) (eq. 1), loss of molecular nitrogen from aryl azides (eq. 2), and loss of carbon monoxide from isocyanates (eq. 3). The second type includes loss

$$Ar\!-\!N\!=\!O \xrightarrow{\text{O-acceptor}} Ar\!-\!N + [O] \qquad (1)$$

$$Ar\!-\!N\!=\!N_2 \xrightarrow{\Delta\ or\ h\nu} Ar\!-\!N + N_2 \qquad (2)$$

$$Ar\!-\!N\!=\!CO \xrightarrow{h\nu} Ar\!-\!N + CO \qquad (3)$$

of halogen from *N,N*-dihaloanilines (eq. 4), and removal of hydrogen from a primary amino group (eq. 5). Other reactions of this type can be

$$Ar\!-\!NX_2 \xrightarrow{\Delta} Ar\!-\!N + X_2 \qquad (4)$$

$$Ar\!-\!NH_2 \xrightarrow{[O]} Ar\!-\!N \qquad (5)$$

easily designed, but they have not been investigated because of difficult handling or even inaccessibility of the requisite starting materials.

A third type of reaction that may lead to arylnitrenes involves the removal of a singly and a doubly bound group from a nitrogen in the ammonium condition. The only representative of such a process appears to be the photolysis of certain nitrones (eq. 6), which may, however, first be isomerized to oxaziridines.

$$(6)$$

A very small number of cases of possible formation of arylnitrenes by fragmentation of a heterocyclic ring forms a fourth type (eqs. 7 and 8).

$$(7)$$

$$R_2C\!=\!O + ArN \qquad (8)$$

Since deoxygenation as a route to nitrenes (2) is covered in detail in Chapter 5, we can turn directly to loss of nitrogen from azides. Most aryl azides decompose smoothly in solution or in the vapor phase at temperatures between 140 and 200°. Occasional examples, especially *o*-nitro

azides, decompose readily at lower temperatures (3a), but there is reason
to suspect that these may not be nitrene-forming decompositions. Temperatures above 200° are required only in exceptional instances. Decomposition (i.e., loss of nitrogen) is generally unimolecular (3–5), as
demanded by equation 2. The reaction does not in general appear to be
subject to surface catalysis (5), although one example of it (palladium
on charcoal) has been reported (6); 2-azido-3,4-dimethoxy-2'-ethylbi-
phenyl was found to be exceptionally stable, and did not evolve nitrogen
even at 235–240° until the catalyst was added.

Thermolysis of aryl azides has usually been carried out in dilute solution, in order to moderate the exothermic reaction, and sometimes in
order to minimize subsequent bimolecular reactions of the nitrene.
Decomposition does not appear to depend greatly on the nature of the
solvent (as long as it is not acidic), although the ultimate products may
involve reaction with the solvent. Those commonly used seem to have been
chosen largely for the convenience of their boiling points, most of which
fall in the range 140–220°. Among hydrocarbon solvents, decalin, tetralin,
hexadecane, and kerosine have seen much use. Resorcinol dimethyl ether
and m-dichlorobenzene are useful when a solvent that is a poor source
of hydrogen is wanted, but when the latter is used, the product may be
contaminated with small amounts of nicely crystalline trichlorobenzenes
sometimes present in commercial samples. n-Hexyl ether has been used
with success. More polar solvents, such as dimethyl sulfoxide, have
sometimes been used for purposes of mechanistic investigation, and even
such reactive substances as acetoacetic ester, malonic ester, and benzyl
mercaptan have been utilized.

Although the influence of solvents on the ease of thermolysis of azides
is not in most cases significant, it is worth mentioning that olefinic solvents
have a marked accelerating effect (3,4).

The thermolysis of aryl azides in indene has an activation energy about
15 kcal/mole lower than in other solvents, and has a negative entropy of
activation (ca. −5 eu) instead of a positive one (ca. 19 eu). These facts
strongly suggest a different mechanism for decomposition in olefinic
solvents. These effects might be expected to arise through initial formation
of a triazoline (eq. 9), and indeed, triazolines are known to be formed
from olefins and azides, and to lose nitrogen easily. However, Walker and
Waters (4) have noted that the rate of disappearance of p-anisyl azide in

solution in indene is the same as the rate of evolution of nitrogen, and they could not detect formation of triazoline. They suggested that the acceleration is due to a concerted process, in which the formation of the aziridine occurs simultaneously with the release of nitrogen from the azido group, without intermediate triazoline formation. If triazoline formation were rapidly reversible, however, it could nevertheless be a step in the process without requiring more than a very low concentration of triazoline at equilibrium, and the observed rate constant would be the product of the equilibrium constant, the rate constant for triazoline decomposition, and the nonvarying concentration of indene solvent.

Thermolysis of some azides has been accomplished neat (7), but this procedure is not generally to be recommended, owing to the danger of explosion. Small azides in particular are likely to be dangerous. When it is wanted to avoid the presence of a solvent, the attenuated vapor phase can be resorted to, utilizing a stream of nitrogen at low pressure to carry the vapors into the thermolysis zone (8).

Photolysis of aryl azides has received somewhat less attention than thermolysis, but enough has been done to illustrate that it is a fairly general process. Phenyl azide itself appears to be an exception, inasmuch as it has been reported to be isomerized to benzotriazole without loss of nitrogen (4). Most aryl azides, however, are readily decomposed by sunlight (9,10), and many show surface discoloration even from exposure to diffuse daylight. A difficulty frequently encountered is that the wall of the photolysis vessel becomes coated with a firmly adhering brown plaque that is opaque to the radiation needed to decompose the azide (11). Photolysis must thus be interrupted several times to allow cleaning or one must be satisfied with incomplete reaction. The wavelength required for effective absorption by an aryl azide varies with the presence of other substituents, inasmuch as the azido group may interact with them electronically through the ring (12); the range of the maxima of longest wavelength is about 250–310 mμm.

Photolysis of isocyanates has received very little attention, and only one example of the possible formation of an arylnitrene has been reported. o-Isocyanatobiphenyl was converted to carbazole in 15% yield by incomplete photolysis through a Vycor filter in dilute ethereal solution (13) (eq. 10). An attempt to generate phenylnitrene by pyrolysis of phenyl isocyanate failed (14), even at temperatures as high as 700°.

(10)

Loss of halogen from N,N-dihaloanilines is a route of very limited potentiality for the production of arylnitrenes, owing to the difficulty of preparation and the ease of rearrangement to ring-halogenated anilines. Three examples have been reported, in which the ring is protected by electron-withdrawing substitution: p-nitro, o-nitro, and 2,3,4,5,6-penta-chloro-N,N-dichloroaniline (15). These compounds lose free halogen spontaneously near room temperature, and are converted to azobenzenes in unspecified yields (eq. 11).

$$p\text{-}O_2NC_6H_4\text{—}NCl_2 \xrightarrow{\sim 20°} Cl_2 + p\text{-}O_2NC_6H_4\text{—}N{=}N\text{—}C_6H_4\text{-}p\text{-}NO_2 \quad (11)$$

Oxidation of primary aromatic amines usually gives complex products involving attack on the aromatic ring as well as the amino group; alternatively, oxygen may be added to the amino group, converting it to a nitroso group (16). Although it is highly unlikely that nitrenes are involved in such reactions, there are a few examples of different behavior which suggest that simple removal of hydrogen from the amino group may occur under certain circumstances. Fischer and Heiler reported (17) in 1893 that oxidation of o-aminodiphenylamine by litharge brought about simple cyclization to phenazine (eq. 12), a different result from oxidation

$$(12)$$

with ferric chloride. This reaction was extended by McCombie, Scarborough, and Waters many years later (18), but they reported that yields were very low, generally only about 5%. The oxidation of o-amino azo compounds (with ammoniacal copper sulfate, copper acetate in pyridine, or lead dioxide) has been shown (19–23) to give benzotriazoles, the same type of product as thermolysis of o-azido azo compounds (eq. 13). Another

$$(13)$$

example of the formation of the same products from oxidation of amines as from thermolysis of azides is the conversion of o-phenylenediamines to cis-1,4-dicyanobutadienes (24) (eq. 14). As in the oxidation of o-amino-

$$(14)$$

diphenylamines, the oxidizing agent is important here; ferric chloride gives phenazines, and silver oxide or lead dioxide give o-quinone diimines.

The conversion of aniline to azobenzene by exposure to ultrasonic vibrations, perhaps through the agency of dissolved oxygen, may perhaps also involve nitrene formation (25).

Other examples of possible oxidation of amines to nitrenes are discussed in the section on heteroaryl nitrenes.

Photolysis of a group of nitrones derived from p-quinone has led to formation of azobenzenes in substantial amounts (26). The two nitrone functions break up in sequence, and the two stages can be separately accomplished as a result of the fact that the half-nitrones produced in the first stage require radiation about 20 mμm shorter in wavelength. The reaction has been interpreted (27) as involving arylnitrene intermediates (eq. 15). Since irradiation commonly converts nitrones to oxaziridines,

$$\text{(structure)} \xrightarrow{h\nu,\ 300\text{--}450\ m\mu m} \text{(structure)}$$

$$+\ [\text{Ar—N}] \longrightarrow \text{ArN}{=}\text{NAr} \qquad (15)$$

which in turn have been photolyzed to give products presumed to arise from nitrenes (28), the same pathway may be followed in the present case. The wavelength of the radiation required for these transformations may show a considerable sensitivity to structure, but the dependence has not been adequately delineated.

Several different types of heterocyclic rings have been found to break up with the possible formation of arylnitrenes, but none of them has seen wide use for the purpose. The photolysis of N-aryl oxaziridines has been mentioned. 3-Arylanthranils, which can be prepared by mild thermolysis of o-azidobenzophenones (7,29), isomerize on strong heating to acridones (7,30) in good yields, or on photolysis in methanol to methoxyazepines (31) (eq. 16), probably through intermediate nitrene formation (eq. 7). The formation of azepines does not, of course, require a 3-phenyl substituent, and has in fact been observed from anthranils bearing hydrogen or methyl at the 3-position.

The thermolysis of 4-phenyl-5,6-benzo-1,2,3-triazine-3-oxide to phenylanthranil may be considered as a case of formation of a nitrene from a heterocyclic ring, although it has been postulated that an azide is first formed (32) (eq. 17).

A possible further case of formation of an arylnitrene is seen in the thermolysis of hexaphenylborazine (33), which gives rise to a small amount of hydrazobenzene (eq. 18). However, B,B',B''-trimethyl-N,N',N''-triphenylborazine gave no indication of nitrene formation.

$$(16)$$

$$(17)$$

$$\xrightarrow{440-490°} \ [\phi N] \ \xrightarrow{[H]} \ \phi—NH—NH—\phi \qquad (18)$$

B. Intermolecular Reactions

The criteria for the intermediacy of a nitrene include a number of reactions, each of which is more or less difficult to explain without invoking a nitrene, but none of which by itself constitutes absolute proof. In the case of arylnitrenes, the presumed intermolecular reactions are: abstraction of hydrogen (eq. 19), addition to a multiple bond (eq. 20), insertion into a single bond (usually C—H) (eq. 21), ring expansion accompanied by reaction with a suitable nucleophile (eq. 22), bond formation at an unshared electron pair (usually on phosphorus) (eq. 23), and dimerization (eq. 24).

$$\text{Ar—N} + \text{RH} \longrightarrow \text{R·} + \text{Ar—NH·} \xrightarrow{\text{RH}} \text{Ar—NH}_2 \qquad (19)$$

$$\text{Ar—N} + \underset{\substack{\diagdown \diagup \\ C \\ \| \\ C \\ \diagup \diagdown}}{} \longrightarrow \text{Ar—N} \underset{\substack{\diagup \\ C— \\ \diagdown \\ C— \\ |}}{\overset{\substack{| \\ C— \\ |}}{}} \tag{20}$$

$$\text{Ar—N} + \text{RH} \longrightarrow \text{Ar—NH—R} \tag{21}$$

$+ \ \text{R}_2\text{NH} \longrightarrow$ $\tag{22}$

$$\text{Ar—N} + (\text{EtO})_3\text{P:} \longrightarrow \text{Ar—N}{=}\text{P(OEt)}_3 \tag{23}$$

$$2\text{Ar—N} \longrightarrow \text{Ar—N}{=}\text{N—Ar} \tag{24}$$

The isolation of a specific reaction product does not necessarily allow the unambiguous deduction of the reaction by which it arose, of course. There are usually alternative pathways to be considered, some of which may not involve a nitrene at all. Interpretation may be further complicated by the fact that the initial products may be so labile that the products actually isolated are the result of further transformation.

Abstraction of hydrogen to form primary amines is perhaps the most general reaction when arylnitrenes are produced in solution. The source of hydrogen may be the solvent or the nitrene precursor. Hydrogen abstraction must always compete with the other reactions, but may nevertheless become virtually the sole reaction. No thorough study of the relation between structure and products has yet been published, but the yields of anilines from a series of aryl azides decomposed under strictly comparable conditions are shown in Table I (5).

There are scattered indications that the extent of hydrogen abstraction by decomposing azides is also dependent on the medium, but no study has

TABLE I
Hydrogen Abstraction by Phenyl Azides
Thermolyzed in Decalin at 141.3°

Substituent	ArNH_2, %
Unsubstituted	44
m-Bromo	85
m-Methoxy	97
m-Nitro	1
p-Nitro	Small
p-Bromo	78
2,4-Dichloro	41

been reported. Phenyl azide thermolyzed in p-xylene at 138° or in benzene at 150–160° gives markedly less aniline (34) (20–25%) than reported for thermolysis in decalin (5) (44%). The yield of aniline for thermolysis in n-pentane (30%) is twice as great as in isopentane (15%) (35). In thiophenol as solvent, the highest yield of aniline (52%) has been reported (8). In the vapor phase, intermolecular hydrogen abstraction is at a minimum, as shown by the formation of only 1% of o-n-butylaniline from o-n-butylphenyl azide, compared to 29% for thermolysis in hexadecane (36). (Much of the difference is made up in this instance by intramolecular abstraction, which increases from 25% in hexadecane to 46% in the vapor phase.) Similar results have been obtained with o-n-propylaniline (37). However, even pyrolysis of phenyl azide in the vapor phase leads to some hydrogen abstraction (14).

Abstraction of hydrogen is not confined to nitrenes generated from azides, but has, in fact, been observed in many instances in the deoxygenation of nitro compounds with triethyl phosphite (37). In only rare instances with deoxygenation by ferrous oxalate has abstraction of hydrogen been reported (38); 2-nitro-2′,4′,6′-trimethylbiphenyl gave a 27.4% yield of $ArNH_2$ neat, 42.1% in hexadecane, and o-ethyl and o-cyclohexylnitrobenzene also gave significant amounts of anilines. Oxidative generation of arylnitrenes would not, of course, be conducive to observation of hydrogen abstraction. In two instances of potential generation of an arylnitrene by photolysis of an oxaziridine (28), 2-phenyl-3,3-pentamethyleneoxaziridine gave roughly equal quantities of aniline and of diethylaminoazepine when irradiated in solution in diethylamine (28a), and 2,3-diphenyloxaziridine gave a mixture of benzaldehyde, its anil, nitrosobenzene, and benzanilide (28b).

The nature of the hydrogen abstraction process is indicated by the fact that the hydrogen-supplying solvent undergoes dehydrogenative coupling to a considerable extent (although generally not enough to account for all the hydrogen required for amine formation). Cumene is converted to bicumyl in 11% yield by the deoxygenation of 2-nitro-2′,4′,6′-trimethylbiphenyl with triethyl phosphite (accompanied by the amine in 32% yield) (39), for example, and the thermolysis of phenyl azide in p-xylene gave about 4% of sym-di-p-tolylethane (along with 43% of aniline) (34). With benzene as the solvent, however, no biphenyl was detected even though aniline was produced to the extent of 20%. Thermolysis of p-anisyl azide in cumene also gives bicumyl, but in insufficient quantities to account for the large quantity of p-anisidine formed (along with $p,p′$-dimethoxyazobenzene and 2,7-dimethoxyphenazine) (4).

A strong preference for abstraction of hydrogen from the nitrene precursor rather than from alkane solvents is indicated by the results of thermolysis of $C_6D_5N_3$ in pentane (35b). The two principal products were aniline, which was found to contain 40% of N—D, and N-pentylanilines, which contained 18% of N—D. The converse situation, thermolysis of $C_6H_5N_3$ in $(CH_3)_3CD$, produced aniline containing only 7% of N—D, in spite of the fact that the tertiary position of isobutane is the most readily attacked.

The evidence suggests that hydrogen atoms are abstracted one at a time, forming amino and alkyl radicals initially (eq. 19). (For dehydrogenation reactions of carbonylnitrenes, see Chapter 6, Section IV-D). This is a reaction to be expected of a triplet state (35b,36b), and presumably must wait upon decay of the singlet-state nitrene initially formed.

There is evidence that formation of amines by thermolysis or photolysis of aryl azides may in some circumstances take place without formation of an intermediate nitrene. Shingaki (40a–c) has measured the kinetics of thermolysis of phenyl azide in a variety of solvents, and found that the rate was slightly higher in disulfides, markedly higher in thiols. Free-radical sources, such as tetraphenylhydrazine, hexaphenylethane, or triphenylmethyl hydroperoxide, greatly accelerated the reaction. The products of both thermolytic and photolytic decomposition in thiophenol as solvent were aniline (principally), o-aminodiphenyl sulfide, and diphenyl disulfide. He concluded that thiyl radicals, R—S·, were involved in attack on the azido group (40d). These observations raise the question of whether apparent abstraction of hydrogen by a nitrene may in some cases be caused by traces of peroxides in normally inert solvents.

Intermolecular insertion of an arylnitrene into a C—H bond has been reported only recently (35). Yields are low in comparison to those of competing reactions as well as to those of intramolecular insertion. Bertho (34) detected no products of insertion when phenyl azide was thermolyzed in p-xylene or benzene, and Doering and Odum (41) reported none from the photolysis of phenyl azide in benzene or cyclohexane. Although Horner, Christmann, and Gross (11) photolyzed a selection of aryl azides in a wide variety of solvents, they likewise reported no insertion products (azobenzenes were generally the major product).

Hall and co-workers (35) thermolyzed phenyl azide in cyclohexane, n-pentane, and isopentane. Insertion products were isolated in yields up to 10%, along with aniline as the major product. It seems likely that insertion products may also have been formed in the earlier experiments, but escaped notice because of the low yields and the lack of modern methods of separation, or because the earlier investigators were not concerned with minor products.

A 5% solution of phenyl azide in cyclohexane at 180° gave 8% of N-cyclohexylaniline (35). With n-pentane, a 10% yield of a mixture of

N-(2-pentyl)- and N-(3-pentyl)-aniline, free of the 1-pentyl isomer, was formed, and with isopentane, a mixture of N-pentyl anilines consisting mostly of the *tert*-amyl isomer was formed (eq. 25). The selectivity of

$$\phi N_3 + CH_3CH_2CH(CH_3)_2 \xrightarrow{180°} \phi NH_2$$

$$+ \ \phi NH\overset{\underset{\displaystyle CH_2CH_3}{|}}{\underset{\displaystyle}{C}}(CH_3)_2 + \phi NH\overset{\underset{\displaystyle CH_3}{|}}{\underset{\displaystyle}{C}}HCH(CH_3)_2$$

$$\text{ratio } 10\text{--}20{:}1$$

$$+ \ \phi NH\!-\!CH_2CH_2CH(CH_3)_2 \text{ and } \phi NH\!-\!CH_2\overset{\underset{\displaystyle CH_3}{|}}{C}HCH_2CH_3 \quad (25)$$

$$\text{ca. } 14\% \text{ of } \phi NH\!-\!C_5H_{11}$$

phenylnitrene for tertiary C—H vs. secondary C—H, (a factor of 20–40 after statistical correction) is thus much greater than that of carboethoxynitrene (42), 3.8. This fact suggests that phenylnitrene may be longer lived and probably of lower energy than carboethoxynitrene.

Thermolysis of phenyl azide in optically active 2-phenylbutane gave the insertion product, 2-anilino-2-phenylbutane, with 40% retention of optical activity (35b). Thermolysis in isobutane gave aniline, *tert*-butylaniline, and isobutylaniline; the ratio of insertion into the tertiary C—H to insertion into a primary C—H was 88:1 (after the 9:1 statistical correction). In 2-deuterio-2-methylpropane, the ratio dropped to 21:1, which corresponds to an isotope effect k_H/k_D of 4.15. This high value is interpreted to mean that the C—H bond breaks before the N—C bond forms, and thus that the insertion products arise from a two-step process (eq. 19). By contrast carboethoxynitrene (from the azide) showed $k_H/k_D = 1.3$, which is more consistent with a direct insertion process, and it also inserts at an optically active site with full retention of configuration. A reasonable interpretation of the evidence is that carboethoxynitrene reacts in the singlet state, in contrast to phenylnitrene.

Further information about the nature of the insertion and hydrogen-abstraction processes comes from the work of Hall, Hill, and Fargher (35b) with thermolysis of aryl azides in mixed solvents. Thermolysis of phenyl azide in cyclohexane gave aniline and N-cyclohexylaniline in a ratio of about 4:1. Dilution with neopentane, which is virtually inert to insertion, caused no perceptible change in the ratio of abstraction to insertion, over a wide range of concentrations. Similar results were obtained with p-tolyl azide and with p-chlorophenyl azide (which incidentally, showed a much higher ratio, ~70). The effect of an inert diluent should be to increase the deenergizing of the initially formed nitrene species, and thereby to increase the ratio of triplet nitrene to singlet nitrene available for reaction with cyclohexane. The insensitivity of the product composition to such an influence indicates that both types of products derive from the same state of the nitrene, rather than, as might have been anticipated, derivation of the insertion product from singlet phenylnitrene and the hydrogen-abstraction product from triplet phenylnitrene. (There remains a certain ambiguity regarding the phase involved in the experiments, owing to the fact that they were of necessity conducted under pressure above the critical temperature of the solvent.)

These results in conjunction with the other evidence are most reasonably encompassed by the concept that the initial step for both abstraction and insertion is a reaction between a triplet nitrene and the hydrocarbon to form an amino radical, ArNH·, and an alkyl radical, R·. If these should combine before leaving the solvent cage, an N-alkyl aniline, ArNHR, would be produced, with a good probability of some retention of configuration if attack was at an optically active site. Separation of the two radicals, however, would lead eventually to the amine, ArNH₂, by further abstraction (from Ar— or RH) as well as to R—R and other products.

Insertion in N—H and S—H bonds is also very unfavorable. Investigations of thermolysis (43) or photolysis (41) of aryl azides in the presence of primary and secondary amines have been made, and in no case have phenylhydrazine derivatives been reported; thermolysis of phenyl azide in thiophenol (8) converted it to aniline only. Phenylnitrene generated by deoxygenation of nitrosobenzene (45) by triethyl phosphite or photolysis of oxaziridines (28) in the presence of amines also gave no phenylhydrazine derivatives. The principal reaction of nitrenes generated in the presence of amines is instead ring expansion to form azepines (43,45) (discussed later), or perhaps dimerization.

One observation of probable insertion into an N—H bond has nevertheless been made (46). 3,5-Dimethyl-4-azidopyrazole when thermolyzed in the presence of aniline gives 6% of the mixed azo compound, 3,5-dimethyl-4-phenylazopyrazole (eq. 26), which can most simply be accounted for if one assumes that an initially formed hydrazo compound underwent dehydrogenation by more of the nitrene (the corresponding amine was also formed).

(26)

Insertion of a nitrene into other types of bonds has not been firmly established, but there is one reaction that appears to involve insertion into a C—O bond. Smalley and Suschitzky (47) thermolyzed a series of aryl azides in boiling acetic anhydride, and obtained the usual nitrene products (azobenzenes and anilines) plus 18–46% of acetylated o-amino-phenols. These may have arisen through simple insertion to give N-acetoxyacetanilides, which are known to rearrange to o-acetoxyacetan-ilides (48) when heated (eq. 27). Only the three nitrophenyl azides failed to give this type of product.

In support of the view that an insertion product is involved is the fact that the ordinary acid-catalyzed decomposition of aryl azides leads to p-aminophenol derivatives. Furthermore, when the opportunity for intramolecular insertion into a C—H bond is also present, no o-amino-phenols are produced, and cyclization takes place instead (49). The process of insertion may well be different from that into C—H bonds, however, in view of the availability of unshared electron pairs on the oxygen atoms. A reasonable path would be electrophilic attack by the nitrene on the central oxygen, followed (or accompanied) by shift of an acetyl group (eq. 28).

Addition of arylnitrenes to a multiple bond of another molecule has seen little study and is not well understood. Reaction with olefins to form aziridines (or their subsequent transformation products) has been postu-lated, by analogy with carbene reactions, but investigation of such reactions has been complicated by the fact that azides, the usual source of nitrenes, themselves react with olefins, undergoing 1,3-dipolar addition to form triazolines. Since triazolines lose nitrogen rather easily (3,4,50) (thermally or photolytically) to form the same product as might have been expected from a nitrene reaction (eq. 28a), the formation of an aziridine from an azide cannot be taken as a criterion of the involvement of a nitrene.

$$\text{Ar—N}_3 + \underset{}{-\text{C}=\text{C}-} \longrightarrow \underset{\underset{\text{Ar}}{\overset{\displaystyle}{\text{N}}\diagdown_{\text{N}}\diagup^{\text{N}}}}{-\text{C}---\text{C}-} \xrightarrow[\text{or } h\nu]{\Delta} \underset{\underset{\text{Ar}}{\overset{\displaystyle}{\text{N}}}}{-\text{C}--\text{C}-} + \text{N}_2 \qquad (28a)$$

Triphenylethylene does not react detectably with aryl azides, either on prolonged standing or heating below the decomposition point of the azide, as shown by ultraviolet spectra, which remain strictly the additive spectra of the components. The rates of decomposition of aryl azides in triphenylethylene as solvent do not show the enhancement observed with other olefinic solvents due to triazoline formation (the rates are actually lower than in decalin). Thermolysis of aryl azides in this solvent might therefore reasonably be expected to generate some arylnitrene. The products from p-bromophenyl azide at 145° were found to be complex, consisting very largely of intractable tars and highly colored materials; no aziridine was detected, and the only identifiable product was an isomer of it, the p-bromoanil of diphenylacetophenone, in only 5% yield (51).

Very little work has been done with deoxygenation of nitro or nitroso compounds in the presence of olefins. Nitrobenzene heated with ferrous oxalate in the presence of hexadecene gave an apparent nitrene–olefin adduct whose structure was tentatively deduced to be N-2-hexadecenyl-aniline (38) rather than 1-phenyl-2-tetradecylaziridine. However, deoxygenation with carbon monoxide and iron pentacarbonyl in the presence of 2-methylpentene-1 gave only aniline (30%) and no adduct with the olefin (51). Photolysis of nitrobenzene in cyclohexene or 2-methylbutene-2 also gave no nitrene–olefine adduct, although azobenzene was formed ($\sim 10\%$) and the intermediacy of phenylnitrene was suggested (53). The principal products were dimeric and contained oxygen, and obviously did not arise from a nitrene. Thus there is as yet no compelling evidence that arylnitrenes form aziridines by reaction with any olefin.

Even less information is available regarding possible reactions of arylnitrenes with other multiple bonds. Aryl azides have been thermolyzed or photolyzed in a variety of solvents containing multiple bonds, such as acetoacetic ester (11), acetonitrile (11), dimethyl sulfoxide (11), and triethyl phosphate (37), without evidence of unusual reaction; azobenzenes remain the principal product.

A reaction which might qualify as an intermolecular reaction of a nitrene with a carbonyl group occurs when phenyl azide is heated at 90° in the presence of iron pentacarbonyl (54a,b). A dimeric addition product is formed whose structure appears to contain the bidentate diphenylurey-lene moiety, ϕN—CO—Nϕ. (The temperature used was unusually low for decomposing aryl azides, which ordinarily require a temperature of about 140°, but the long duration of the experiment (24 hr) might have sufficed for generating the quantity of nitrene required, assuming the rate at 90° to be 0.01–0.02 that at 140°.) However, the decomposition of phenyl azide occurs at only 20° in the presence of diiron nonacarbonyl (54c), which suggests that metal carbonyls actually induce decomposition (and that free nitrenes are thus not involved).

The foregoing result should be compared to that of Kmiecik (52), who obtained only azobenzene (up to 80% yield) by heating nitrobenzene with iron carbonyl at 210° under a pressure of 2000 psi of carbon monoxide. Nitrenes may not be involved in the latter reaction, and indeed, Kmiecik demonstrated that azoxybenzenes and not nitrenes were the immediate precursors of the azobenzenes. The absence of a nitrene is further supported by the lack of formation of isocyanates, which by contrast have been obtained in substantial yields when aryl azides are thermolyzed under a high pressure of carbon monoxide (55) (eq. 29).

$$\phi N_3 + CO \xrightarrow[160-180°]{300 \text{ atm}} N_2 + \phi NCO \qquad (29)$$
$$(47\%)$$

Addition to an unshared electron pair on sulfur or phosphorus might be expected as a manifestation of electrophilic character in a nitrene. The possible occurrence of such a reaction when an azide is the source of the nitrene is obscured by the fact that azides readily form 1:1 addition products, phosphazines, with trivalent phosphorus compounds. The phosphazines in turn lose nitrogen to form phosphine imines (eq. 30), the product that would have resulted from direct attack of a nitrene.

$$ArN_3 + R_3P \longrightarrow Ar\text{—}N\text{=}N\text{—}N\text{=}PR_3 \longrightarrow N_2 + ArN\text{=}PR_3 \qquad (30)$$

This complication does not arise when deoxygenation is used to generate the nitrene. A number of nitrobenzene derivatives have been deoxygenated by heating in excess triethyl phosphite (37,39,56); triethyl phosphorimidate was the major product (eq. 31) except where adjacent unsaturation favored

$$ArNO_2 + (EtO)_3P(\text{excess}) \longrightarrow [ArN] \xrightarrow{(EtO)_3P} ArN\text{=}P(OEt)_3 \qquad (31)$$

competing cyclization. In all cases, hydrogen abstraction or cyclizing insertion were significant reactions, however. Similar results were obtained with nitrosobenzenes (37).

There is little information on the possibility of reaction between aryl-nitrenes and tetravalent or divalent sulfur. The formation of azoanisole in essentially undiminished yield when p-anisyl azide was photolyzed in dimethyl sulfide (91% yield) or dimethyl sulfoxide (82.5% yield) (11) is presumptive evidence that the nitrene is not trapped by the sulfide or sulfoxide function. Phenyl azide has also been photolyzed in the presence of sulfur dioxide (57); there was no evidence for the formation of a nitrene–sulfur dioxide adduct, $\phi N{=}SO_2$. In benzene solution, the products were phenylsulfuric acid and benzene-sulfonanilide, and in methanol solution, phenylsulfuric acid and ammonium sulfate, whereas in liquid sulfur dioxide, only an unidentified green solid was obtained.

The formation of azo compounds, the dimers of nitrenes, during reactions that are presumed to generate arylnitrenes, is widely documented, but whether they are, in fact, formed by a direct dimerization process is still uncertain. By far the majority of instances of formation of azo compounds consists of thermolysis or photolysis of azides. Deoxygenation of nitro-benzenes and nitrosobenzenes has only rarely been reported to give azo-benzenes, perhaps because most of the compounds studied have been chosen with the object of promoting cyclization to an *ortho*-substituent. Such examples as there are must be considered ambiguous at best, for deoxygenation of the nitroso dimer stepwise through the azoxy compound (eq. 32) is always a possible route, and does not involve nitrenes. Deoxy-

$$2ArNO \;\rightleftharpoons\; \underset{\underset{O}{\parallel}}{Ar{-}\overset{+}{N}{-}Nar} \;\overset{-O}{\longrightarrow}\; Ar{-}N{=}\overset{\underset{|}{O^-}}{N^+}{-}Ar \;\overset{-O}{\longrightarrow}\; ArN{=}NAr \qquad (32)$$

genation of azoxybenzenes has been demonstrated with both carbon mon-oxide (52) and triethyl phosphite (58), and formation of azoxybenzenes from the reaction of nitrosobenzenes with a limited amount of triethyl phosphite has also been reported (59).

Less ambiguous is the formation of azobenzenes from N,N-dichloro-anilines (15) (p-nitro, o-nitro, pentachloro) (eq. 11), although the lack of reported yields makes assessment uncertain. The formation of high yields of azobenzenes from photolysis of quinone nitrones (26) (eq. 15) in benzene solution also implies the intermediacy of nitrenes. Although N,N'-diphenyl-quinonedioxime gave only 35% of azobenzene, the N-phenyl-N'-β-naph-thyl derivative gave an 81% yield of a mixture consisting of the three possible azo compounds in close to the statistical proportions (eq. 33). A mixture of the N,N'-diphenyl and N,N'-di-p-anisyl derivatives also gave a mixture of the three possible azo compounds. Although pathways alternative to formation and dimerization of nitrenes can be postulated, all seem improbable.

$$\phi-\overset{+}{\underset{O^-}{N}}\left\langle\right\rangle=\overset{+}{\underset{O^-}{N}}-\alpha\text{-Nap} \xrightarrow{h\nu}$$

$$\phi N=N\phi + \phi N=N-\alpha\text{-Nap} + \alpha\text{-Nap}-N=N-\alpha\text{-Nap} \qquad (33)$$

$$\text{ratio } 1.00 : 1.78 : 0.96$$

The formation of azobenzenes from aryl azides can be envisaged by several pathways, only some of which involve nitrenes. The simplest path is, of course, cleavage of the azide to nitrene, followed by dimerization (eq. 34).

$$\text{ArN}_3 \longrightarrow \text{N}_2 + \text{ArN}; \qquad 2\,\text{ArN} \longrightarrow \text{ArN}=\text{NAr} \qquad (34)$$

Since nitrenes are not known to be stable, the first step may be presumed to be rate-determining, and the rate law would accordingly be first-order.

It is also possible that the arylnitrene might attack an undecomposed molecule of azide, a statistically more probable process, to produce a tetrazadiene (eq. 35),

$$\text{ArN} + \text{ArN}_3 \longrightarrow \text{Ar}-\text{N}=\text{N}-\text{N}=\text{N}-\text{Ar}$$

or

$$\text{Ar}-\text{N}=\overset{+}{\underset{\text{Ar}}{\text{N}}}-\text{N}=\text{N}^- \longrightarrow \text{N}_2 + \text{ArN}=\text{NAr} \qquad (35)$$

The 1,2-diaryltetrazadiene would give an azo compound directly by loss of nitrogen, whereas the 1,4-isomer would presumably have to decompose first to two molecules of arylnitrene (which might combine within the solvent cage, or might separate first). Tetrazadienes have not been described and we can thus only speculate on their behavior. If the slowest step in the sequence is the initial cleavage of the azide, first-order kinetics would follow, as they also would if the breakdown of the tetrazadiene should be the slow step (except in the special circumstance that the formation of tetrazadiene should be reversible, and slower than the initial cleavage of the azide). A third possibility is that the aryl azide dimerizes before losing nitrogen (eq. 36),

$$2\text{ArN}_3 \longrightarrow (\text{ArN}_3)_2 \longrightarrow 2\text{N}_2 + \text{ArN}=\text{NAr} \qquad (36)$$

which would lead to second-order kinetics. Lastly, hydrogen abstraction might be involved so as to produce amino radicals, which would dimerize to hydrazo compounds and then in turn be dehydrogenated (eq. 37) in a chain process. The kinetics would be expected to show an inhibition period, and a pronounced dependence on solvent, unless cleavage of the azide to a nitrene must precede the hydrogen transfer.

$$ArNH-NHAr \xrightarrow[\text{or 2ArN}]{2ArN_3} ArN{=}NAr + 2ArNH \tag{37}$$

The facts as known point strongly to either of the first two pathways mentioned, and thus to involvement of nitrenes. Kinetics of thermolysis of variously substituted phenyl azides in different solvents have been measured (2–4,60); all are first-order, whether the products are azobenzenes or other substances. No evidence for dimerization of azide before loss of nitrogen has been found, although the possibility cannot be completely ruled out that azo compounds might arise through reaction of a photo-excited azide with one in the ground state (11).

The possibility of a route involving amino radicals (eq. 37) is supported by the fact that hydrazobenzene and benzidine have been found accompanying azobenzene and aniline in the thermolysis of phenyl azide in decalin. However, hydrazo compounds or benzidines are not formed in the reduction of azides below their ordinary decomposition temperature, and azobenzenes can be formed in high yield in the gas phase (8). These facts suggest that dimerization by way of amino radicals may require prior formation of arylnitrene and may function as an alternative path to direct dimerization of arylnitrene under suitable conditions.*

Evidence for the amino radical route to azo compounds is found in the oxidation of a group of 5-aminopyrazoles, which are converted to azo compounds by treatment with permanganate (61) whereas thermolysis of the corresponding azides gives no azo compound, but only a monomer (eqs. 38 and 39) which does not dimerize under the conditions of the

$$\tag{38}$$

$$\tag{39}$$

* In one instance, thermolysis of p-methoxyphenyl azide in cumene, formation of a phenazine, an oxidative dimer, has been reported (4); amino radicals may well be involved.

oxidation experiments. If the monomeric substance can be considered to arise from the nitrene (if it is not, indeed, the electronically stabilized singlet nitrene), then the azo compound must thus have arisen through some intermediate other than the nitrene. An amino radical, formed in the first stage of oxidation, satisfies the requirements, for it could either dimerize to a hydrazo compound before further oxidation to the azo compound, or could be oxidized directly to nitrene (eq. 40), depending on conditions (as observed).

$$Ar—NH_2 \xrightarrow{[O]} Ar—NH· \longrightarrow Ar—NH—NH—Ar \xrightarrow{[O]} Ar—N{=}N—Ar \quad (40)$$

$$\downarrow [O]$$

$$Ar—N \longrightarrow \text{products}$$

The formation of azo compounds from aryl azides is considerably influenced by the medium, in a way that suggests a competition in reactions of an arylnitrene between conversion to azo compound vs. abstraction of hydrogen and unimolecular reactions. A thoroughly systematic study has not been made, but a comparison of various published results for formation of azobenzene from phenyl azide is interesting (Table II). The significance of these data should be considered in the light of the fact that some substituted phenyl azides give much higher yields of azo compound, and others give much lower yields, under the same conditions. p-Anisyl azide, for example, consistently gives more azo compound, and p-chlorophenyl azide gives very little (Table III). Horner, Christmann, and Gross (11), noting these facts along with the observation that photolysis of p-anisyl azide gave only 18% of azo compound in benzene solution, but 82–91% in solvents, such as tetrahydrofuran, having unshared electron pairs, suggested that formation of azo compound is favored by factors that stabilize the nitrene. These factors are coordination of the electrophilic nitrene with unshared electron pairs of solvent molecules, and electron donation by substituents. One can deduce that it is the singlet state of the arylnitrene that would be stabilized by these effects (5). The effect of a p-methoxy group, for example, can be envisaged in terms of conjugation with an unshared pair on the oxygen, represented by the limiting structure 2 (or more generally, by 3).

TABLE II
Formation of Azobenzene from Phenyl Azide

Medium	Thermolysis (T) or photolysis (P)	Temperature, °C	Yield, %	Other products	Ref.
Vapor phase	T	330–900	+	ϕNH_2	14
Vapor phase	T	350–360	72	—	8
ϕSH	T	165	0	ϕNH_2	8
Ac_2O	T		9.3	ϕNH_2, $o\text{-}AcOC_6H_4NHAc$	47
DMSO	P		Trace		11
DMSO	P		0		11
AcOH	P		0		11
$Cl_2FC\text{—}CF_2Cl(+CO)$	T	160–180	50	ϕNCO	55
$n\text{-}C_5H_{12}$	T	180	0	ϕNH_2, ϕNHR	35
Decalin	T		5	ϕNH_2, $\phi NHNH\phi$	4
Benzene	T	150–160	~10	ϕNH_2	34
p-Xylene	P	138	Small	ϕNH_2	41
	T				34
ϕNH_2	T	~165	0	Anilinoazepine	43

TABLE III
Yields of Azo Compound by Photolysis of Azides in Benzene

Aryl azide	Yield of ArN=NAr, %	Ref.
Phenyl	5	41
p-Anisyl	18	11
p-Biphenylyl	81	11, 62
p-Chlorophenyl	0	11

Distinction between direct dimerization of arylnitrenes (eq. 34) and attack of arylnitrene on aryl azide to form an intermediate tetrazadiene (eq. 35) cannot be expected from available kinetic evidence. Dimerization of highly reactive species formed in low concentration is inherently improbable; therefore, if direct dimerization of arylnitrenes is indeed the source of azo compounds, arylnitrenes must be of sufficiently low reactivity to survive many collisions with solvent molecules before encountering another arylnitrene with which to combine. This view is consistent with the observed sensitivity of the yield of azo compound to the reactivity of the solvent, and with the maximum yield in the vapor phase. Reaction of an arylnitrene with undecomposed aryl azide is statistically much more probable; although it, too, would be susceptible to interference by reaction of arylnitrene with solvent, the reactivity of the arylnitrene would not need to be so low as in the case of direct dimerization.

The only experimental evidence on the subject lies in the thermolysis of 1-methyl-3,5-diphenyl-4-azidopyrazole (46), which takes place readily at 80°. When thermolyzed alone in cyclohexane, it gives the azopyrazole in 23–36% yield. When it was thermolyzed in the presence of a large excess of p-anisyl azide, which is stable at 80°, it still gave azopyrazole (25% yield) (eq. 41) and no detectable mixed azo compound, which would have been expected if the anisyl azide had been attacked by a pyrazolylnitrene.

(plus tar and fragmentation products)

$$\tag{41}$$

Ring expansion to form a 2-substituted azepine (eq. 22) has been observed when arylnitrenes are generated in the presence of ammonia (41), primary and secondary amines (28,41,43–45a) hydrogen sulfide (41), or

triethyl phosphite (45b). The presumption that this is a reaction of a
nitrene is supported by the fact that azepines have been obtained by
thermolysis (43,44) or photolysis (41,44) of azides, by deoxygenation of
nitrosobenzene with tervalent phosphorus (45), and by photolysis of an
N-phenyloxaziridine (28), and the fact that the kinetics of azide decomposition are not significantly different in aniline compared to other solvents (60). The 2-position of the azepine arises from C-1 of the phenyl
azide (43c) (eq. 42).

$$\text{(diagram)} \quad \text{C}^{14}\text{—N}_3 + \phi\text{NH}_2 \xrightarrow{\Delta} \text{(diagram)} \quad \text{C}^{14}\text{—NH}\phi \tag{42}$$

This reaction was first observed in 1912 by Wolff (63), who named the
product from phenyl azide and aniline "dibenzamil"; the correct structure
was not determined until 46 years later, by Huisgen, Vossius, and Appl (43).
The highest yield reported is only 62% (nitrosobenzene, dimethylamine)
(45a), and most are much lower. A very large excess of amine as solvent
promotes azepine formation (43) although ether (45a), trimethylamine (41),
and chlorobenzene (44) have been used successfully. (See Appendix.)

An azirine intermediate has been proposed (8,41,43) (eq. 43). It has
been concluded that its formation is not concerted with loss of nitrogen
from the azide, in view of the fact that substituents on the phenyl azide
ring have almost no effect on the kinetics of thermolysis in aniline solution
(60) (at 174.1°, $k = 8.47 \times 10^{-4}$ sec^{-1} for phenyl azide, 8.88×10^{-4} for
m-tolyl azide, and 8.84×10^{-4} for m-anisyl azide). A concerted reaction,
which would have the character of electrophilic attack at an $ortho$-position,
should have been markedly influenced. However, this conclusion is
somewhat weakened by the fact that the kinetic investigations did not
include an examination of the products; none of the foregoing substituted
azides has been shown to yield an azepine. That there is a real uncertainty
is shown by the fact that neither p-anisyl azide (11), p-chlorophenyl
azide (43) the naphthyl azides (43), nor p-nitrophenyl azide (43) give rise

$$\text{(diagram)}\text{—N}_3 \longrightarrow \text{(diagram)}\text{—N:} \rightleftharpoons \text{(diagram)} \xrightarrow{\text{R}_2\text{NH}} \text{(diagram with } \text{NR}_2) \tag{43}$$

to detectable azepine. A symmetrical (1H)-azirine (4) is ruled out by the
fact that there is no scrambling in the product between C-1 and C-2 of the
azide (43c); furthermore, p-tolyl azide gives only a single azepine. (The

(4)

earliest proposal of a fused azirine appears to be that of Michaelis and Schäfer (64), who were concerned with the dehydrogenation of the amino group of 5-amino-3-methyl-1-phenylpyrazole by hydrogen peroxide). (See Appendix.)

There is evidence that azepine formation in the decomposition of aryl azides may be more widespread than was initially realized. Smalley and Suschitzky (44) observed that thermolysis of various fluoro-, chloro-, and bromophenyl azides gave rise to small amounts of sublimate consisting of the corresponding aniline hydrohalide (7.8% in the case of p-chlorophenyl azide). The amount was markedly increased when an added aniline was also present. They convincingly demonstrated that the hydrogen halide did not derive from reaction of aniline with undecomposed azide or with other thermolysis products of the aryl azides, and showed that the amino haloazepines were easily attacked by aniline to form aniline hydrohalide. Presumably the substituted aniline formed by hydrogen abstraction during thermolysis is capable of reacting with more arylnitrene to form an anilinoazepine, which is then dehydrohalogenated.

Postulation of an azirine intermediate raises the question of why ring expansion is not commonly observed when arylnitrenes are formed in the absence of amines (or hydrogen sulfide). The simplest explanation is that the nitrene and azirine are in equilibrium with each other, and only when a suitably reactive nucleophile is present can the azirine react to form an azepine. In this connection it is perhaps significant that hydrogen sulfide gives only a 5% yield of azepine (41) (photolysis of azide), and thiophenol gives none (8) (thermolysis of azide). An attempt to increase the yield of azepine from photolysis of phenyl azide in liquid ammonia by carrying out the reaction in the presence of potassium amide failed, owing to the occurrence of a rapid, nonphotolytic reaction leading to aniline (50% yield) (41).

Most of the rather limited work on the formation of azepines has been concerned with sources of unsubstituted phenylnitrene (Table IV). Other potential instances have been p-tolyl azide (43), p-chlorophenyl azide (44), m-tolyl azide (60), m-anisyl azide (60), m-nitrophenyl azide (60), p-anisyl azide (11), p-nitrophenyl azide (43), and the naphthyl azides (43). Of these, only from the first two has an azepine been characterized, whereas the last three were found to give no azepine (p,p'-dimethoxyazobenzene and p-nitroaniline, respectively, were identified from two of them). p-Chlorophenyl azide gives no isolable azepine on thermolysis in aniline (43,44), but gave 4% of 2-anilino-4-chloroazepine on photolysis (44). These observations suggest that the chloro azepine may in fact have been formed on thermolysis, but was destroyed by further reaction with aniline; at the much lower temperature of the photolysis, some chloroazepine could survive subsequent destruction.

Splitter and Calvin (28c) have recently adduced evidence that formation of azepines is a reaction of the singlet state of an arylnitrene. Photolysis of two N-phenyloxaziridines disclosed competition between a rearrangement leading to an anilide, and fragmentation leading to phenylnitrene. Because fragmentation was repressed in experiments carried out in the presence of oxygen, an efficient quencher for triplet states, they deduced that fragmentation must arise through a triplet-state oxaziridine, which would therefore lead to the triplet state of phenylnitrene. Since photolysis of the oxaziridines produced largely aniline and azobenzene, and hardly any azepine, even in the presence of diethylamine, triplet-state phenylnitrene could not be the forerunner of azepine.

Further support for the view that it is the singlet state of an arylnitrene that gives rise to azepine was adduced from photolysis of phenyl azide in diethylamine in the presence of p-dimethylaminobenzaldehyde, a triplet sensitizer. The yield of diethylaminoazepine dropped markedly (to 7%), and that of aniline rose correspondingly (to 70–80%). As was expected, azepine formation was not quenched by oxygen.

In no case of azepine formation has a material balance been reported; in fact, the accompanying products have not usually been identified (although the azepine must in many cases be considered to be a minor product). The only information available is the report that azobenzene is formed to the extent of 50% in the photolysis of p-anisyl azide in aniline (11) (no other product could be detected by chromatography), and the report that an approximately equal quantity of aniline is formed along with the azepine when phenylnitrene was generated by photolysis of an oxaziridine in diethylamine (28). In all these reactions, large quantities of tarry or resinous material are formed.

The last of the intermolecular reactions has been observed in two instances only—the deoxygenation of o-nitrosotoluene and o-butyl-nitrosobenzene with triethyl phosphite (37). The product, a pyridyl nitrone (5), is formally derived from a molecule each of nitrene and nitroso compound. Sundberg has proposed that it may have arisen by reaction of an azirine (eq. 44), which undergoes attack by the nitroso group at the alkylated ring-carbon perhaps concerted with migration of the aziridine nitrogen to the position originally *meta* to the nitroso group. The accompanying products in the case of o-nitrosotoluene were the corresponding

TABLE IV
Azepine Formation from Sources of Phenylnitrene

Source	Means[a]	Amine used	Azepine, %	Ref.
Azide	P	Diethyl	34	41
Oxaziridine	P	Diethyl	10	28
Nitroso	$\phi_3 P$	Diethyl	62	45a
Nitroso	$\phi_3 P$	Methyl	4	45a
Nitroso	$\phi_3 P$	Dimethyl	50	45a
Azide	P	Diisopropyl	14	60
Nitroso	$\phi_3 P$	Butyl	60	45a
Nitroso	$\phi_3 P$	Ammonia	0	45a
Azide	P	Ammonia	25	41
Azide	T	Cyclohexyl	21	43
Oxaziridine	P	Cyclohexyl	+	28
Azide	T	p-Tolyl	+	43
Azide	T	o-Tolyl	9	60
Azide	T	p-Ethoxyphenyl	8.5	43
Azide	T	Benzyl	41	43
Azide	T	Methyl phenyl	21	43

[a] P = photolysis; T = thermolysis.

$$(44)$$

(5)
(15%)

o-methylanil (deoxygenated 5) (20–34%), triethyl N-o-tolylphosphorimidate (10%), and o-toluidine (6%). The ring-alteration products were not observed in deoxygenation of the corresponding nitro compounds.

C. Intramolecular Reactions (Also see Appendix)

The simplest intramolecular reaction attributable to an aryl nitrene is fragmentation. The formation of mucononitriles by thermolysis of o-phenylene diazides (24b–d) (eq. 14) is a possible example, although the evidence does not distinguish between intermediate formation of a nitrene and fragmentation concerted with rupture of the azido groups. A benzotriazolyl nitrene has been proposed as a possible intermediate. The mucononitriles are formed initially in an all-cis configuration and in good yields; they isomerize readily.

We have already considered reactions of arylnitrenes that result in expansion or alteration of the benzene ring. In contrast, intramolecular contraction of the benzene ring has been observed in the high-temperature vapor-phase pyrolysis of phenyl azide; the product is cyclopentadiene-carbonitrile (14) (eq. 45). This substance was formed only when the rate of introduction of the azide vapor into the hot zone was fast; at slow rates,

$$\langle\!\!\!\bigcirc\!\!\!\rangle\!-N_3 \xrightarrow{\ 300-700^\circ\ } \langle\!\!\!\square\!\!\!\rangle\!-CN + N_2 \qquad (45)$$

the products were instead azobenzene and aniline. To explain this unexpected circumstance, Crow and Wentrup (14) suggested that a vibrationally excited singlet nitrene may be formed first, and undergoes ring contraction if it does not diffuse out of the hot zone fast enough; otherwise it decays to a triplet state, which can take part in dimerization and hydrogen abstraction. Benzotriazole, which is isomeric with phenyl azide, also gave rise to cyclopentadienecarbonitrile on pyrolysis, but did not form azobenzene under any conditions. It was suggested that the benzotriazole may give rise to a symmetrical azirine (4) directly, and not via phenylnitrene, thus obviating formation of azobenzene or aniline. (See Appendix.)

Intramolecular insertion of a nitrene into a saturated C—H bond, which forms a ring, is much favored over the corresponding bimolecular reaction. This type of cyclization shows a strong preference for forming a five-membered ring, and a selectivity for *prim*-C—H < *sec*-C—H (< *tert*-C—H). The initial product of such an insertion, an indoline (or analog), is often oxidized during the reaction to an indole (eq. 46).

$$\begin{array}{ccc} \overset{Y}{\underset{N}{\bigcirc}}\!\!\!\text{CH}_2R & \longrightarrow & \overset{Y}{\underset{NH}{\bigcirc}}\!\!\!\text{CHR} & \longrightarrow & \overset{Y}{\underset{N}{\bigcirc}}\!\!\!\text{C—R} \end{array} \qquad (46)$$

Intramolecular insertion into a saturated C—H has been encountered mostly with thermolysis of *ortho*-substituted aryl azides (8,36,49,65–67). Deoxygenation of nitroso compounds (37) and nitro compounds (37,39,68) with triethyl phosphite also gives rise to intramolecular insertion, but with ferrous oxalate as the deoxygenating agent, insertion into a purely aliphatic side chain has not been reported. The only direct comparison of the two reagents is the work of Smolinsky and Feuer (68) with *o*-(2-methylbutyl)-nitrobenzene, however (eq. 47). They have suggested that the occurrence of insertion of this type might be taken as diagnostic of the production of a true nitrene as an intermediate, and have accordingly inferred that the ferrous oxalate deoxygenations do not produce nitrenes.

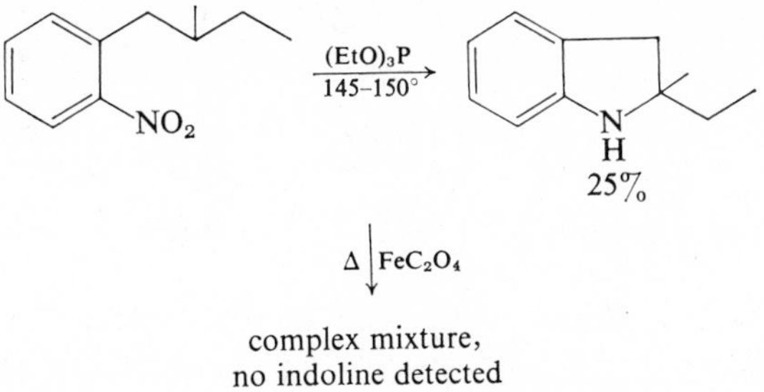

(47)

25%

$\Delta \Big| FeC_2O_4$

complex mixture,
no indoline detected

There are other instances of deoxygenation by ferrous oxalate in which cyclization to a saturated C—H nevertheless takes place. These are N-benzyl-o-nitroaniline (69), which is converted to 2-phenylbenzimidazole (eq. 48), 2-nitro-2',4',6'-trimethylbiphenyl (38,68), which gives dimethyl-

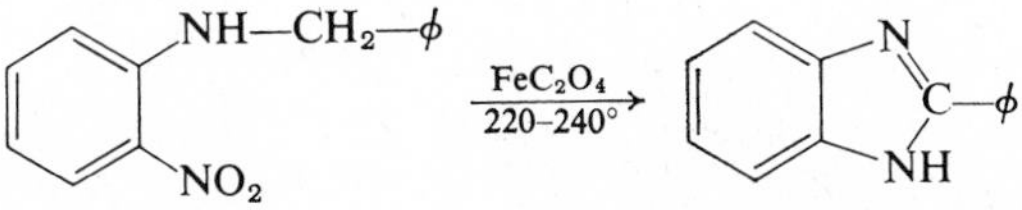

(48)

phenanthridine (eq. 49), and 1-cyclohexyl-2-nitrobenzene (38,68), which gives carbazole (eq. 50). These examples are not definitive, however, for the same products are formed by simple heating without a reducing agent, albeit not always so cleanly. The first two cases are believed to involve isomerization to a nitronic acid (eq. 51), which may then cyclize

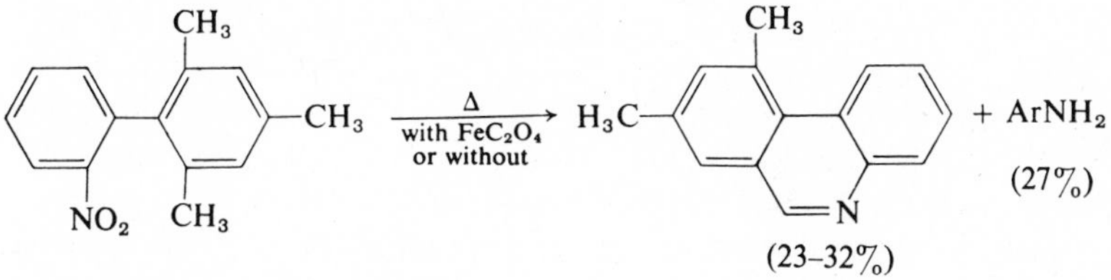

+ ArNH$_2$ (49)

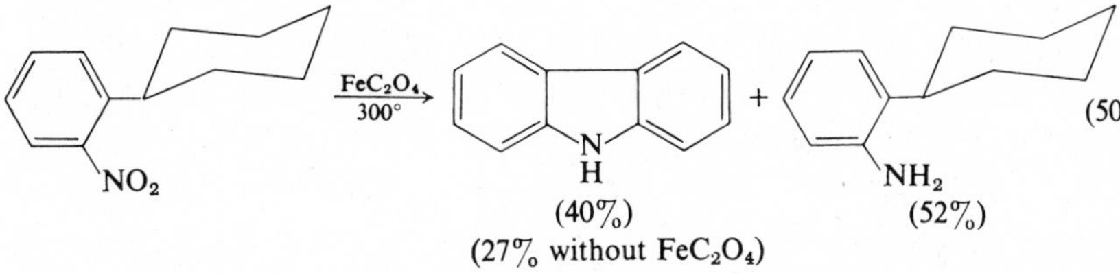

(27%)

(23–32%)

(50)

(40%)   (52%)
(27% without FeC$_2$O$_4$)

by another mechanism, not involving prior deoxygenation (68,69) (the precise details of such an alternative pathway are not yet clear, but are not needed in the present discussion). In the last case, oxidation of the cyclohexyl group to a phenyl group probably precedes cyclization at the high

$$\text{(51)}$$

temperatures used (68) (*o*-aminobiphenyl is known to undergo thermal conversion to carbazole).

Reduction of a nitro group by tin and hydrochloric acid, or by sodium bisulfite, reagents that are not usually considered to be simple deoxygenating agents for nitro groups, has led to cyclization, formally by insertion into a saturated C—H, in the case of 2-nitro-*N*,*N*-dimethyl-*p*-toluidine (70), which gives 1,5-dimethylbenzimidazole. The original investigators believed that they were observing cyclization at the stage of an intermediate nitroso compound; indeed, the fact that the same reagents do not produce cyclization products from simple *o*-alkyl nitrobenzenes suggests that a nitrene is not involved.

The preference for cyclization to form a five-membered ring is most clearly illustrated by Smolinsky and Feuer's results with vapor-phase thermolysis of *o*-butylphenyl azide (36a) (eq. 52). The same products in similar ratio (but lower yield) were obtained from deoxygenation of

$$\text{(52)}$$

ratio 43 : 11 : 46; total yield 70%

o-butylnitrobenzene or *o*-butylnitrosobenzene with triethyl phosphite (37). *o*-(2-Methylbutyl)phenyl azide also undergoes cyclization principally to the β-carbon (50–60% yield) (36b). *o*-Cyclohexylnitrobenzene behaves analogously (37) when treated with triethyl phosphite (eq. 53); the results are distinct from those of deoxygenation by ferrous oxalate. No dehydrogenation takes place. *o*-Cyclohexylphenyl azide apparently cyclizes only to hexahydrocarbazole (*cis*/*trans* = 1; 86% yield) (71).

Barton and Morgan (72) have proposed in connection with alkyl azides that the triplet nitrene attacks at hydrogen through a cyclic transition

$$\text{(53)}$$

(*cis* and *trans*) 37%   14%

state, which would be six-membered in the situation leading to eventual closure at carbon to form a five-membered ring (eq. 54). The seven-membered transition state required for eventual closure to a six-membered ring would be less favorable, thus accounting for the observed preference. Such a mechanism is very similar to that of the Hoffman-Loeffler-Freytag cyclization of N-chloro amines, and is in accord with the order of reactivity

$$\text{(54)}$$

tert-C—H > *sec*-C—H > *prim*-C—H observed in intermolecular insertions. Evidence for this order in cyclizations of aryl azides is seen in the fact that o-propylnitrosobenzene, o-propylnitrobenzene (eq. 55), and o-propylphenyl azide give only 2-methylindoline and no tetrahydroquinoline on cyclization (37). Although these results demonstrate a clear reluctance to cyclize to a methyl group, it should be noted that o-isopropylphenyl azide has been cyclized to 3-methylindoline in 15% yield by heating in diphenyl ether; the major product was o-aminocumene, however. Cyclization in the vapor phase (8), where no solvent is present to serve as a source of hydrogen, raised the yield to 55%.

$$\xrightarrow[156°]{(EtO)_3P}$$

(7%)   (6.5%)

$$+ \text{Ar—N}{=}\text{P(OEt)}_3 + \text{Ar—NH}_2 \quad \text{(55)}$$

(40%)   (3%)

Although much evidence favors the diradical mechanism for cyclization of a nitrene to a saturated site, there is contrary evidence in the form of Smolinsky and Feuer's demonstration (36b,68) that cyclization to an optically active site occurs with about 50% retention of configuration. The system used was o-(2-methylbutyl)phenylnitrene, obtained either from the nitro compound (eq. 47) or the azide. The degree of retention was greater in the vapor phase than in solution. The conclusion was drawn that the nitrene is initially formed in the singlet state, in which it takes part in direct insertion with retention of configuration. Insertion is in competition with decay to the triplet state, in which abstraction of hydrogen (inter- or intramolecular) occurs. Such decay would be faster in solution.

If abstraction of hydrogen is intramolecular, the resulting diradical (eq. 54) may undergo racemization before cyclization. In addition, the amino radical moiety may abstract a second hydrogen, so as to generate a double bond (eq. 56). Intermolecular abstraction of hydrogen would lead through an amino radical to the amine (eq. 56).

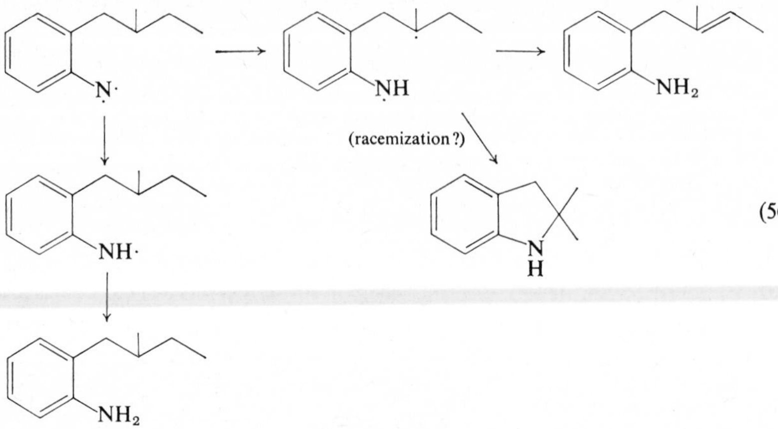

(56)

(racemization?)

It should be noted that a mechanism by which the singlet nitrene abstracts hydride from the adjacent alkyl group to form a discrete intermediate zwitterion (eq. 57) is negated by the foregoing evidence, for the carbonium ion would become racemic if it had an independent existence.

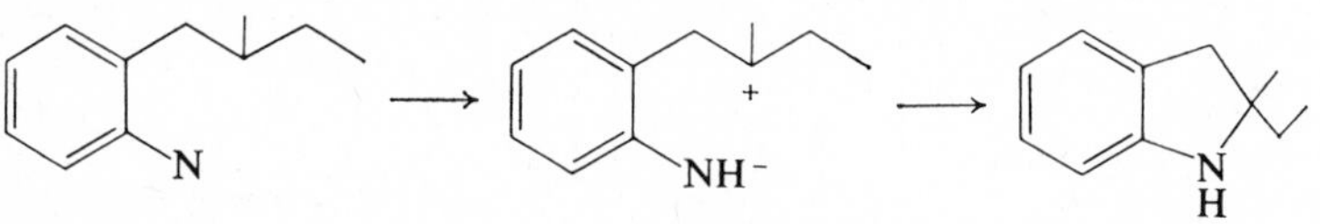

(57)

Before leaving the subject of insertion into a saturated C—H, the special behavior of some o-dialkylaminophenylnitrenes should be considered. Saunders has reported (67) the formation of benzimidazoles in moderate to high yields from heating various o-hexahydroazepyl-, o-piperazinyl-, and o-morpholino-phenyl azides in nitrobenzene (eq. 58). Benzimidazolines were presumably formed first, but were oxidized by the

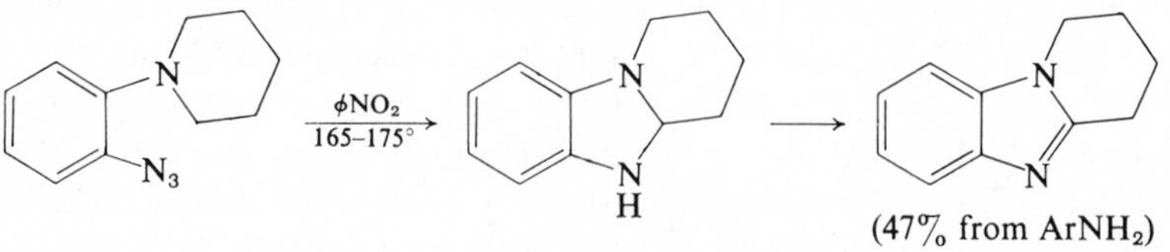

$$\frac{\phi NO_2}{165-175°}$$

(58)

(47% from ArNH$_2$)

nitrobenzene; aniline was obtained in one instance in 84% yield. Further examples have been reported by Schmutz and Kunzle (65), and by Meth-Cohn, Smalley, and Suschitzky (49), who were able to intercept the benzimidazoline stage in one instance by conducting the thermolysis in acetic anhydride. They also carried out analogous cyclizations with azidopyridines (see below). Smith and Suschitzky (69) attempted similar reactions by deoxygenation of o-nitroaniline derivatives with ferrous oxalate. They obtained the expected 2-phenylbenzimidazole with N-benzyl-o-nitroaniline (40% yield), but the N-cyclohexyl and N-phenethyl compounds behaved unexpectedly differently (eqs. 59 and 60). The same products were obtained by heating without a deoxygenating agent (in

$$(59)$$

$$(21\%)$$

$$(60)$$

$$(7\%) \qquad (6\%)$$

lower yields), and the reactions do not appear to involve arylnitrenes (cf. the reduction of 2-nitro-N,N-dimethyl-p-toluidine (70)).

The possibility of cyclization by insertion of a nitrene into an O—H bond has been investigated in the case of o-azidobenzyl alcohol (8). The only product isolated was o-aminobenzaldehyde (65% yield).

Intramolecular insertion of a nitrene (or nitrene source) into an aromatic C—H occurs with much higher yields than insertion into saturated C—H, and has become the basis of a general synthesis of carbazoles from ortho-substituted biphenyls. The majority of the instances involve the azide, but nearly every other source of aryl nitrenes has been applied as well. Table V summarizes the reported examples.

Several cases of cyclization to a pyridine (59,73,76,77) (eq. 61), or thiophene (73) (eq. 62) ring have been reported. The ratio of the two isomeric carbolines obtained through o-β-pyridylphenylnitrene differs considerably when the source is thermolysis of the azide (73) (α-carboline = 2:1) or deoxygenation of the nitroso compound with triethyl phosphite (59) (α-carboline/γ-carboline = 4.4:1), but the difference is probably due to the large difference in temperature of the two reactions. The α-pyridyl analogs lead not to carboline, but to pyrid[1,2-b]indazole (59,76) (eq. 63), if conversion to the amine does not predominate (73).

TABLE V
Carbazoles from Biphenyl Derivatives

| Biphenyl | | Method | Carbazole | | |
Substituents	Nitrene precursor		Substituents	Yield, %	Ref.
None	$-N_3$	T	None	76	9
		P		98	5
		$(EtO)_3P$		77	9
	$-NO_2$	$Fe(CO)_5 + CO$		82.5	58
		FeC_2O_4		37.5	52
	$-NCO$	P		63	74
	$-NO$	$(EtO)_3P$		15	13
				76	59
5-Chloro	$-N_3$	T	3-Chloro	60.5	75
4-Bromo	$-N_3$	T	2-Bromo	87	5
5-Bromo	$-N_3$	T	3-Bromo	83	5, 9
	$-N_3$	P		23	9
3,5-Dibromo	$-N_3$	T	1,3-Dibromo	74	9
	$-N_3$	P		57	9
3-Nitro	$-N_3$	T, P	(4-Phenylbenzofuroxan)	0	9
4-Nitro	$-N_3$	T	2-Nitro	98	5
5-Nitro	$-N_3$	T	3-Nitro	88–93	5, 9
5-Nitro	$-N_3$	P	(Azide recovered)	0	9
3,5-Dinitro	$-N_3$	T, P	(Furoxan formed)	0	9
4-Methoxy	$-N_3$	T	2-Methoxy	98	5
5-Methoxy	$-N_3$	T	3-Methoxy	75	5
4-Methyl	$-N_3$	T	2-Methyl	90	5

Substituent	Group	Method	Product / Remarks	Yield	Ref.
5-Methyl	$-N_3$	T	3-Methyl	80	5
2'-Chloro	$-NO_2$	(EtO)₃P	4-Chloro	36	58
2'-Bromo	$-NO_2$	(EtO)₃P	4-Bromo	42	58
2'-Methyl	$-NO_2$	(EtO)₃P	4-Methyl	72	58
2'-Methoxy	$-N_3$	T	4-Methoxy	95	71
2'-Hydroxy	$-N_3$	T	(Amine, 30%, and tar)	0	71
2'-Hydroxy, anion	$-N_3$	T	(Amine, 10%, and tar)	0	71
2'-Nitro	$-N_3$	T		0	9
2'-Cyano	$-N_3$	T	(Tetrazole formed)	0	75
4'-Methyl	$-NO_2$	(EtO)₃P	2-Methyl	83	58
4'-Nitro	$-N_3$	T	2-Nitro	87	5, 9
	$-N_3$	P		0	9
4'-Methoxy	$-N_3$	T	2-Methoxy	71.8	75
4'-Hydroxy	$-N_3$	T	2-Hydroxy	87.5	75
2',4',6'-Trimethyl	$-N_3$	T	2,4,9-Trimethyl	4.5	66
2'-Methoxy-6'-Methyl	$-NO_2$	(EtO)₃P		0	58
	$-N_3$	T	Amine, 29%; phenanthridine, 16%	0	71
2',5'-Dimethoxy	$-N_3$	T	1,4-Dimethoxy	71–99	5, 75
2',4',6'-Trimethoxy	$-N_3$	T	(Amine, 3%)	0	71
4,4'-Dinitro	$-N_3$	P	2,7-Dinitro	91	9
	$-N_3$	T		65	9
5,4'-Dinitro	$-N_3$	T	3,7-Dinitro	94	9
5,2'-Dimethoxy	$-N_3$	T	1,4-Dimethoxy	71	75
4,5-Dimethoxy-2-ethyl	$-N_3$	T	2,3-Dimethoxy-5-ethyl	58	6
2',3'-Benzo	$-N_3$	T	3,4-Benzo	99/100	5, 75
3',4'-Benzo	$-N_3$	T	1,2-Benzo	93–4	5, 75
3,4-Benzo	$-N_3$	T	1,2-Benzo	94	75

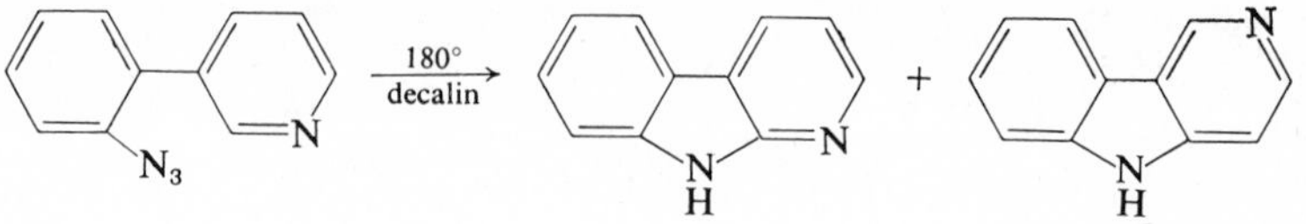

$$\text{(61)}$$

(ratio 2:1; 94% total yield)

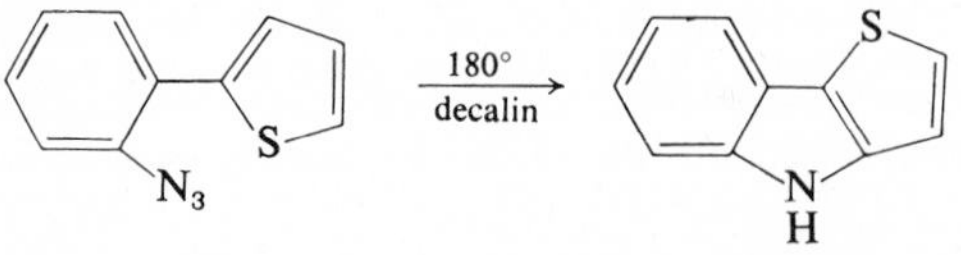

$$\text{(62)}$$

(93%)

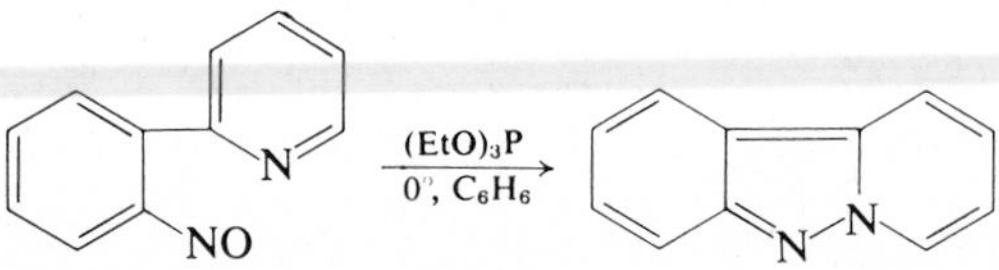

$$\text{(63)}$$

Cyclization to form a six-membered ring by insertion into an aromatic C—H takes place only if formation of a five-membered ring is interdicted in the manner shown in equation 64, and is even then generally a poor

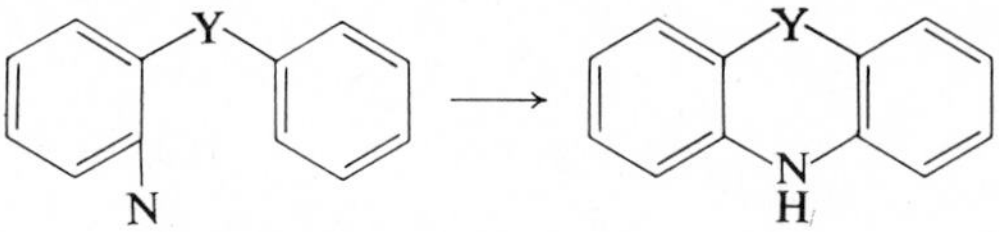

$$\text{(64)}$$

reaction. Yields are commonly low, and in many instances no cyclization product whatsoever can be isolated.

The best results arise when Y = S, leading to phenothiazines; the unsubstituted compound has been obtained by thermolysis of the azide in 32% yield (29), and a variety of substituted phenothiazines have been prepared in this way in apparently respectable yields (78).

Rearrangement has been observed in the cyclization of 4′-substituted 2-nitro and 2-azido diphenyl sulfides; the 4′-substituent (chloro or methyl) appears in the 3-position of the phenothiazine produced, rather than the expected 2-position (78c). This result can be explained by assuming that the nitrene first attacks the 1-carbon of the second benzene ring, giving rise to an unstable intermediate with a five-membered ring with a *spiro* structure (compare equation 68 and the accompanying discussion of rearrangement in the formation of acridones from substituted *o*-azido-benzophenones). It has been suggested (78c) that cyclization of all *o*-nitro and *o*-azido-diphenyl sulfides takes place through such an intermediate; rearrangement would only be observed when the second phenyl group was substituted.

Phenothiazine dioxide has been prepared from o-azidodiphenyl sulfone in 42% yield (29).

Only one report exists of an attempt to bring about cyclization when $Y = O$; o-azidodiphenyl ether apparently gave a small amount of phenoxazine by thermolysis (29).

A considerable number of examples of formation of phenazines by nitrenoid cyclizations are known (eq. 64, $Y = NH$), although yields are very low. Only one case of an azide is included; o-azido-N-acetyldiphenyl-amine gave only colored gum on thermolysis or photolysis (29). Deoxygenation of o-nitrodiphenylamines with iron or ferrous oxalate at 250–290° has been the most widely used method (74,79–81); it is sometimes referred to as the Waterman-Vivian phenazine synthesis (eq. 65). Yields as high as 50% have been reported but many are apparently very low. There is no evidence that dihydrophenazines are intermediates, and there is reasonable doubt that this type of deoxygenation actually involves

$$\begin{array}{c}\text{(structure)}\end{array} \quad \underset{240\text{--}290°}{\xrightarrow{FeC_2O_4}} \quad \begin{array}{c}\text{(structure)}\end{array} \qquad (65)$$

nitrenes (cf. the earlier part of this chapter). An attempt to extend the reaction to o-nitroanilinopyridines gave only tars (76).

Phenazines have also been obtained by oxidation of o-aminodiphenyl-amines with lead oxide (Pb_3O_4) (17,18) at 200–240° (eq. 12). Yields are of the order of only 5%. Although various substituents (p'-Cl, p'-Br, p'-CH_3O) remain unaffected by the conditions, o-amino-o'-methoxydiphenyl-amine lost the entire methoxy group and gave only unsubstituted phenazine (18).

o-Azidobenzophenones cyclize easily to anthranils, without attack of the nitrene on the aromatic ring (7,29) (eq. 65). When the anthranils are brought to higher temperatures, however, isomerization takes place, giving acridones (7,30,31), which are, at least formally, the result of nitrenoid insertion (eq. 67). Yields are only moderate, but still high enough to be of preparative value. Photolysis of anthranils appears to give poorer results, judging by the single reported comparison (31) (3-pentafluoro-anthranil).

Rearrangement has been reported during thermolysis of three anthranils bearing an alkoxyl group in the p'-position (7). The structure of the products, which are position isomers of the expected ones, is accounted for by the hypothesis that the nitrene attacks the 1'-position, thus closing a five-membered ring (eq. 68).

$$(66)$$

$$(67)$$

$$(68)$$

(50%)

    o-Azidodiphenylmethane undergoes thermolytic cyclization of a different type (82), which illustrates in still another way the preference for formation of a five-membered ring. No detectable dihydroacridine is formed; instead, an isomeric substance corresponding to insertion of a nitrene into an aromatic C—C bond is obtained (eq. 69). This compound is obtained in good yield, and its structure is unequivocally established by NMR and by reduction to known compounds.

$$(69)$$

    The question of whether true nitrenes are actually involved in the foregoing cyclizations to an adjacent aromatic ring has been investigated in the instance of *o*-azidobiphenyl (83). Irradiation in a frozen matrix at 77°K produces a single species, stable at that temperature, whose ultraviolet spectrum is independent of the nature of the solvent-matrix; the quantum yield is 0.43. Softening of the matrix causes this new species to disappear. Prolonged irradiation of the new species at wave lengths above 320 mμm, and thus in the range where only the new species absorbs (λ_{max} 342, 400 mμm), causes slow replacement of the spectrum of the intermediate by that of carbazole.

The spectrum of the intermediate closely resembles those of other arylnitrenes that have been obtained by photolysis in frozen matrices, and which are apparently ground-state triplets. The second step, photolytic cyclization, has a low-quantum yield (0.01–0.02), which indicates that it is a process requiring activation, which would not be available by thermal collisions in the frozen matrix, but which would become available upon melting of the matrix, allowing spontaneous cyclization to carbazole. The simplest explanation of the activation would be formation of an excited triplet, but the evidence does not rule out the possibility that the excited triplet crosses to a singlet state before cyclization. On the other hand, photolysis of o-azidobiphenyl ordinarily gives only carbazole, but in the presence of triplet sensitizers, the principal product is 2,2'-diphenylazobenzene (84).

The nature of the insertion reaction is uncertain. In addition to the paths of direct insertion and of hydrogen abstraction followed by combination of radicals (cf. eqs. 54 and 56), available for cyclization to a saturated site, there is the possibility of addition to an aromatic C—C bond. The resulting bicyclic aziridine might isomerize to the observed products directly or by way of an azepine (eq. 70). The intermediacy of azepines in intermolecular attack of nitrenes on aromatic compounds has recently been demonstrated by Abramovitch and Uma (85). The ability of this pathway to provide an explanation for anomalous cyclization products (eqs. 68 and 69) is obvious.

(70)

A number of examples of cyclization of arylnitrene precursors to an ortho-vinyl system (eq. 71) have been reported; most of them have been o-nitrostyrene derivatives (38,56,58). o-Nitrostyrene itself gives only a trace of indole with either triethyl phosphite (58) or ferrous oxalate (38), but yields as high as 85% (*cis-o*-nitrostilbene, triethyl phosphite) have been reported (58).

The known examples of this type of reaction are summarized in Table VI. Usefulness as a synthetic method for indoles is, of course, limited by the availability of the appropriate *o*-substituted styrene derivatives, most of

$$ \tag{71} $$

TABLE VI

Indoles from *o*-Alkenylphenylnitrene Sources (eq. 71)

R	R'	NX	Method	Indole yield, %	Ref.
H	H	NO_2	$(EtO)_3P$	Trace	58
H	H	NO_2	FeC_2O_4	Trace	38
H	CH_3	NO_2	$(EtO)_3P$	51	56
H	Pr	NO_2	$(EtO)_3P$	60	56
H	ϕ (*trans*)	NO_2	$(EtO)_3P$	58	58
				71	56
H	ϕ (*cis*)	NO_2	$(EtO)_3P$	85	58
H	ϕ (*trans*)	N_3	Boiling decalin	88	86
H	ϕ (*cis*)	N_3	Boiling decalin	18[a]	86
CH_3	CH_3[b]	N_3	Boiling decalin	24	87
H	CH_3CO	NO_2	$(EtO)_3P$	16	56
H	ϕCO	NO_2	$(EtO)_3P$	16	56
H	Ethylene ketal of above	NO_2	$(EtO)_3P$	21	88
H	$\phi CON\!\!<\!\!>\!\!-CH_2CO$	NO_2	$(EtO)_3P$	13	88
H	Ethylene ketal of above	NO_2	$(EtO)_3P$	12	88
H	—COOH	NO_2	$(EtO)_3P$	7.5	58
	—COOH	NO_2	FeC_2O_4	Low	38
H	COOEt	NO_2	$(EtO)_3P$	19	56
H	$o\text{-}O_2N\phi$	NO_2	$(EtO)_3P$	2[c]	58

[a] The accompanying product was a nonbasic, nonhydrolyzable gum.
[b] Unknown stereochemistry, probably mixed.
[c] Indolo-[3,2-*b*]-indole.

which are rather tedious to obtain. *o*-Azidostyrenes are perhaps best obtained from the *o*-bromostyrenes through the reaction of the Grignard reagent with toluenesulfonyl azide (86); most of the *o*-nitrostyrenes (56) have been made from *o*-nitrobenzyl bromide and aldehydes by the Wittig reaction. Yields of indole are much higher when the β-substituent is alkyl or aryl rather than a functional group.

There is always a quantity of the corresponding β,β-biindolyl formed along with the indole during deoxygenation of *o*-nitrostyrenes with triethyl phosphite (56). The formation of indoles by this method may not involve nitrenes, for Sundberg (56) has shown that *N*-hydroxy indoles are reduced to indoles under the reaction conditions, and are thus possible intermediates (which might arise from nitrosostyrenes).

The cyclizations of *o*-nitrostyrenes that take place in low yields are usually accompanied by extensive tar formation. The two examples bearing a dioxolane substituent (ethylene ketal) underwent fragmentation and rearrangement as well (eq. 72); Sundberg (88) has proposed that easy formation of 2-phenyldioxolanyl cation may be the cause.

The mechanism of cyclization of *o*-alkenylphenylnitrenes, apart from the general ambiguity about deoxygenation reactions, might reasonably involve attack of the nitrene on the π-electrons of the olefinic bond, or might involve insertion into the C—H bond in one of the ways described in connection with the cyclization of *o*-alkylphenylnitrenes (cf. eqs. 54, 56, and 57). There is evidence that attack on the π-bond may be possible. Sundberg and Yamazaki (89) studied a series of *o*-nitrostyrenes bearing two substituents on the β-position, and thus having no hydrogen at the required point of attack for indole formation. In every case they obtained an indole (yields 15–77%) with concomitant migration of a substituent to the adjacent carbon (eq. 73).

$$(73)$$

2-Methyl-*o*-nitro-*trans*-stilbene underwent rearrangement exlusively by phenyl migration, a result which could perhaps be due to steric as well as electronic effects. A bipolar intermediate of the type shown is purely speculative, but it would account for the occurrence of such rearrangement. This reaction may, in fact, not involve a nitrene at all, of course.

One example of an *o*-azidostyrene bearing two substituents on the β-carbon has been studied (86): β-methyl-β-phenyl-*o*-azidostyrene (stereochemistry mixed). 2-Methyl-3-phenylindole, a product resulting from migration of the phenyl group, was the only pure substance isolable from thermolysis in decalin (∼10% yield). Deoxygenation of the corresponding nitro compound (89) gave the same product, but in much higher yields.

The effect of geometrical configuration on the formation of indoles from styrenes deserves separate mention (see Table VI). Cyclization by attack of a nitrene on hydrogen or the C—H bond would presumably be favored by a *trans*-stilbene configuration and be very difficult for the *cis*-stilbene. Attack on the β-carbon, however, as shown in equation 73, should be relatively little influenced by configuration. The fact that *trans*-*o*-azidostilbene undergoes cyclization in high yield, whereas the *cis* isomer gives only a low yield of 2-phenylindole under identical conditions, is thus in accord with attack at H or C—H. The quite different behavior of the *o*-nitrostilbenes, in which both isomers give 2-phenylindole in good yields, with the *cis* giving the most, indicates a difference in mechanism. One possibility is that reaction of the azido group with the olefinic double bond to form a triazoline precedes cyclization, or competes with it. This seems unlikely in view of the general unreactivity of stilbene toward 1,3-dipolar additions, and the high-strain energy that would be involved in intramolecular triazoline formation (resulting in fusion of a four-membered ring with the triazoline system). Another possibility is that formation of indoles from *o*-nitrostyrenes does not actually involve nitrenes, but, as

Sundberg has suggested (89), proceeds by cyclization before completion of deoxygenation.

Cyclization of a nitrene source to an adjacent azomethine function is represented by two reports of thermolysis of o-azidoanils of aldehydes (eq. 74) (90,91). Yields vary from 45 to 96%, except in the instance R = benzhydryl, which gave no cyclization product (91).

$$
\text{(structure)} \xrightarrow{130-150°} \text{(structure)} \tag{74}
$$

Cyclization of arylnitrenes to unsaturated atoms other than carbon is formally represented by four systems, all of which involve closure to oxygen or nitrogen (eq. 75). The temperatures required for the azides are generally much lower than those required for fragmentation of other aryl azides, a fact that raises the question whether these cyclizations may actually be concerted with loss of nitrogen. No general evidence is available, but the negative entropy of activation for decomposition of o-nitrophenyl azide (2) suggests a cyclic transition state.

$$
\text{(structure)} \xrightarrow{-G_2} \text{(structure)} \qquad (G_2 = N_2, O_2, \text{ or } H_2) \tag{75}
$$

o-Azidophenyl ketones, including 1-azidoanthraquinones (92, 93), form anthranils (7,29,32) (eq. 75, Y = C, Z = O), generally in high yields. There are a number of examples (93) of formation of anthranils by reduction of o-nitro- and o-nitrosophenyl ketones or oxidation of o-aminophenyl ketones, but it seems doubtful that nitrenes are involved, particularly in view of the fact that it has been demonstrated (94) that in the reduction of o-nitrobenzaldehyde to anthranil, the oxygen atom in the product is not that of the carbonyl group, but comes from the nitro group.

Benzaldimine derivatives behave analogously, giving rise to indazoles (eq. 75, Y = C, Z = N) by thermolysis of azides (95) or deoxygenation of nitro compounds (58) with triethyl phosphite; yields are high by the former method, moderate by the latter. The azine of o-azidobenzaldehyde reacts in two distinct stages (eq. 76).

Apparent cyclization of a nitrene to an azo group has been observed with o-azido- (96), o-nitro- (58), and o-aminoazobenzenes (19–23,96,97) (eq. 75; Y = Z = N). The products are benzotriazoles. o,o'-Diazidoazobenzene loses nitrogen in two discrete stages (eq. 77) (96). The second stage is, of course, no longer of the type shown in equation 75, and the much higher temperature required is the usual one for ordinary aryl azides. The

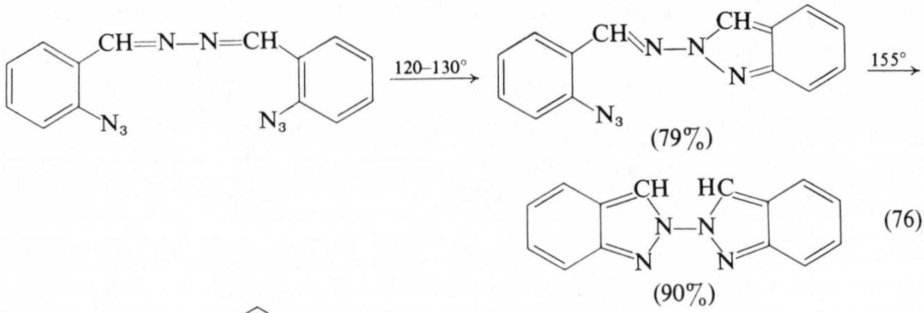

(76)

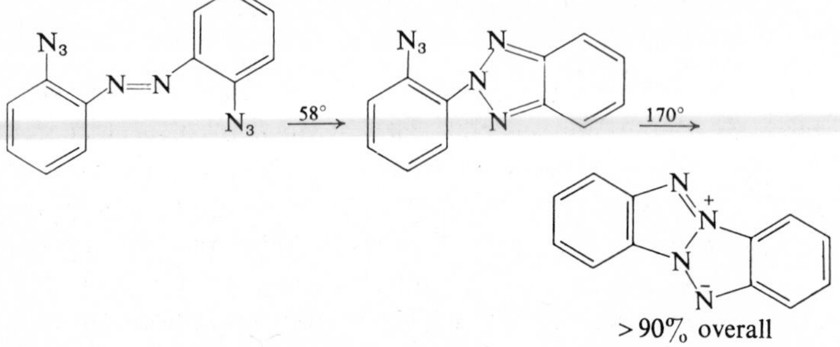

(77)

> 90% overall

correctness of the assignment of structure to the final product as a tetra-azapentalene rather than tetraazacyclooctatetraene is demonstrated by the behavior of o-azidophenylnaphthotriazole, which produces two isomeric tetraazapentalenes (eq. 78) (there would be only one tetraazacyclooctatetraene).

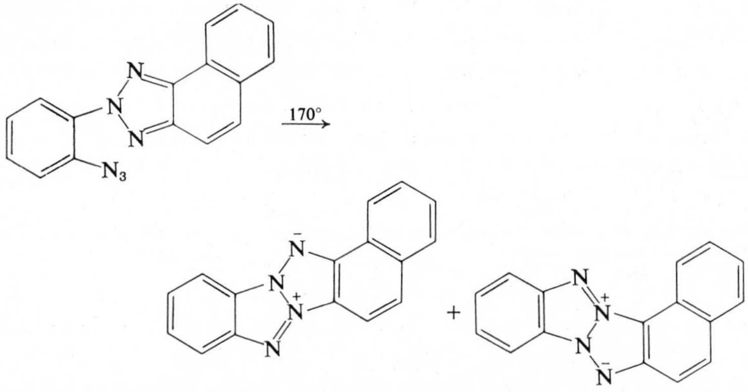

(78)

Yields in cyclization to an azo group seem to be distinctly higher (>90%) with azides than with deoxygenation of nitro compounds (31–83%). Oxidation of o-amino azobenzenes appears to give good yields of benzotriazoles, but the reports are not specific. Oxidation has usually been carried out with cupric salts, but chromic acid has also been used (97).

Kinetics are third order (23), rate = $k[\text{ArNH}_2][\text{Cu}^{\text{II}}]^2$, in most, but not all, instances. There is evidence that a copper complex is involved, perhaps with chelation to the azo group and the erstwhile amino nitrogen (21,23).

The second stage of cyclization of o,o'-diazidoazobenzenes apparently involves attack of a nitrene at an unsaturated nitrogen atom whose double bond is in the β,γ-position with respect to the phenyl group. Many examples have been reported (93,98). Deoxygenation of o-nitrophenyl-triazoles with triethyl phosphite brings about the same result in slightly lower yields (58,98), but ferrous oxalate failed in the one instance reported (98). This type of cyclization is more than formally analogous to the formation of pyrido[1,2-b]indazoles by cyclization to a pyridine nitrogen (75) (eq. 63).

Cyclization of an arylnitrene to the oxygen of an o-nitro group would lead to a benzofuroxan (eq. 75, Z = O, Y = NO). o-Nitroaryl azides in general do, indeed, give benzofuroxans in high yield on mild thermolysis (10,99). This type of reaction takes place readily below 100°. In 2-azido-3-nitrobiphenyls, the nitrene that would be derived from the azido group would have the choice of attacking the ortho position of the neighboring group or forming a phenylbenzofuroxan; the latter compounds are the exclusive products (9) (eq. 79). These facts suggest that the loss of nitrogen from o-nitroaryl azides is concerted with cyclization, an

$$(79)$$

$$(83\%)$$

interpretation that is also in agreement with the negative entropy of activation of o-nitrophenyl azide (2).

Deoxygenation can also lead to benzofuroxans, as shown in the instance of o-nitronitrosobenzene (58) (yield 18%). Oxidation of o-nitroaniline by sodium hypochlorite is reported to give benzofuroxan in basic solution, but to give o,o'-dinitroazobenzene in methanolic acetic acid (100). Although there is insufficient information for deducing the mechanism of these reactions, it is possible that o-nitro-N,N-dichloroaniline may be an intermediate, inasmuch as that compound is formed from the same reagents at lower temperatures (15).

An attempt to prepare iodine analogs of benzofurazan and benzo-furoxan from o-iodosophenyl azide and o-iodoxyphenyl azide failed (101). The iodoso compound underwent disproportionation before loss of nitrogen, and the iodoxy compound exploded upon attempted thermolysis (thermolysis in solution was apparently not tried, however).

II. QUASI-ARYL SYSTEMS

Azidotropone rearranges thermally or photolytically to salicylonitrile (102) (eq. 80). There is insufficient evidence to decide if a nitrene is involved.

$$(80)$$

An azidoquinone, 2,3-diazido-1,4-naphthoquinone, loses nitrogen in boiling toluene and forms phthaloyl cyanide (eq. 81); the kinetics are first-order (103). Photolysis in benzene gives the same product, with a quantum yield of 1. Several other azidoquinones were found to lose

$$(81)$$

nitrogen by first-order kinetics, but unfortunately the products were not reported. This reaction bears an obvious similarity to the fragmentation of o-diazidobenzenes to dinitriles, discussed earlier. As with them, there is no evidence to distinguish whether the fragmentation is a concerted process or proceeds through a discrete nitrene stage.

III. HETEROCYCLIC NITRENES

A. Six-Membered Rings

Relatively very little work has been done of a type that might involve a nitrene function attached to a six-membered aromatic heterocyclic ring, but the information available suggests that these systems are quite similar to their homocyclic analogs. 2-Phenyl-3-azidopyridine, for example, is converted to δ-carboline by thermolysis (104) (eq. 82). Intramolecular insertion into a saturated carbon has also been reported (49) (eq. 83). The yields from 4-substituted 3-azidopyridines (41–72%) were much higher than those from the 2-substituted isomers (10–15%). The reason is not apparent, but since the reported results are overall figures for conversion

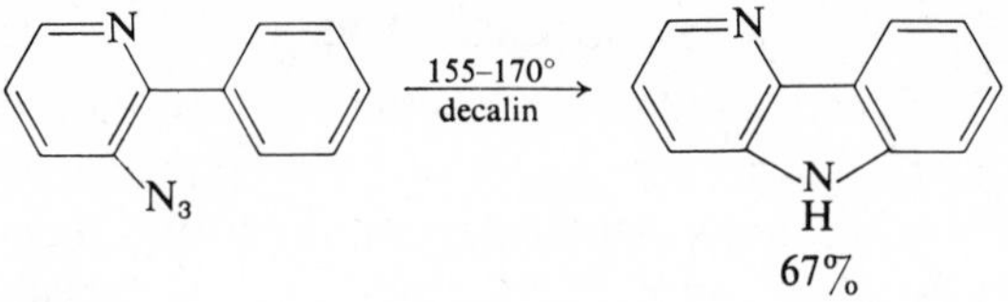

$$\text{(82)}$$

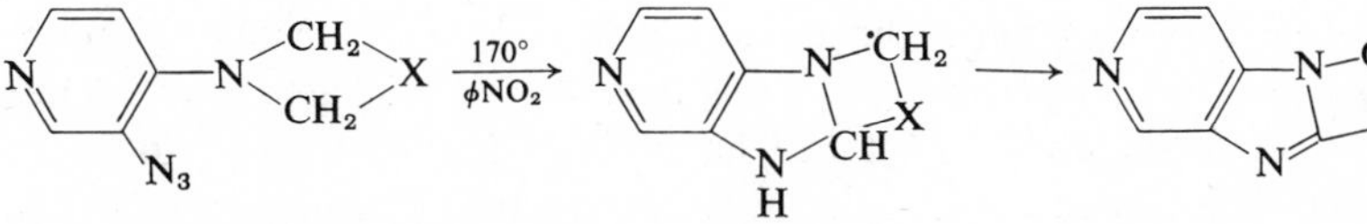

$$\xrightarrow[\text{decalin}]{155-170°}$$

67%

$$\xrightarrow[\phi NO_2]{170°}$$

$$\text{(83)}$$

X = $(CH_2)_2$, $(CH_2)_3$, $(CH_2)_4$, $(CH_2—O—CH_2)$

of amine to azide to product, the difference is not necessarily attributable to the stage involving the nitrenes.

The oxidative cyclization of *o*-aminoazobenzenes to benzotriazoles also has its parallel in the pyridine series (20) in the successful conversion of 2,6-diamino-3-phenylazopyridine to 2-phenyl-6-aminopyridoimidazole by oxidation with ammoniacal copper sulfate.

One example of thermolysis of an azidophthalazine has been reported (105) (eq. 84). Decomposition in tetralin results in abstraction of hydrogen to form the amine in high yield; in xylene or amyl ether, the same product was obtained, but in lower yields.

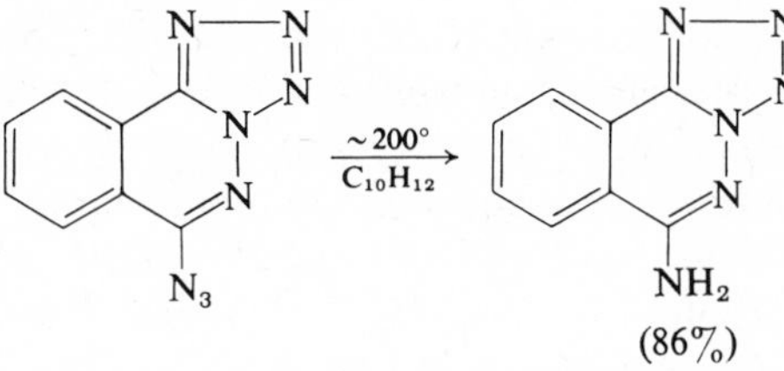

$$\xrightarrow[C_{10}H_{12}]{\sim 200°}$$

$$\text{(84)}$$

(86%)

Cyanuryl azide explodes on heating above 100°; the products, entirely gaseous, consist of nitrogen and cyanogen (106a) (eq. 85). The formation of these substances requires more than a simple fragmentation, but there is ESR evidence for initial formation of a nitrene intermediate (106b).

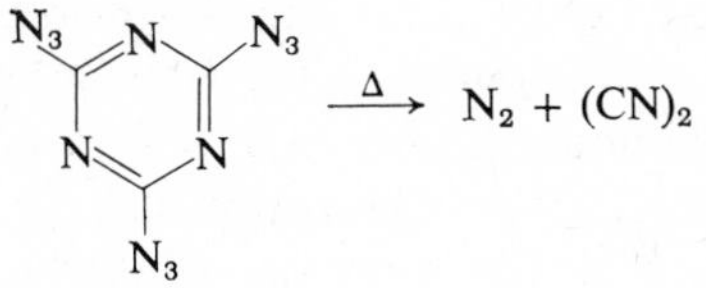

$$\xrightarrow{\Delta} N_2 + (CN)_2$$

$$\text{(85)}$$

B. Five-Membered Rings

The chemistry of heteroaryl nitrenes with five-membered rings is largely confined to the behavior of azides, principally pyrazoles and triazoles. The usual nitrene reactions must compete, often not very successfully, with favorable possibilities for fragmentation of the nitrene (discrete or incipient).

The earliest example is the oxidation of 1-phenyl-3-methyl-5-aminopyrazole (64) with hydrogen peroxide, from which Michaelis and Schäfer obtained a light-brown substance having two less hydrogen atoms, to which they assigned a fused-ring azirine structure ("azipyrazole"). The compound was much later demonstrated (107) to be in all probability a product of fragmentation: β-phenylazocrotononitrile in one of its four possible geometrically isomeric forms (eq. 86). There was, however, a

$$\text{H}_3\text{C}\overbrace{\underset{\phi}{\underset{N\text{-}N}{\bigvee}}}\text{-}NH_2 \xrightarrow{\text{H}_2\text{O}_2} \left[\text{H}_3\text{C}\overbrace{\underset{\phi}{\underset{N\text{-}N}{\bigvee}}}\text{-}N\right] \longrightarrow \phi\text{-}N=N\text{-}\underset{\underset{CH_3}{|}}{C}=CH\text{-}C\equiv N \quad (86)$$

considerable amount of difficulty in duplicating the original preparation. The corresponding azide, on the other hand, thermolyzes very easily (~80°) with formation of a different substance (61), isomeric with the oxidation product, in high yield. The spectroscopic properties of the second isomer (a red compound) also correspond to a phenylazocrotononitrile structure. A third isomeric phenylazocrotononitrile is produced when the phenylhydrazone of acetoacetonitrile is oxidized with N-bromosuccinimide or when 2-chloroacetoacetonitrile reacts with phenylhydrazine (108). The isomer from the azide is converted to that from the phenylhydrazone by exposure to dilute acid, but no interconversion with the isomer of Michaelis and Schäfer has so far been achieved. The isomer from the azide, evidently the less stable, presumably has the geometry imposed by the ring system from which it came (eq. 87). There is no evidence concerning the possibility of an equilibrium between this isomer and a mesoionic cyclic structure corresponding to the singlet nitrene.

This pattern of behavior is shown by several other 5-amino and 5-azido pyrazoles, and appears to be general, although not in all instances so clearly defined.

Of special interest is 1,3,4-triphenyl-5-azidopyrazole, which is converted by heating above 50° into what may be an azoacrylonitrile, which, however, cannot be converted into any isomer (61). Spectroscopic information is ambiguous; a very weak infrared absorption at 2150 cm^{-1} might be due to an inherently inactive nitrile group, or might instead indicate a small

$$(87)$$

amount of nitrile in equilibrium with a mesoionic cyclic structure (cf. eq. 87). The same substance is formed by oxidizing the aminopyrazole in acid solution, and even more easily by oxidation of the phenylhydrazone of benzoylphenylacetonitrile. The best evidence for the existence of the mesoionic nitrenoid structure (most likely as the minor partner in an equilibrium) is the fact that the substance is dimerized to an azopyrazole on mild heating (eq. 88) or by treatment with strong sodium hydroxide

$$(88)$$

solution. This is an unprecedented behavior for nitriles, but is a reasonable reaction of a singlet nitrene.

One 5-azidopyrazole lacking a 1-substituent has been studied (46); thermolysis of 3-methyl-4-phenyl-5-azidopyrazole requires a temperature of ca. 110°, at which four atoms of nitrogen are lost, producing α-phenyl-crotononitrile in 39% yield (eq. 89). The higher temperature required in this instance is more consistent with the initial formation of a nitrene

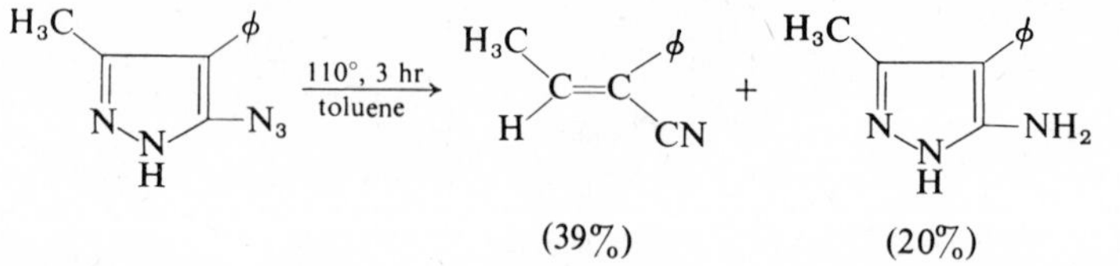

(89)

(39%)   (20%)

than are the low temperatures that suffice for analogs bearing a 1-aryl group. It may be significant in this respect that the amine, a product of hydrogen abstraction, is also formed in appreciable amounts.

The positionally isomeric 4-azidopyrazoles of necessity behave differently. The anticipated nitrene cannot be so well stabilized in a meso-ionic singlet state, nor can it open to form an azoacrylonitrile. Instead, the general behavior (46) on thermolysis or photolysis is a competition between formation of azopyrazoles (dimerization of a nitrene?) and a more deep-seated fragmentation (eq. 90). 1,3,5-Triphenyl-4-azidopyrazole and

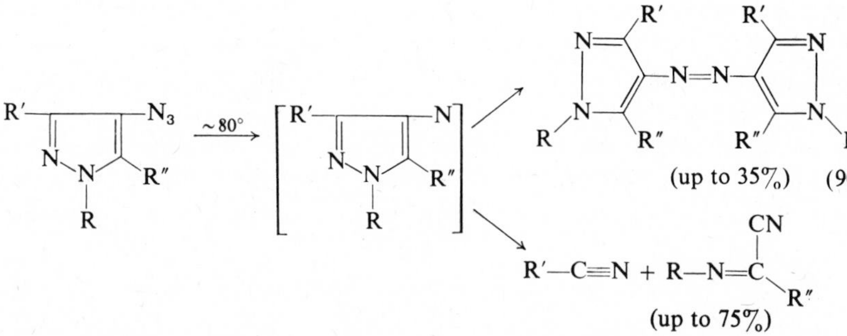

(up to 35%)   (90)

(up to 75%)

1-methyl-3,5-diphenyl-4-azidopyrazole give products of both types; 4-azidopyrazoles bearing 3,5-dimethyl-, 3,5-dimethyl-1-phenyl-, 1,3,5-trimethyl-, 1,3-diphenyl-5-methyl-, and 1,5-diphenyl-3-methyl substituents give fragmentation exclusively. The same reactions were observed with deoxygenation of 4-nitrosopyrazoles with triethyl phosphite, but in lower yields.

The formation of the observed fragmentation products from 4-azidopyrazoles can be quite readily accommodated in terms of a concerted process not involving formation of an intermediate nitrene. Two observations (46) suggest, however, that nitrenes are, in fact, intermediates. 1-Methyl-3,5-diphenyl-4-azidopyrazole (eq. 90, R = CH_3, R' = R" = ϕ) gives more azopyrazole and less fragmentation as the concentration is increased. 3,5-Dimethyl-4-azidopyrazole (eq. 90, R = H, R' = R" = CH_3) gives only fragmentation products and tars when decomposed in hydrocarbon solvents, but in the presence of aniline, 3,5-dimethyl-4-phenylazopyrazole is formed (6% yield), and in the presence of phenylhydrazine, 3,5-dimethyl-4-aminopyrazole is formed (43% yield). Although other explanations are obviously possible, the formation of these products can be regarded not unreasonably as resulting from interception of an intermediate nitrene before fragmentation ensues (eq. 91). (Insertion of aryl nitrenes into N—H bonds is not commonly observed, but at least one example, the insertion of phenylnitrene into dimethylamine, has been reported (45).)

$$\text{ArN}_3 \longrightarrow \text{Ar—N} \longrightarrow \text{fragmentation products}$$

Ar—NH$_2$ Ar—NH—NH—ϕ $\xrightarrow{[0]}$ Ar—N=N—ϕ (91)

An azidopyrazolone, "azidoantipyrine," was reported by Forster and Müller (109) in 1909 to decompose on standing or heating, leaving a red, amorphous solid, to which they assigned an azopyrazolone structure on the basis of analysis for nitrogen and an approximate molecular weight. Reinvestigation (46) confirmed the general observations, but thin-layer chromatography (as well as spectroscopy) indicated the product to be a mixture. No crystalline substance could be obtained from it, and it could not be reduced to an aminopyrazolone. Thermolysis of the azide in solution (cumene) gave similar results, except for the isolation (after chromatography) of a ring-cleavage product, N-oxamyl-N-phenyl-N'-methyl-N'-acetylhydrazine (eq. 92), which might conceivably have arisen from an intermediate nitrene and subsequent reaction with the air.

     (92)

10%

A possible example of the formation of a nitrene attached to a 1,2,4-triazole ring has been reported (110). Thermolysis of 3-azido derivatives bearing an adjacent benzylidenamino group leads to ring closure of the type to be expected of a nitrene insertion reaction (eq. 93). It is interesting that simple ring-cleavage at the 3,4-bond, which would have led to a conjugated azonitrile, was not observed.

5-Azido-1,2,3-triazoles decompose at much lower temperatures (40–50°) than do most aryl azides, and are converted in high yields to substances having two fewer nitrogen atoms (111). The evolved nitrogen comes

(93)

entirely from the azido group. The red products have a somewhat ambiguous chemistry in solution, which suggests the possibility of an equilibrium between a conjugated, open-chain nitrile and a mesoionic nitrenoid structure (eq. 94). X-Ray crystallography of the example with Ar = p-tolyl, Ar' = phenyl indicates the open-chain structure for the solid (112),

(94)

although the cyano group is bent 6.5° from the normal 180° angle.

Thermolysis of the 5-azidotriazoles follows unimolecular kinetics. No intermediate (such as a possible triplet nitrene) more reactive than the isolated product has been intercepted, and there is no evidence for intramolecular insertion at either the 1-aryl or the 5-aryl group. The isolated product does not show the reactivity ordinarily associated with nitrenes, except for a mild ability to abstract hydrogen not normally encountered with either triazenes or nitriles (cold ethereal mercaptans convert the substance to a 5-aminotriazole in a few minutes, for example). Dimerization to an azo compound by heating, as encountered with the isosteric pyrazole system (61), has not been observed, presumably owing to the ease with which nitrogen is lost on heating. The available evidence thus leaves uncertain the question of whether ring opening is concerted with rupture of the azido group, or whether a nitrene is involved in any way. Support for the view that a nitrene may be formed is provided by the behavior of the isoelectronic system, 1,4-diphenyl-5-diazomethyltriazole, which gives the typical reactions of a carbene (113) (which, however, cannot be isolated and apparently does not cleave to form a conjugated open-chain system).

Oxidation of 5-amino-1,4-diphenyltriazole by permanganate (or other oxidizing agents) gives the same product as from thermolysis of the azide, but it is accompanied by a dimer, the azotriazole (61). In view of the fact that the dimer is not formed from the azide, one can conclude either that only the oxidation produces a true nitrene, or that the azotriazole is formed in a manner that does not involve a nitrene. The latter possibility seems the most likely, and dimerization at the intermediate oxidation stage of an amino radical provides a rational path.

The possibility of forming nitrenes by oxidation of aminobenzotriazoles has been explored by Rees and co-workers (114–116). 2-Aminobenzotriazole gave the fragmentation product dicyanobutadiene when treated with lead tetraacetate (115) (eq. 95), but it is not known if the fragmentation is concerted or whether a nitrene is an intermediate. The position isomer,

$$\text{(95)}$$

1-aminobenzotriazole, gives rise to trappable benzyne under the same conditions (114) (eq. 96). An analogous reaction has been observed with the 1-amino-1,2,3-triazine system (116) (eq. 97), but if a nitrene was an

$$2 N_2 + \text{(96)}$$

$$\longrightarrow \text{products} \quad \text{(97)}$$

intermediate, it was not detected (the isolated products were those derived from perinaphthalyne).

One example of the possible formation of a nitrene from an oxadiazole has been reported (46); thermolysis of 2-phenyl-5-azido-1,3,5-oxadiazole. Decomposition occurs slowly in boiling toluene to yield principally benzoyl cyanide, accompanied by a small amount of a substance of enigmatic structure, formally a trimer of benzoyl isocyanide (117) (eq. 98). The major product involves formation of a new carbon-carbon bond, which might occur intramolecularly, concerted with or following

$$\phi-\overset{\text{O}}{\underset{\|}{C}}-CN + (\phi CONC)_3 \quad \text{(98)}$$

$$(35\%) \qquad (2\%)$$

disruption of the ring. There is no evidence on the possible intermediacy of a nitrene; *cis*-benzoylazocarbonitrile, which might also be an intermediate, could not be intercepted by reaction with anthracene.

IV. NITRENES FROM OPENING OF HETEROCYCLIC RINGS

A substantial number of examples of the possible formation of arylnitrenes by opening of a heterocylic ring have been mentioned in the earlier parts of this chapter. These will be recapitulated here only insofar as is necessary to present a complete picture of the subject under one heading.

There are in principle two types of ring opening that might lead to nitrenes: those in which the ring is broken in but one place, leading to a single initial product, and those in which the ring is broken in more than one place, leading to multiple fragments. The first type is represented by only two systems: anthranils and azirines. Anthranils by their nature can only give rise to arylnitrenes (by photolysis or thermolysis), for which reason all reported examples have already been discussed (7,30,31,92a) (eqs. 16 and 67). There is no compelling evidence that nitrenes are in actual fact formed from anthranils, but the temperature required (250–260°) for the formation of the products presumed to arise through nitrenes is credible for such a reaction path. The alternative products are also those to be expected of a nitrene: intramolecular insertion to form acridones (7,30,92a); abstraction of hydrogen to form amines (30); reaction with the solvent accompanied by ring-expansion to an azepine (31); and formation of azobenzenes (92a).

The possibility that the azirine ring might equilibrate with a nitrene structure (eq. 43) has been invoked by various investigators to explain the formation of azepines when arylnitrenes are generated in the presence of nucleophilic solvents (8,41,43,60). Unfortunately, no unequivocal synthesis of an azanorcaratriene is known by which this hypothesis might be tested. A similar interconversion has been used to explain the photolytic isomerization of some 3-benzoylazirines to oxazoles (118) (eq. 99).

$$\phi-\underset{O}{\underset{\|}{C}}\overset{}{\diagdown}\underset{N}{\diagup}-Ar \xrightarrow{h\nu} \left[\phi-\underset{O}{\underset{\|}{C}}-CH=C\overset{Ar}{\diagdown_{N}}\right] \longrightarrow \underset{O\diagdown N}{\boxed{}}-Ar \qquad (99)$$

The most important types of fragmentation of a heterocyclic ring into two (or more) parts, one of which is a nitrene, can be represented in a general way by equations 100 and 101. The only established instance of

$$R-X-Y \longrightarrow R-X=Y + R-N$$
$$\underset{\underset{R}{|}}{\diagdown N \diagup} \qquad\qquad\qquad (100)$$

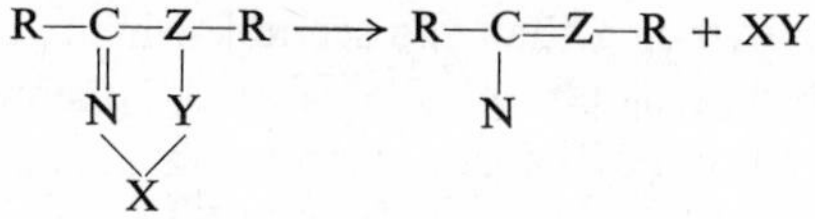

$$R-\underset{\underset{\displaystyle X}{\underset{\displaystyle \diagdown\diagup}{N-Y}}}{\overset{\displaystyle \|}{C}}-R \longrightarrow R-\underset{\underset{\displaystyle N}{\|}}{C}=Z-R + XY$$

(101)

equation 100 appears to be photolysis of oxaziridines (28), in which R—X=Y is a ketone (see eq. 8). Another possible example (highly speculative) is the photolytic reaction of nitrobenzene with olefins (53), which might involve initial formation of a dioxazolidine, which might fragment to a carbonyl compound and an oxaziridine, followed by further fragmentation of the latter compound (eq. 102).

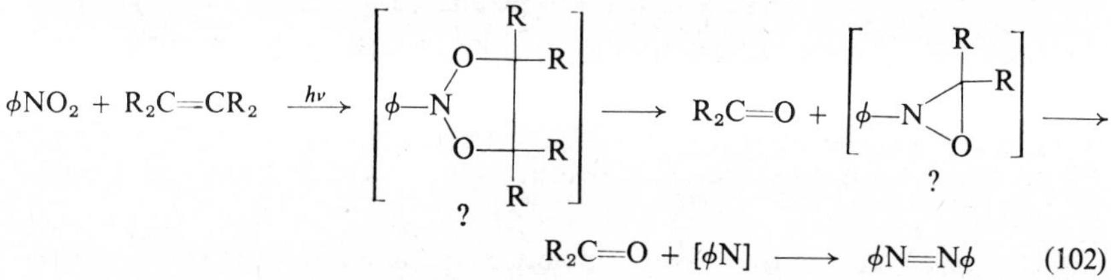

$$\phi NO_2 + R_2C\!=\!CR_2 \xrightarrow{h\nu} \left[\phi-N\underset{\underset{?}{\displaystyle O}}{\overset{\displaystyle O}{\diagup\diagdown}}\underset{R}{\overset{R}{\underset{\displaystyle \diagdown\diagup}{\big|}}}\right] \longrightarrow R_2C\!=\!O + \left[\phi-N\underset{O}{\overset{R}{\diagdown\diagup}}\right] \longrightarrow$$

$$R_2C\!=\!O + [\phi N] \longrightarrow \phi N\!=\!N\phi \qquad (102)$$

Equation 101 is represented by loss of carbon dioxide from 1,2,4-oxadiazolones (119,120) (eq. 103), thioxazolones (121) (eq. 104), and dioxazolones (122) (eq. 105). The first reaction, which involves an

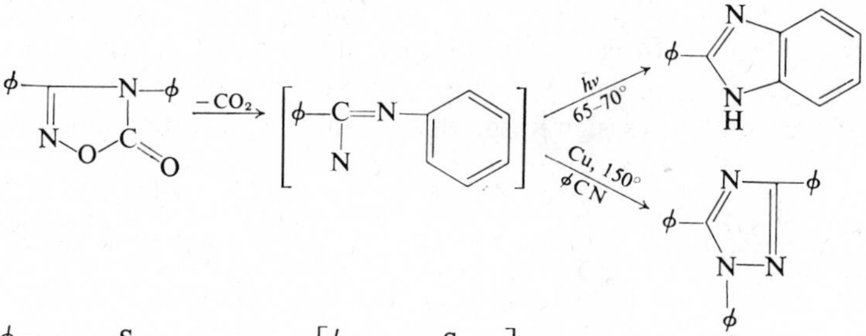

(103)

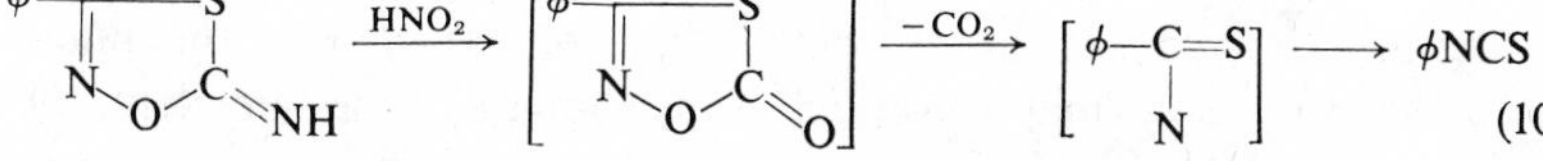

(104)

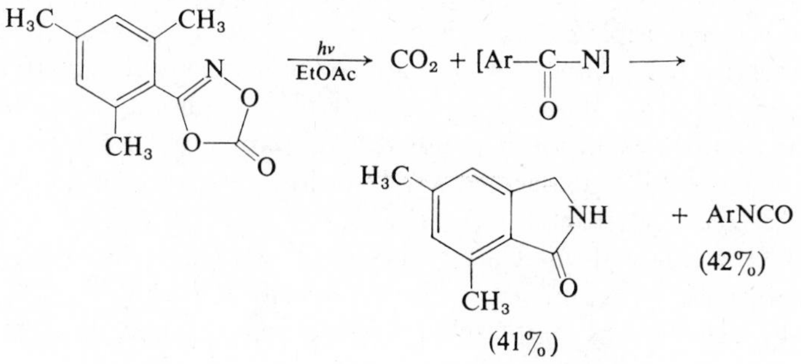

(105)

insertion, is more convincing than the second, which could well be a concerted process. Whereas photolysis of 1,2-diphenyloxadiazolone (in dioxane) gave only 2-phenylbenzimidazole, and a nitrene could not be intercepted with benzonitrile, carbon disulfide, or acrylonitrile (120), thermolysis (150°) in the presence of copper allowed interception of the presumed imidoyl nitrene with benzonitrile in 77% yield. With phenylacetonitrile, however, the intermediate abstracted hydrogen and formed N-phenylbenzamidine.

The third reaction (122) appears to give benzoylnitrenes, whose presence is deduced from the cyclic insertion reaction shown in equation 105, by intermolecular insertion into cyclohexane to give N-cyclohexylbenzamide (8% yield), by interception with dimethyl sulfoxide to form a sulfoximine or with dimethyl sulfide to form a sulfimine, and by hydrogen abstraction to form benzamide. Thermolysis (130–150°) gave no evidence for a nitrene intermediate, and resulted only in rearrangement to aryl isocyanate.

Closely related are the loss of carbon oxysulfide from the thione corresponding to the thioxazolone of equation 104, which gives rise to phenyl isothiocyanate (121), the loss of carbon oxysulfide from dioxazolone thiones (122), which gives products similar to those from dioxazolones, and the loss of phenyl isocyanate from the corresponding anil, which gives the same result (121).

Loss of molecular nitrogen according to equation 101 is represented by the photolysis of 1,5-diphenyltetrazole, which forms 2-phenylbenzimidazole (123) (eq. 106). It is significant that no diphenylcarbodiimide, the

$$\phi\overset{}{\underset{N\diagdown N\diagup N}{\boxed{}}}N-\phi \quad\overset{h\nu}{\longrightarrow}\quad \left[\phi-\underset{N}{\overset{|}{C}}{=}N-\boxed{}\right] \quad\longrightarrow\quad \phi-C\overset{N}{\underset{\underset{H}{N}}{\boxed{}}} \qquad (106)$$

product of C-to-N migration of a phenyl group, is formed by this photolysis, although it is the major product of thermolysis of the same tetrazole. The roughly 200° difference in the temperature of the two types of experiment might be responsible for the change in products as a result of a difference in activation energies for two competing pathways for reaction of the same nitrene, but it seems more likely that equilibration of the tetrazole with the open-chain imidoyl azide structure is involved. Rearrangement of the imidoyl azide, analogous to the Curtius rearrangement of acyl azides, may be a concerted process, not involving a nitrene (eq. 107), as appears to be the case with the Curtius rearrangement (124).

Thiatriazoles also lose nitrogen, thermolytically or photolytically. The thermal process results in multiple fragmentation to sulfur, nitrogen,

$$\phi \overset{N-\phi}{\underset{N_3}{\rightleftharpoons}} \phi-\underset{N_3}{\overset{|}{C}}=N-\phi \xrightarrow{\text{concerted}} \phi-N=C=N-\phi + N_2 \qquad (107)$$

and a nitrile (125,126), and there is no reason to believe that a nitrene is an intermediate. Photolysis, however, produces isothiocyanates (5–10%) as well (126) (eq. 108). It is, of course, purely speculative whether a thio-

$$R \overset{S}{\underset{N-N}{\diagup}} \xrightarrow{h\nu} \left[R-\underset{N}{\overset{|}{C}}=S \right] \longrightarrow R-NCS \qquad (108)$$

carbonyl nitrene might be an intermediate, but the alternative, that irradiation first brings about ring-opening to the thioacyl azide structure, which subsequently rearranges in a concerted manner, is on no better foundation.

The remaining case of loss of nitrogen from a heterocyclic ring is of a different type. The oxidative fragmentation of 2-amino-benzotriazole to form a dinitrile (115) (eq. 95) might conceivably involve fragmentation of a triazolylnitrene to an o-phenylenedinitrene, which then undergoes valence isomerization to the observed product.

Loss of X=Y in equation 101 as O=SO is represented by thermolysis of a group of diarylthiaoxadiazoles (127), which give diarylcarbodiimides in high yield (eq. 109). Efforts to intercept an intermediate nitrene were

$$Ar \overset{N-Ar'}{\underset{N-O-S=O}{\diagup}} \xrightarrow{72-150°} SO_2 + ArN=C=NAr' \qquad (109)$$

unsuccessful, and the authors concluded that the reaction was a concerted process. No 2-arylbenzimidazoles were detected, unlike the thermolysis of 1,5-diaryltetrazoles (128). The corresponding compounds with an ethyl group instead of phenyl in the 5-position were investigated by Eloy (129).

The last example of equation 101 is the loss of triphenylphosphine oxide from oxadiazaphospholines and oxazaphospholines (130). The former produce diarylcarbodiimides (eq. 110); the latter produce the analogous product, ketene anils, plus a ring-contraction product, 2-arylazirines (eq. 111). Whether a nitrene is involved in either reaction is speculative.

$$\phi CNO + \phi N=P\phi_3 \xrightarrow{80°} \overset{\phi}{\underset{N-O-P\phi_3}{\diagup}}\overset{N^{\diagup \phi}}{} \longrightarrow \phi_3P=O + \phi N=C=N\phi \qquad (110)$$
$$(66\%)$$

$$\phi \overset{}{\underset{N-O-P\phi_3}{\diagup}} \xrightarrow{130-140°} \phi_3P=O + \phi N=C=CH_2 + \phi \overset{}{\underset{N}{\diagup \diagdown}} \qquad (111)$$
$$(39\%)$$

The fact that only diarylcarbodiimide and no 2-arylbenzimidazole is formed from the oxadiazaphospholines suggests that migration of the aryl group to nitrogen may be concerted with a stage of the ring rupture.

The possible formation of phenylnitrene by thermolysis of hexaphenylborazine (33) has been discussed earlier (eq. 18).

APPENDIX

The possible formation of phenylnitrene from sulfinylaniline is suggested by two recent observations. Pyrolysis of sulfinylaniline at 800° gave aniline in 10% yield (but no azobenzene), but at 1000°, cyanocyclopentadiene was formed (4% yield) (131) (eq. 112). This is a product formed in the high-temperature pyrolysis of phenyl azide, and is perhaps derived from isomerization of phenylnitrene. This view is supported by the fact that the mass spectrum of sulfinylaniline shows a strong peak corresponding to $C_6H_5N^+$ (also seen in the mass spectrum of phenyl azide). Azo-

$$C_6H_5\!-\!N\!\!=\!\!S\!\!=\!\!O \xrightarrow{\Delta} C_6H_5\!-\!NH_2 \text{ and } \qquad\qquad (112)$$

$$\xrightarrow[110-140°]{Cu} C_6H_5\!-\!N\!\!=\!\!N\!\!-\!\!C_6H_5$$

benzene is formed in low yields when sulfinylaniline is heated in the presence of copper powder, but there is no further evidence about the mechanism (132).

The formation of cyanocyclopentadienes by vapor-phase pyrolysis of phenyl azides has recently been independently confirmed (133). At 600° the yield is only 9%, but at 700° it levels off near 50%; the accompanying products are aniline (4%), azobenzene (1–2%), benzene (7%), benzonitrile (3%), diphenylamine (2-3%), and carbazole (>2%). The reactions leading to some of these products, such as benzene and benzonitrile, are evidently complex, and may involve aryl radicals.

Ring contraction of this type has been demonstrated to be a general reaction (134). The same mixture of 2- and 4-substituted 1-cyanocyclopentadienes (ratio 1:2) was formed when m- or p-tolyl azide was heated at 450° (1.0 mm); similar results were obtained with o-, m-, and p-fluorophenyl azides (eq. 113). The isomer composition of the products is apparently not significant, inasmuch as the isomers undergo thermal equilibration.

$$R\!-\!C_6H_4\!-\!N_3 \xrightarrow{450°} \qquad + \qquad \text{etc.} \qquad (113)$$

o-Tolyl azide gave only *o*-toluidine and *o,o'*-azotoluene under the same conditions, and 2,4,6-trichlorophenyl azide gave only hexachloroazobenzene (68% yield). These results are attributed to rapid crossing of an initially formed singlet nitrene to a triplet state in these two instances, the ring-contraction reaction being assumed to occur through the singlet nitrene.

Those azides that gave cyanocyclopentadienes also produced small amounts (10%) of pyridines (eq. 114). This behavior parallels the formation of pyridine derivatives from deoxygenation of nitrobenzenes with

$$+ \text{Ar—CN} + \text{HCN, etc.} \qquad (114)$$

triethyl phosphite under photolytic conditions. In both instances, the ring transformation may be a reaction of the arylnitrene, but the overall process is evidently complex; the fate of the carbon atom lost in the process is not known.

Ring contraction under nitrene-forming conditions occurs particularly readily with heteroaromatic systems. Crow and Wentrup (135) have reported conversion of tetrazolo[1,5-*a*]pyridine to 2-cyanopyrrole, accompanied by 2-aminopyridine (yield decreased by higher temperature) and glutacononitrile (yield increased by higher temperature) (eq. 115). These substances presumably arose through competitive isomerization, hydrogen abstraction, or ring cleavage of α-pyridylnitrene. γ-Pyridyl azide

$$(115)$$

reacted much like phenyl azide, giving 4,4'-azopyridine, 2- and 3-cyanopyrrole, and 4-cyanopyridine.

In the pyrimidine series, tetrazolo[1,5-*a*]pyrimidines were said to produce mixtures of 1-cyanopyrazoles and 2-aminopyrimidines on pyrolysis, presumably through isomerization to the 2-pyrimidyl azides and fragmentation to the nitrenes (eq. 116). The isomeric tetrazolo[1,5-*c*]pyrimidines gave 1-cyanoimidazoles quantitatively, in what may actually be a

$$\qquad\qquad\qquad\qquad\qquad\qquad\qquad (116)$$

concerted process not involving a nitrene (eq. 117). This view is supported by the observation that analogs that exist in the 4-azidopyrimidine form owing to its stabilization by substitution give much lower yields (e.g., 2,6-dimethoxy-4-azidopyrimidine gives 1-cyano-2,4-dimethoxyimidazole

$$\qquad\qquad\qquad\qquad\qquad\qquad\qquad (117)$$

in only 12% yield). 2-Methylthio-4-azido-6-methylpyrimidine, on the other hand, is converted to the 1-cyanoimidazole in 75% yield even at 200°. The authors attribute this effect to stabilization of a bicyclic intermediate by the methylthio group (eq. 118).

$$\qquad\qquad\qquad\qquad\qquad\qquad\qquad (118)$$

Similar ring contractions have been observed in pyrolysis of benzotriazoles (136) and triazolopyridines (135). Cyanocyclopentadienes or cyanopyrroles are formed nearly quantitatively, unaccompanied by the other products associated with nitrene reactions. These pyrolyses are therefore believed not to involve formation of nitrene intermediates; they may involve an intermediate, perhaps diradical in nature, which collapses directly to a bicyclic intermediate tautomeric with the one pro-posed to form from arylnitrenes. The same ring contraction has been observed in the pyrolysis of isatins, which lose carbon monoxide instead of nitrogen (136).

Of particular importance is the report of experiments that imply that phenylnitrenes and α-pyridylcarbenes equilibrate, or react through a common intermediate (137). The formation of small amounts of pyridine derivatives, already mentioned, from phenylnitrenes has since been paralleled by the formation of benzene derivatives from presumed sources of α-pyridylcarbenes. v-Triazolo[1,5-a]pyridine when pyrolyzed at 500° gave azobenzene in 77% yield and aniline (4%), the principal products of mild pyrolysis of phenyl azide (eq. 119). Loss of nitrogen directly, or after prior ring opening to form α-diazomethylpyridine, would be expec-ted to produce α-pyridylcarbene; equilibration in favor of the nitrene would account for the observed products. At 800°, the products were

$$Ar—N{=}N—Ar + Ar—NH_2 \qquad (119)$$

benzene (2.8%), benzonitrile (3.7%), pyridine (8.3%), α-picoline (1.1%), and aniline (7.6%).

The methyl analog, which should form 5-methyl-2-pyridylcarbene, gave the same products as m-tolyl azide (137): largely m,m'-azotoluene on mild pyrolysis, and a complex mixture at 800°, consisting largely of toluene (10.2%), β-picoline (15.3%), and m-toluidine (4.3%), resembling that from the violent pyrolysis of m-tolyl azide. The formation of m-tolui-dine exclusive of the ortho and para isomers has been taken to preclude those intermediates which could reasonably be expected to give the other isomers, such as an open-chain nitrile or a [3.1.1] bicyclic system. The authors favored a seven-membered ring (eq. 120), which might also be the source of the aminoazepines formed when nitrenes are generated in the presence of amines. If such an intermediate is formed, it cannot be the

$$(120)$$

source of the cyanocyclopentadienes produced in the pyrolysis of phenyl azides, for they were formed in only barely detectable amounts from the triazolopyridines.

α-Pyridylmethylcarbene apparently tautomerizes to α-vinylpyridine, the only detected product, too fast for the ring transformation to compete (137). The phenyl analogs are exceptionally interesting in that carbazoles are formed essentially quantitatively, implying complete transformation to the nitrene (eq. 121).

(121)

REFERENCES

1. H. J. Shine, *Aromatic Rearrangements*, Elsevier, New York, 1967, p. 182–190.
2. J. I. G. Cadogan, *Quart. Rev.*, **22**, 222 (1968).
3. (a) T. F. Fagley, J. R. Sutter, and R. L. Oglukian, *J. Am. Chem. Soc.*, **78**, 5567 (1956); (b) K. E. Russell, *J. Am. Chem. Soc.*, **77**, 3487 (1955).
4. P. Walker and W. A. Waters, *J. Chem. Soc.*, **1962**, 1632.
5. P. A. S. Smith and J. H. Hall, *J. Am. Chem. Soc.*, **84**, 480 (1962).
6. R. E. Moore and H. Rapaport, *J. Org. Chem.*, **32**, 3335 (1967).
7. R. Kwok and P. Pranc, *J. Org. Chem.*, **33**, 2880 (1968).
8. G. Smolinsky, *J. Org. Chem.*, **26**, 4108 (1961).
9. P. A. S. Smith and B. B. Brown, *J. Am. Chem. Soc.*, **73**, 2438 (1951).
10. J. H. Boyer and F. C. Canter, *Chem. Rev.*, **54**, 1 (1954).
11. L. Horner, A. Christmann, and A. Gross, *Chem. Ber.*, **96**, 399 (1963).
12. P. A. S. Smith, J. H. Hall, and R. O. Kan, *J. Am. Chem. Soc.*, **84**, 485 (1962).
13. J. S. Swenton, *Tetrahedron Letters*, **1967**, 2855.
14. W. D. Crow and C. Wentrup, *Tetrahedron Letters*, **1967**, 4379.
15. S. Goldschmidt and L. Strohmeyer, *Ber.*, **55**, 2450 (1922).
16. Summarized in P. A. S. Smith, *Chemistry of Open-Chain Organic Nitrogen Compounds*, Vol. I, Benjamin, New York, 1965, pp. 108–114.
17. O. Fischer and O. Heiler, *Ber.*, **26**, 378 (1893).
18. H. McCombie, H. A. Scarborough, and W. A. Waters, *J. Chem. Soc.*, **1928**, 353.
19. M. P. Schmidt and A. Hagenböcker, *Ber.*, **54**, 2191, 2201 (1921).
20. G. Charrier and M. Joris, *Gazz. Chim. Ital.*, **68**, 640 (1938).
21. V. I. Mur, *Zh. Obshch. Khim.*, **25**, 374 (1955).
22. J. Poskočil and Z. J. Allan, *Chem. Listy*, **50**, 111 (1956).
23. I. Čepčiansky, V. Slavík, L. Němec, H. Fingerová, and I. Németh, *Coll. Czechoslov. Chem. Commun.*, **33**, 100 (1968).
24. (a) K. Nakagawa and H. Onoue, *Tetrahedron Letters*, **1965**, 1433; *Chem. Commun.*, **1965**, 396; (b) J. H. Hall, *J. Am. Chem. Soc.*, **87**, 1147 (1965); (c) J. H. Hall and E. Patterson, *ibid.*, **89**, 5856 (1967); (d) J. H. Hall, J. G. Stephanie, and D. K. Nordstrom, *J. Org. Chem.*, **33**, 2951 (1968).
25. W. Wawryczek, *Naturwissenschaften*, **48**, 160 (1961).

26. C. J. Pedersen, *J. Am. Chem. Soc.*, **79**, 5014 (1957).

27. W. Kirmse, *Angew. Chem.*, **71**, 537 (1959).

28. (a) E. Meyer and G. W. Griffin, *Angew. Chem.*, **79**, 648 (1967); (b) H. Shindo and B. Umezawa, *Chem. Pharm. Bull (Tokyo)*, **10**, 492 (1962); *Chem. Abstr.*, **57**, 14598e (1962); (c) J. S. Splitter and M. Calvin, *Tetrahedron Letters* **1968**, 1445.

29. P. A. S. Smith, B. B. Brown, R. K. Putney, and R. R. Reinisch, *J. Am. Chem. Soc.*, **75**, 6335 (1953).

30. (a) A. Kliegl, *Ber.*, **42**, 591 (1909); (b) P. L. Coe, A. E. Jukes, and J. C. Tatlow, *J. Chem. Soc.*, Part C, **1966**, 2020.

31. M. Ogata, H. Kanō, and H. Matsumoto, *Chem. Commun.*, **1968**, 397.

32. J. Meisenheimer, O. Senn, and P. Zimmermann, *Ber.*, **60**, 1736 (1927).

33. H. C. Newsom, W. D. English, A. L. McCloskey, and W. G. Woods, *J. Am. Chem. Soc.*, **83**, 4134 (1961).

34. A. Bertho, *Ber.*, **57**, 1138 (1934).

35. (a) J. H. Hall, J. W. Hill, and H. C. Tsai, *Tetrahedron Letters*, **1965**, 2211; (b) J. H. Hall, J. W. Hill, and J. Fargher, Paper 39, Organic Division, 156th National Meeting, American Chemical Society, Atlantic City, New Jersey, September, 1968; *J. Am. Chem. Soc.*, **90**, 5315 (1968).

36. (a) G. Smolinsky and B. I. Feuer, *J. Org. Chem.*, **29**, 3097 (1964); (b) G. Smolinsky and B. I. Feuer, *J. Am. Chem. Soc.*, **86**, 3085 (1964).

37. R. J. Sundberg, *J. Am. Chem. Soc.*, **88**, 3781 (1966); *Tetrahedron Letters*, **1966**, 477.

38. R. A. Abramovitch, Y. Ahmad, and D. Newman, *Tetrahedron Letters*, **1961**, 752.

39. J. G. G. Cadogan and M. J. Todd, *Chem. Commun.*, **1967**, 178.

40. (a) T. Shingaki, *Sci. Rep. Coll. Gen. Educ., Osaka Univ.*, **11**, 67 (1963); (b) *ibid.*, **11**, 81 (1963); (c) *ibid.*, **11**, 93 (1963) (*Chem. Abstr.*, **60**, 6733d–6734b (1963)); (d) cf. the reaction of phenyl azide with trichloromethyl radicals: J. E. Leffler and H. H. Gibson, Jr., *J. Am. Chem. Soc.*, **90**, 4117 (1968).

41. W. E. Doering and R. A. Odum, *Tetrahedron*, **22**, 81 (1966).

42. W. Lwowski and T. L. Maricich, *J. Am. Chem. Soc.*, **86**, 3164 (1964).

43. (a) R. Huisgen, *Angew. Chem.*, **67**, 756 (1955); (b) R. Huisgen, D. Vossius, and M. Appl, *Chem. Ber.*, **91**, 1 (1958); (c) R. Huisgen and M. Appl, *Chem. Ber.*, **91**, 12 (1958).

44. R. K. Smalley and H. Suschitzky, *J. Chem. Soc., Suppl.*, **1964**, 5922.

45. (a) R. A. Odum and M. Brenner, *J. Am. Chem. Soc.*, **88**, 2074 (1966); (b) J. I. G. Cadogan, R. K. Mackie, and M. J. Todd, *Chem. Commun.*, **1968**, 736.

46. P. A. S. Smith and H. Dounchis, unpublished results; H. Dounchis, Ph.D. dissertation, University of Michigan, Ann Arbor, Mich., 1967.

47. R. K. Smalley and H. Suschitzky, *J. Chem. Soc.*, **1963**, 5571.

48. L. Horner and H. Steppan, *Ann.*, **606**, 24 (1957).

49. O. Meth-Cohn, R. K. Smalley, and H. Suschitzky, *J. Chem. Soc.*, **1963**, 1666.

50. P. Scheiner, *J. Am. Chem. Soc.*, **90**, 988 (1968).

51. P. A. S. Smith and J. H. Carter, unpublished results; J. H. Carter, Ph.D. dissertation, University of Michigan, Ann Arbor, Mich., 1962.

52. J. E. Kmiecik, *J. Org. Chem.*, **30**, 2014 (1965).

53. G. Büchi and D. E. Ayer, *J. Am. Chem. Soc.*, **78**, 689 (1956).

54. (a) T. A. Manuel, *Inorg. Chem.*, **3**, 1703 (1964); (b) J. A. Jarvis, B. E. Job, B. T. Kilbourn, R. H. B. Mais, P. G. Owston, and P. F. Todd, *Chem. Commun.*, **1967**, 1149; (c) M. Dekker and G. R. Knox, *Chem. Commun.*, **1967**, 1243.

55. R. P. Bennett and W. B. Hardy, *J. Am. Chem. Soc.*, **90**, 3295 (1968).
56. R. J. Sundberg, *J. Org. Chem.*, **30**, 3604 (1965).
57. T. Nagai, K. Yamamoto, and N. Tokura, *Bull. Chem. Soc., Japan*, **40**, 408 (1967).
58. J. I. G. Cadogan, M. Cameron-Wood, R. K. Mackie, and R. J. G. Searle, *J. Chem. Soc.*, **1965**, 4831.
59. P. J. Bunyan and J. I. G. Cadogan, *J. Chem. Soc.*, **1963**, 42.
60. M. Appl and R. Huisgen, *Chem. Ber.*, **92**, 2961 (1959).
61. P. A. S. Smith, G. J. W. Breen, and M. Hajek, *Abstracts* (Paper No. 55), Division of Organic Chemistry, 155th National Meeting, American Chemical Society, San Francisco, California, April, 1968.
62. L. Horner and A. Gross, in *Präparative Organische Photochemie*, A. Schönberg, Ed., Springer, Berlin, 1958, p. 192.
63. L. Wolff, *Ann.* **394**, 59 (1912).
64. A. Michaelis and A. Schäfer, *Ann.*, **397**, 119 (1913); **407**, 234 (1915).
65. J. Schmutz and F. Kunzle, *Helv. Chim. Acta*, **39**, 1144 (1956).
66. G. Smolinsky, *J. Am. Chem. Soc.*, **82**, 4717 (1960).
67. K. H. Saunders, *J. Chem. Soc.*, **1955**, 3275.
68. G. Smolinsky and B. I. Feuer, *J. Org. Chem.*, **31**, 3882 (1966).
69. R. H. Smith and H. Suschitzky, *Tetrahedron*, **16**, 80 (1961).
70. (a) W. M. Lauer, M. M. Sprung, and C. M. Langkammerer, *J. Am. Chem. Soc.*, **59**, 225 (1936), (b) J. Pinnow, *J. Prakt. Chem.*, **65**, 579 (1902).
71. G. Smolinsky, *J. Am. Chem. Soc.*, **83**, 2489 (1961).
72. D. H. R. Barton and L. R. Morgan, Jr., *J. Chem. Soc.*, **1962**, 622.
73. P. A. S. Smith and J. H. Boyer, *J. Am. Chem. Soc.*, **73**, 2626 (1951).
74. H. C. Waterman and D. L. Vivian, *J. Org. Chem.*, **14**, 289 (1949).
75. P. A. S. Smith, J. M. Clegg, and J. H. Hall, *J. Org. Chem.*, **23**, 1595 (1958).
76. (a) R. A. Abramovitch and K. A. H. Adams, *Can. J. Chem.*, **39**, 2516 (1961); (b) R. A. Abramovitch, *Chem. Ind.*, **1957**, 422.
77. R. A. Abramovitch and J. G. Saha, *J. Chem. Soc.*, **1964**, 2175.
78. (a) M. Nakanishi and A. Naraki, Japan. Pat. 16,283 (1962); *Chem. Abstr.*, **59**, 15516e (1963); (b) M. Nakanishi, A. Naraki, and C. Tashiro, Japan. Pat. 16,284 (1962); *Chem. Abstr.*, **59**, 15516g (1963); (c) J. I. G. Cadogan, S. Kulik, and M. J. Todd, *Chem. Commun.*, **1968**, 736.
79. D. L. Vivian, J. T. Greenberg, and J. L. Hartwell, *J. Org. Chem.*, **16**, 1 (1951).
80. D. L. Vivian and J. L. Hartwell, *J. Org. Chem.*, **18**, 1065 (1953).
81. P. Z. Slack and R. Slack, *Nature*, **160**, 437 (1947).
82. L. O. Krbechek and H. Takimoto, *J. Org. Chem.*, **33**, 4286 (1968); P. A. S. Smith and C. R. Rowe, unpublished results (C. R. Rowe, Ph.D. dissertation, University of Michigan, Ann Arbor, Mich., 1968).
83. A. Reiser, H. Wagner, and G. Bowes, *Tetrahedron Letters*, **1966**, 2635; A. Reiser, G. Bowes, and R. J. Horne, *Trans. Faraday Soc.*, **62**, 3162 (1966).
84. J. S. Swenton, *Tetrahedron Letters*, **1968**, 3421.
85. R. A. Abramovitch and V. Uma, *Abstracts of Papers, Fifty-first Annual Conference of The Chem. Inst. of Canada, Vancouver, B.C., June, 1968.*
86. P. A. S. Smith and C. D. Rowe, unpublished results (C. D. Rowe, Ph.D. dissertation, Univ. of Michigan, Ann Arbor, Mich., 1968).
87. P. A. S. Smith and L. O. Krbechek, unpublished results.
88. R. J. Sundberg, *J. Org. Chem.*, **33**, 487 (1968).
89. R. J. Sundberg and T. Yamazaki, *J. Org. Chem.*, **32**, 290 (1967).
90. L. O. Krbechek and H. Takimoto, *J. Org. Chem.*, **29**, 3630 (1964).
91. J. Hall and D. R. Kamm, *J. Org. Chem.*, **30**, 2092 (1965).

92. (a) L. Gattermann and R. Ebert, *Ber.*, **49**, 2117 (1916); (b) L. Gattermann and H. Rolfes, *Ann.*, **425**, 135 (1921); (c) H. Schaarschmidt, A. Constandachi, and M. Thiele, *Ber.*, **49**, 1632 (1916); (d) K. Brass and F. Albrecht, *Ber.*, **61**, 983 (1928).

93. Altaf-ur-Rahman and A. J. Boulton, *Tetrahedron, Suppl.* No. 7, 49 (1966); K. H. Wünsch and A. J. Boulton, *Advances in Heterocyclic Chemistry*, **8**, 303 (1967).

94. I. T. Kukhtenko, *Dokl. Akad. Nauk*, **132**, 609 (1960); *Chem. Abstr.*, **54**, 24619 (1960).

95. L. O. Krbechek and H. Takimoto, *J. Org. Chem.*, **29**, 1150 (1964).

96. (a) R. A. Carboni and J. E. Castle, *J. Am. Chem. Soc.*, **84**, 2453 (1962); (b) R. A. Carboni, J. C. Kauer, J. E. Castle and H. E. Simmons, *ibid.*, **89**, 2618 (1967); (c) R. A. Carboni, J. C. Kauer, W. R. Hatchard, and R. J. Harder, *ibid.*, **89**, 2626 (1967).

97. (a) T. Zincke and A. T. Lawson, *Ber.*, **19**, 1452 (1886); (b) T. Zincke and H. Jaenke, *Ber.*, **21**, 540 (1888).

98. J. C. Kauer and R. A. Carboni, *J. Am. Chem. Soc.*, **89**, 2633 (1967).

99. P. A. S. Smith and J. H. Boyer, *Org. Synth.*, **31**, 14 (1951).

100. (a) W. Meigen and W. Normann, *Ber.*, **33**, 2711 (1900); (b) A. G. Green and F. M. Rowe, *J. Chem. Soc.*, **101**, 2443 (1912).

101. M. O. Forster and J. H. Schaeppi, *J. Chem. Soc.*, **101**, 1359 (1912).

102. J. D. Hobson and J. R. Malpass, *Chem. Commun.*, **1966**, 141.

103. J. A. Van Allan, W. J. Priest, A. S. Marshall, and G. A. Reynolds, *J. Org. Chem.*, **33**, 1100 (1968).

104. R. A. Abramovitch, K. A. H. Adams, and A. D. Notation, *Can. J. Chem.*, **38**, 2152 (1960).

105. R. Stollé and H. Storch, *J. Prakt. Chem.*, **135**, 128 (1932).

106. (a) E. Ott and E. Ohse, *Ber.*, **54**, 179 (1921); (b) R. M. Moriarty, M. Rahman, and G. J. King, *J. Am. Chem. Soc.*, **88**, 842 (1966).

107. S. Searles, Jr., and W. R. Hine, Jr., *J. Am. Chem. Soc.*, **79**, 3175 (1957).

108. A. Quilico and R. Justoni, *Rend. Ist. Lomb.*, **69**, fasc. XI–XV (1936); A. Quilico, R. Fusco, and V. Rosnati, *Gazz. Chim. Ital.*, **76**, 30 (1946).

109. M. O. Forster and R. Müller, *J. Chem. Soc.*, **95**, 2072 (1909).

110. H. H. Takimoto, G. C. Denault, and S. Hotta, *J. Heterocyclic Chem.*, **3**, 119 (1966); *J. Org. Chem.*, **30**, 711 (1966).

111. P. A. S. Smith, L. O. Krbechek, and W. Resemann, *J. Am. Chem. Soc.*, **86**, 2025 (1964).

112. C. E. Nordman and J. W. Schilling, private communication; J. W. Schilling, Ph.D. dissertation, University of Michigan, Ann Arbor, Mich., 1968.

113. P. A. S. Smith and J. G. Wirth, *J. Org. Chem.*, **33**, 1145 (1968).

114. C. D. Campbell and C. W. Rees, *Proc. Chem. Soc.*, 296 (1964).

115. C. D. Campbell and C. W. Rees, *Chem. Commun.*, **1965**, 192.

116. C. W. Rees and R. C. Storr, *Chem. Commun.*, **1965**, 193.

117. O. Diels and H. Stein, *Ber.*, **40**, 1658 (1907).

118. B. Singh and E. F. Ullman, *J. Am. Chem. Soc.*, **89**, 6911 (1967).

119. T. Bacchetti and A. Alemagna, *Atti Accad. Nazl. Lincei, Rend. Classe Sci. Fis., Mat., Nat.*, **28**, 824 (1960); *Chem. Abstr.*, **56**, 7304 (1962).

120. J. Sauer and K. K. Mayer, *Tetrahedron Letters*, **1968**, 325.

121. R. Fusco and C. Musante, *Gazz. Chim. Ital.*, **68**, 665 (1938).

122. J. Sauer and K. K. Mayer, *Tetrahedron Letters*, **1968**, 319.

123. R. M. Moriarty, J. M. Kliegman, and C. Shovlin, *J. Am. Chem. Soc.*, **89**, 5958 (1967).

124. G. T. Tisue, S. Linke, and W. Lwowski, *J. Am. Chem. Soc.*, **89**, 6303 (1967); S. Linke, G. T. Tisue, and W. Lwowski, *ibid.*, **89**, 6308 (1967).
125. (a) K. A. Jensen and A. Holm, *Acta Chem. Scand.*, **18**, 826 (1964); (b) D. Martin and W. Mucke, *Ber.*, **98**, 2059 (1965); (c) P. A. S. Smith and D. H. Kenny, *J. Org. Chem.*, **26**, 5221 (1961).
126. W. Kirmse, *Ber.*, **93**, 2353 (1960).
127. R. Rajagopalan and B. G. Advani, *J. Org. Chem.*, **30**, 3369 (1965); R. Rajagopalan and H. U. Daeniker, *Angew. Chem.*, **75**, 91 (1963).
128. P. A. S. Smith and E. Leon, *J. Am. Chem. Soc.*, **80**, 4647 (1958); J. Vaughan and P. A. S. Smith, *J. Org. Chem.*, **23**, 1909 (1958).
129. F. Eloy, *Helv. Chim. Acta*, **48**, 380 (1965).
130. R. Huisgen and J. Wulff, *Tetrahedron Letters*, **1967**, 917, 921.
131. W. D. Crow and C. Wentrup, private communication.
132. T. Minami and T. Agawa, *Tetrahedron Letters*, **1968**, 4109.
133. E. Hedaya, M. E. Kent, D. W. McNeil, F. P. Lossing, and T. McAllister, *Tetrahedron Letters*, **1968**, 3415.
134. W. D. Crow and C. Wentrup, *Tetrahedron Letters*, **1968**, 5569.
135. W. D. Crow and C. Wentrup, *Chem. Commun.*, **1968**, 1082.
136. W. D. Crow and C. Wentrup, *Chem. Commun.*, **1968**, 1026.
137. W. D. Crow and C. Wentrup, *Tetrahedron Letters*, **1969**, in press.

CHAPTER 5

Deoxygenation of Nitro and Nitroso Groups*

J. H. BOYER

Department of Chemistry, University of Illinois, Chicago Circle Campus, Chicago, Illinois

I. INTRODUCTION

The generation of a nitrene by the deoxygenation of a nitroso group may follow, in principle, either electrophilic or nucleophilic attack at nitroso oxygen and examples of both types of reaction have been reported.

$$RNO \xrightarrow{X} RN\overset{\oplus}{=}\overset{\ominus}{O}—X \longrightarrow RN + O{=}X$$
$$RNO \xrightarrow{Y} R\overset{\ominus}{N}—O—\overset{\oplus}{Y} \longrightarrow RN + \overset{\ominus}{O}—\overset{\oplus}{Y}$$

The probability of nitrene intermediacy in the deoxygenation of nitro and nitroso groups has been most frequently encountered in reactions with aromatic C-nitro and nitroso derivatives. Complementary information on deoxygenation of N- and O-nitro and nitroso derivatives is not abundant. Similar information is also not available for saturated aliphatic C-nitro and nitroso compounds; however, three reports are pertinent insofar as deoxygenation in these examples may not require alkylnitrene or nitrenoid intermediates.

* The financial assistance for this work from a NASA grant, NGR 14-012-004, is gratefully acknowledged.

First, in a deoxygenation of a *gem*-nitroso chloride, it was proposed that the ring expansion brought about by treatment with triphenyl phosphine in benzene proceeded from an oxime salt (1) (eq. 1). The second example is

found in the observation that alkali salts of primary nitroparaffins react with diethyl phosphorochloridite to form nitriles and diethyl hydrogen phosphate and the sodium salt of 2-nitropropane under similar treatment was transformed into diethyl isopropylideneaminophosphate (2). The third example consists of the transformation of β-nitroethylbenzene into phenylacetonitrile upon treatment with triethyl phosphite which was explained by initial deoxygenation to β-nitrosoethylbenzene and tautomerization to an oxime followed by dehydration (3). Deoxygenation of a tertiary nitroalkane has not been reported.

Nitrenes and nitrenoid intermediates from nitroso compounds via isolated derivatives are considered outside the scope of this chapter; but an occasional example such as the fragmentation of an oxaziridine will be described. Benzaldehyde, benzaldehyde anil, nitrosobenzene and a trace of benzanilide obtained from the irradiation of 2,3-diphenyloxaziridine (4) (eq. 2) suggests that both phenylcarbene and phenylnitrene were generated initially. More recently phenylnitrene has been an assumed intermediate in the photofragmentation of 2-phenyl-3,3-pentamethyleneoxaziridine (5). When carried out in the presence of a primary or secondary amine the expected azepine was produced. Each example represents a possible

$$C_6H_5CH{=}N(O)C_6H_5 \underset{}{\overset{h\nu}{\rightleftarrows}} C_6H_5\overset{\overset{\displaystyle O}{\diagup\diagdown}}{CH{-}N}C_6H_5 \xrightarrow{h\nu} [C_6H_5N] + C_6H_5CHO$$

$$+ [C_6H_5CH] + C_6H_5NO + C_6H_5CH{=}NC_6H_5 + C_6H_5CONHC_6H_5 \text{ (trace).} \quad (2)$$

overall deoxygenation of nitrosobenzene to phenylnitrene since nitrones, the valence tautomers of oxaziridines, have been obtained from reactions between nitroso compounds and either diazoalkanes or sulfur ylids (6).

II. DIAGNOSTIC PROPERTIES FOR NITRENES FROM DEOXYGENATIONS

That deoxygenation of a nitro or nitroso group may generate a nitrene or nitrenoid intermediate is an assumption often supported by similarities with the same intermediate which may be derived from the structurally related azide on elimination of molecular nitrogen, from a hydroxylamine derivative by α-elimination, or from some other source. These similarities include inter- and intramolecular insertion and hydrogen abstraction with C—H bonds, less frequently both insertion and hydrogen abstraction from other bonds, and other molecular rearrangements (3,7–9). Arylnitrene intermediacy in the pyrolysis of aryl azides is supported by kinetic evidence (10) (see also reference 17), but corresponding information on nitrenes from deoxygenation is not available. Both ultraviolet (11) and ESR spectroscopy (12–14) depict a triplet ground state for a variety of nitrenes generated from azides. At least one attempt to detect an intermediate triplet nitrene in the reaction between an aromatic nitroso compound and triethyl phosphite by observing an ESR spectrum was unsuccessful (15).

At the present time the similarities do not include addition of a nitrene or nitrenoid intermediate generated by deoxygenation to an unsaturated bond. For diagnostic purposes nitrene addition to the olefinic (14) and acetylenic (16) bond, demonstrated with carbethoxynitrene derived from ethyl azidoformate, has limited applicability insofar as azide addition to the olefinic linkage followed by extrusion of nitrogen from a triazoline ring may be an energetically preferred route to an aziridine (17). Since a nitrene derived from a nitroso or a nitro compound receives no such competition from its precursor, aziridine formation may become an additional chemical test for these intermediates when generated in the presence of a suitable olefin or acetylene.

It should also be pointed out that transformation of a nitrene to an azo compound during deoxygenation by a tervalent phosphorous reagent has not been clearly demonstrated; however, this is a reaction often attributed to a nitrene derived from an azide with the general recognition that combination of a nitrene and an azide may bring about the evolution of nitrogen and formation of the azo compound (18). This apparent dimerization cannot be employed as a chemical test for the presence of a nitrene from a nitroso group unless it is established that the azo compound has not been generated through deoxygenation of the corresponding nitroso dimer. Azobenzene has been obtained from nitrobenzene during photolysis with an olefin (Section IV), during a reaction with iron pentacarbonyl and carbon monoxide (page 168) and from nitrosobenzene and certain metal salts (Section III). The observation that deoxygenations of nitro compounds generally fail to produce either azo or azoxy derivatives is based on the assumption that reduction of certain aromatic nitro compounds by phosphine in alkali to azoxy and other products proceeds by another mechanism (19,20).

Support for the formation of an azoxy compound by direct combination of a nitrene and a nitroso derivative is found in an example of its reversal whereby azoxybenzene upon dry distillation afforded nitrosobenzene, aniline and azobenzene (21) (eq. 3). The formation of aniline suggests the intermediate presence of phenylnitrene which abstracted hydrogen.

$$C_6H_5N(O){=}NC_6H_5 \xrightleftharpoons{\text{heat}} C_6H_5NO + C_6H_5N$$

$$C_6H_5N + C\!-\!H \longrightarrow C_6H_5NH_2$$

$$C_6H_5NH_2 + C_6H_5NO \xrightarrow{-H_2O} C_6H_5N{=}NC_6H_5 \longleftarrow 2C_6H_5N \qquad (3)$$

An intramolecular reaction between a nitrene and a nitroso group probably accounts for the formation of a benzocinnoline oxide from o-nitroso-o'-azidobiphenyl during photolysis or pyrolysis (22) (eq. 4).

$$(4)$$

III. DEOXYGENATION BY METALS, METAL SALTS, OTHER ELEMENTS, AND FREE RADICALS

Nitrosobenzene is obtained from nitrobenzene with iron powder in the presence of carbon dioxide at 220° (23). Similar deoxygenations have been carried out in dry organic solvents with other metals (24), and metal oxides (24). Certain metals attack the nitroso group at oxygen with subsequent dimerization (25,26) (eq. 5). Deoxygenation by magnesious iodide apparently proceeds by similar mechanism and transforms nitrosobenzene into azobenzene (26,27).

$$C_6H_5NO \xrightarrow{Na} C_6H_5\dot{N}ONa \longrightarrow \underset{\underset{ONa}{|}}{\overset{\overset{ONa}{|}}{C_6H_5N}}\!-\!NC_6H_5 \qquad (5)$$

A phenazine synthesis designed by Vivian and Waterman (28) consists of heating 2-nitrodiphenyl amine with a mixture of ferrous oxalate dihydrate and lead at 270–280°. Pyrophoric ferrous oxide is claimed to be the active cyclizing agent (29) and an intermediate nitrene has been postulated not only for this reaction but also for the related transformation of o-nitrobiphenyl into carbazole (9) (eq. 6). Support for the postulate

$$(6)$$

seemed to be found in the formation of pyrido[1,2-b]indazole not only from a similar treatment of 2-o-nitrophenylpyridine but also from the pyrolysis of 2-o-azidophenylpyridine (30) (eq. 7), in the conversion of

$$\text{(7)}$$

o-ethylnitrobenzene into cis-o,o'-diethylazobenzene and o-ethylaniline (31) (eq. 8), and in the formation of 4-methylphenoxazine from 2-methyl-2'-nitrodiphenyl ether in the presence of ferrous oxalate (32,33).

$$\text{(8)}$$

These reactions apparently take place on the surface of the catalyst (ferrous oxalate or oxide) where it is not unlikely that iron is bonded to nitroso oxygen following initial deoxygenation of the nitro group. Dimerization (see eq. 5), elimination of ferric oxide and deoxygenation of an azoxy group accounts for the formation of an azo compound (eq. 9). Supporting evidence for this sequence is found in the isolation of p,p'-

$$\overset{\cdot}{\text{ArNOFeO}} \longrightarrow \overset{\text{ArNOFeO}}{\underset{\text{ArNOFeO}}{|}} \xrightarrow{-\text{FE}_2\text{O}_3} \text{ArN(O)}{=}\text{NAr} \xrightarrow{\text{Fe}^{2+}} \text{ArN}{=}\text{NAr}$$

$$\text{(9)}$$

dichloroazoxybenzene from the reaction between p-chloronitrobenzene and ferrous oxalate (oxide) (9). Cyclization by attack at a CH bond and hydrogen abstraction to form the primary amino group may follow the formation of intermediates of the type ArNOFeO. Cyclization to pyridine nitrogen may occur by a concerted process (eq. 10).

$$+ \text{Fe}_2\text{O}_3 \qquad \text{(10)}$$

It has been shown that heating 2,4,6-trimethyl-2'-nitrobiphenyl at 350° does not require the presence of the catalyst (ferrous oxalate or oxide) for conversion into 1,3-dimethylphenanthridine presumably via an aci-form of the starting material (3) (eq. 11). A similar explanation has been offered to account for phenazine formation from o-nitrodiphenylamine (34). Another related reaction is found in the pyrolysis, in the absence of a

$$(11)$$

catalyst, of o-nitrophenylcyclohexane at 350°; tetrahydrocarbazole (12%), carbazole (27%), and 2-cyclohexylaniline (10%) are formed, again presumably via an intermediate aci-form of the starting material (3).

A nitrene intermediate has also been considered (9) in the transformation of o-dimethylaminonitrobenzene into 1-methylbenzimidazole upon heating with ferrous oxalate and lead (eq. 12). This same transformation apparently

$$(12)$$

occurs when 2-dimethylamino-5-methylnitrobenzene is reduced with sodium bisulfite (Piria reaction) or with tin and hydrochloric acid (35) (eq. 13). It has been shown that the Piria reduction to an aminosulfonic acid proceeds by way of the corresponding nitroso and hydroxylamino compounds. Apparently imidazole ring closure in each example occurs probably through an intramolecular condensation between a nitroso group and an N-methyl group. An insertion reaction of a nitrenoid intermediate subsequently generated from the nitroso compound appears unlikely.

$$(13)$$

The interesting transformation of o-cyclohexylaminonitrobenzene into 1,2-pentamethylenebenzimidazole on pyrolysis with ferrous oxalate (oxide) (34) may also be an example of cyclization through either intramolecular condensation (eq. 14) or nitrene insertion (9). The last two steps are identical with steps previously suggested (9).

Azobenzene and carbon dioxide are produced when nitrobenzene is treated with iron pentacarbonyl and excess carbon monoxide (36). In similar reactions o-nitrobiphenyl and o-ethylnitrobenzene are transformed into carbazole and o-ethylaniline, respectively (36). Nitrenoid (36) and/or radical intermediates, as described in equation 9, may be required in each

$$(14)$$

of these reactions as well as in the transformation of nitrobenzene into phenyl isocyanate by treatment with carbon monoxide in the presence of a Lewis acid and a noble metal (37) (eq. 15). Phenyl isothiocyanate has been

obtained from phenyl azide and carbon disulfide in the presence of aluminum chloride by a parallel but probably ionic reaction (38) (eq. 16).

In contrast with metals, free radicals attack a nitroso group first at nitrogen. A second radical combines with the initially formed nitroxide at oxygen to complete the formation of a trisubstituted hydroxylamine (39,40). As measured by methyl affinities (41) the nitro group is less powerfully activating than is the nitroso group for free-radical attack and is less restrictive insofar as the radical may react first at carbon or at either nitro nitrogen or oxygen (42). Cyanoisopropyl radicals apparently react at oxygen with subsequent deoxygenation to a nitroso derivative (43). A related mechanism permits nitrosobenzene to be considered an intermediate in the free-radical polymerization of methyl methacrylate employing nitrobenzene as a chain transfer agent (44). Deoxygenation of a nitroso group by free-radical attack has not been reported.

An end-on collision mechanism has been ascribed to the deoxygenation of nitric and nitrous oxide by atomic carbon (45). Deoxygenation of nitroso derivatives by atomic carbon does not appear to have been observed.

IV. DEOXYGENATION BY OLEFINS

Irradiation of a mixture of nitrobenzene and 2-methyl-2-butene at room temperature with a mercury resonance arc produced azobenzene along with at least four other products. An explanation (46) (eq. 17 reproduces the pertinent portion) for the reaction included phenylnitrene as the precursor for azobenzene.

$$C_6H_5NO_2 + (CH_3)_2C{=}CHCH_3 \xrightarrow{h\nu} C_6H_5N \begin{array}{c} O{-}C(CH_3)_2 \\ \diagup \\ \diagdown \\ O{-}CHCH_3 \end{array} \longrightarrow$$

$$C_6H_5\overset{\ominus}{N}\overset{\oplus}{OC}(CH_3)_2 + C_6H_5\overset{\ominus}{N}\overset{\oplus}{OC}HCH_3 + CH_3COCH_3 + CH_3CHO$$

$$C_6H_5\overset{\ominus}{N}\overset{\oplus}{OC}(CH_3)_2 \longrightarrow C_6H_5N + (CH_3)_2CO$$

$$C_6H_5N \longrightarrow C_6H_5N{=}NC_6H_5 \tag{17}$$

V. DEOXYGENATION BY TERVALENT PHOSPHOROUS REAGENTS

A. Defensible Nitrene Intermediacy

The strongest defense for the generation of an arylnitrene by deoxygenation of a nitro or a nitroso group appears to be based on certain reactions with tervalent phosphorous reagents. Aromatic nitro compounds react more slowly than nitroso compounds do, and it has been proposed that stepwise deoxygenation first transforms the nitro group into the nitroso group (3,7,15) (eq. 18).

$$ArNO_2 + R_3P \longrightarrow ArN\begin{array}{c}\overset{\ominus}{O}\\ |\\ \overset{\oplus}{}\\ \diagdown\\ O\end{array}PR_3 \xrightarrow{-R_3PO} ArNO \xrightarrow{R_3P}$$

$$ArN\begin{array}{c}\\ \diagdown\\ O\end{array}PR_3 \longrightarrow ArN + R_3PO \tag{18}$$

Convincing evidence for nitrene intermediacy from 2'-nitro-2,4,6-trimethylbiphenyl partially consists of isolating the products of hydrogen abstraction, CH bond insertion and coupling with the tervalent phosphorous reagent (8) (eq. 19). When the reaction was carried out in isopropylbenzene as solvent the formation of bi-α-cumyl suggested intermolecular hydrogen abstraction by a triplet nitrene. Similar results were obtained from the pyrolysis of 2,4,6-trimethyl-2'-azidobiphenyl at 230° which produced 2,4,6-trimethyl-2'-aminobiphenyl, 2,4,9-trimethyl carbazole, and 8,10-dimethylphenanthridine (47) (eq. 20).

$$\text{[o-nitrobiphenyl derivative]} \xrightarrow[-(C_2H_5O)_3PO]{(C_2H_5O)_3P}$$

$$\text{[phenanthridine derivative]} CH_3 + ArNH_2 + \underset{(C_2H_5O)_3P}{Ar\overset{\parallel}{N}} \quad (19)$$

$$\text{[o-azidobiphenyl derivative]} \xrightarrow{-N_2}$$

$$\text{[phenanthridine derivative]} CH_3 + \text{[carbazole derivative]} CH_3 + ArNH_2 \quad (20)$$

A strong similarity has been observed for the reaction between phenyl azide and diethyl amine (48) (eq. 21) with the reaction between o-nitro-biphenyl, diethyl methyl phosphonite (preferred reagent) and diethyl amine (8,49) (eq. 22). Since an azepine is formed in both reactions and the

$$C_6H_5N_3 \xrightarrow{-N_2} \text{[bicyclic azirine]} \xrightarrow{R_2NH} \text{[azepine NR}_2\text{]} \quad (21)$$

$$\text{[o-nitrobiphenyl]} \xrightarrow[R_2NH]{R'_3P} \text{[azepine NR}_2\text{]} + \text{[carbazole]} \quad (22)$$

latter also produces a carbazole, three possibilities have been recognized: (1) both types of product are formed from a nitrene or some other electron deficient species without the intervention of a bicyclic azirine, (2) both are formed from a bicyclic azirine intermediate, and (3) the bicyclic azirine is formed reversibly from a nitrene thereby permitting carbazole formation from the nitrene and azepine from the bicyclic azirine. It is a persuasive argument for the contention that in these examples nitrene intermediacy is as acceptable for deoxygenation of a nitro compound as it is for pyrolysis and photolysis of an azide (8). This hypothesis that phenylnitrene is in equilibrium with a bicyclic azirine may be deficient by not accounting for the absence of formation of either hydrazobenzene or diaryl amines by insertion reactions between phenylnitrene and CH or NH bonds unless it can be assumed that these are sufficiently kinetically less probable. Phenylnitrene has also been suggested as an intermediate in the formation of azepines from nitrosobenzene, triphenyl phosphine and dialkyl amine (49).

For the more general example there is a fourth possibility, which could be considered a corollary to the third. Through reversible valence tautomerization the bicyclic azirine is in equilibrium with azabicyclopropenes and, in turn, a pyridylcarbene* (eq. 23). A clean-cut demonstration established the reality of this possibility in the transformation of o-

$$(23)$$

(a total of 6 bicyclic valence tautomers)

nitrosotoluene, treated with triethyl phosphite, into o-toluidine, ethyl N-o-tolylphosphorimidate, the o-methylanil of α-acetylpyridine and the corresponding nitrone (7) (eq. 24). In a formal sense the anil represents

$$(24)$$

direct combination of the nitrene and the carbene (equations 23 and 24, R = CH₃); however, it is more reasonable that the nitrone was formed first by a similar combination of either the carbene or its bicyclic precursor with nitrosotoluene (7) and that triethyl phosphite deoxygenated the nitrone to the anil. The last step was verified in a separate experiment (7).

* See also Chapter 4.

An equilibrium between a benzocyclopropene and the corresponding diradicals appears to be a closely related process (50) (eq. 25).

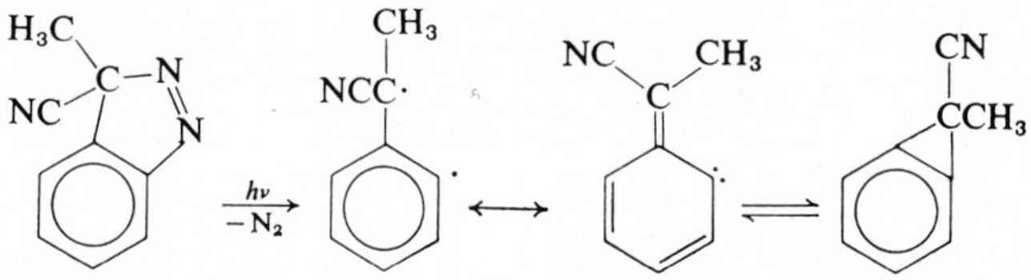

(25)

Numerous cyclizations by intramolecular insertion have been reported from deoxygenation of o-alkyl- (or aryl) nitro-, or nitrosobenzene. From o-nitropropylbenzene and triethyl phosphite 2-methyl-2,3-dihydroindole (7%) is obtained along with o-allylaniline (6%), o-propylaniline (3%), and ethyl N-o-tolyphosphorimidate (40%) (7) (eq. 26). From o-ethylnitroso-benzene there is apparently no cyclization to dihydroindole, instead an

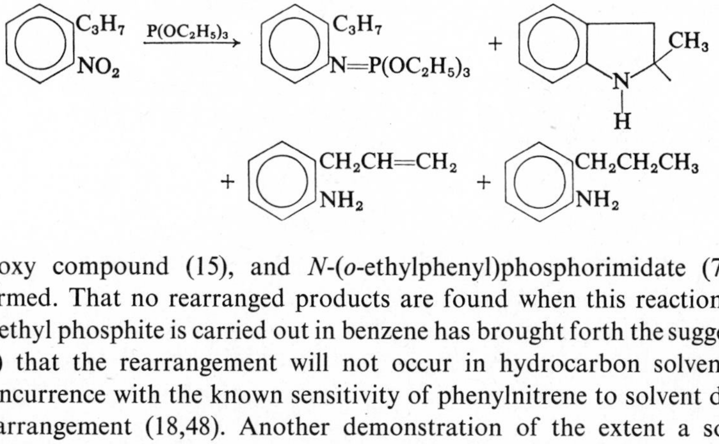

(26)

azoxy compound (15), and N-(o-ethylphenyl)phosphorimidate (7) are formed. That no rearranged products are found when this reaction with triethyl phosphite is carried out in benzene has brought forth the suggestion (7) that the rearrangement will not occur in hydrocarbon solvents, in concurrence with the known sensitivity of phenylnitrene to solvent during rearrangement (18,48). Another demonstration of the extent a solvent effect may have in controlling the reaction a nitrene may follow is found in the rearrangement of carbalkoxynitrene in methanol but not in hydro-carbons (51,52).

Cyclization from o-n-butyl phenylnitrene produces 2-ethyl-2,3-dihydro-indole, 2-methyl-1,2,3,4-tetrahydroquinoline, and o-n-butylaniline in almost the same ratio for the nitrene generated from an azide (solution phase pyrolysis) or from the corresponding nitro compound (3,7). Simi-larly triethyl phosphite transformed (+)-(S)-2-nitro-1-(2-methylbutyl)-benzene into partially active (about 50%) 2-methyl-2-ethyl-2,3-dihydro-indole (3), the same product (65% optical purity) also obtained from the corresponding azide (3) (eq. 27).

Deoxygenation of an o-nitrosoalkylbenzene gives markedly lower yields

of phosphorimidate esters and amines than are obtained from corresponding nitro compounds, and, as expected, dimeric products (anils, nitrones,

azo, and azoxy compounds) become more important since they may be
formed by reactions of the nitroso group. Presumably the concentration
of the nitroso compound does not become appreciable during the deoxygenation of a nitro compound (7).

A selectivity in abstracting hydrogen from an *o*-alkyl side-chain is
demonstrated by a nitrene obtained both from an azide and a nitro or
nitroso compound. The formation of only the unconjugated olefin from
o-*n*-propyl- or *n*-butylphenylnitrene can be accounted for by a concerted
mechanism for the abstraction process (7) (eq. 28).

B. Intramolecular Cyclization to an Unsaturated Group

A strong tendency toward cyclization accompanies deoxygenation in
triethyl phosphite of aromatic nitro and nitroso compounds in which an
α,β-unsaturated group occupies the *o*-position. Cyclization from a nitrene
may account for the formation of a rearranged indole from a β,β-di-
substituted *o*-nitrostyrene but it apparently cannot be a precursor for a
1-hydroxyindole, a 3-indolinone, or a biindoline. Initial deoxygenation to
the corresponding *o*-nitrosostyrene was considered to be an important
step (53,54) (eq. 29); however, at the present time the extent of further
deoxygenation to a nitrene cannot be evaluated. An investigation of the
pyrolysis and/or photolysis of an *o*-styryl azide as an alternate route to an
o-styrylnitrene would be helpful.

Deoxygenation of both *cis*- and *trans*-*o*-nitrostilbene produced 2-
phenylindole (55), also obtained from α-nitrostilbene (56). Cyclization
from a nitrene in the latter example is unlikely since it would be in contrast
with the isomerization to an azirine or dimerization to a pyrazine of
presumably the same nitrene generated by the photolysis of α-azidostilbene
(57) (eq. 30).

$$3\text{-indoline} + \text{biindoline} \tag{29}$$

$$C_6H_5CH{=}C(N_3)C_6H_5 \xrightarrow{-N_2} C_6H_5CH{=}\overset{N}{\underset{|}{C}}C_6H_5$$

$$\tag{30}$$

Ring-closure reactions from isolated o-nitrosostyrenes have not been investigated since these nitroso derivatives are unknown and attempted synthesis has been unsuccessful (58). A related transformation is found in the reaction between o-nitrosobiphenyl and triethyl phosphite which produces carbazole. Ring closure from both a nitrene and from a zwitterion (which otherwise might be the nitrene precursor) has been proposed (56) (eq. 31,32) in a mechanism which calls for nucleophilic attack by

$$o\text{-}C_6H_5C_6H_4NO \xrightarrow{(C_2H_5O)_3P} o\text{-}C_6H_5C_6H_4\overset{\oplus}{N}O\overset{+}{P}(OC_2H_5)_3 \xrightarrow{-(C_2H_5O)_3PO}$$

$$o\text{-}C_6H_5C_6H_4N \xrightarrow{\hspace{2cm}} \text{carbazole} \tag{31}$$

phosphorus on nitroso oxygen. Deoxygenation of aromatic nitroso compounds has also been accounted for by assuming an electrophilic attack by tervalent phosphorous on nitroso oxygen (59).

Additional persuasive evidence for a similarity between pyrolysis or photolysis of an azide and deoxygenation of a corresponding nitro or

$$ \text{(32)} $$

nitroso compound may be found in intramolecular cyclization by bond formation between nitrogen and an atom other than carbon. Both pyrolysis of o-azidoazobenzene (60) and deoxygenation of o-nitroazobenzene (61,62) produce 2-phenylbenzimidazole (eq. 33). In each conversion, cyclization by a concerted process would account for the absence of other products to be expected from nitrene intermediates. Isolation of 2-phenyl benzimidazole-N-oxide from the reduction of o-nitroazobenzene with sodium sulfide supports the assumption that cyclization of o-nitrosoazobenzene may occur (63).

$$ \text{(33)} $$

Azapentalenes (64), pyrazolobenzotriazoles (65) indazoles (61,62), anthranils (61,62,66) and pyrrolopyrimidines (66a) are examples of other heterocycles produced directly from the appropriate reactions of corresponding nitro compounds and/or the related azides. An important contribution can be anticipated from an investigation of the deoxygenation of 1,4-diaryl-5-nitro(or nitroso)triazole since it is known that the "nitrene" generated from the corresponding azide is in equilibrium with a noncyclic isomeric nitrile (67) (eq. 34).

A different reaction occurs on treating o-dinitrobenzene with triethyl phosphite. A nitro group is replaced and ethyl nitrite is produced (68) (eq. 35). By a similar reaction the first denitration of a pyrimidine compound (69) was claimed.

$$\text{(34)}$$

$$\text{(35)}$$

Deoxygenation of o-nitrosonitrobenzene with triethyl phosphite at room temperature produces o-dinitrosobenzene (benzfuroxan) which, in turn, undergoes further deoxygenation at 150° to form benzfurazan (56). Intermediate nitrenes have been proposed (15); however, concerted ring-closure reactions (70) (eq. 36) appear to account more satisfactorily for the absence of other products to be expected from nitrenoid intermediates.

(2 structures)

$$\text{(36)}$$

It is consistent with the lack of formation of an azo compound during deoxygenation of nitro and nitroso compounds for neither o,o'-dinitro- nor $o,-o'$-dinitrosoazobenzene to be found in the reaction mixtures. In neither of these examples, however, could an azo compound be expected if ring closure is concerted with the elimination of a phosphine oxide. The expected dimerization of o-nitrophenylnitrene, should it have been formed, may have been demonstrated from an independent source in the isolation of o,o'-dinitroazobenzene from the low temperature pyrolysis of N,N-dichloro-o-nitroaniline, a reaction which also brought about the liberation of chlorine. The absence of formation of benzfuroxan in this example has been attributed to the inability of the neighboring nitro group to provide anchimeric assistance for the expulsion of chlorine (9). This neighboring group assistance by the nitro group facilitated the expulsion of nitrogen in the pyrolytic conversion of *o*-nitroazidobenzene into benzfuroxan where a negative entropy of activation (71) probably indicated a cyclic transition state.

Whereas phosphorus tri- or pentachloride fail to deoxygenate *o*-dinitrosobenzene (benzfuroxan) (72), PCl_5 does deoxygenate a furoxan in which the furazan ring is not fused to an aromatic ring (72). Apparently the latter reaction proceeds without ring opening by electrophilic attack on exocyclic furoxan oxygen (eq. 37). Deoxygenation of amine oxides by

$$
\begin{array}{c}
RC-CR' \\
\| \quad \| \\
N \quad N \to O \\
\diagdown O \diagup
\end{array}
\xrightarrow{PCl_3}
\begin{array}{c}
RC-CR' \\
\| \quad \| \\
N \quad N \\
\diagdown O \diagup
\end{array}
\qquad (37)
$$

phosphorus trichloride may also occur by electrophilic attack on oxygen (15).

Furoxans in which the furazan ring is not fused to another aromatic ring are also deoxygenated by trialkyl or triaryl phosphines and by triethyl phosphite (73,74) (eq. 38) but the temperatures required are within the range known to bring about thermal isomerization of furoxans (75,76) (eq. 38). Ring-opening to a dinitroso olefin not only readily accounts for isomerization but also permits deoxygenation by nucleophilic attack on nitroso oxygen.

$$
\begin{array}{c}
RC-CR' \\
\| \quad \| \\
N \quad N \to O \\
\diagdown O \diagup
\end{array}
\longrightarrow
\begin{array}{c}
RC=CR' \\
| \quad | \\
NO \quad NO
\end{array}
\longrightarrow
\begin{array}{c}
RC-CR' \\
\| \quad \| \\
O \gets N \quad N \\
\diagdown O \diagup
\end{array}
$$

$$
\begin{array}{c}
\downarrow R_3P \\[4pt]
R'C=NO\overset{\oplus}{P}R_3 \\
| \\
RC=NO^{\ominus}
\end{array}
\longrightarrow
\begin{array}{c}
R'C \quad N-O \\
\diagup \quad \diagdown \\
\quad \quad PR_3 \\
RC \quad N-O
\end{array}
\longrightarrow
\begin{array}{c}
RC-CR' \\
\| \quad \| \\
N \quad N \\
\diagdown O \diagup
\end{array}
\qquad (38)
$$

A cyclic or zwitterionic intermediate during the deoxygenation of a dinitroso olefin is reminiscent of the reaction between either *cis*- or *trans*-dibenzoylethylene and triethyl phosphine (eq. 39). In this example, the adduct, sensitive to moisture, was transformed into dibenzoylethane (77,78) (eq. 39). A similar hydrolysis of the adduct obtained from *o*-

$$
\begin{array}{c}
C_6H_5 \\
| \\
CHC=O \\
\| \\
CHC=O \\
| \\
C_6H_5
\end{array}
\xrightarrow{(C_2H_5)_3P}
\begin{array}{c}
C_6H_5 \\
| \quad \overset{\oplus}{} \\
CH=C-O-P(C_2H_5)_3 \\
| \\
CH=C-O^{\ominus} \\
| \\
C_6H_5
\end{array}
\xrightarrow{HOH}
$$

$$
(C_2H_5)_3PO + C_6H_5COCH_2CH_2COC_6H_5 \qquad (39)
$$

dinitrosobenzene and triphenyl phosphine was observed (70) (eq. 40). Thermal conversion of the adduct obtained from a phosphine and a

$$
\text{(structure)} \xrightarrow{H_2O} \left[\text{(structure)} \right]^{\oplus} OH^{\ominus} \longrightarrow
$$

$$
\text{(structure)} + (C_6H_5)_3PO + H_2O \qquad (40)
$$

dinitroso derivative into a five-membered furazan ring (eq. 38) is reminiscent of the pyrolytic conversion of a dioxaphospholane into an epoxide (79) (eq. 41).

$$
\begin{array}{c}
C_6H_5(CN)C\!-\!O \\
| \qquad\qquad P(OR)_3 \\
C_6H_5(CN)C\!-\!O
\end{array}
\xrightarrow[80°]{-(RO)_3PO}
\begin{array}{c}
C_6H_5(CN)C \\
\diagdown \\
\diagup \qquad O \\
C_6H_5(CN)C
\end{array}
\qquad (41)
$$

Combining the concept that a cyclic or zwitterionic intermediate may dissociate into product without the intervention of a nitrene with the observation that the best yield of azoxybenzene from nitrosobenzene and triphenylphosphine required a 2:1 molar ratio (15) leads to a modification of equation 18 which may account for at least part of the reaction between a tervalent phosphorous reagent and a nitroso compound (eq. 42). The

$$
ArNO + R_3P \longrightarrow Ar\overset{\ominus}{N}O\overset{\oplus}{P}R_3 \xrightarrow{ArNO} ArN\overset{\oplus}{O}PR_3 \underset{\longleftarrow}{\overset{\longrightarrow}{}}
$$
$$
\underset{ArN\!-\!O^{\ominus}}{|}
$$

$$
\begin{array}{c}
ArNO \\
\diagdown \\
| \qquad PR_3 \longrightarrow ArN\!=\!N(O)Ar + R_3PO \quad (42) \\
\diagup \\
ArNO
\end{array}
$$

transformation of monoketones into dioxaphospholanes may proceed by an analogous mechanism (79,80) (eq. 43).

$$
R_2CO \xrightarrow{R_3P} R_2\overset{\ominus}{C}O\overset{\oplus}{P}R_3 \xrightarrow{R_2CO} R_2\overset{\oplus}{C}OPR_3 \longrightarrow
\begin{array}{c}
R_2CO \\
\diagdown \\
| \qquad PR_3 \quad (43) \\
\diagup \\
R_2CO
\end{array}
$$
$$
\underset{R_2CO^{\ominus}}{|}
$$

Neither *o*-dinitrosoarenes nor dinitroso olefins (furoxans) are transformed into nitriles on deoxygenation. Should both nitroso groups in nitrosobenzene be deoxygenated the assumed dinitrene intermediate, which has been suggested for the pyrolysis of *o*-diazidobenzene (81), would be expected to rearrange into mucononitrile (81) (eq. 44).

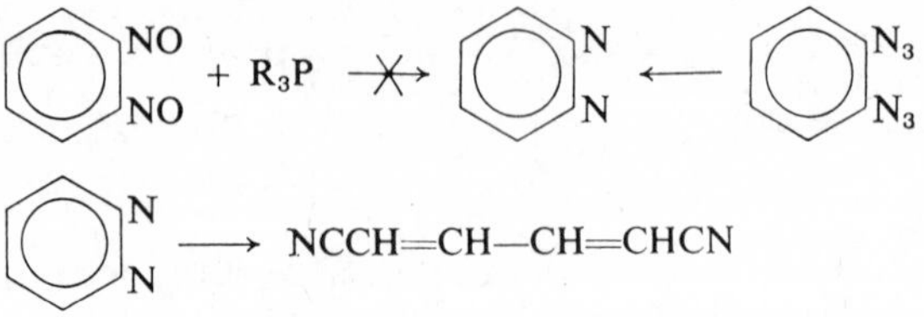

$$C_6H_5CH=CHNO_2 \xrightarrow{R_3P}$$

NCCH=CH—CH=CHCN   (44)

VI. DEOXYGENATION OF NITROOLEFINS

Deoxygenation of β-nitrostyrene is reported to give phenylacetonitrile in 6% yield (9) (eq. 45). Presumably β-nitrosostyrene is produced initially and gives a nitrenoid intermediate on further deoxygenation. Rearrangement of β-styrylnitrene into phenylketene imine, in analogy with the Curtius rearrangement on pyrolysis of an acyl azide,* followed by tautomerization would account for the formation of the product. Supporting evidence for the nitrene intermediate has been obtained from the pyrolysis and photolysis of β-styryl azide into phenylacetonitrile (82) (eq. 45).

$$C_6H_5CH=CHNO_2 \xrightarrow{R_3P}$$

$$C_6H_5CH=CHNO \longrightarrow C_6H_5CH=CHN \xleftarrow{-N_2} C_6H_5CH=CHN_3$$

$$C_6H_5CH_2CN \longleftarrow C_6H_5CH=C=NH \qquad (45)$$

Deoxygenation of α-nitrostilbene has been discussed (Section VB).

A nonterminal vinylnitrene may isomerize into an azirine (57,83) (eq. 46). That azirine formation may not be occurring in the deoxygenation

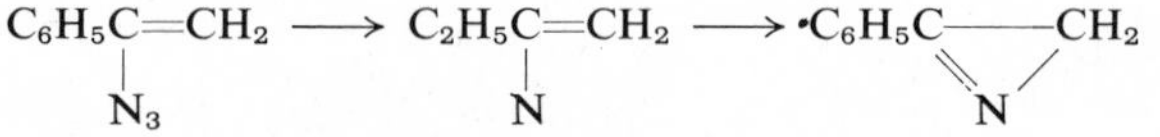

$$C_6H_5C=CH_2 \longrightarrow C_2H_5C=CH_2 \longrightarrow \cdot C_6H_5C{-}{-}{-}CH_2 \qquad (46)$$

of furoxans (eq. 38) is attested to by the formation of 1,2,5-oxadiazoles (furazans) but not of 1,2,4-oxadiazoles (74) (eq. 47); however, azirine ring opening may proceed by preferential cleavage of a CN rather than a CC single bond.

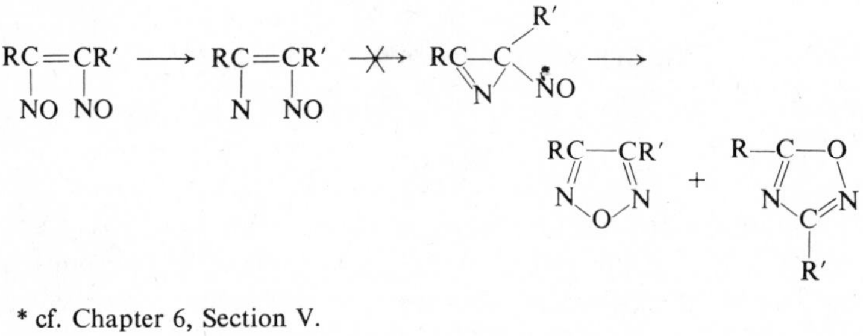

$$RC=CR' \longrightarrow RC=CR' \xrightarrow{} RC{-}C \qquad \longrightarrow$$

(47)

---

* cf. Chapter 6, Section V.

VII. DEOXYGENATION OF N-NITROSO DERIVATIVES

Apparently alkaline sodium hydrosulfite (dithionite) reduction of an N-nitroso derivative of a secondary amine produces an intermediate dialkylaminonitrene (84) (eq. 48) which is also a probable intermediate in the mercuric oxide oxidation of the corresponding N,N-dialkylhydrazine (84). The intermediate has also been described as an azamine or a diazene, $R_2\overset{\oplus}{N}=\overset{\ominus}{N}$ (85). Triphenyl phosphine was found inert to nitrosoamines (78).

$$(C_6H_5CH(CH_3))_2NNO \xrightarrow{Na_2S_2O_4} R_2NNHOH \xrightarrow{\text{base}}$$

$$R_2N\overset{\ominus}{N}OH \xrightarrow{-\overset{\ominus}{O}H} R_2N{-}N \xleftarrow{Hgo} R_2NNH_2$$

$$R_2N{-}N \xrightarrow{-N_2} (C_6H_5CH(CH_3))_2 + C_6H_5CH{=}CH_2 \quad (48)$$

VIII. DEOXYGENATION OF NITRITE ESTERS (86)

In an attempt to generate a nitrene bound to oxygen, the deoxygenation of nitrite esters by tervalent phosphorus reagents was investigated. Both benzyl- and tert-butyl nitrite are transformed into corresponding alcohols by tri-n-butyl- or triphenylphosphine or by triethyl phosphite as the corresponding phosphine oxide or phosphate is formed. The combination of tert-butyl nitrite and tri-n-butyl phosphine may become explosive after an induction period with gentle warming. Insofar as insertion with CH bonds, addition to multiple bonds, and intramolecular rearrangement could not be detected, the possibility of a nitrene intermediate was considered to be doubtful.

A proposed explanation (eq. 49) is based on a combination of nitrite ester and the initially formed zwitterionic adduct of nitrite ester with the phosphorus reagent followed by elimination of a phosphine oxide (compare eq. 42) and generation of a proposed intermediate hyponitrite ester. Mild pyrolysis of each hyponitrite (87) and the generation of alkoxy radicals which may abstract hydrogen from either a phosphine or phosphine oxide was confirmed by independent experiments.

$$R'ONO + R_3P \longrightarrow R'O\overset{\ominus}{N}\overset{\oplus}{O}PR_3 \xrightarrow{R'ONO} R'ON\overset{\oplus}{O}PR_3 \xrightarrow{-R_3PO}$$
$$\overset{|}{R'ON{-}O^{\ominus}}$$

$$\underset{R'ON\to O}{\overset{R'ON}{\|}} \xrightarrow[-R_3PO]{R_3P} \underset{R'ON}{\overset{R'ON}{\|}} \xrightarrow{-N_2} R'O\cdot \xrightarrow{CH} R'OH \quad (49)$$

$$R' = (CH_3)_3C{-},\ C_6H_5CH_2{-}$$
$$R' = nC_4H_9{-},\ C_2H_5O{-},\ C_6H_5$$

Isolation of ethyl nitrite from the reaction between o-dinitrobenzene and triethyl phosphite (eq. 35) apparently requires the escape of gaseous ethyl nitrite as it is formed.

REFERENCES

1. M. Ohno and I. Sakai, *Tetrahedron Letters*, **1965**, 451.
2. T. Mukaiyama and H. Numbu, *J. Org. Chem.*, **27**, 2201 (1962).
3. G. Smolinsky and B. I. Feuer, *J. Org. Chem.*, **31**, 3882 (1966).
4. H. Shindo and B. Umezawa, *Chem. Pharm. Bull.* (*Tokyo*), **10**, 492 (1962); *Chem. Abstr.*, **57**, 14598 (1962).
5. E. Meyer and G. W. Griffin, *Angew. Chem.*, **79**, 648 (1967).
6. A. W. Johnson, *J. Org. Chem.*, **28**, 252 (1963).
7. R. J. Sundberg, *J. Am. Chem. Soc.*, **88**, 3781 (1966).
8. J. I. G. Cadogan and M. J. Todd, *Chem. Commun.*, **1967**, 178.
9. R. A. Abramovitch and B. A. Davis, *Chem. Rev.*, **64**, 149 (1964).
10. P. A. S. Smith and J. H. Hall, *J. Am. Chem. Soc.*, **84**, 480 (1962).
11. A. Reiser and V. Frazer, *Nature*, **208**, 682 (1965). A. Reiser, G. C. Terry, and F. W. Willets, *Nature*, **211**, 410 (1966). A. Reiser, H. Wagner, and G. Bowes, *Tetrahedron Letters*, **1966**, 2635. A. Reiser, G. Bowes, and R. J. Horne, *Trans. Faraday Soc.*, **62**, 3162 (1966).
12. E. Wasserman, G. Smolinsky, and W. A. Yager, *J. Am. Chem. Soc.*, **86**, 3166 (1964). G. Smolinsky, L. C. Snyder, and E. Wasserman, *Rev. Mod. Phys.*, **35**, 576 (1963). G. Smolinsky, E. Wasserman, and W. A. Yager, *J. Am. Chem. Soc.*, **84**, 3220 (1962).
13. R. M. Moriarty, M. Rahman, and G. J. King, *J. Am. Chem. Soc.*, **88**, 842 (1966).
14. W. Lwowski and T. W. Mattingly, Jr., *J. Am. Chem. Soc.*, **87**, 1947 (1965); W. Lwowski and T. W. Mattingly, Jr., *Tetrahedron Letters*, **1962**, 277. J. S. McConaghy, Jr., and W. Lwowski, *J. Am. Chem. Soc.*, **87**, 2357 (1967).
15. P. J. Bunyan and J. I. G. Cadogan, *J. Chem. Soc.*, **1963**, 42.
16. J. Meinwald and D. H. Aue, *J. Am. Chem. Soc.*, **88**, 2849 (1966). R. Huisgen and H. Blaschke, *Tetrahedron Letters*, **1964**, 1409.
17. R. Walker and W. A. Waters, *J. Chem. Soc.*, **1962**, 1632. R. Huisgen and G. Mueller, *Angew. Chem.*, **72**, 371 (1960).
18. G. Smolinsky, *J. Org. Chem.*, **26**, 4108 (1961). W. Lwowski, T. W. Mattingly, Jr., and T. J. Maricich, *Tetrahedron Letters*, **1964**, 1591.
19. S. A. Buckler, L. Doll, F. K. Lind, and M. Epstein, *J. Org. Chem.*, **27**, 794 (1962).
20. A. C. Bellaart, *Tetrahedron*, **21**, 3285 (1965).
21. E. Bamberger, *Chem. Ber.*, **27**, 1182 (1894).
22. L. A. Neiman, V. I. Maimind, and M. M. Shemyakin, *Izv. Akad. Nauk SSSR, Otd. Khim. Nauk.*, **1964** (7), 1357; *Chem. Abstr.*, **61**, 11991 (1964).
23. I. J. Rinks, *Chem. Weekblad*, **2**, 1061 (1914).
24. S. Kobayashi and Y. Aoyama, Japan, 4329; *Chem. Abstr.*, **49**, 4712 (1955). J. Meisenheimer, *Chem. Ber.*, **36**, 4174 (1903).
25. T. Kauffmann and S. M. Hage, *Angew. Chem. Intern. Ed.*, **2**, 156 (1963).
26. T. T. Tsai, W. E. McEwen, and J. Kleinberg, *J. Org. Chem.*, **25**, 1186 (1960).
27. W. E. Bachman, *J. Am. Chem. Soc.*, **53**, 1524 (1931).
28. D. L. Vivian, G. Y. Greenberg, and J. L. Hartwell, *J. Org. Chem.*, **16**, 1 (1951).
29. H. C. Waterman and D. L. Vivian, *J. Org. Chem.*, **14**, 289 (1949).
30. R. A. Abramovitch and K. A. H. Adams, *Can. J. Chem.*, **39**, 2515 (1961).

31. R. A. Abramovitch, Y. Ahmad, and D. Newman, *Tetrahedron Letters*, **1961**, 752.
32. R. Higginbottom and H. Suschitzky, *J. Chem. Soc.*, **1962**, 2367
33. R. A. Abramovitch and B. A. Davis, private communication.
34. R. H. Smith and H. Suschitzky, *Tetrahedron*, **16**, 80 (1961).
35. W. M. Lauer, M. M. Sprung, and C. M. Langkamerer, *J. Am. Chem. Soc.*, **58**, 225 (1936).
36. J. E. Kmiecik, *J. Org. Chem.*, **30**, 2014 (1965).
37. R. P. Bennett, W. B. Hardy, R. K. Madison, and S. M. Davis, 153rd Meeting Am. Chem. Soc., Miami, April 1967, *Abstracts*, O 89. W. B. Hardy and R. P. Bennett, *Tetrahedron Letters*, **1967**, 961.
38. W. Borsche and H. Hahn, *Chem. Ber.*, **82**, 260 (1949).
39. B. A. Gingras and W. A. Waters, *J. Chem. Soc.*, **1950**, 1920.
40. A. Maschke, B. S. Shapiro, and F. W. Lampe, *J. Am. Chem. Soc.*, **86**, 1929 (1964).
41. M. Szwarc, *J. Polymer Sci.*, **16**, 367 (1955).
42. P. A. S. Smith, *Open-Chain Nitrogen Compounds*, Vol. 2, Benjamin, New York, 1966, p. 362, 430.
43. W. P. Norris, *J. Am. Chem. Soc.*, **81**, 4239 (1959).
44. R. L. Dannley and M. Esayian, *Am. Chem. Soc., Div. Polymer Chem., Preprints*, **1**, 337 (1960).
45. J. Dubrin, C. MacKay, M. Pandow, and R. Wolfgang, *J. Inorg. Nucl. Chem.*, **26**, 2113 (1964).
46. G. Buchi and D. E. Ayer, *J. Am. Chem. Soc.*, **78**, 689 (1956).
47. G. Smolinsky, *J. Am. Chem. Soc.*, **82**, 4717 (1960).
48. R. Huisgen and M. Appl, *Chem. Ber.*, **91**, 12 (1958). W. von E. Doering and R. A. Odum, *Tetrahedron*, **22**, 81 (1961).
49. R. A. Odum and M. Brenner, *J. Am. Chem. Soc.*, **88**, 2074 (1966).
50. G. L. Closs, L. R. Kaplan, and V. I. Bendall, *J. Am. Chem. Soc.*, **89**, 3376 (1967).
51. W. Lwowski, R. De Mauriac, T. W. Mattingly, Jr., and E. Scheiffele, *Tetrahedron Letters*, **1964**, 3285.
52. W. Lwowski, A. Hartenstein, C. de Vita, and R. L. Smick, *Tetrahedron Letters*, **1964**, 2497.
53. R. J. Sundberg, *J. Org. Chem.*, **30**, 3604 (1965).
54. R. J. Sundberg and T. Yamazaki, *J. Org. Chem.*, **32**, 290 (1967).
55. J. I. G. Cadogan and M. Cameron-Wood, *Proc. Chem. Soc. (London)*, **1962**, 361.
56. J. I. G. Cadogan, M. Cameron-Wood, R. K. Mackie, and J. G. Searle, *J. Chem. Soc.*, **1965**, 4831.
57. A. Hassner and L. A. Levy, *J. Am. Chem. Soc.*, **87**, 4203 (1965).
58. J. H. Boyer and H. Alul, *J. Am. Chem. Soc.*, **81**, 2136 (1959).
59. R. F. Hudson, *Structure and Mechanism in Organophosphorus Chemistry*, Academic Press, New York, 1965, p. 191.
60. T. Zincke and A. T. Lawson, *Ber.*, **20**, 1176 (1887).
61. J. I. G. Cadogan and R. J. G. Searle, *Chem. Ind.*, **1963**, 1434.
62. J. I. G. Cadogan, R. K. Mackie, and M. J. Todd, *Chem. Commun.*, **1966**, 491.
63. E. Bamberger and R. Hubner, *Chem. Ber.*, **36**, 3822 (1903).
64. J. C. Kauer and R. A. Carboni, *J. Am. Chem. Soc.*, **89**, 2633 (1967).
65. B. M. Lynch and Y-Y. Hung, *J. Heterocyclic Chem.*, **2**, 218 (1965).
66. J. H. Boyer and F. C. Canter, *Chem. Rev.*, **54**, 34, 36 (1954).
66a. E. C. Taylor and E. E. Garcia, *J. Org. Chem.*, **30**, 655 (1965).
67. P. A. S. Smith, L. O. Krbechek, and W. Resemann, *J. Am. Chem. Soc.*, **86**, 2025 (1964).

68. J. I. G. Cadogan, D. J. Sears, and D. M. Smith, *Chem. Commun.*, **1966**, 491.
69. E. C. Taylor, F. Sowinski, T. Yll, and F. Yoneda, *J. Am. Chem. Soc.*, **89**, 3369 (1967).
70. J. H. Boyer and S. E. Ellzey, Jr., *J. Org. Chem.*, **26**, 4684 (1961).
71. T. F. Fagley, J. R. Sutter, and R. L. Oglukian, *J. Am. Chem. Soc.*, **78**, 5567 (1956).
72. J. H. Boyer, "Oxadiazoles," in *Heterocyclic Compounds*, Vol. 7, R. C. Elderfield, Ed., Wiley, New York, 1961, p. 483.
73. T. Mukaiyama, H. Nambu, and M. Okamoto, *J. Org. Chem.*, **27**, 3651 (1962).
74. C. Grundmann, *Ber.*, **97**, 575 (1964).
75. Reference 72, p. 498.
76. F. B. Mallory and A. Cammarata, *J. Am. Chem. Soc.*, **88**, 61 (1966).
77. L. Horner and K. Klupfel, *Ann.*, **591**, 69 (1955).
78. L. Horner and H. Hoffmann, *Angew. Chem.*, **68**, 473 (1956).
79. J. H. Boyer and R. Selvarajan, unpublished results.
80. I. J. Borowitz and M. Anschel, *Tetrahedron Letters*, **1967**, 1517.
81. J. H. Hall, *J. Am. Chem. Soc.*, **87**, 1147 (1965).
82. J. H. Boyer, W. E. Krueger, and G. J. Mikol, *J. Am. Chem. Soc.*, **89**, 5504 (1967).
83. G. Smolinsky, *J. Org. Chem.*, **27**, 3559 (1962).
84. C. G. Overberger, N. P. Marullo, and R. G. Hiskey, *J. Am. Chem. Soc.*, **83**, 1374 (1961).
85. Reference 41, p. 136. See also D. M. Lemal, T. W. Rave, and S. D. McGregor, *J. Am. Chem. Soc.*, **85**, 1944 (1963).
86. J. H. Boyer and J. D. Woodyard, unpublished results.
87. H. Keifer and T. G. Traylor, *Tetrahedron Letters*, **1966**, 6163.

CHAPTER 6

Carbonylnitrenes

WALTER LWOWSKI

Research Center, New Mexico State University,
Las Cruces, New Mexico 88001

I. INTRODUCTION

Carbonylnitrenes, —CO—N, comprise the best-known class of nitrenes. Those that have been studied in detail, so far, are quite similar to each other in their chemical reactions. That is, nitrenes of the types

185

$>$C—CO—N and —O—CO—N show similar chemical reactions, and the order of reactivity with different C—H bonds is the same. The selectivity of one nitrene $>$C—CO—N was found to be greater than that of R—O—CO—N. However, the data comparing these two groups of nitrenes are quite limited, and much more work will have to be done before a general comparison of the two can be given.

While the chemical reactions of the types $>$C—CO—N and —O—CO—N seem to be analogous, the known ways for generating them are not. Both species can be made by photolyzing the corresponding azides, but thermolysis of these azides leads to the nitrene in the case of RO—CO—N_3 and to Curtius rearrangement (to give R—NCO) in the case of $>$C—CO—N_3. Carbalkoxynitrenes, RO—CO—N, can be made by α-elimination, e.g., from EtO—CO—NH—OSO_2Ar, but the corresponding acylhydroxylamines, R—CO—NH—Oacyl, undergo Lossen rearrangement (to give R—NCO) in all cases known so far.

To separate the unanalogous generation of the nitrenes while keeping together the analogous reactions, this chapter deals first with the making of each group, and then with the reactions of both together.

II. CARBALKOXYNITRENES

Carbethoxynitrene was first reported as an intermediate in the photolysis of ethyl azidoformate (1). Photolysis of EtOOC—N_3 in cyclohexene leads to the corresponding aziridine, 7-carbethoxy-7-azabicyclo[4.1.0]heptane (1) and to insertion of EtOOC—N into the allylic and other C—H bonds to give 2 and 3. Photolysis of the azide in cyclohexane gives N-cyclohexylurethan (4). A priori, these products could be formed by azide mechanisms not involving a nitrene intermediate (see Chapter 1), but carbethoxynitrene was shown to be the actual intermediate by generating it by an alternate route (2), α-elimination from N-(p-nitrobenzenesulfonyloxy)-urethan (5). Identical products are formed, and the intermediates generated by photolysis (1), thermolysis (3–5), and α-elimination (6,7) show very similar selectivities.

A. Carbalkoxynitrenes from Azidoformates

Ultraviolet irradiation of ethyl azidoformate in a rigid matrix at liquid helium temperatures gives carbethoxynitrene, as shown by the ESR spectrum of its triplet ground state (8). Earlier attempts to obtain the

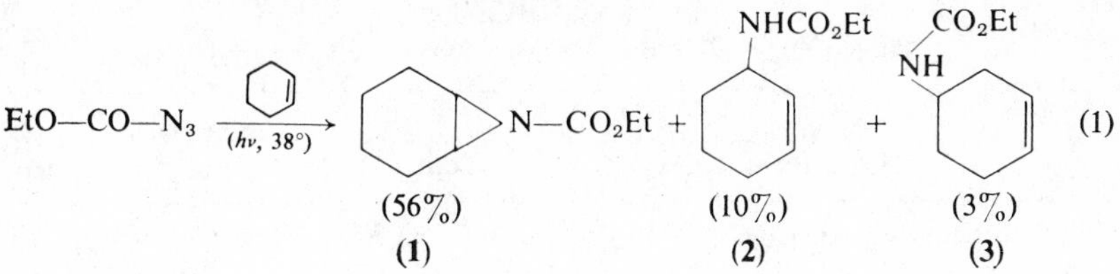

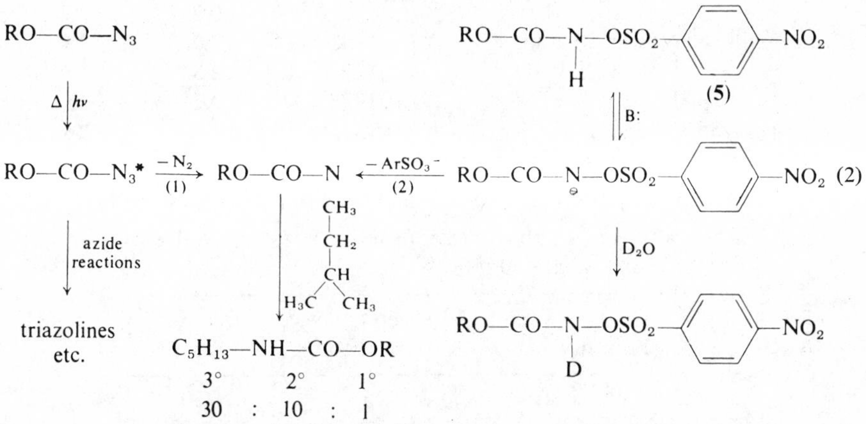

EtO—CO—N$_3$ $\xrightarrow[(hv, 38°)]{}$ [bicyclic structure] N—CO$_2$Et + [cyclohexadiene with NHCO$_2$Et] + [cyclohexadiene with NH and CO$_2$Et] (1)

(56%)   (10%)   (3%)
(1)   (2)   (3)

RO—CO—N$_3$ RO—CO—N—OSO$_2$——[benzene ring]——NO$_2$
                                                              |
Δ $\Big|$ hv   H   (5)

$\Big\downarrow$ $\Big\Uparrow$ B:

RO—CO—N$_3$* $\xrightarrow[(1)]{-N_2}$ RO—CO—N $\xleftarrow[(2)]{-ArSO_3^-}$ RO—CO—N—OSO$_2$——[benzene ring]——NO$_2$ (2)
                                                              |
                                                              ⊖

$\Big|$ azide $\Big|$ CH$_3$ $\Big\downarrow$ D$_2$O
$\Big|$ reactions $\Big|$ CH$_2$
                                              $\Big|$ CH
                                    H$_3$C$\diagdown$ $\diagup$CH$_3$

triazolines   RO—CO—N—OSO$_2$——[benzene ring]——NO$_2$
etc.   C$_5$H$_{13}$—NH—CO—OR |
                  3°   2°   1°   D
                  30 : 10 : 1

ESR spectrum, at higher temperatures, had not been successful (9) (see Chapter 13). Irradiation of ethyl azidoformate in the gas phase also generates the nitrene, and flash photolysis in the presence of cyclohexene gave the aziridine (1). In the absence of a reaction partner, carbethoxynitrene decomposed to NCO and EtO radicals (10,11) (see Chapter 2).

Best documented is the decomposition of alkyl azidoformates in solution. The kinetics of solution thermolysis (5,12) have been studied by D. S. Breslow (5,13–15) and Huisgen (16). The results are shown in Tables I and II. Excellent first-order kinetics were obtained in a variety of solvents with n-octadecyl azidoformate, tetramethylene bis(azidoformate), and 2,2-bis(4-azidocarbonyloxyphenyl)propane. For example, the thermolysis of n-octadecyl azidoformate in diphenyl ether at 120° did not significantly deviate from first-order kinetics up to 95% completion. The rate was not affected by a variety of additives, such as potassium tert-butoxide, p-toluenesulfonic acid, fatty acid salts of divalent calcium, zinc, cadmium, copper, lead, manganese, cobalt, and nickel, titanium(IV) naphthenate, and aluminum and magnesium acetyl acetonates at concentrations of 10 mole %. Weights equal to that of the azide of carbon black and calcium carbonate also did not affect the decomposition rate. The thermolysis rates of n-propyl azidoformate at 130° (16) of n-octadecyl azidoformate at 120° (5,15) have been measured in various solvents. Despite large

TABLE I

Thermolysis of Alkyl Azidoformates in Diphenyl Ether

Azidoformate	Temp., °C	$k \times 10^4$, sec^{-1}	Ref.
n-Octadecyl azidoformate	133.3	9.00	15
	120.0	2.40	15
	100.0	0.267	15
Tetramethylene bis(azidoformate)	133.3	9.00	15
	120.0	2.34	15
	100.0	0.261	15
n-Propyl azidoformate	130.0	6.0	16

TABLE II

Thermolysis of n-Propyl Azidoformate in Various Solvents
at 130° (16)

Solvent	Nitrogen evolved, % of theor.	$k \times 10^4$, sec^{-1}
Anethole	96	9.9
Ethyl undecenylate	97	8.4
Benzontirile	100	6.9
Diphenylacetylene	87	6.0
Diphenyl ether	95	6.0
Phenylacetylene	56	4.6
Mesitylene	91	3.4
Paraffin, mp 65–70°	82	2.4

differences in the chemical nature of the solvents used, the thermolysis rates of either azide changed only by a factor of 4. Thus, the solvent does not seem to be intimately involved in the rate-determining step, and a nitrene intermediate is indicated. The activation parameters for both n-octadecyl azidoformate and tetramethylene bis(azidoformate) are (5): $\Delta H\ddagger = 29.9$ kcal/mole and $\Delta S\ddagger = +4.7$ eu.

The photolysis of ethyl azidoformate (1,4) is best carried out using low-pressure mercury lamps and fused silica or Vycor 7912 vessels, because the light absorption of the azidoformate is quite low at wavelengths above 2800 Å (Table III) (4,17).

Approximate quantum yields for the photolysis (at 2537 Å) of ethyl azidoformate are 0.2 in cyclohexane, and about 1 in methanol (18) (for nitrogen evolution). Photosensitization is possible with acetophenone ($E_T = 73.6$ kcal/mole), but benzophenone ($E_T = 68.5$ kcal/mole) does not

TABLE III
The Ultraviolet Spectrum of EtOOC—N_3

Wavelength, Å	Extinction coefficients in		
	Hexane	Methanol	Cyclohexene
2500	88.1	85.1	—
2600	66.9	63.1	—
2700	40.6	36.9	41.9
2800	17.4	15.3	19.0
2900	6.9	6.0	8.1
3000	2.7	2.1	3.6
3100	1.2		
3200	[0.1]		
3300	[0.04]		
3400	[0.02]		

sensitize the photolysis of ethyl azidoformate. The acetophenone sensitization is not 100% efficient (4), presumably because the lowest triplet state of ethyl azidoformate lies somewhat above that of acetophenone. The photolysis of some other alkanoyl azides also cannot be sensitized by benzophenone (19) which, however, has been found effective with an aroyl azide, benzazide (20). However, with ethyl azidoformate and with benzazide, the sensitized reaction leads only to the amide and dehydrogenation products from the solvent. Ethyl azidoformate gives little or no carbethoxynitrene on acetophenone sensitized decomposition in cyclohexene (4) or in 4-methylpentene-2 (21). Instead, the reaction in cyclohexene gave ethyl carbamate, EtOOC—NH_2 (73.5% yield) and 3,3′-biscyclohexenyl (62.7% yield). Since it is known (21–24) that triplet carbethoxynitrene adds readily to C=C double bonds, the formation of substantial quantities of the triplet nitrene seems to be excluded. The dehydrogenation of the solvent, and the hydrogenation of the starting material to urethan, might be explained by an intermediate triplet azide, or a (triplet) sensitizer–azide complex. With benzazide, Horner obtained a 63% yield of benzamide in the benzophenone sensitized photolysis. However, Nozaki (25) reported the formation of (formal) C—H insertion products when he sensitized the photolysis of ethyl azidoformate in 1,4-dioxane. Perhaps the triplet nitrene is capable of reacting with the α-CH bonds of ethers, by hydrogen abstraction and recombination, to

give R—O—C—NH—COOEt. However, a dehydrogenation process not

involving singlet or triplet nitrenes might possibly lead to radicals
$-\overset{\displaystyle .}{\underset{\displaystyle |}{C_\alpha}}-O-$ and $H\overset{\displaystyle .}{N}-COOEt$, which could then combine to the observed,

formal, insertion product $-O-\underset{\displaystyle |}{C_\alpha}-NHCOOEt$.

B. Carbalkoxynitrenes by α-Elimination

Treating N-(p-nitrobenzenesulfonyloxy)-carbamates with base forms the N-anion, which eliminates p-nitrobenzenesulfonate anion to form singlet carbalkoxynitrenes (eq. 2). In many cases, triethylamine is a convenient base, since it is miscible with most solvents, and its p-nitrobenzenesulfonate salt is insoluble in ether, allowing its easy removal (2,7). The triethylammonium salt of the precursor $Ar-SO_2O-NH-COOEt$ has been isolated but tends to decompose spontaneously (7,26). In solution, it eliminates the sulfonate ion and forms the same products (in the same ratios) as does a mixture of precursor 5 and triethylamine. The anion of 5 (R = Et) decomposes, in dichloromethane solution, with first-order kinetics, and the reaction rate does not depend on the concentration of added cyclohexene. This indicates a nitrene mechanism, rather than a process in which 5 (or its anion) combines with the olefin in some way, and the adduct then eliminates sulfonate ion (27). The position of the triethylamine + precursor vs. triethylammonium ion + 5-anion equilibrium is not accurately known, hence the rate of decomposition of the anion of 5 (R = Et) could only be estimated to be about $10^{-1}\ sec^{-1}$ in dichloromethane at room temperature (27).

Carbomethoxynitrene can be made from a precursor 5 (R = Me) in a manner entirely analogous to that published (7) for R = Et (28).

C. Other Routes to Carbalkoxynitrenes

An attempt to make carbethoxynitrene by oxidation of ethyl carbamate with lead tetraacetate (29) did not lead to the isolation of any products that would indicate the presence of carbethoxynitrene.

D. Singlet and Triplet Carbethoxynitrenes

For nitrenes in general, two (or perhaps three) electronic states of low energy may be expected. All of them have a normal bond to the ligand R in R—N, a set of two spin-paired electrons, and two electrons that can be arranged in two (or perhaps three) different ways: The singlet state commonly discussed in the literature has these last two electrons paired in one orbital, and has an empty bonding orbital. One might, however, think of another singlet state, having the two electrons spin-paired, but in two different orbitals, and having no empty bonding orbital of low energy. This would be something like a singlet 1,1-diradical. Should this be the actual configuration of the commonly observed singlet carbonylnitrenes, then their apparent inability to undergo the Curtius rearrangement (see Section V) might be explained. The triplet carbethoxynitrene would also not have an empty orbital, the two electrons discussed here would be in separate orbitals, with their spins parallel. Organic chemists have long speculated that such electronic configurations should behave as di-radicals. On the basis of Hund's rule, the triplet state would be expected as the ground state. Substituent interactions, however, might well upset the situation and lead to a singlet ground state in some special nitrenes.

Wasserman (8) has recently shown carbethoxynitrene to have a triplet ground state (see Chapter 13). However, the nitrene can be generated in the singlet state, which undergoes many types of selective, intermolecular, reactions before decaying to the triplet (21–24). To fully describe a nitrene reaction, one must then determine the electronic state of the nitrene involved.

In the addition of carbethoxynitrene to olefins, two separate species are involved (21–23,30). One of them adds stereospecifically, the other adds non-stereospecifically and can be trapped by α-methylstyrene. The first species is considered the singlet, the latter the triplet, in part because the system N—COOEt–olefin behaves according to a model introduced by Skell (31) for carbenes. This model correlates the data in the carbene field rather well (32), even if some theoretical details of the interpretation might still have to be revised (33,34). Written for nitrenes, Skell's scheme is shown in Figure 1. The addition of the singlet species to the C=C double bond is supposed to occur in a single step and stereospecifically (cf. reference 34). The addition of the triplet species, on the other hand, is to occur in two discrete steps, via a triplet 1,3-diradical intermediate. The rate of ring closure of the latter is supposed to be considerably smaller than that for rotation about the C—C bond between the two former olefin carbons. Consequently, stereospecificity is lost, and a mixture of cis and trans disubstituted three-membered rings formed from either cis or trans disubstituted olefin. Scheiner (35) has generated N—C—C— 1,3-diradicals independently, by photosensitized decomposition of triazolines, and has confirmed that they will cyclize nonstereospecifically.

In cases in which much or all of the triplet nitrene is formed by inter-system crossing from the singlet, and where the singlet undergoes inter-molecular reactions, one can control the amount of triplet formed by adjusting the concentration of the reactant that consumes the singlet. The addition of singlet and triplet carbethoxynitrene to cis- and trans-4-methylpentene-2 was studied in this manner (21–23), over a range of olefin

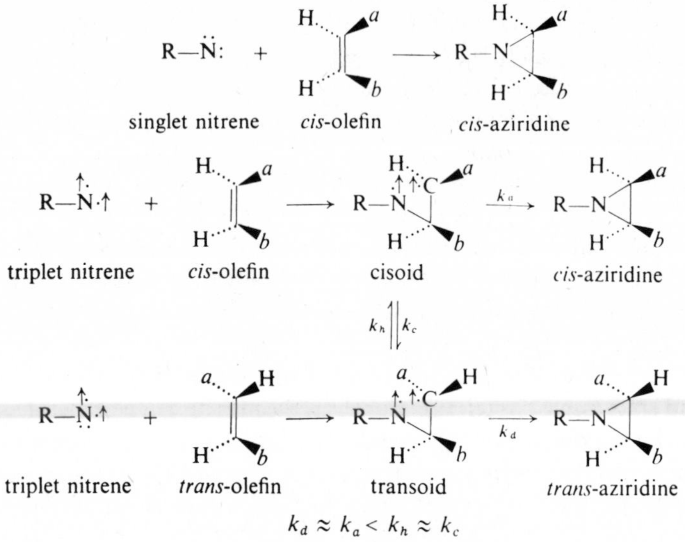

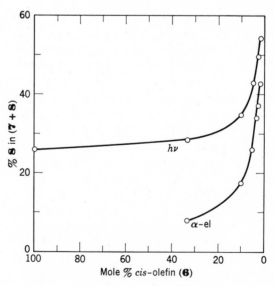

Fig. 1. Skell's scheme for the addition of singlet and triplet species to olefins, adapted for the addition of a nitrene.

Fig. 2. Fraction of *trans*-aziridine (8) from *cis*-olefin (6) as a function of olefin concentration.

concentrations from 1 to 100 mole % in dichloromethane solution. The ratio of N-carbethoxy-2-methyl-3-isopropylaziridines formed stereospecifically (by the singlet) and nonstereospecifically (by the triplet) was measured. Equation 3 gives the reaction, Figure 2 the experimental results for generating the carbethoxynitrene by photolysis and by α-elimination (see above), and Figure 3 the kinetic scheme used to reduce the data. The

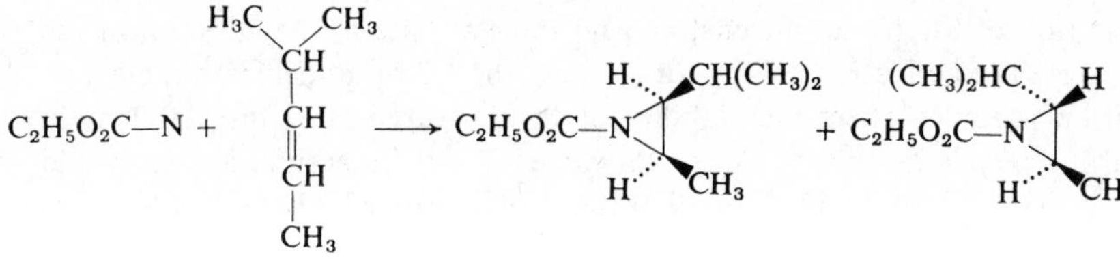

$$
C_2H_5O_2C-N \; + \;
\begin{matrix} H_3C \quad CH_3 \\ CH \\ | \\ CH \\ \| \\ CH \\ | \\ CH_3 \end{matrix}
\;\longrightarrow\;
C_2H_5O_2C-N\underset{H}{\overset{H}{\diagdown}}\!\!\begin{matrix} CH(CH_3)_2 \\ \\ CH_3 \end{matrix}
\; + \;
C_2H_5O_2C-N\underset{H}{\overset{(CH_3)_2HC}{\diagdown}}\!\!\begin{matrix} H \\ \\ CH_3 \end{matrix}
\qquad (3)
$$

$$\qquad\qquad\qquad (6) \qquad\qquad\qquad\qquad (7) \qquad\qquad\qquad\qquad\qquad (8)$$

scheme of Figure 3 includes a rate constant (k_5) for all the side reactions of the triplet nitrene, because such side reactions reduce the amount of aziridines formed nonstereospecifically. Side reactions of the singlet do not enter into the scheme as long as all the triplet is formed from the singlet. The triplet nitrene gives an aziridine mixture, the composition of which is determined by the equilibrium between the cisoid and the transoid 1,3-diradical intermediates (Fig. 1) and the rate constants k_a and k_d. Thus, the composition of this mixture is constant and can be found by analyzing the results of runs at several concentrations of *cis* and *trans* olefins (23), as long as the triplet addition is completely nonstereospecific. This was found to be true for the addition of ^{3}N—COOEt to 4-methylpentene-2, and the fraction of stereospecific reaction path (called S) and the nonstereospecific reaction path (called T) could be determined for each olefin concentration used. The scheme in Figure 3 corresponds to equation 4, derived using a steady-state approximation for the concentration of triplet nitrene.

$$
T/S = \frac{k_2}{k_3 \times [\text{olefin}] + k_5/k_4} \qquad (4)
$$

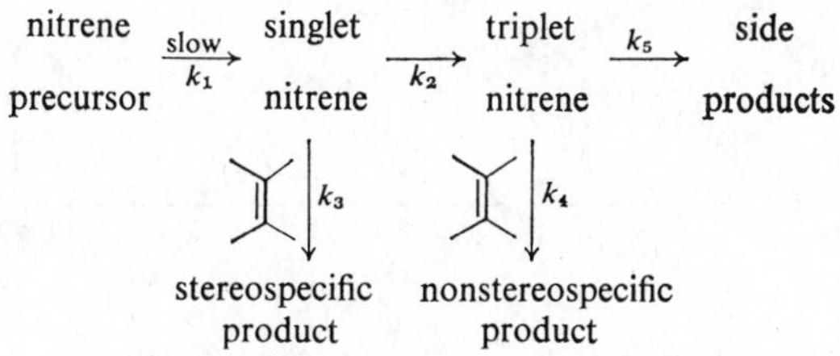

nitrene precursor $\xrightarrow[k_1]{\text{slow}}$ singlet nitrene $\xrightarrow[k_2]{}$ triplet nitrene $\xrightarrow{k_5}$ side products

$$\Big\downarrow k_3 \qquad\qquad \Big\downarrow k_4$$

stereospecific product   nonstereospecific product

Fig. 3. Kinetic scheme for the addition of carbethoxynitrene to olefins.

Thus, a plot of T/S vs. $1/\{[\text{olefin}] + k_5/k_4\}$ should give a straight line going through the origin. This is indeed found when carbethoxynitrene is generated by α-elimination (23) (Fig. 4) or by thermolysis of ethyl azidoformate (21). When the nitrene is made by photolysis, the line is straight, but has an intercept at $T/S = 0.44$. Analysis of the data (21) shows that about 30% of the photolytically formed nitrene is in the triplet state.

The scheme of Figure 1 shows a 1,3-diradical intermediate for the addition of the triplet nitrene, and no intermediate in the addition of the singlet. The activation energy for forming the 1,3-diradical from the triplet nitrene should reflect the stability of the 1,3-diradical. This stability, in turn, depends on the type of olefin chosen for the reaction, and it should be possible to divert much or all of the triplet nitrene by having two olefins present, one of which gives a highly stabilized 1,3-diradical. This was done, using α-methylstyrene added as a triplet trap to the carbethoxynitrene–cis-4-methylpentene-2 system (23). Making carbethoxynitrene (by α-elimination) in a dichloromethane solution containing 3.3 mole % cis-4-methylpentene-2 gives a 25% yield of cis-aziridine (7), plus a 13% yield of trans-aziridine (8). Adding α-methylstyrene (in 3.3 mole % concentra-

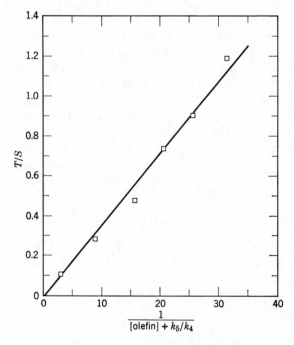

Fig. 4. Concentration dependence of the stereospecificity of the reaction of carbethoxynitrene with cis-4-methylpentene-2, plotted according to equation 4.

tion) to the same solution changes the yield of **7** to 16% and that of **8** to a trace. Evaluation of such data, using lower α-methylstyrene concentrations, shows the triplet nitrene to react 86 times faster with the α-methylstyrene than with cis-4-methylpentene-2 (23). These results agree well with Skell's contention that triplet electron-deficient intermediates, such as triplet carbenes or our triplet nitrene, are much like radicals in their chemistry. The product formed from the α-methylstyrene supports his assumption of a triplet diradical intermediate: The 1:1 adduct is not an aziridine but 3-carbethoxyamino-2-phenyl-propene-1 (**9**). It might well be formed by a 1,4-hydrogen atom transfer in an initial 1,3-diradical intermediate (eq. 5).

$$C_6H_5C{=}CH_2 + NCOOEt \longrightarrow C_6H_5C\overset{CH_2\diagdown H}{\underset{CH_2NCOOEt}{\diagup}} \longrightarrow C_6H_5C\overset{CH_2}{\underset{CH_2NHCOOEt}{\diagup}} \qquad (5)$$
$$\underset{CH_3}{|} \qquad\qquad\qquad\qquad\qquad\qquad\qquad\qquad\qquad (\mathbf{9})$$

Thermolysis of ethyl azidoformate in 4-methylpentene-2-dichloromethane mixtures gave results perfectly analogous to those obtained using the α-elimination route to the nitrene (21). All the nitrene is initially formed as the singlet.

Beckwith and Redmond have studied the temperature dependence of the singlet–triplet crossing of carbethoxynitrene, using its reactions with anthracene and with the 2-butenes (36). Relative to the rate of addition of singlet carbethoxynitrene to cis-2-butene, the intersystem crossing rate increases with temperature.

The change of singlet to triplet carbethoxynitrene, competing with intermolecular singlet reactions, also explains the concentration dependences of product ratios in the reaction with cyclohexene (24), cyclic ethers (25), anthracene (36), and other condensed aromatic hydrocarbons (37). Concentration dependence of stereospecificity has been analogously explained in reactions of cyanonitrene (38) (see Chapter 9), and carbenes (39,40). Quantitative correlations have, as yet, been reported only for the carbethoxynitrene addition to the 4-methylpentenes. In this case, the nature of the solvent (dichloromethane, neopentane) does not seem to materially affect the rate of singlet–triplet crossing (21).

E. The Ultimate Fate of Carbalkoxynitrenes

In the absence of a suitable reactant, singlet carbethoxynitrene crosses to the triplet state, which may add to double bonds (see Section IV-E) or abstract hydrogen atoms from suitable substrates (see Section IV-D). If neither reaction is possible, triplet carbethoxynitrene seems to dissociate to ethoxy and cyanato radicals, at least in the gas phase (10)(see Chapter 2).

III. ALKANOYL- AND AROYLNITRENES

A. Formation

Carbonylnitrenes of the type $\diagdown C$—CO—N were first postulated by Tiemann in 1891 (41) as intermediates in the Lossen rearrangement. It has now become unlikely that they are actually involved in this rearrangement (see Section V), and reactions of R—CO—N (R = alkyl or aryl) have been observed only recently. Cyclizations of carbonylnitrenes by intramolecular insertion into C—H bonds, first reported by Edwards (42), are discussed in Chapter 7. Horner (43) found intermolecular nitrene products in the photolysis (but not the thermolysis) of benzazide. In all cases investigated so far, only the photolysis, and not the thermolysis, of azides of the type $\diagdown C$—CO—N_3 gives the corresponding nitrenes. Thermolysis leads exclusively to Curtius rearrangement, giving the isocyanates R—N=C=O (see Section V). Attempts to generate nitrenes, $\diagdown C$—CO—N, by α-elimination have, so far, only led to Lossen rearrangement. Horner (20,43) trapped benzoylnitrene with water, acetic acid, and aniline, which gave the corresponding O—H or N—H insertion products, and with dimethyl sulfoxide, which gave the sulfoximine C_6H_5—CO—N=S(O)(CH_3)_2$. Photosensitization led only to the formation of benzamide. Huisgen (44) trapped acetylnitrene with benzonitrile and with phenylacetylene. The photolysis of pivaloyl azide gives pivaloylnitrene, trapped by olefins and saturated hydrocarbons in yields approaching 50%, based on pivaloyl azide. The rest of the azide is converted to the isocyanate, tBu—NCO, in about 40% yield (45–47). Intermolecular insertion products were also found in the photolysis of 1,2-diphenyl-3-azidocarbonyl-cyclopropene (48). Most interestingly, Just (49) found C—H insertion and cyclization products in the photolysis of nitrile oxides — isomers of the corresponding carbonylnitrenes. He discusses a mechanism involving isomerization to the nitrene, followed by C—H insertion (eq. 6).

The thermolysis of trimethylammonium dodecanoylimide, $Me_3\overset{+}{N}$—N—CO—$C_{11}H_{23}$, gave only the isocyanate, $C_{11}H_{23}$—NCO, but photolysis in dimethyl sulfoxide gave the amide, $C_{11}H_{23}CO$—NH_2 (50). The authors ascribe this to the intermediate formation of a triplet nitrene. The assignment of the triplet state is because they find that photolysis of the corresponding azide, $C_{11}H_{23}$—CO—N_3 in dimethyl sulfoxide gives (besides the isocyanate) sulfoximine, $C_{11}H_{23}$—CO—N=S(O)Me_2$. Because of the analogy with Horner's (20,43) work, this latter reaction is ascribed to the singlet nitrene. Formation of the amide in the photolysis of the amine-

(6)

imide might, however, be due to an intermediate other than triplet nitrene.

The formation of nitrenes by fragmentation of heterocyclic compounds is more systematically treated in Chapter 4. Sauer (51) has investigated such fragmentations of 3-substituted 1,4,2-dioxazolidin-5-ones and their thio analogs. Some of them seem to dissociate, on thermolysis or photolysis, into a carbonylnitrene fragment and CO_2 or COS. For example, the photolysis of 3-phenyl-1,4,2-dioxazolidin-5-one in dimethyl sulfide gave the corresponding sulfimine in 34% yield, and photolysis in cyclohexane gave N-cyclohexyl-benzamide in 8% yield. It is possible to write mechanisms that avoid postulating an aroylnitrene intermediate for these reactions, but the analogy to the known (20,43) behavior of benzoylnitrene is striking.

$R = C_6H_5$; pMeO—C_6H_4; pO$_2$N—C_6H_4; oMeO—C_6H_4; 2,4,6-Me$_3$C$_6$H$_4$; C_6H_5—CH=CH; mMeO—C_6H_4

$R = C_6H_5$; pMeO—C_6H_4; pO$_2$N—C_6H_4

Carbonyl azide, N_3—CO—N_3, when thermolyzed in benzene, gives N-azidocarbonyl-azepine and products of its further conversion (52), indicating a nitrene intermediate N_3—CO—N (eq. 7). Prevented by its

$$N_3\text{—}CO\text{—}N_3 \ + \ \langle \text{benzene} \rangle \ \xrightarrow{\Delta} \ \langle \text{azepine} \rangle \text{N}\text{—}CON_3$$

$$\langle \text{azepine} \rangle \text{N—CONH}_2 \qquad \langle \text{phenyl} \rangle\text{—NHCON}_3 \qquad \langle \text{azepine} \rangle \text{N—CO OEt}$$

with NH₃, H⁺, EtO⁻/EtOH branches   (7)

structure from undergoing Curtius rearrangement, this carbonyl azide gives a nitrene on thermolysis, as do the alkoxycarbonyl azides.

Diethylcarbamoyl azide, upon thermolysis or photolysis, does not give the corresponding nitrene, $Et_2N\text{—}CO\text{—}N$, but rearranges to the iso-cyanate $Et_2N\text{—}NCO$ **(10)**, which reacts with nucleophilic solvents and dimerizes in inert solvents (53) (eq. 8). Also, a carbamoylnitrene was not

$$Et_2N\text{—}CO\text{—}N_3 \ \longrightarrow \ Et_2N\text{—}N\text{=}C\text{=}O$$
$$\textbf{(10)}$$

with ROH and inert solvent branches:

$$Et_2N\text{—}NH\text{—}CO\text{—}OR \qquad \overset{+}{Et_2N}\text{——}\overset{-}{N}$$

$$O\text{=}C\underset{N}{\diagdown}\diagup C\text{=}O$$
$$\underset{NEt_2}{|}$$
     (8)

found in the thermolysis of 3-phenyl-2-anilinocarbonyl-oxaziridine (54).

Treating carboxylic acid amides with lead tetraacetate leads to re-arrangement to the corresponding isocyanates. Intra- or intermolecular nitrene products have not been isolated from such reactions, and it seems that a Lossen rearrangement (not involving a nitrene C—CO—N) takes place (29,55–57).

B. Side Reactions

In the photolysis of alkanoyl and aroyl azides, Curtius rearrangement always accompanies nitrene formation. The isocyanate seems to be formed not from the nitrene, but from the excited azide directly (see Section V). Substantial yields of isocyanate are usually observed (see also Chapter 7), decreasing the maximum yield of nitrene products that can be obtained. Pivaloyl azide, for example, gives a 40% yield of isocyanate (when de-composed at 0° with light of 2537 Å wavelength), regardless of solvent (47).

Other side reactions include the (nonphotolytic) addition of azides to unsaturated functions, and the displacement of azide ion from carbonyl azides by nucleophiles.

C. Dissociation of the Nitrenes

Little is known about the ultimate fate of nitrenes, $\diagup$C—CO—N, which are not consumed by intra- or intermolecular reactions. Pivaloylnitrene, when generated in unreactive solvents (dichloromethane, neopentane), decomposes to give isobutene and a polymeric gum. Based on the nitrene presumably formed (half of the azide gives products other than isocyanate, and yield of about 50% of nitrene trapping products were found in other experiments), a 20% yield of isobutene can be obtained. The gum seems to be (on the basis of its IR and NMR spectra), a polymer of HNCO and isobutene (47), formed perhaps as shown in equation 9.

$$(H_3C)_3C—CO—N^1 \xrightarrow[\text{crossing}]{\text{intersystem}} (H_3C)_3C—CO—N^3$$

$$(H_3C)_3C—CO—N^3 \longrightarrow (H_3C)_3C^{\cdot} + {}^{\cdot}NCO \longrightarrow \text{polymer} \qquad (9)$$

$$\downarrow$$

$$(H_3C)_2C{=}CH_2$$

IV. REACTIONS OF CARBONYLNITRENES

A. Reactions with C—H Bonds to give —CO—NH—C (Insertion)

Reactions of carbonylnitrenes with substances containing C—H bonds commonly lead to products of the type —CO—NH—C$\diagup$. This process is often referred to as "insertion," irrespective of the reaction mechanism. Perhaps it would be better to restrict the term "insertion" to processes leading from R—N to R—NH—C$\diagup$ in one single step. A variety of mechanisms can be written for the formation of R—NH—C$\diagup$, including one-step processes with various geometries of the transition state, as in equations 10–14.

$$R—N + H—C \longrightarrow R—\overset{\cdot}{N}H + {}^{\cdot}C\diagdown \longrightarrow R—NH—C \qquad (10)$$

$$R—N + H—C \longrightarrow R—\overset{+}{N}H + {}^{-}C\diagdown \longrightarrow R—NH—C \qquad (11)$$

$$R—N + H—C \longrightarrow R—\overset{-}{N}H + {}^{+}C\diagdown \longrightarrow R—NH—C \qquad (12)$$

Special mechanisms could operate in aromatic systems (37) (eq. 15), and with olefins (eq. 16).

$$R\!-\!N + H\!-\!C \longrightarrow R\!-\!N\text{-}\text{-}H\text{-}\text{-}\text{-}C\!\!-\!\! \longrightarrow R\!-\!NH\!-\!C \qquad (13)$$

$$R\!-\!N + H\!-\!C \longrightarrow \quad H\text{-}\text{-}\text{-}\text{-}\text{-}\text{-}C\!\!-\!\! \longrightarrow R\!-\!NH\!-\!C \qquad (14)$$

$$(15)$$

$$(16)$$

1. Unactivated C—H Bonds

In the absence of functional groups and unsaturation, the choice of mechanisms for the formation of R—NH—C$\diagup$ is from equations 10–14, above. In open-chain hydrocarbons, the order of reactivity for all nitrenes investigated is tertiary > secondary > primary C—H bonds. This order is compatible with the intermediacy of a radical pair, or a carbonium ion–nitrogen anion pair, or a concerted mechanism, but it excludes the mechanism in equation 11. The experimental data for carbethoxynitrene are given in Table IV, and are compared with those of pivaloylnitrene (46), phenylnitrene (58), the reactivity of C—H bonds toward hydrogen abstraction by *tert*-butyloxy radical (59), and the solvolysis rates (in formic acid) of some bromides (60) in Table V.

The data in Tables IV and V correlate better with those of hydrogen atom abstraction by tBuO$\cdot$ than with solvolytic reactivity. One might thus tentatively conclude that the transition state leading to the insertion

TABLE IV
Selectivities of Carbethoxynitrene in the Insertion into C—H Bonds

N—COOEt generated by	Reactivity per C—H bond								
	In 2-methylbutane[a]			In 44.5% 2-methylbutane in CH_2Cl_2 soln.[b]			In 3-methylhexane[c]		
	3°	2°	1°	3°	2°	1°	3°	2°	1°
Photolysis of EtOOC—N₃	34	9	1	36	10	1	16.0	5.3	1
α-Elimination from 5	27	11	1	25	8	1	13.5	4.7	1[d]
Thermolysis of EtOOC—N₃	32	10	1	—	—	—	17.0	6.0	1

[a] Photolysis and elimination values from reference 7; thermolysis values from reference 5.
[b] From reference 7.
[c] From references 61 and 62.
[d] 26.8 mole % of 3-methylhexane in CH_2Cl_2.

TABLE V
Comparison of the Selectivity of Carbethoxynitrene with That of Various Other Systems

System	Type of carbon atom involved		
	Tertiary	Secondary	Primary
EtOOC—N from photolysis of EtOOC—N₃ (7)[a]	34	10	1
tBu—CO—N from photolysis of tBuCON₃ (46)[a]	160 ± 40	9 ± 1	1
Phenylnitrene from thermolysis of $C_6H_5N_3$ (58)[a]	140–280	>7	1
Hydrogen atom abstraction by tBuO· from tBuOCl (59)[b]	44	12	1
Solvolysis in formic acid of some bromides (60)	tBuBr, 10^8	iPrBr, 26	EtBr, 1

[a] 2-Methylbutane.
[b] Averaged from data for various hydrocarbons.

product (or the intermediate preceding it) has little or no positive charge on the carbon atom. This tentative conclusion is supported by the observation that insertion into bridgehead C—H bonds takes place readily. Carbethoxynitrene will insert into the bridgehead C—H bonds of

TABLE VI

Thermal Decomposition of Ethyl Azidoformate in Various Hydrocarbons

Hydrocarbon	Position	Type of H	Relative reactivity per hydrogen	
Norbornane	1	3°	0.32	0.31
	2-*exo*	2°	1.00	
	2-*endo*	2°	0.34	0.34
	7	2°	0.20	0.23
Bicyclo[2.2.2]octane	1	3°	1.94	1.74
	2	2°	1.00	
Adamantane	1	3°	6.65	6.74
	2	2°	1.00	
Tricyclo[3.3.0^{2,6}]octane[a]	1	3°	1	
	2	2°	1	
2-Methylbutane	2	3°	3.2	
	1	2°	1	

[a] From reference 65.

norbornane (63), and the relative reactivities (64) are given in Table VI. Competition experiments with cyclohexane show the secondary C—H bonds in adamantane, bicyclo[2.2.2]octane, and norbornane (*endo* and *exo*) to be about as reactive as those in cyclohexane, with the exception of the 7-norbornane CH$_2$ group. The latter is only 19% as reactive as the cyclohexane CH$_2$ (64). Meinwald (65) reports the same reactivities for the bridgehead C—H bonds of tricyclo[3.3.0^{2,6}]octane and for its secondary C—H bonds in the ethylene bridges. No general explanation for the bridgehead reactivities, as compared with each other, has emerged so far. Partial charge, ease of homolytic bond breaking (bond dissociation energy), adjustment of angle strain, might all be involved. The best guide to the reactivity of C—H bonds seems to be their ease of homolysis, despite the fact (see below) that the formation of C—H insertion products is not a radical process, but a one-step reaction of the singlet nitrene.

The deuterium isotope effect for the insertion of carbethoxynitrene into the C—H bonds of cyclohexane and cyclohexane-d_{12} was found to be $k_H/k_D = 1.5 \pm 0.2$, indicating a rather unsymmetrical or nonlinear transition state (7). Insertion into C—H bonds at optically active carbons proceeds with retention of configuration. Smolinsky obtained optically active 4-ethyl-4-methyl-oxazolidin-2-one from the vapor-phase pyrolysis of S(+)-2-methylbutyl azidoformate (66). Yamada and collaborators correlated the absolute configurations of starting material and product in Smolinsky's reaction, and repeated it in solution (67). They found nearly

100% retention of configuration in this intramolecular insertion. The intermolecular insertion of carbethoxynitrene into the tertiary C—H bond of optically active 3-methylhexane produces the corresponding urethan with 100% retention of configuration (61,62,68,69). This result is independent of the concentration of the (+)3-methylhexane, and of the method used to generate the nitrene, as shown in Table VII. The relative reactivi-

TABLE VII

Stereospecificity of the Insertion of Carbethoxynitrene into the Tertiary C—H Bond of $S(+)$-3-Methylhexane (61)

Nitrene made by	Concentration of (+)3-methylhexane in dichloromethane, mole %	Retention of configuration, %
Azide photolysis	100	97 ± 5
Azide photolysis	27	100 ± 6
Azide photolysis	1.2	102 ± 3
Azide thermolysis	100	99 ± 5
α-Elimination	1.2	98 ± 7

ties of the various C—H bonds of 3-methylhexane are also independent of the concentration of 3-methylhexane. This is different from the concentration dependence of the stereospecificity of the olefin addition. In the latter, both singlet and triplet nitrene add to the C=C double bond, but by different mechanisms (see above, Section II-D). If both the singlet and the triplet would insert into C—H bonds, one would expect a concentration dependence of selectivity (as the transition states would necessarily be different), and one might expect the triplet to insert with loss of stereospecificity. Neither is found, so that one must conclude that only the singlet carbethoxynitrene inserts into unactivated C—H bonds to an extent detectable by the experiments that have been carried out so far. This conclusion is supported by experiments with cyclohexene, at concentrations from 0.2 to 100 mole %. Both singlet and triplet nitrene add to the double bond, but only the singlet inserts. Consequently, the fraction of insertion product in total product decreases with decreasing concentration of the substrate, the more of the nitrene that finds time to go to the triplet state, the less insertion product is obtained. At 100% olefin concentration, 17% of the products are insertion products, at 0.2 mole % cyclohexene (in CH_2Cl_2) the figure is only 1.7% (24).

The angular transition state of equation 14 explains best the observations made on the mechanism of the C—H insertion. Especially the sterospecificity of the insertion is much more easily understood on the basis of the angular arrangement than by an analogon of the linear transition state proposed by DeMore and Benson (70) for the carbene insertion (R—N $\cdots$ H $\cdots$ C—). If one wants to explain the stereospecifiity while adopting a radical-pair mechanism (eq. 10), one has to assume that either the radical pair collapses without any of the R—$\dot{\text{N}}$H moiety getting to the back face of the $R^1R^2R^3C^{\cdot}$, or that radicals which escape the cage will never recombine.

Care should be taken not to generalize the findings made with carbonylnitrenes: A nonstereospecific reaction of N—CN with C—H bonds, leading to C—NH—CN, has been observed (38) (see Chapter 9).

The assumption of an angular transition state also agrees with the results of intramolecular C—H insertions of alkanoyl azides. In these cyclizations, δ-lactams are formed in preference to γ-lactams (71–73), and six-membered rings are also the preferred products in the cyclization of long-chain alkyl azidoformates (5). Models show a less strained transition state for the formation of six-membered rings, when the triangular arrangement of the atoms C, H, and N is used (see Chapter 7).

Intermolecular insertions of nitrenes C—CO—N seem to occur with as much facility as those of RO—CO—N, but cyclization competes strongly

TABLE VIII

Competition of Intramolecular and Intermolecular C—H Insertion of Various Carbonylnitrenes

Nitrene	Ref.	Intramol. C—H insertion (in CH_2Cl_2), % yield	In cyclohexane solution	
			Intramol.	Intermol.
nC_3H_7—CO—N	73	3.5	1	8.1
nC_4H_9—CO—N	73	23.5	7.8	4.7
nC_5H_{11}—CO—N	73	33.5	15.8	1.9
	72	—	21	3
nC_7H_{15}—CO—N	73	32.0	16	1.4
iPr—CH_2—CO—N	73	5.3	1.9	8.0
tBu—CH_2—CO—N	73	20.1	5.4	8.2
Et—CMe_2—CO—N	73	35.6	11.2	7.4
tBu—CMe_2—CO—N	73	56.2	58.3	0
$C_{18}H_{37}O$—CO—N	14	—	10.8	62.6
C_2H_5O—CO—N	74	45 (gas phase)	—	—
	4	—	—	51
$tBuO$—CO—N	81	60 (in $tBuOH$)	—	—
	81	80 (in MeCN)	—	—

as soon as the chain attached to the carbonyl group has three or more members (72–74), and cyclization can be strongly promoted by a geminal dimethyl effect (73), as shown in Table VIII.

Typical yields of C—H insertion products of various carbonylnitrenes (RO—CO—N and C—CO—N) are summarized in Table IX.

Since only the singlet carbethoxynitrene inserts into C—H bonds with any efficiency, the yield of insertion products depends on the reactivities of the C—H bonds of the substrate. The lower their reactivity, the more of the nitrene crosses over to the triplet state and is lost for the insertion reactions. The ultimate fate of the triplet nitrene is not known in detail. In the presence of suitable hydrogen donors, hydrogen abstraction may occur. In the gas phase, carbethoxynitrene dissociates to give EtO˙ and NCO˙ (11), and pivaloylnitrene decomposes in an analogous fashion in solution (46).

A most interesting effect on the yield of C—H insertion products and hydrogen abstraction products (the latter presumably arising from the

TABLE IX

Some Yields of Intermolecular C—H Insertion Products of Carbonylnitrenes

Nitrene source	Mode of de-compo-sition	Substrate	% Yield (based on precursor)	Ref.
EtO—CO—N$_3$	Light	Cyclohexane	51	4
			78	16
	Heat	Cyclohexane	52	15
	Heat	Cyclohexane + additive	74	15
	Light	3-Methylhexane	28	62
	Heat	3-Methylhexane	25	62
EtOOC—NH—OSO$_2$Ar	Base	3-Methylhexane	16	62
EtO—CO—N$_3$	Light	Cyclohexene	18	4
MeO—CO—N$_3$	Light	Tricyclo[3.3.0.0^{2,6}]-octane	51	65
nC$_{18}$H$_{37}$O—CO—N$_3$	Heat	Cyclohexane	60	5
3β-Acetoxy-androst-5-en-17β-yl-azido-formate	Light	Cyclohexane	48	75
tBu—CO—N$_3$	Light	Cyclopentane	13	46
		Cyclohexane	20	46
		2-Methylbutane	25	46
		Cyclohexene	1.5	46

triplet nitrene) has been uncovered by D. S. Breslow (5,15). Substances like nitrobenzene, dinitrobenzene, sulfur, hydroquinone, and others (all potential radical inhibitors) increase the absolute yield of insertion product, while decreasing the absolute yield of hydrogen abstraction product, $RO-CO-NH_2$. Thus, when ethyl azidoformate was thermolyzed in cyclohexane, the addition of 0.27 g/100 ml of m-dinitrobenzene raised the yield of cyclohexylurethan from 52 to 74%, while decreasing the yield of urethan from 25 to 17%. The reaction rate, as measured by the nitrogen evolution, remained unchanged. One possible explanation is (15) that radicals such as $RO-CO-\overset{.}{N}H$ catalyze the singlet–triplet change of the nitrene. Intercepting the radicals thus would increase the amount of singlet nitrene available for insertion. Using tetrachloroethylene as a triplet trap, Breslow was able to reduce the yield of urethan, $RO-CO-NH_2$, to 0.9%.

2. Insertion into Activated C—H Bonds

The single-step stereospecific insertion mechanism that seems to operate in insertions of carbonylnitrenes into unactivated C—H bonds is not generally applicable to all types of C—H bonds or all types of nitrenes. Cyanonitrene triplet reacts with tertiary C—H bonds to give the apparent insertion product nonstereospecifically (38). Nozaki's results (25) with cyclic ethers and carbethoxynitrene suggest a hydrogen abstraction–recombination process at C_α in ethers to give formal insertion products—the process of equation 10. Masamune's 2,3-diphenylcycloprop-2-ene-1-carbonylnitrene inserts into the methylene group of diethyl ether (48), in accord with the exclusive reaction at C_α—H Nozaki observed for carbethoxynitrene with cyclic ethers.

Carbethoxynitrene (76) and n-octadecyloxycarbonylnitrene (13) insert into the CH_3 groups of dimethylaniline, perhaps via an aminimide intermediate, $Ar-\overset{+}{N}(CH_3)_2-\overset{-}{N}-COOR$ (76).

The allylic C—H bonds in cyclohexene react faster with carbethoxynitrene than the nonallylic ones by a factor of 3 to 4 (4,7). This ratio varies very little with the reaction conditions, so that rearrangement of the aziridine is probably not an important source for the allylic "insertion" product in this case. However, thermolysis at 230° of the aziridine gave some of the allylic urethan (4), and the possibility of such a rearrangement as a source for allylic insertion products should not be generally discounted (cf. 46). The insertion of carbethoxynitrene into the allylic methyl group of an enol-acetate has been reported (77).

The reaction of carbethoxynitrene with benzene gives N-carbethoxyazepine and some N-carbethoxyaniline (78,2,76,12). It is not clear whether all of the latter is formed by rearrangement of the azepine (2), or of a valence tautomer of the azepine, or whether some of the acylaniline is formed directly by an independent process. Studies of the relative abundance of products from carbethoxynitrene with anthracene, phenanthrene, and pyrene (37), and of the concentration dependence of the products from anthracene (36) indicate that at least some direct formation of *N*-aryl urethans takes place—certainly from triplet carbethoxynitrene, and possibly also from the singlet. For the reactions of singlet and triplet carbethoxynitrene with anthracene, Beckwith suggests mechanism 17 (36).

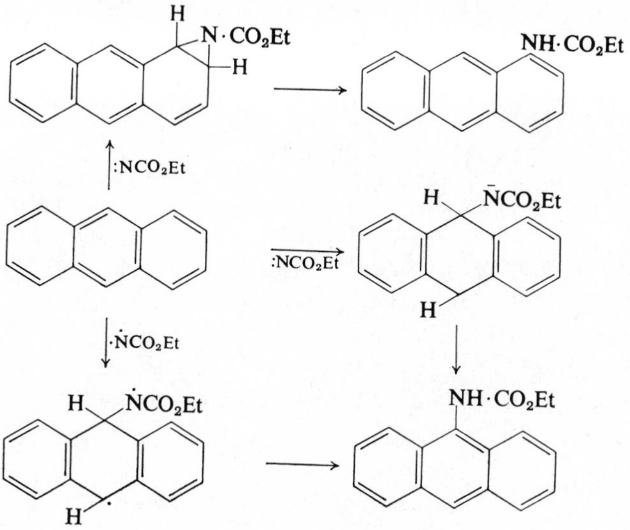

$$
(17)
$$

B. Insertion into O—H Bonds

Decomposition of azidoformates in alcohols gives aldehydes or ketones plus urethans (13,79), by hydrogen atom transfer, *N*-alkoxy urethans by O—H insertion (80,81), and the addition products of the alcohols to alkoxy-isocyanates, formed by Curtius rearrangement of the azides (80) (eq. 18). Formation of both O—H insertion product and addition product

$$
\begin{array}{lll}
\text{EtOOC—N}_3 & \longrightarrow \text{EtOOC—N} & + & \text{EtO—N=C=O} \\
\text{In}\cdot\downarrow & \text{R—CH}_2\text{OH}\downarrow & & \downarrow\text{R—CH}_2\text{OH} \\
\text{EtOOC—NH}_2 & \text{EtOOC—NH—O—CH}_2\text{R} & \text{EtO—NH—COO—CH}_2\text{R} & (18) \\
+ & + & \\
\text{R—CHO} & \text{EtOOC—NH}_2 & \\
& + & \\
& \text{R—CHO} &
\end{array}
$$

to EtO—NCO are also observed when the nitrene is made by α-elimination (27). The ratio of insertion product to rearrangement product, however, is very different. Thus, it seems that the alkoxy-isocyanates are formed from the precursors (by Curtius and Lossen rearrangements, respectively), rather than by rearrangement of RO—CO—N.

An azide reaction, leading to amide and ketone or aldehyde, and proceeding by a radical-chain mechanism, has also been observed (82) and might be partially responsible for some of the dehydrogenation reactions attributed to carbonylnitrenes (eq. 19).

$$R—CO—\dot{N}H + Me_2CHOH \longrightarrow Me_2\dot{C}—OH \xrightarrow[-N_2]{R-CON_3}$$

$$Me_2CO + R—CO—\dot{N}H \quad (19)$$

This chain reaction is analogous to that reported earlier for sulfonyl azides (83).

Perhaps because of the competing radical-chain decomposition of the azides, the yields of O—H insertion products depend much on reaction conditions. Typical yields of O—H insertion products are: EtOOC—NH—OCH$_3$ (44%) (80), EtOOC—NH—O—tBu (60%) (81), MeOOC—NH—OtBu (55%) (79), tBuOOC—NH—OtBu (29%) (81).

The photolysis of benzazide in the presence of water gave benzhydroxamic acid in 9% yield, in acetic acid O-acetyl-N-benzoyl hydroxylamine was formed in 30% yield (20), together with the addition products of water or acetic acid to phenyl isocyanate.

C. Insertion into N—H Bonds

Thermolysis of ethyl azidoformate in aniline gave a 52% yield of ethyl phenylhydrazoformate, C$_6$H$_5$—NH—NH—COOEt. N-Methylaniline gave a 49% yield of the N—H insertion product, plus some of the product of displacement of azide ion by the aniline (76). Photolysis of benzazide in aniline gave N-phenyl-N'-benzoylhydrazine in 14% yield, together with N,N'-diphenylurea (20). More nucleophilic amines displace azide ion rapidly from azidoformates. N-p-nitrobenzenesulfonoxy urethans (formula 5), however, are suitable starting materials for nitrene–amine reactions (28,84). Using the α-elimination route to the nitrene, tBu—NH—NH—COOEt was obtained in 51% yield, and nBu—NH—NH—COOEt in 49% yield. The synthetic utility of the N—H insertion reaction with primary amines is limited by the dehydrogenation of the primary product, R—NH—NH—COOR', to the azo compound R—N=N—COOR', by the nitrene (84). Secondary amines give insertion products R$_2$N—NH—COOR', which are stable towards carbalkoxynitrenes (85).

D. Dehydrogenation Reactions

Dehydrogenations by carbonylnitrenes have been reported by several authors for a variety of systems. The reaction is not completely understood, and it is sometimes difficult to distinguish from radical induced dehydrogenations by the parent azide (cf. 82,83). The triplet nitrene seems to be primarily responsible.

Cumene is dehydrogenated to dicumyl (3), cyclohexane gives cyclohexene (5,14) and it seems that both hydrogens at the adjacent carbons are removed simultaneously (5). The photosensitized decomposition of n-hexanoyl azide in cyclohexane gives both cyclohexene and biscyclohexyl (72).

Alcohols are readily dehydrogenated to aldehydes or ketones (13,20, 79,81).

Hydrazo compounds are converted to azo compounds (84).

E. Reactions with C=C Double Bonds

Carbethoxynitrene adds readily to carbon–carbon double bonds to give N-carbethoxyaziridines, both in solution (1,4,7,22–24) and in the gas phase (10,11). The addition is stereospecific for the singlet (22–24) and nonstereospecific for the triplet (see Section II-D). With 1,3-dienes, such as butadiene (86), isoprene, cyclopentadiene, and cyclohexadiene (30) and cyclooctatetraene (26), only 1,2-addition is observed. However, the primary products (vinylaziridienes) can be thermally rearranged to the apparent 1,4-addition products (pyrrolines) (30,87) (eq. 20). In the reaction of

carbethoxynitrene with isoprene the singlet nitrene discriminates very little between the disubstituted and the monosubstituted double bonds, but the triplet prefers the methyl-substituted double bond by a factor of about 2 (30).

Pivaloylnitrene adds to the double bond of cyclohexene in 75% yield, based on the nitrene actually generated (46). Table X gives examples of the yields of aziridines produced from carbonylnitrenes and unactivated olefins.

The photolysis of ethyl azidoformate in dihydropyran (88) gives the corresponding aziridine, most likely via the nitrene. Thermolysis of ethyl azidoformate in dihydropyran liberates nitrogen twenty times as fast (at 80°) as in inert solvents, and the primary reaction appears to be the

TABLE X
Yields of Aziridines from Carbonylnitrenes and Olefins

Nitrene	Mode of generation	Olefin	Aziridine yield, %	Ref.
EtOOC—N	Azide photol.	Cyclohexene	56	4,7
		Cyclohexane (at −75°)	75	4
	α-Elim.	Cyclohexane	57	7
		4-Methylpentene-2	57	23
	Azide photol.	4-Methylpentene-2 (pure olefin)	67	21
		4-Methylpentane-2 (3.3 mole %)	60	21
		Cyclohexene	52	24
		Cyclohexene in dichloromethane (7%)	53	24
		Cyclohexene in dichloromethane (0.2%)	35	24
		Isoprene (100%)	95	30
		Isoprene (0.5%)	85	30
	α-Elim.	Cyclopentadiene	30	30
		Cyclooctatetraene	40	26
tBu—CO—N	Azide photol.	Cyclohexene	45	46

addition of the azide, to form a triazoline (88). Photolysis of ethyl azido-formate in the presence of isopropenyl acetate or 1-acetoxy-cyclohexene gave the corresponding aziridines (77).

A side reaction that can be fast enough to prevent any nitrene formation is the addition of the parent azide to the olefin. This becomes important as soon as there is any activating feature present in the olefin molecule, such as an electron-donating group or ring strain. The triazoline formation is, of course, a well-known reaction (89,90). Ethyl azidoformate does not add, at room temperature, to unactivated double bonds (such as in cyclo-hexene), but azidoformates add to enamines (91), norbornene (92,93), norbornadiene (94), and the like. Benzazide also adds to the strained double bond of norbornene (95), and the triazoline so formed loses nitrogen at 40° to give the corresponding aziridine (95). The decomposi-tion of triazoline can give aziridines and imines (92,93,96,97). Huisgen has studied the addition of organic azides (mostly aryl azides) to enol ethers and enamines (98,99). In the absence of more extensive data on the addi-tion of carbonyl azides, this work can be used as a guide to predicting the reactivity of certain olefins in the triazoline formation.

F. Reactions with Aromatic and Heteroaromatic Systems

Carbethoxynitrene, when generated in benzene solution by photolysis (76,78), thermolysis (12), or by α-elimination (2), expands the benzene ring to give N-carbethoxyazepine (eq. 7). Analogously, carbonyl azide, N_3—CO—N_3, when thermolyzed in benzene gives N-azidocarbonyl-azepine (50). Two independent studies suggest that the singlet carbethoxy-nitrene is responsible for the azepine formation. Baldwin (100) correlated the rates of disappearance of benzene and of substituted benzenes in the presence of thermolytically decomposing ethyl azidoformate and ethyl diazoacetate. Both the rates of the nitrene and of the carbene reactions correlate with σ_ρ. For carbethoxynitrene, ρ is -1.32; for carbethoxy-carbene, ρ is -0.38. These substituent effects are much smaller than those observed in electrophilic aromatic substitution, and a concerted mechanism is thus indicated. In turn, a concerted mechanism is much more likely for the singlet electron-deficient species. In the other study (101), competition experiments were run between benzene and cyclohexane, and between benzene and cyclohexene. Equimolar mixtures of the competitors were diluted with dichloromethane and allowed to react with a small amount of carbethoxynitrene, made by α-elimination. In the benzene–cyclohexane reactions, the yields of both azepine and cyclohexylurethan fell by a factor of 10 over a concentration range from 32 to 0.5 mole % of each substrate, but the fraction of the azepine in the product mixture changed little. Both products must be formed by the same intermediate, the singlet nitrene (cf. Section IV-A-1). This is borne out by the benzene–cyclohexene experiments. Over a concentration range from 32 to 0.1 mole % of each substrate, the azepine yield decreased from 29% to a trace, while the yield of the aziridine (1) decreased only from 56 to 16.4%, in accord with the known ability of both singlet and triplet carbethoxynitrene to add to the double bond of cyclohexene (21,23,24). The azepine (and cyclohexyl-urethan) yields fall with the amount of singlet nitrene undergoing reactive collision with these substrates, the aziridine yield does not.

Alkyl-substituted benzenes form azepines with little directive effect from the alkyl groups (76,100). Condensed aromatic hydrocarbons, such as anthracene, phenanthrene, and pyrene (36,37) do not give azepines but N-arylurethans (see Section IV-A-2).

Some five-membered heteroaromatic compounds react with carbeth-oxynitrene to form N-carbethoxypyrrole derivatives (102). Pyrrole gives a 14% yield of 1-carbethoxy-2-aminopyrrole, plus the (formal) insertion product in the 2-position, 2-carbethoxyaminopyrrole, in 24% yield. These products might be formed via a primary 1,4-adduct, but they could also be explained by an intermediate 1,2-adduct (and the latter prod-uct by simple insertion), as shown in equation 21. Thiophene gave N-carbethoxypyrrole in 21% yield, and 2,6-dimethylthiophene gave

$$(21)$$

2,6-dimethyl-N-carbethoxypyrrole in 18% yield. The latter product shows the intermediacy of a symmetrical stage, such as the 1,4-bridged thiophene in equation 22 (102). The direct rearrangement of a 1,2-adduct to the 2,6-dimethylthiophene would give 2,3-dimethyl-N-carbethoxypyrrole, which was not observed.

$$(22)$$

G. Reactions with Allenes

Thermolysis of ethyl azidoformate in tetramethylallene gives a mixture of triazoline (from 1,3-dipolar addition of the azide) and oxazoline. The latter is not formed from the triazoline, and is most likely formed by either 1,3-cycloaddition of the nitrene, or rearrangement of an initially formed aziridine. Photolysis of ethyl azidoformate in 1,1-dimethylallene leads to reaction at the more substituted double bond (eq. 23)(103).

H. Reactions with Alkynes

Thermolysis of ethyl azidoformate in diphenylacetylene gives a mono- and a diadduct (eq. 24) (16,104). Monoadducts were also obtained with ethyl propiolate, phenylacetylene, dimethyl acetylenedicarboxylate, and methyl propiolate. In the latter three cases, addition of the azide to form triazoles also took place, but the triazoles are stable under the reaction

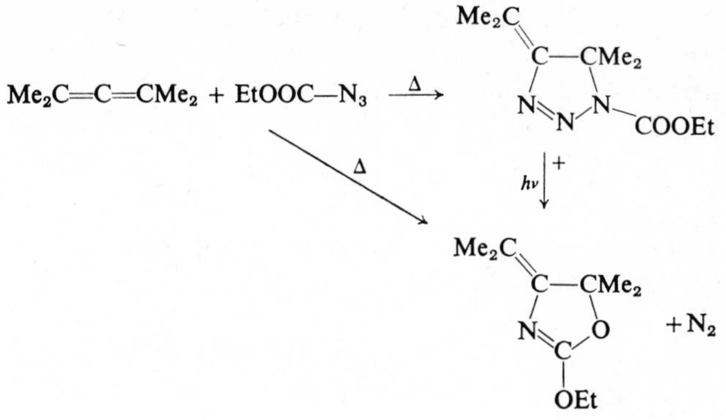

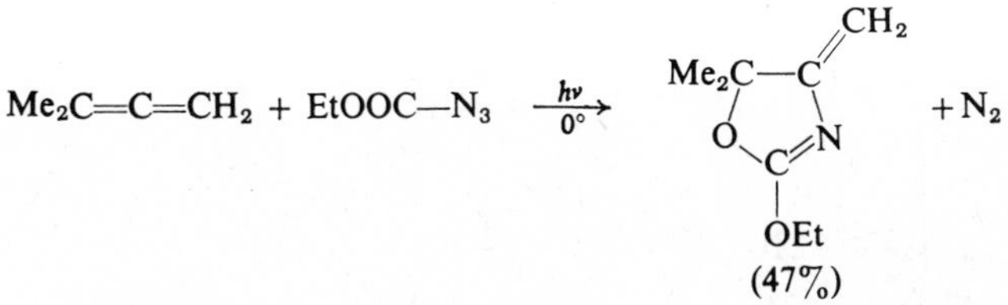

$$Me_2C=C=CMe_2 + EtOOC-N_3 \xrightarrow{\Delta}$$

$$Me_2C=C=CH_2 + EtOOC-N_3 \xrightarrow[0°]{hv}$$

(47%)

(23)

conditions, which leaves carbethoxynitrene (of undetermined electronic state) as the intermediate in the formation of the oxazoles. This conclusion is strengthened by the observation that the decomposition rate of propyl azidoformate, at 130°, is the same within a factor of two in diphenyl-acetylene, diphenyl ether, benzonitrile, mesitylene, and molten paraffin (104).

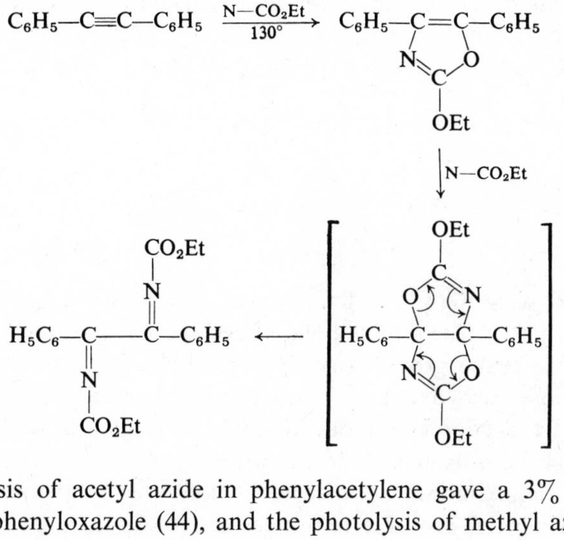

$$C_6H_5-C\equiv C-C_6H_5 \xrightarrow[130°]{N-CO_2Et} C_6H_5-C=C-C_6H_5$$

(24)

Photolysis of acetyl azide in phenylacetylene gave a 3% yield of 2-methyl-5-phenyloxazole (44), and the photolysis of methyl azidoformate in 2-butyne gave a 12% yield of the oxazole, plus a 30% yield of an adduct of two molecules of 2-butyne and one nitrene molecule. Meinwald (105) considers a carbenoid nitrene–acetylene adduct as an explanation of his results (eq. 25).

$$\text{(25)}$$

I. Reactions with Nitriles

Carbethoxynitrene (79,106,107) and acetylnitrene (44) add to nitrile groups to give 1,3,4-oxadiazoles. Carbalkoxynitrenes give the 2-alkoxy-1,3,4-oxadiazoles described earlier by Bacchetti (108). Carbethoxynitrene, made by α-elimination in the presence of acrylonitrile gave a 14% yield of the oxadiazole, together with a 73% yield of the addition product to the C=C double bond, 1-carbethoxy-2-cyanoaziridine (106) (eq. 26). Thus,

$$\text{(26)}$$

in $H_2C=CH-CN$, the C=C double bond is about five times as reactive as the cyano group. The double bond of cyclohexene is about 12 times as reactive as the cyano group in acetonitrile (109). Dilution experiments, similar to those carried out with benzene, again with cyclohexane and cyclohexene (cf. Section IV-E) show that it is the singlet carbethoxynitrene that forms the 1,3,4-oxadiazole from nitriles.

The reaction mechanism is not known. A concerted 1,3-dipolar addition, as well as a diazir-2-ene or a nitrilimine intermediate, seem reasonable.

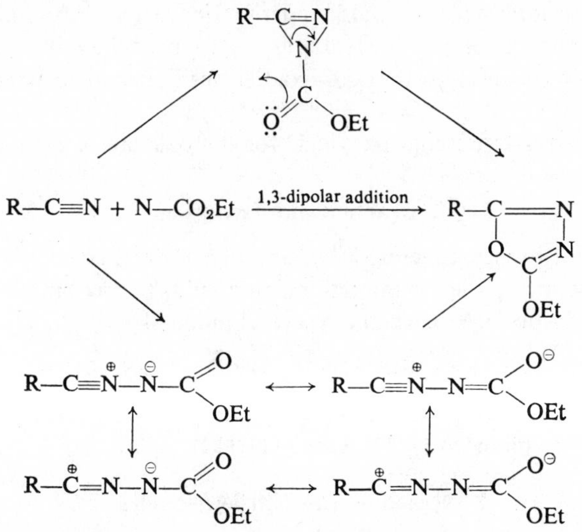

$$
\begin{array}{c}
R-C\equiv N \\
\diagdown N \\
\ddot{O}\diagup C \diagdown OEt
\end{array}
$$

$$R-C\equiv N + N-CO_2Et \xrightarrow{\text{1,3-dipolar addition}} \quad R-C \begin{array}{c} N \\ | \quad \| \\ O \quad N \\ \diagdown C \diagup \\ | \\ OEt \end{array}$$

$$R-C\equiv \overset{\oplus}{N}-\overset{\ominus}{N}-C\overset{O}{\underset{OEt}{\diagdown}} \longleftrightarrow R-C\equiv \overset{\oplus}{N}-N=C\overset{O^{\ominus}}{\underset{OEt}{\diagdown}}$$

$$R-\overset{\oplus}{C}=N-\overset{\ominus}{N}-C\overset{O}{\underset{OEt}{\diagdown}} \longleftrightarrow R-\overset{\oplus}{C}=N-N=C\overset{O^{\ominus}}{\underset{OEt}{\diagdown}} \tag{27}$$

TABLE XI
Yields of 2-Alkoxy-1,3,4-oxadiazoles from Nitriles

Nitrile	Mode of nitrene generation[a]	Yield, %	Ref.
H_3C-CN	$h\nu$	55	106
	$h\nu$	60	107
	αE	30	106
iPr$-CN$	$h\nu$	64	106
EtO$-CH_2-CH_2CN$	$h\nu$	73	106
C_6H_5-CN	$h\nu$	6.8	107
	Heat	14.5	107
pCl$-C_6H_4-CN$	Heat	8.3	107
$p-O_2N-C_6H_4-CN$	Heat	4.0	107
MeO$-CO-CH_2CN$	Heat	16	107
EtO$-CO-CN$	Heat	1.7	107
	$h\nu$	0.2	107
$H_2C=CH-CN$	αE	14[b]	106

[a] $h\nu$, azide decomposed by light;
Heat, azide decomposed by heat;
αE, $p-O_2N-C_6H_4-SO_2O-NH-COOEt$ decomposed by Et_3N.
[b] Plus 73% cyanoaziridine.

The diazirene intermediate could undergo ring expansion analogous to the known conversion of N-acylaziridines to oxazolines (cf. 6,30), the cyclization of nitrilimines to 1,3,4-oxadiazoles is a known process (108) (eq. 27).

Table XI gives representative yields for the oxadiazole formation.

J. Reaction with Isonitriles

Carbethoxynitrene, generated by photolysis or thermolysis of ethyl azidoformate, or by the α-elimination route, adds to *tert*-butyl isocyanide to form N-*tert*-butyl-N'-carbethoxycarbodiimide (110):

$$tBu—NC + N—COOEt \longrightarrow tBu—N=C=N—COOEt$$

The compound was identified by comparing its physical and spectral data with those reported by Neidlein (111,112).

K. Reaction with Nitrile N-oxides

The decomposition of N-p-nitrobenzenesulfonyloxyurethan (5) by triethylamine in the presence of nitrile oxides led to the formation of 10 (113). The structure of 10 was proved by X-ray crystallography, and by comparison with compounds obtained by Huisgen (114). It is doubtful whether 10 is actually the product of a nitrene reaction, because Huisgen obtained his compounds from benzonitrile oxides and azodicarboxylates. The latter are formed, in the standard α-elimination procedure, by attack of the nitrene on precursor-anion (115). Under the usual conditions, the azodicarboxylates are then destroyed by triethylamine, but in the presence of nitrile oxide, they might well be captured by the latter.

$$
\begin{matrix}
& Ar & & Ar & \\
& | & & | & \\
EtOOC—O—N=C—N=N—C=N—O—COOEt \\
& & (10) & &
\end{matrix}
$$

L. Reactions with Sulfoxides

Horner observed the addition of benzoylnitrenes (generated by photolysis of benzazide and substituted benzazides) to sulfoxides (20,43):

$$
C_6H_5—CO—N + O=SR_2 \longrightarrow \begin{matrix} C_6H_5—CO—N \\ \diagdown \\ SR_2 \\ \diagup \\ O \end{matrix}
$$

Sulfoximine yields range from less than 1% (p-nitrobenzoylnitrene) to 32% (p-methylbenzoylnitrene), perhaps due to partial photodecomposition of the products. D. S. Breslow (13) observed a fairly complex reaction of n-octadecyloxycarbonylnitrene with dimethylsulfoxide. Analysis of the gas evolved in this reaction indicates an oxygen transfer from the sulfoxide to the nitrene, and radical decomposition of the nitrosoformate formed:

$$RO—CO—N + O{=}S(CH_3)_2 \longrightarrow RO—CO—NO + S(CH_3)_2$$

isolated

$$\downarrow$$

$$R{\cdot} + CO_2 + NO$$

isolated

Parallel to the oxygen transfer reaction, addition of the nitrene to form a sulfoximine also occurs.

Robson and Speakman (50) observed sulfoximine formation when they photolyzed dodecanoyl azide in the presence of dimethyl sulfoxide.

M. Reactions with Trivalent Phosphorus Compounds

Triethyl phosphite and triphenyl phosphine react with ethyl azidoformate at room temperature (13) to give the corresponding phosphazanes, $EtO—CO—N{=}PX_3$. This reaction takes place far below the decomposition temperature of ethyl azidoformate, thus it cannot be a nitrene reaction. The question of how a nitrene would react with phosphines and phosphites remains open.

N. Reaction with Carbon Monoxide

When ethyl azidoformate is thermolyzed in the presence of high pressures of carbon monoxide in benzene or cyclohexane solution, only nitrene products derived from the solvent were found, but when an inert solvent, 1,1,2-trichloro-1,2,2-trifluoroethane was used, a 35% yield of carbethoxy isocyanate, $EtOOC—NCO$, was obtained (116). The apparent scavenging role of the solvent seems to indicate that this is a nitrene reaction.

V. THE CURTIUS, HOFMANN, AND LOSSEN REARRANGEMENTS

The common feature of the Curtius, Hofmann, and Lossen rearrangements is migration of a group—alkyl, aryl, alkylamino (51), alkoxy (51), and sulfur (118)—to a potentially electron-deficient nitrogen, to form an isocyanate:

$$Mig—CO—N—X \longrightarrow Mig—N{=}C{=}O + X$$

The three rearrangements are mainly different in their leaving groups. They have been extensively reviewed (117). At this place, we shall consider only a somewhat puzzling mechanistic problem: Are nitrenes involved in all or some of the rearrangements, or not? This question has been debated since Tiemann (41) and Stieglitz (119) postulated nitrene mechanisms for the Lossen and Curtius rearrangements, respectively (eq. 28). The most

$$R\text{—}CO\text{—}\bar{N}\text{—}OAcyl \underset{-\ ^-OAcyl}{\underline{\hspace{3cm}}}\qquad \text{(Tiemann)}$$

$$R\text{—}\overset{|}{C}O\text{—}N \longrightarrow R\text{—}N{=}C{=}O \qquad (28)$$

$$R\text{—}CO\text{—}N_3 \underset{-\,N_2}{\overline{\hspace{3cm}}}\qquad \text{(Stieglitz)}$$

obvious alternative to the stepwise process of equation 28 is a concerted one, in which the leaving group departs while the migrating group makes the new bond to nitrogen.

Since carbonylnitrenes can be trapped efficiently (see Section IV-A to H), it should be possible to test the nitrene mechanism of equation 28 by trapping experiments. On the other hand, a concerted mechanism should lead to a dependence of the reaction rate (rate of disappearance of starting material) on the electronic properties of the migrating group. Both types of experiment have been done, both with negative results.

Trapping experiments in the Hofmann (120) and the (thermally induced) Curtius rearrangements all failed (20,43,44,47,71,120–122; also see Chapter 7), disposing of the nitrene mechanism. Somewhat disconcertingly, the concerted mechanism is only very feebly supported by substituent effect studies trying to correlate rate constants with the nature of the migrating groups. The kinetics of the Curtius rearrangement was studied in several ways. Brower (123,124) found a small, positive volume change of activation in the decomposition of benzazide, and concluded that a nitrene intermediate is formed. However, one can just as easily assume that bond-breaking in the transition state is sufficiently advanced to account for this result, without postulating a free nitrene intermediate. No clear correlation emerges from kinetic data comparing the activation parameters of the decomposition of carbonyl azides with various migrating groups (125–133).

The same is true for the Hofmann (134–136) and the Lossen rearrangements (137–140). There are some correlations, covering a few migrating groups of similar structure (e.g., 139), but the scope of these correlations

$$\underset{\substack{|\\ \text{Co}\\ |\\ \text{Ar}}}{\overset{R}{\underset{|}{\text{Me}_2\text{C}\text{—}O}}} \overset{O}{\underset{}{}} \longrightarrow \underset{\substack{|\\ \text{Co}\\ |\\ \text{Ar}}}{\overset{R}{\text{Me}_2\text{C}\text{-----}O}} \overset{O}{\underset{}{}} \longrightarrow \underset{\substack{^-O_2\text{CAr}\\ \downarrow\\ \text{products}}}{\overset{+}{\text{Me}_2\text{C}\text{--}O\text{—}R}} \qquad (29)$$

is narrow, and the magnitude of the effects is rather small, compared to other rearrangements which proceed with anchimeric assistance, for example the rearrangement of 2-substituted 2-propyl p-nitroperbenzoates (141) (eq. 29). To illustrate the point, selected data are assembled in Table XII.

TABLE XII
Selected Activation Parameters for Four Rearrangements

Migrating group, R	Rate or relative rate	$\Delta H\ddagger$ kcal/mole	$\Delta S\ddagger$ eu	Ref.
1. Me_2C—O—O—CO—C_6H_4—$NO_2(p)$				
R = CH_3	1.00	27.6	−6.7	141
C_2H_5	45	25.6	−0.3	141
$CH(CH_3)_2$	2,940	24.0	−3.0	141
$3(3H_3)_3$	228,000	21.9	−1.3	141
C_6H_5	118,000	—	—	141
2. p-X—C_6H_4—CO—N_3				
R = C_6H_5	$3.03 \times 10^{-5\,a}$	27.8	+4.1	133
p-O_2N—C_6H_4	2.14×10^{-5}	28.3	+4.7	133
p-H_3C—O—C_6H_4	2.065×10^{-5}	28.0	+3.9	133
3. R—CO—NH—Br				
		E_A		
R = C_2H_5	$0.29 \times 10^{-4\,b}$	30.1	+23.3	135
$CH(CH_3)_2$	7.26×10^{-4}	28.4	+23.8	135
$C(CH_3)_3$	12.20×10^{-4}	28.4	+24.9	135
C_6H_5—CH_2	0.78×10^{-4}	28.8	+21.0	135
$(C_6H_5)_2CH$	3.89×10^{-4}	28.6	+23.2	135
4. R—CO—NH—O—CO—C_6H_5				
		H		
R = C_6H_5	$2.30 \times 10^{-5\,c}$	26.5	+5.9^d	137, 138
p-H_3C—C_6H_4	5.98×10^{-5}	26.5	+7.7	137, 138
p-H_3C—O—C_6H_4	15.72×10^{-5}	25.8	+7.2	137, 138
C_2H_5	$1.13 \times 10^{-5\,c}$	29.2	+15.1	139
$CH(CH_3)_2$	17.8×10^{-5}	28.0	+16.1	139
$C(CH_3)_3$	14.9×10^{-5}	27.1	+13.3	139

[a] In sec^{-1}, at 60°.
[b] In sec^{-1}, at 14.9°.
[c] In sec^{-1}, at 30°.
[d] Calculated from the author's data (135).

Fry and Wright (142) measured kinetic isotope effects for the migrating group, the carbonyl group, and the migration terminus for all three rearrangements. Substantial isotope effects for the migrating carbon, the migration terminus (nitrogen), and—in the Curtius rearrangement—the leaving nitrogen indicate that all of them are involved in the rate determining step. Thus, the concerted mechanism is strongly supported (Table XIII). The paucity of the substituent effects can perhaps be attrib-

TABLE XIII

Carbon and Nitrogen Kinetic Isotope Effects in the Curtius, Hofmann, and Lossen Rearrangements (142)%

uted to a rather lopsided transition state, the energy of which is largely determined by the formation (in the case of nitrogen) or solvation (in the Hofmann rearrangement) of the leaving group.

The photolytic decomposition of carbonyl azides gives the corresponding carbonylnitrenes (see Section III-A) along with the corresponding isocyanates. Several mechanisms seem reasonable for this photoinduced Curtius rearrangement (143). The isocyanate could be formed by rearrangement of the nitrene, or directly from the carbonyl azide, without intervention of a nitrene. For the case of pivaloyl azide, this has been tested by running the photolysis in the presence and in the absence of good nitrene traps. If the pivaloylnitrene is an intermediate on the way to the *tert*-butyl isocyanate, trapping it should decrease the yield on isocyanate.

This was found not to be the case—the yield on isocyanate remained unchanged at 40%, whether nitrene products were formed (in cyclohexene solution) in 45% yield, or whether the nitrene was not trapped at all (in dichloromethane solution, where it finally dissociates; see section III-C). Thus, the nitrene tBu—CO—N does not rearrange at a rate competitive with its other reactions (and dissociation), and cannot be an intermediate in the photolytic Curtius rearrangement (47).

REFERENCES

1. W. Lwowski and T. W. Mattingly, Jr., *Tetrahedron Letters*, **1962**, 277.
2. W. Lwowski, T. J. Maricich, and T. W. Mattingly, Jr., *J. Am. Chem. Soc.*, **85**, 1200 (1963).
3. M. F. Sloan, T. J. Prosser, N. R. Newburg, and D. S. Breslow, *Tetrahedron Letters*, **1964**, 2945.
4. W. Lwowski and T. W. Mattingly, Jr., *J. Am. Chem. Soc.*, **87**, 1947 (1965).
5. D. S. Breslow, T. J. Prosser, A. F. Marcantonio, and C. A. Genge, *J. Am. Chem. Soc.*, **89**, 2384 (1967).
6. W. Lwowski and T. J. Maricich, *J. Am. Chem. Soc.*, **86**, 3164 (1964).
7. W. Lwowski and T. J. Maricich, *J. Am. Chem. Soc.*, **87**, 3630 (1965).
8. E. Wasserman, private communication.
9. G. Smolinsky, E. Wasserman, and W. A. Yager, *J. Am. Chem. Soc.*, **84**, 3220 (1962).
10. R. S. Berry, D. W. Cornell, and W. Lwowski, *J. Am. Chem. Soc.*, **85**, 1199 (1963).
11. D. W. Cornell, R. S. Berry, and W. Lwowski, *J. Am. Chem. Soc.*, **87**, 3626 (1965).
12. R. J. Cotter and W. F. Beach, *J. Org. Chem.*, **29**, 751 (1964).
13. T. J. Prosser, A. F. Marcantonio, and D. S. Breslow, *Tetrahedron Letters*, **1964**, 2479.
14. T. J. Prosser, A. F. Marcantonio, C. A. Genge, and D. S. Breslow, *Tetrahedron Letters*, **1964**, 2483.
15. D. S. Breslow and E. I. Edwards, *Tetrahedron Letters*, **1967**, 2123.
16. R. Huisgen and H. Blaschke, *Chem. Ber.*, **98**, 2985 (1965).
17. Yu. N. Sheinker, *Dokl. Akad. Nauk USSR*, **77**, 1043 (1951).
18. W. Lwowski and E. Scheiffele, unpublished results.
19. C. H. Robinson, private communication; cf. *Abstracts*, 151th Natl. Mtg. Am. Chem. Soc., March 1965, *Abstract K 019*.
20. L. Horner, G. Bauer, and J. Dörges, *Chem. Ber.*, **98**, 2631 (1965).
21. J. S. McConaghy, Jr., and W. Lwowski, *J. Am. Chem. Soc.*, **89**, 4450 (1967).
22. W. Lwowski and J. S. McConaghy, Jr., *J. Am. Chem. Soc.*, **87**, 5490 (1965).
23. J. S. McConaghy, Jr., and W. Lwowski, *J. Am. Chem. Soc.*, **89**, 2357 (1967).
24. W. Lwowski and F. P. Woerner, *J. Am. Chem. Soc.*, **87**, 5491 (1965).
25. H. Nozaki, S. Fujita, H. Takaya, and R. Noyori, *Tetrahedron* **23**, 45 (1967).
26. S. Masamune and N. T. Castellucci, *Angew. Chem.*, **76**, 569 (1964).
27. T. J. Maricich, Thesis, Yale University, New Haven, Conn., 1965.
28. W. Lwowski and L. Selman, unpublished; L. Selman, Thesis, Yale University, New Haven, Conn., 1966.
29. B. Acott, A. L. J. Beckwith, A. Hassalani, and J. W. Redmond, *Tetrahedron Letters*, **1965**, 4093.

30. A. Mishra, S. N. Rice, and W. Lwowski, *J. Org. Chem.*, **33**, 481 (1968).
31. P. S. Skell and R. C. Woodworth, *J. Am. Chem. Soc.*, **78**, 4496 (1956).
32. Cf. P. P. Gaspar and G. S. Hammond, in *Carbene Chemistry*, W. Kirmse, Ed., Academic Press, New York, 1964.
33. R. Hoffmann, *Trans. N.Y. Acad. Sci.*, *2*, **28**, 75 (1966).
34. R. Hoffmann, *J. Am. Chem. Soc.*, **90**, 1475 (1968).
35. P. Scheiner, *J. Am. Chem. Soc.*, **88**, 4759 (1966).
36. A. L. J. Beckwith and J. W. Redmond, *Chem. Commun.*, **1967**, 165; *J. Am. Chem. Soc.*, **90**, 1351 (1968).
37. A. L. J. Beckwith and J. W. Redmond, *Australian J. Chem.*, **19**, 1859 (1966).
38. A. G. Anastassiou, *J. Am. Chem. Soc.*, **89**, 3184 (1967).
39. M. Jones and K. R. Rettig, *J. Am. Chem. Soc.*, **87**, 4013 (1965).
40. M. Jones, A. Kulczycki, Jr., and K. F. Hummel, *Tetrahedron Letters*, **1967**, 183.
41. F. Tiemann, *Ber.*, **24**, 4162 (1891).
42. J. ApSimon and O. E. Edwards, *Proc. Chem. Soc.*, **1961**, 461.
43. L. Horner and A. Christmann, *Chem. Ber.*, **96**, 388 (1963).
44. R. Huisgen and J.-P. Anselme, *Chem. Ber.*, **98**, 2998 (1965).
45. W. Lwowski and G. T. Tisue, *J. Am. Chem. Soc.*, **87**, 4022 (1965).
46. G. T. Tisue, S. Linke, and W. Lwowski, *J. Am. Chem. Soc.*, **89**, 6303 (1967).
47. S. Linke, G. T. Tisue, and W. Lwowski, *J. Am. Chem. Soc.*, **89**, 6308 (1967).
48. N. C. Castellucci, M. Kato, H. Zenda, and S. Masamune, *Chem. Commun.*, **1967**, 473.
49. G. Just and W. Zehetner, *Tetrahedron Letters*, **1967**, 3389.
50. P. Robson and P. H. R. Speakman, *J. Chem. Soc.* (*B*)., **1968**, 463.
51. J. Sauer and K. K. Mayer, *Tetrahedron Letters*, **1968**, 319.
52. L. E. Chapman and R. F. Robbins, *Chem. Ind.*, **1966**, 1266.
53. W. Lwowski, R. DeMauriac, T. W. Mattingly, Jr., and E. Scheiffele, *Tetrahedron Letters*, **1964**, 3285; W. Lwowski and R. DeMauriac, *Abstracts*, 155th Meeting, ACS, April 1968, *Abstract P 056*; R. DeMauriac, Ph.D. thesis, Yale University, New Haven, Conn., 1967. Cf. W. S. Wadsworth and W. D. Emmons, *J. Org. Chem.*, **32**, 1279 (1967).
54. E. Schmitz, S. Schramm, and R. Ohme, *J. Prakt. Chem.*, **36** (4), 86 (1967).
55. B. Acott and A. L. J. Beckwith, *Chem. Commun.*, **1965**, 161.
56. H. E. Baumgarten and A. Staklis, *J. Am. Chem. Soc.*, **87**, 1141 (1965).
57. J. Bollinger, Thesis, Marburg, 1937; as quoted by R. Criegee, in *Oxidation in Organic Chemistry*, Part A, K. Wiberg, Ed., Academic Press, New York, 1965, p. 353.
58. J. H. Hall, J. W. Hill, and H.-C. Tsai, *Tetrahedron Letters*, **1965**, 2211.
59. C. Walling and W. Thaler, *J. Am. Chem. Soc.*, **83**, 3877 (1961).
60. L. C. Bateman and E. D. Hughes, *J. Chem. Soc.*, **1940**, 945.
61. W. Lwowski and J. Simson, *Abstracts*, 153rd Meeting, Am. Chem. Soc., April 1967, *Abstract O 163*.
62. J. Simson and W. Lwowski, *J. Am. Chem. Soc.*, in press.
63. J. Reed and W. Lwowski, unpublished results.
64. D. S. Breslow, E. I. Edwards, R. Leone, and P. von R. Schleyer, *J. Am. Chem. Soc.*, **90**, 7097 (1968).
65. J. Meinwald and D. H. Aue, *Tetrahedron Letters*, **1967**, 2317.
66. G. Smolinsky and B. I. Feuer, *J. Am. Chem. Soc.*, **86**, 3085 (1964).
67. S.-I. Yamada, S. Terashima, and K. Achiwa, *Chem. Pharm. Bull.* (*Japan*), **13**, 751 (1965).

68. See W. Lwowski, *Angew. Chem.*, **79**, 922 (1967); *Intern. Ed.*, **6**, 897, (1967).
69. J. Simson, Ph.D. thesis, Yale University, New Haven, Conn., 1967.
70. W. B. DeMore and S. W. Benson, *Advan. Photochem.*, **2**, 252 (1964).
71. J. W. ApSimson and O. E. Edwards, *Can. J. Chem.*, **40**, 896 (1962).
72. I. Brown and O. E. Edwards, *Can. J. Chem.*, **45**, 2599 (1967).
73. S. Linke and W. Lwowski, unpublished results; cf. *Angew. Chem.*, **79**, 922 (1967); *Intern. Ed.*, **6**, 897 (1967).
74. R. Kreher and D. Kühling, *Angew. Chem.*, **77**, 43 (1965).
75. J. Hora, *Collection Czech. Chem. Commun.*, **29**, 1079 (1964).
76. K. Hafner, D. Zinser, and K. L. Moritz, *Tetrahedron Letters*, **1964**, 1733.
77. J. F. W. Keana, S. B. Keana, and D. Beetham, *J. Org. Chem.*, **32**, 3057 (1967).
78. K. Hafner and C. König, *Angew. Chem.*, **75**, 89 (1963); *Intern. Ed.*, **2**, 96 (1963).
79. R. Puttner and K. Hafner, *Tetrahedron Letters*, **1964**, 3119.
80. W. Lwowski, R. DeMauriac, T. W. Mattingly, Jr., and E. Scheiffele, *Tetrahedron Letters*, **1964**, 3285.
81. R. Kreher and G. H. Bockhorn, *Angew. Chem.*, **76**, 681 (1964).
82. L. Horner and G. Bauer, *Tetrahedron Letters*, **1966**, 3573.
83. M. Reagan and A. Nickon, *Abstracts*, 150th Natl. Meeting, ACS, Atlantic City, Sept. 1965, *Abstract S 23*; *J. Am. Chem. Soc.*, **90**, 4096 (1968).
84. W. Lwowski and L. Selman, *Abstracts*, 150th Natl. Meeting, ACS, Atlantic City, Sept. 1965, *Abstract S 25*.
85. W. Lwowski, L. Hartsell, and P. Worstell, unpublished results.
86. K. Hafner, W. Kaiser, and R. Puttner, *Tetrahedron Letters*, **1964**, 3953.
87. R. S. Atkinson and C. W. Rees, *Chem. Commun.*, **1967**, 1232.
88. I. Brown and O. E. Edwards, *Can. J. Chem.*, **43**, 1266 (1965).
89. L. Wolff, *Ann. Chem.*, **394**, 30, 69 (1912).
90. R. Huisgen, R. Grashey, and J. Sauer, in *The Chemistry of the Alkenes*, S. Patai, Ed., Interscience, New York, 1964, p. 835ff.
91. R. Fusco, G. Bianchetti, and D. Pocar, *Gazz. Chim.*, **91**, 933 (1961); *Chem. Abstr.*, 14020h (1962).
92. A. C. Oehlschlager, P. Tillman, and L. H. Zalkow, *Chem. Commun.*, **1965**, 596.
93. K. Tori, K. Kitahonoki, Y. Takano, H. Tanida, and T. Tsuji, *Tetrahedron Letters*, **1965**, 869.
94. Shinogi and Co., Neth. Pat. Appl. 6,514,295 (May 6, 1966); *Chem. Abstr.*, **65**, 16941a (1966).
95. R. Huisgen and G. Müller, *Angew. Chem.*, **72**, 371 (1960).
96. P. Scheiner, *J. Org. Chem.*, **30**, 7 (1965).
97. P. Scheiner, *J. Org. Chem.*, **32**, 2022 (1967).
98. R. Huisgen, L. Möbius, and G. Szeimies, *Chem. Ber.*, **98**, 1138 (1965).
99. R. Huisgen and G. Szeimies, *Chem. Ber.*, **98**, 1153 (1965).
100. J. E. Baldwin and R. A. Smith, *J. Am. Chem. Soc.*, **89**, 1886 (1967); *J. Org. Chem.*, **32**, 3511 (1967).
101. W. Lwowski and R. L. Johnson, *Tetrahedron Letters*, **1967**, 891.
102. K. Hafner and W. Kaiser, *Tetrahedron Letters*, **1964**, 2185.
103. R. F. Bleiholder and H. Shechter, *J. Am. Chem. Soc.*, **90**, 2131 (1968).
104. R. Huisgen and H. Blaschke, *Tetrahedron Letters*, 1409 (1964).
105. J. Meinwald and D. H. Aue, *J. Am. Chem. Soc.*, **88**, 2849 (1966).
106. W. Lwowski, A. Hartenstein, C. DeVita, and R. L. Smick, *Tetrahedron Letters*, **1964**, 2497.
107. R. Huisgen and H. Blaschke, *Ann.*, **586**, 145 (1965).

108. T. Bacchetti, *Gazz. Chim.*, **91**, 866 (1961); *Chem. Abstr.*, **56**, 8708 (1962).
109. W. Lwowski and Pen-Li Kao, unpublished results.
110. W. Lwowski, M. Grassman, and T. Shingaki, unpublished results.
111. R. Neidlein and E. Heukelbach, *Tetrahedron Letters*, **1965**, 149.
112. R. Neidlein and E. Heukelbach, *Arch. Pharm.*, **299**, 709, (1966); *Chem. Abstr.*, **66**, 37600j (1967).
113. P. Rajagopalan and C. N. Talaty, *Tetrahedron Letters*, **1966**, 4877.
114. R. Huisgen, H. Blaschke, and E. Brunn, *Tetrahedron Letters*, **1966**, 405.
115. W. Lwowski, J. Simson, and J. McConaghy, unpublished results.
116. R. P. Bennett and W. B. Hardy, *J. Am. Chem. Soc.*, **90**, 3295 (1968).
117. P. A. S. Smith, in *Molecular Rearrangements*, Vol. 1, P. DeMayo, Ed., Wiley, New York, 1963, p. 528ff; P. A. S. Smith, *The Chemistry of Open-Chain Organic Nitrogen Compounds*, Vol. 2, Benjamin, New York, 1966, p. 217ff.
118. M. Prince and C. M. Orlando, *Chem. Commun.*, **1967**, 818.
119. J. Stieglitz, *Am. Chem. J.*, **18**, 751 (1896).
120. C. R. Hauser and S. W. Kantor, *J. Am. Chem. Soc.*, **72**, 4284 (1950).
121. H. Lindemann and W. Schultheis, *Ann. Chem.*, **464**, 237 (1928).
122. G. Powell, *J. Am. Chem. Soc.*, **51**, 2436 (1929).
123. K. R. Brower, *J. Am. Chem. Soc.*, **85**, 1401 (1963).
124. K. R. Brower, *J. Am. Chem. Soc.*, **83**, 4370 (1961).
125. M. S. Newman, S. H. Lee, and A. B. Garrett, *J. Am. Chem. Soc.*, **69**, 113 (1947).
126. R. A. Coleman, M. S. Newman, and A. B. Garrett, *J. Am. Chem. Soc.*, **76**, 4534 (1954).
127. Y. Yukawa and Y. Tsuno, *J. Am. Chem. Soc.*, **79**, 5530 (1957).
128. Y. Tsuo, *Mem. Inst. Sci. Res. Osaka U.*, **15**, 183 (1958); *Chem. Abstr.*, **53**, 1116c (1959).
129. Y. Yukawa and Y. Tsuno, *J. Am. Chem. Soc.*, **80**, 6346 (1958).
130. Y. Otsuji, M. Furakawa, and E. Imoto, *J. Chem. Soc. Japan, Pure Chem. Sect.*, **80**, 1307 (1959).
131. Y. Yukawa and Y. Tsuno, *J. Am. Chem. Soc.*, **81**, 2007 (1959).
132. W. S. Hamilton, Ph.D. thesis, Tulane University, New Orleans, La., 1963.
133. R. Huisgen and G. Müller, private communication; G. Müller, Dissertation, München, 1962.
134. C. R. Hauser and W. B. Renfrow, *J. Am. Chem. Soc.*, **59**, 121 (1937).
135. W. P. Judd, Ph.D.Thesis, Univ. of Auckland, New Zealand, 1967.
136. K. M. Joshi and K. K. Shah, *J. Indian Chem. Soc.*, **43**, 481 (1966).
137. W. B. Renfrow and C. Hauser, *J. Am. Chem. Soc.*, **59**, 2308 (1937).
138. R. D. Bright and C. R. Hauser, *J. Am. Chem. Soc.*, **61**, 618 (1939).
139. D. C. Berndt and H. Shechter, *J. Org. Chem.*, **29**, 916 (1964).
140. D. C. Berndt and W. J. Adams, *J. Org. Chem.*, **31**, 976 (1966).
141. E. Hedeya and S. Winstein, *J. Am. Chem. Soc.*, **89**, 1661 (1967).
142. A. Fry and J. C. Wright, *Chem. Eng. News*, **Jan. 1, 1968**, 28; J. C. Wright, Ph.D. thesis, Univ. of Arkansas, Fayetteville, Ark., 1966.
143. L. Horner, E. Spietschka, and A. Gross, *Ann. Chem.*, **573**, 17 (1951).

CHAPTER 7

Acylnitrene Cyclizations

O. E. EDWARDS

Division of Pure Chemistry, National Research Council of Canada, Ottawa, Canada

I. INTRODUCTION

Cyclizations which can, at least formally, be considered to involve nitrenes (monosubstituted nitrogen with six-valence electrons) have enabled synthesis of a number of hitherto unknown systems and ready access to others. In addition they have proven useful in the structure proof and synthesis of natural products.

The first of these cyclizations appears to have been observed by Zincke and colleagues (1) in 1887. Thermal decomposition of an *o*-azidoazobenzene (1) gave them a 2-arylbenzotriazole (2). Soon benzofuroxan (3)

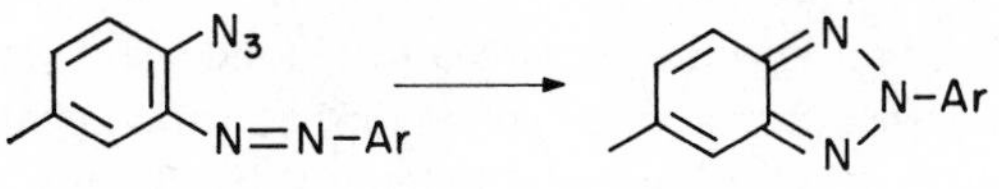

(1)   (2)

was prepared (2) by pyrolysis of *o*-nitrophenylazide, but it was 1953 before Smith and co-workers (3) added anthranils (e.g., **4**) to this list of compounds that may be formed by nitrene attack on a heteroatom. Boyer and Ellzey partially deoxygenated *o*-dinitrosobenzene using triphenylphosphine (4) and obtained benzofurazan in what may be a nitrene cyclization.

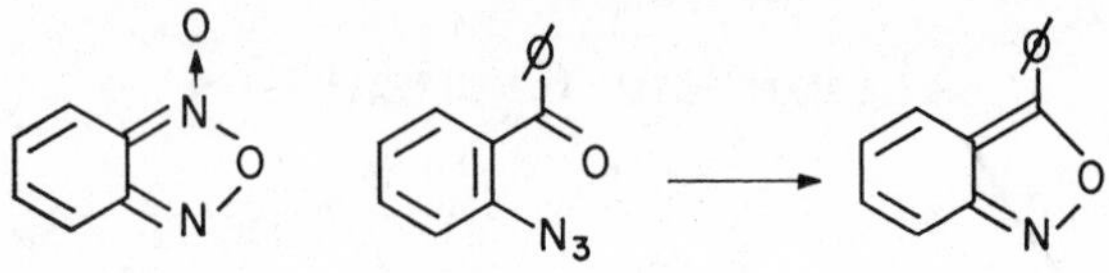

(3)   (4)

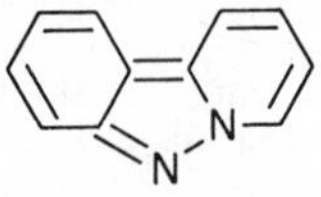

(5)

Bunyan and Cadogan (5) deoxygenated 3-(2′-nitrosophenyl)pyridine and obtained a high yield of pyrid[1,2-*b*]indazole (5).

Starting in 1951 Smith and his co-workers (6) described production of an array of heterocyclic systems apparently formed by nitrene attack on aromatic rings, for example, carbazole (6), carboline (7), indolothiophene (8), and phenothiazine-5-dioxide (9).

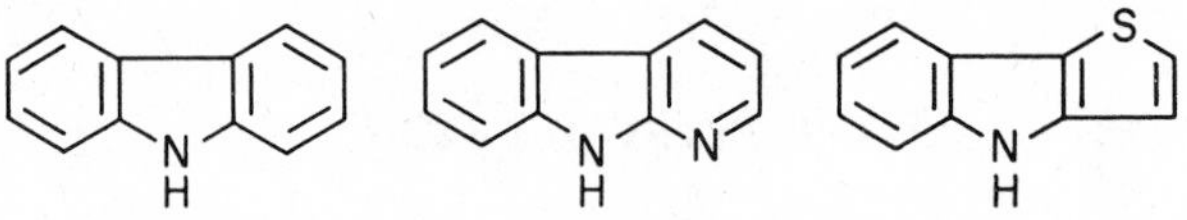

(6)   (7)   (8)

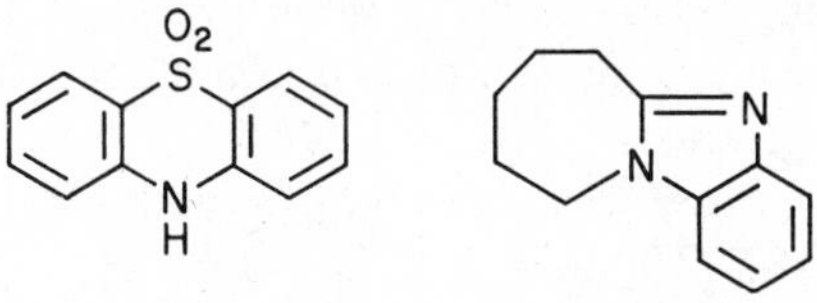

(9)   (10)

The insertion of an arylnitrene into a CH bond was first described by Saunders (7) who used the reaction to form piperido and hexahydroazepino-benziminazoles (e.g., 10). This was extended by Smolinsky (8) to the formation of phenanthridines (e.g., 11), indolines (e.g., 12) and 1,2,3,4-tetrahydroquinolines. The above and related reactions are described in detail in Chapters 4 and 5.

Cyclization of aliphatic azides on irradiation was first reported by Barton and Morgan (9) but this work has not been confirmed (10). The thermal decomposition of vinyl azides was used by Smolinsky (11) to make azirenes (e.g., 13), and Horner and co-workers (12) and Hassner and

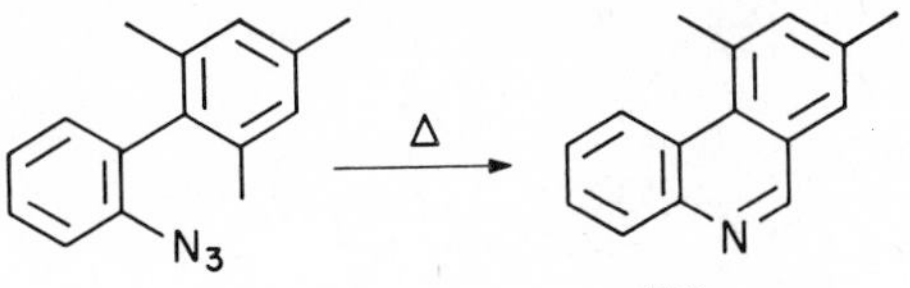

(11)

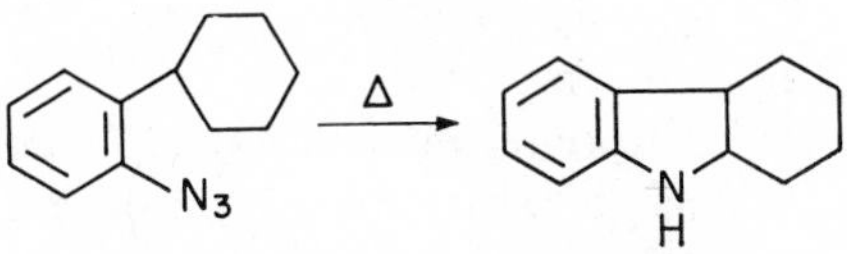

(12)

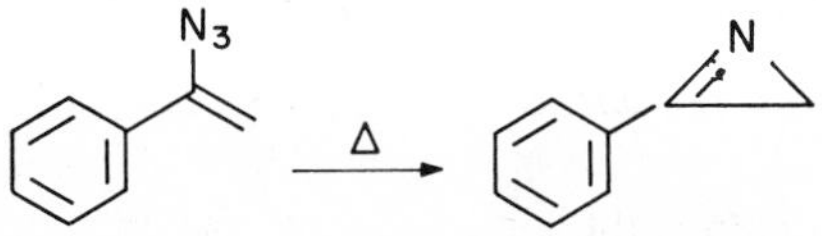

(13)

Fowler (13) later showed that this could also be accomplished photolytically. Finally, Logothetis (14) was the first to show that aliphatic azides would cyclize by addition to more remote double bonds to give aziridines (e.g., **14**). However, the latter reactions are best accounted for as involving heterolysis of intermediate triazolines. Aliphatic nitrenes are fully discussed in Chapter 3.

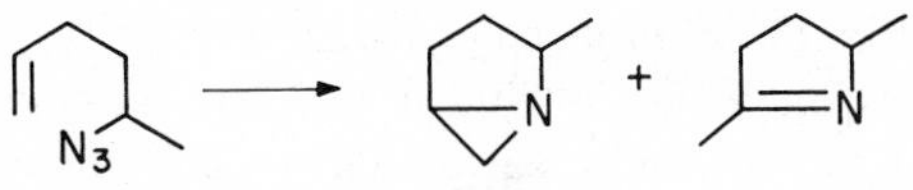

(14)

The thermal decomposition of acyl azides (Curtius reaction) had long been considered to involve an acylnitrene intermediate (eq. 1). Two lines

$$RCON_3 \longrightarrow N_2 + RC\overset{..}{O}N: \longrightarrow RNCO \qquad (1)$$

of evidence seemed to support this contention. The volume of activation of the Curtius reaction ($\Delta V^* = 2\text{--}5$ cc/mole) corresponds to a large measure of N—N bond rupture in the transition state (15). The comparable magnitude of ΔV^* in polar and nonpolar solvents indicates a lack of polarity in the transition state. This was considered to be incompatible with participation of R in the transition state (see **15**). The small difference in the rate of decomposition between *p*-methoxy- and *p*-

$$\underset{\overset{\displaystyle |}{R}}{\overset{\overset{\displaystyle O}{\diagdown}}{C}} \cdots N \cdots N_2$$

(15)

nitrobenzazide (16) also appeared inconsistent with this type of participation in the loss of nitrogen.

If acylnitrene intermediates in the thermal Curtius rearrangement had any significant lifetime, however, the above nitrene reactions would lead one to expect that it could be trapped. But so far no evidence for attack of an intermediate on the solvent or functions in the same molecule has been observed. To be consistent with the above evidence, one would have to postulate that the movement of R starts after the transition state is passed, but while the nitrene nitrogen is still shielded by the departing nitrogen molecule. Hence the nitrene would never be free to react with its surroundings.

However, more recent evidence contradicts the postulate of a nitrene intermediate. Fry and Wright (17) observed large kinetic isotope effects in the thermal Curtius rearrangement with the illustrated ^{13}C and ^{15}N substitution. This requires the

$$-\,^{13}C - CO - \,^{15}N - \,^{15}N \equiv N$$

migration of the carbon to be concerted with the rupture of the N—N bond. Lwowski and colleagues (18) showed that the percentage of isocyanate produced in the photolysis of pivaloyl azide was independent of the substrates present. They concluded that the photochemical Curtius reaction involved synchronous rearrangement and nitrogen elimination from excited azide. (See Chapter 6 for a thorough discussion of this and related problems.)

In contrast to the thermal decomposition, photolysis of acyl azides and nitrile oxides does give a reactive intermediate, apparently the acylnitrene, and reactive alkoxycarbonyl- and cyanonitrenes can be generated by thermal, photochemical, and chemical processes (see Section V and Chapters 6 and 9). This chapter will be concerned with intramolecular reactions of these acylnitrenes.

II. ACYL AZIDE PHOTOLYSIS

A. Insertion into CH bonds

The demonstrated reactivity of carbenes and arylnitrenes in CH bond insertion encouraged work in the author's laboratory on acylnitrenes. It was hoped that with suitable proximity and energy, acylnitrenes might insert into unactivated CH bonds, providing a method for synthesis of some natural products.

For the primary objective, synthesis of the heterocyclic system of diterpenoid alkaloids, podocarpic acid (16) was very suitable. The carboxyl group and angular methyl group are in close proximity. It was considered that the buttressing effect of ring C in the *trans*-anti-*cis* perhydrophenanthrene carboxylic acid (17) (R = COOH) would make this promixity even more marked. The acyl azides 17 (R = CON₃) and 18 were prepared. Despite the unusual proximity but in full agreement with previous observations on the Curtius rearrangement, no substitution of the angular methyl group occurred when these azides were thermally decomposed.

(16)   (17)   (18)

Photolysis of these azides (19) produced, in agreement with the earlier work of Horner et al. (20), around 70% of isocyanate. In addition however, azide 17 (R = CON₃) gave 25% of δ-lactam (19) and a trace of a γ-lactam. Similarly azide 18 gave 20% of the δ-lactam (20) and 5% of a γ-lactam presumed to be 21 or 22. The δ-lactam had the desired azabicyclononane system of the diterpenoid alkaloids, and 20 was used to confirm the structure and stereochemistry of atisine. In this and a later paper (21) the photolysis of the equatorial acyl azides 23 and 24 derived from dihydropimaric and dihydroisopimaric acids was shown to give the γ-lactams 25 (26%) and 26 (9%), respectively. In this case only the equatorial 6-hydrogen was available for attack in the stable conformation. The added rigidity of ring B in the dihydroisopimaric case may make the geometry for C—H insertion less favorable, resulting in the lower yield of lactam. In none of the above cases was attack on solvent (hexane) detected.

In order to provide information about factors determining the relative amounts of γ- and δ-lactam when both can be formed, hexanoyl azide was photolyzed (22). As expected from the above work, δ-lactam 27 was produced in higher yield (13%) than the γ-lactam 28 (8%). Attack on the

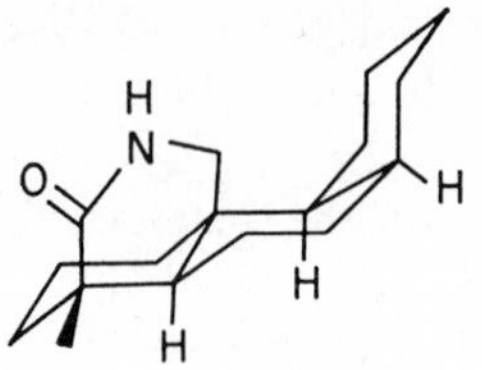

(19) 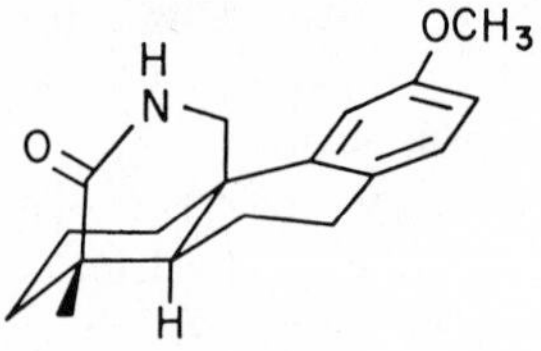   (20)

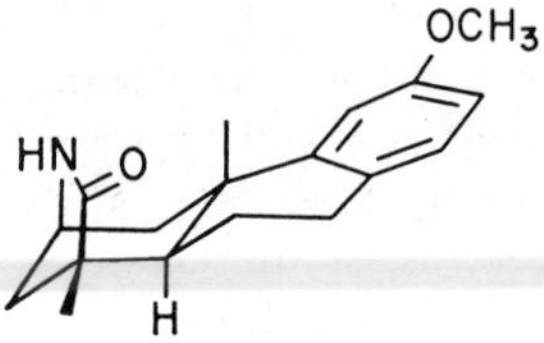

(21) 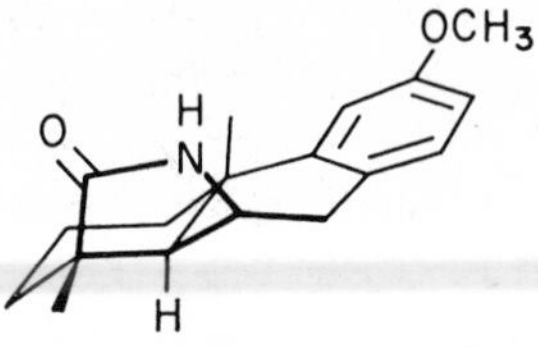   (22)

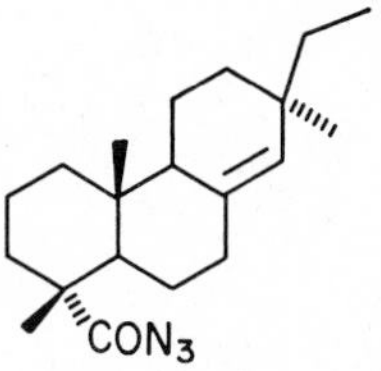

(23) 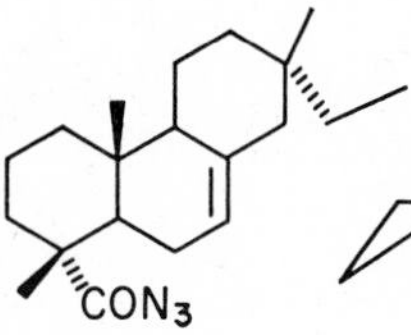   (24) 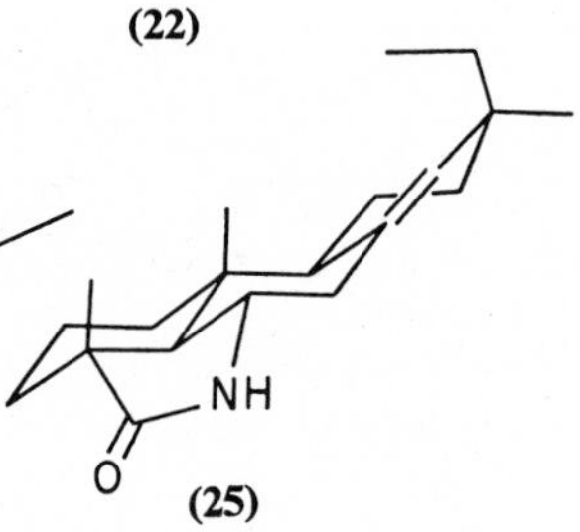   (25)

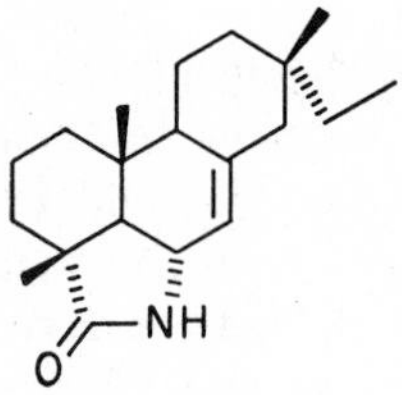

(26)

solvent (cyclohexane) also occurred, giving 3% of N-cyclohexylhexan-amide (29). Linke and Lwowski (23) made a more extensive study of the photolysis of alkylcarbonyl azides in both cyclohexane and methylene chloride. For hexanoyl and heptanoyl azides in cyclohexane they too found the δ-lactams to predominate over the γ-lactams (roughly 2:1 ratio) while less than 2% of N-cyclohexyl amides were formed. In methylene

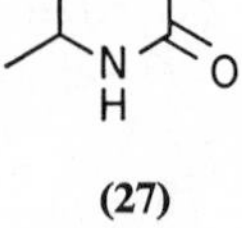

(27) 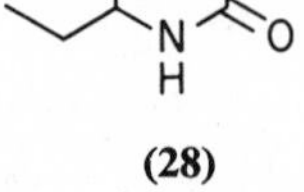   (28) 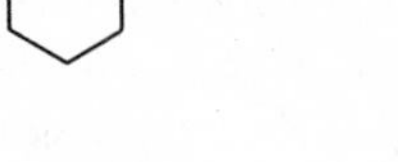   (29)

chloride the total yield of lactams doubled while the ratio of $\delta:\gamma$ remained nearly constant. Pentanoyl azide gave 23.5% of 5-methyl-2-pyrrolidone in methylene chloride but butyryl azide gave only 3.5% of 2-pyrrolidone. Branching in the chain however, dramatically increased the internal attack on the primary C—H bonds as shown by the yields of lactams **30–33** from the corresponding azides (in CH_2Cl_2) (increasing from 2 to 58%).

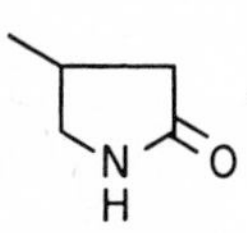

(30) 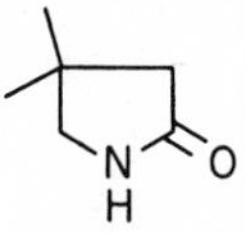   (31) 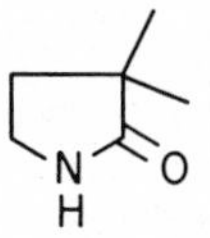   (32) 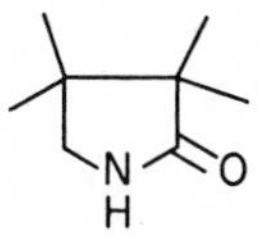   (33)

Masamune also explored the photolysis of acyl azides, and used this in his synthesis of diterpenoid alkaloids (24) (**34–35**). In an extension of the above approach to the azabicyclononane system, Meyer and Levinson (25) studied the photolysis of both the *cis-* and *trans-*decalin derivatives **36** and **37**. Brown (26) also photolyzed these azides. The δ-lactam **38** predominated (ca. 18%) in the case of **36**, but the γ-lactam (probably **39**) predominated (14%) in the products from **37** (see Section V).

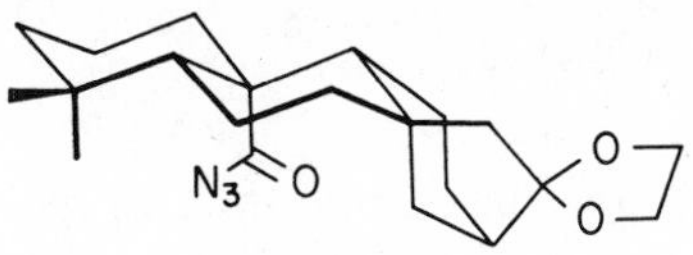

(34) 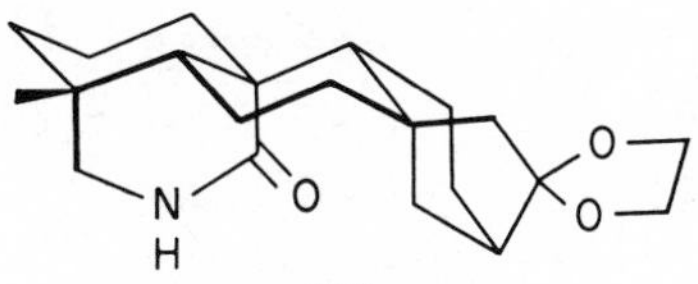   (35)

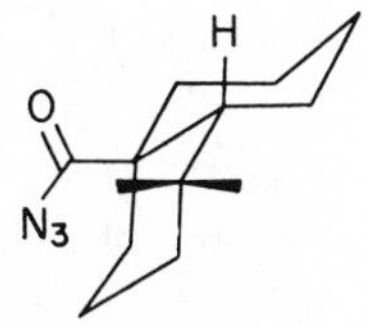

(36) 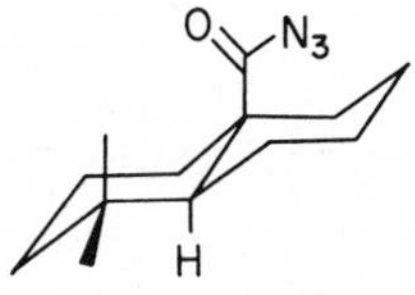   (37)

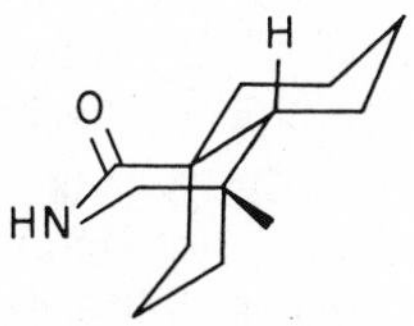

(38) 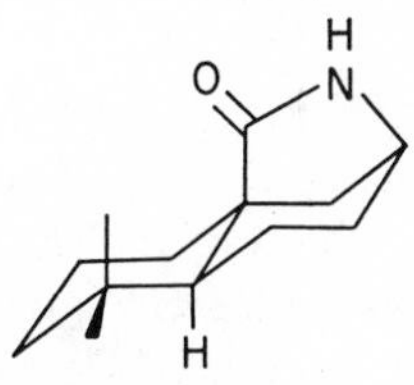   (39)

Robinson and Hofer (27) studied C—H insertion in the photolysis of lithocholic azide (40) in cyclohexane. Some amide (41) probably resulting from triplet nitrene attack on solvent (see Section V), was produced but no N-cyclohexyl amide was identified. The γ- and δ-lactams 42 and 43 were obtained, each in a yield of about 15%. As usual the major product was isocyanate.

(40)   (41)

(42)   (43)

Another example of lactam formation in the natural product field was provided by Huneck (28). Irradiation of the azide of 3α-acetoxyurs-12-en-24-oic acid (44) gave 45 as the major amide product.

(44)   (45)

A most interesting recent development was the photolysis of nitrile oxides by Just and Zehetner (29). The cyclized product (25%) from the nitrile oxide 46, derived from the methyl ether of podocarpic acid, was identical with the δ-lactam 20. The products from irradiation of a methanol solution of mesitonitrile oxide (47) were urethane (48) and lactam (49) in nearly equal amounts. The authors suggested that acylnitrenes were formed via the oxazirene (eq. 2) and that these reacted as expected.

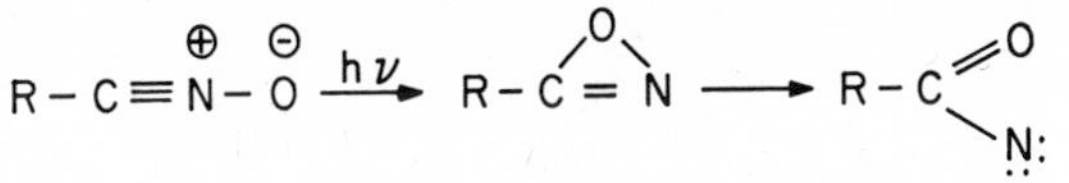

$$R - C \equiv \overset{\oplus}{N} - \overset{\ominus}{O} \xrightarrow{h\nu} R - \overset{\overset{O}{\diagup}\diagdown}{C} = N \longrightarrow R - C\overset{\diagup O}{\underset{\diagdown N:}{}} \qquad (2)$$

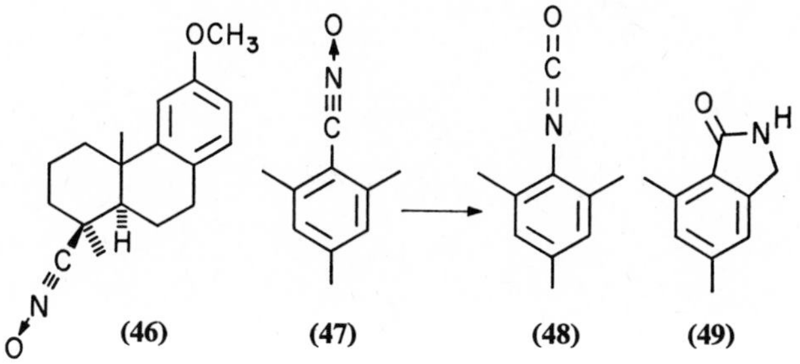

(46) (47) (48) (49)

B. Photolysis of Unsaturated Acyl Azides

Work has been done in the author's laboratory (30) on the internal addition of irradiated acyl azide groups to olefinic double bonds. These are probably additions of acylnitrenes, but the possibility that excited triazolines are intermediates has not been excluded. Again isocyanates are major products, but in the most favorable case studied, up to 20% yields of cyclization product have been obtained.

Despite the probable unfavorable geometry for cyclization of the normal conformation (see **50**) 4-cycloheptenecarbonyl azide gave 30% of polymeric amide and acyl aziridine (**51**). The latter was very readily hydrolyzed to the hydroxylactam **52** and traces of water probably initiated its polymerization. This prompted examination of the photolysis of **50** in aqueous dioxane. Polymerization was suppressed and quite pure hydroxylactam (**52**) was obtained in around 12% yield.

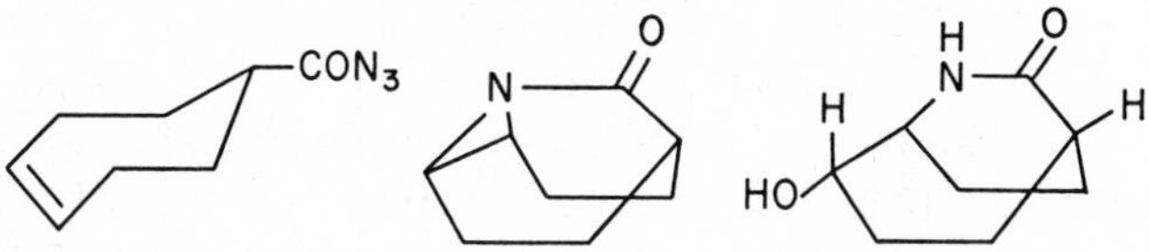

(50) (51) (52)

The *endo*-norbornenylcarbonyl azide **53** also cyclized on irradiation but despite the rigidly enforced proximity of the two functions the yield of the aziridine **54** was only 20%. As in the case of **51** this was only observed spectroscopically because of its high reactivity. It hydrolyzed rapidly in water to the hydroxylactam **55**.

cis-2-Vinylcyclopropylcarbonyl azide **56** was photolyzed at $-78°$. A weak infrared band at 1765 cm^{-1} may have indicated the presence of the

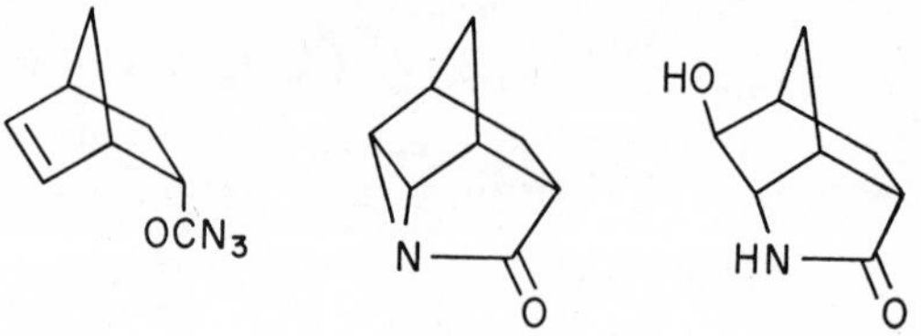

(53)   (54)   (55)

acylaziridine **57**, but attempts to isolate its hydration product were not successful.

(56)   (57)

A third rather rigid example studied was o-vinylbenzazide (**58**). When this was irradiated isocyanate and the acylaziridine **59** (ν_{max} 1750 cm^{-1}) were produced. Water hydrolyzed this to a mixture of hydroxy amides **60** (4%) and **61** (1%).

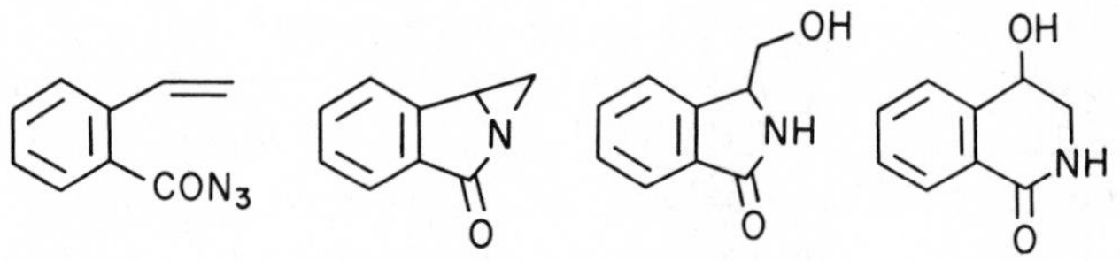

(58)   (59)   (60)   (61)

Simultaneously, Masamune and co-workers (31) were studying the photolysis of 2,3-diphenylcycloprop-2-enylcarbonyl azide (**62**) and the homologous azide (**63**). The former when irradiated in ethyl ether gave isocyanate (50–60%), diphenylacetylene (4%), and amide (**64**) (15%) arising from insertion into the α-CH bond of the ether. Photolysis of **63** followed by heating with ethanol gave the expected urethane (15%), diphenylacetylene (1%), 3,4-diphenyl-2-pyridone (**65**) (21%), 2,5-diphenylpyrrole (**66**) (6%), and 5,6-diphenyl-2-pyridone (**67**) (5%). The authors showed that **66** and **67** did not arise by photolysis of the isocyanate, and they suggested that they arose by initial formation of **68** followed by rearrangement. Possible mechanisms for their formation are illustrated in equations 3 and 4. It was shown that **65** was formed by heating the isocyanate in protic solvents.

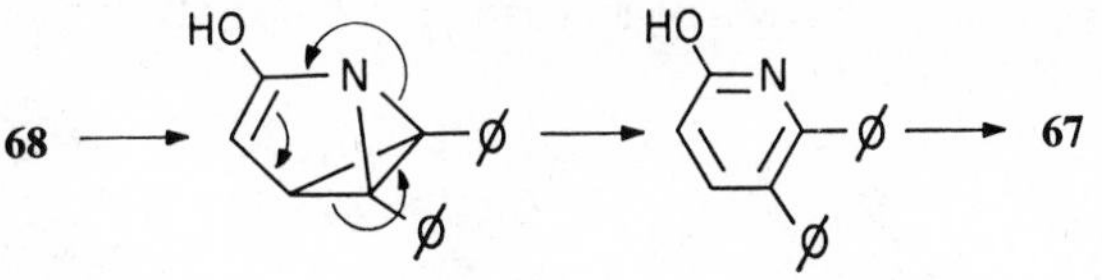

$$68 \longrightarrow \text{[structure]} \longrightarrow \text{[structure]} \longrightarrow 67 \quad (3)$$

$$\text{[structure]} \longrightarrow CO + \text{[structure]} \longrightarrow 66 \quad (4)$$

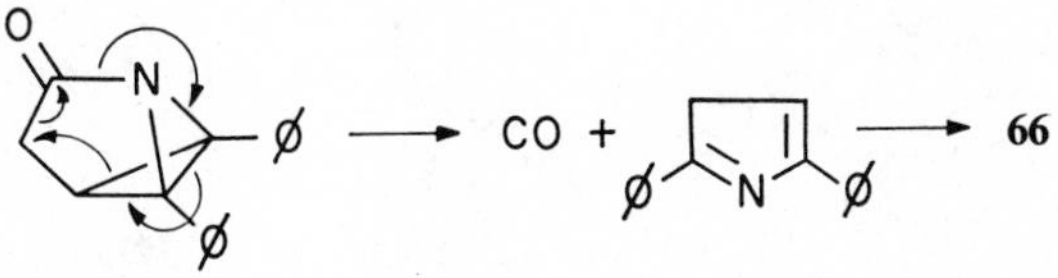

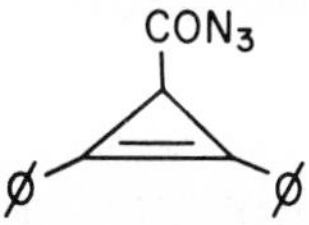

(62) 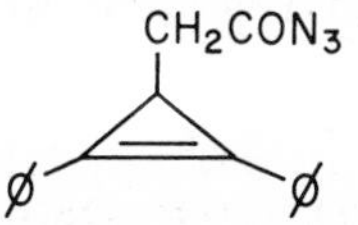   (63) 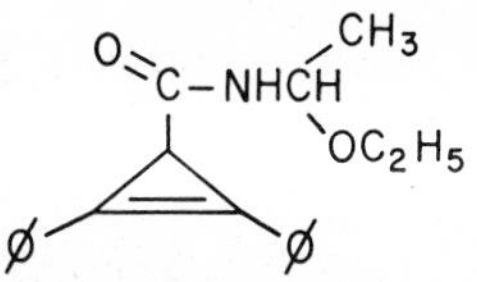   (64)

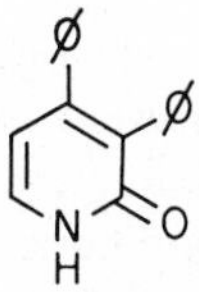

(65) 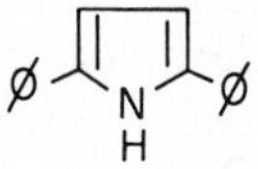   (66) 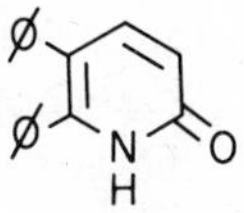   (67) 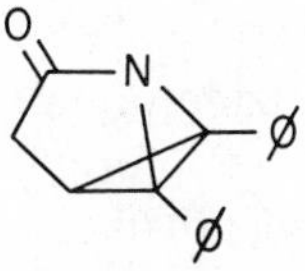   (68)

III. PHOTOLYSIS OF AZIDOFORMATES

The acyl azide reactions described in the preceding section are rather low-yield reactions because of predominant rearrangement to isocyanate. Lwowski and colleagues pioneered the study of photolysis of azidoformates (which do not give much isocyanate) (32), showing that new carbon–nitrogen bonds could be formed by C—H insertion (see Chapter 6). The use of this reaction in cyclization has been developed recently by Kreher and Bockhorn (33) and Puttner and Hafner (34). They showed that photolysis of *t*-butylazidoformate in acetonitrile gave an 80% yield of 5,5-dimethyloxazolidone (69).

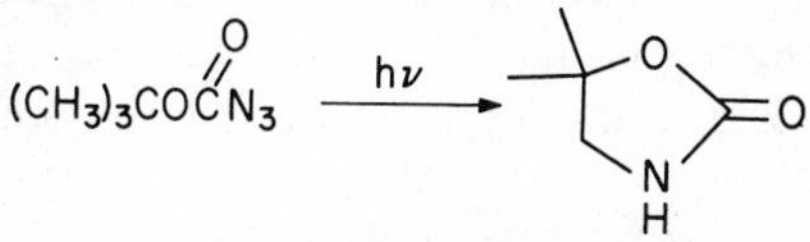

$$(CH_3)_3CO\overset{O}{\overset{\|}{C}}N_3 \xrightarrow{h\nu} \text{[structure]}$$

(69)

Curran and Angier (35) photolyzed 1-adamantyl azidoformate and obtained a 45% yield of adamanto[2,1-*d*]oxazolidin-2-one (70). A small amount (4%) of the cyclohexylcarbamate **71** was also produced by attack on solvent cyclohexane.

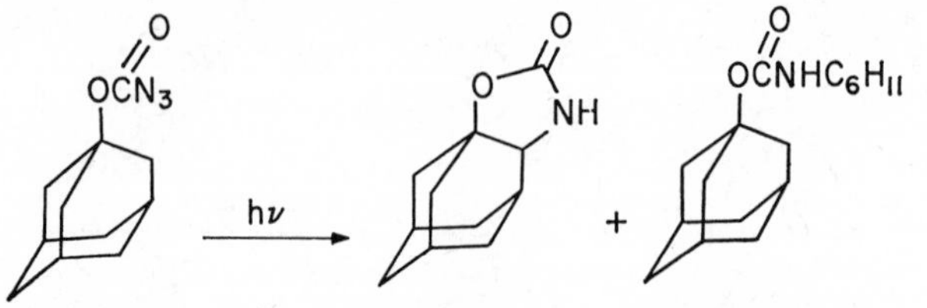

(70)   (71)

These reactions provide good routes to the corresponding amino-alcohols.

IV. THERMAL DECOMPOSITION OF AZIDOFORMATES

A. Cyclization by CH Insertion

Breslow and colleagues (36) studied the thermal decomposition in solution of azidoformates. They found that at 130° in solvent cyclohexane *n*-octadecyl azidoformate gave the corresponding carbamate (23%) and *N*-cyclohexylcarbamates (60%), the *n*-hexadecyloxazolidinone (**72**) (5%) and an isomer of the latter (8%). Contrary to their conclusions, it is probable that the isomer is the 4-*n*-pentadecyltetrahydro-1,3-oxazin-2-one (**73**). It seems likely that the conformational rigidity imparted to **73** by the

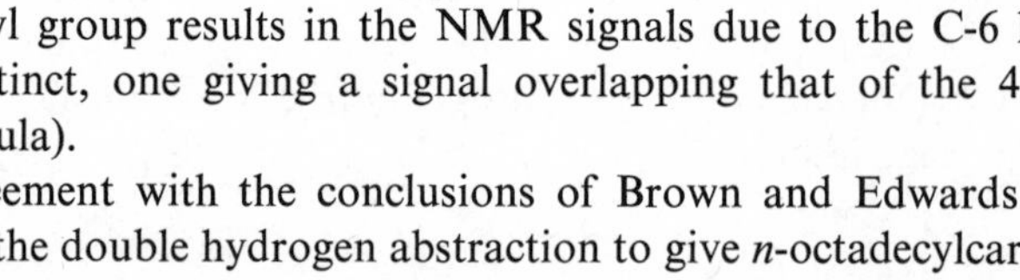

(72)   (73)

large alkyl group results in the NMR signals due to the C-6 hydrogens being distinct, one giving a signal overlapping that of the 4-hydrogen (see formula).

In agreement with the conclusions of Brown and Edwards (22) they attribute the double hydrogen abstraction to give *n*-octadecylcarbamate to triplet nitrene. This is presumably formed by intersystem crossing from the singlet.

Smolinsky and Feuer (8b) were simultaneously studying the stereochemistry at the reacting carbon in thermal sulfonyl azide and azidoformate cyclizations. In the vapor phase decomposition of 2-methylbutyl azidoformate (**74**) they obtained a 68% yield of the oxazolidinone (**75**) besides four minor products. They did not identify any of the tetrahydro-1,3-oxazin-2-one **(76)** which one would expect to be the next most abundant

product. The stereochemistry is discussed in Section V. Huisgen and Blaschke (37) observed the formation of 16% of 2-oxazolidone during thermolysis of ethyl azidoformate in methyl cyanoacetate.

In similar gas-phase thermal decompositions of the corresponding azidoformates Kühling (38) obtained yields of around 75% of cyclic products 77–80. In the case of *n*-propyl azidoformate both 4-methyl-oxazolidinone **(81)** (63%) and tetrahydro-1,3-oxazin-2-one **(82)** (8%) were produced. Hexahydrobenzyl azidoformate **(83)** gave the oxazolidinone **(84)** and the hexahydro-1,3-oxazin-2-one **(85)** in a 3:1 ratio.

B. Substitution of an Aromatic Ring

Kühling (38) found that thermolysis of phenyl azidoformate in the gas phase gave 57% of benzoxazolin-2-one **(86)**. This may be the product of

(86)

direct CH insertion, or of addition to the aromatic ring followed by rearrangement.

There seem to be no recorded cases of cyclizations using alkenyl azidoformates or of cyclizations using chemically generated alkyloxy-carbonyl nitrenes.

V. MECHANISM AND FACTORS DETERMINING RING SIZE

Among the fundamental questions raised by the above work are:

1. Are vibrationally and electronically excited azides reacting directly with the C—H or C=C bonds or aromatic rings or are acylnitrenes intermediates?

2. If nitrenes are intermediates, what is their electronic state at the time of reaction?

3. What factors in acylnitrene reactions are involved in determining the rates of insertion into CH bonds on carbons γ or δ to the carbonyl?

This section reviews the evidence bearing on each of these questions.

(*a*) The similarity between products from reactions of alkoxycarbonyl-nitrene produced by α-elimination, and those from thermally or photo-chemically decomposed azidoformates (32), the fact that the rate of thermolysis of azidoformates is almost independent of the solvents despite their widely-different reactivities (see Chapter 6), the proof of a symmetrical intermediate in cyanogen azide reactions (see Chapter 9) and the ESR evidence for triplet ground state nitrenes, all support the contention that in most cases nitrenes are intermediates in the thermolysis and photolysis of the alkoxycarbonyl and cyanogen azides. The finding of Lwowski and colleagues (18) that in the photolysis of pivaloyl azide the percentage of isocyanate formed does not vary with solvent, can also be construed to mean that excited acyl azides do not react with these solvents and hence that acylnitrenes are the reactive intermediates in acyl azide photolysis.

The reaction of acyl azides with double bonds is less clearly a nitrene process. Indeed, triazolines have been implicated in at least two thermal reactions of azidoformates (39–42). The triazoline **87** gave a mixture of aziridine **88** and imine **89** on photolysis (41), while the photolysis of the mixture of norbornylene and ethyl azidoformate (41) gave no imine. This

(87)   (88)   (89)

suggests that triazoline is not intermediate in the photolysis of the mixture. The cyclization of **50** seemed surprisingly efficient if the normal conformation is as illustrated. This led us to think that excited azide had a reasonable lifetime, leading to triazoline then to aziridine. We hence expected that **53**, which appears from models to give a strain-free triazoline **90**, would give an unusually good yield of aziridine because of the enforced proximity. However, the yield was only 20%. Thus both sets of observations lend some support to the intermediacy of nitrenes in the photochemical decomposition of acyl azide.

(90)

(*b*) Evidence has accumulated that singlet nitrenes are the main reacting entities in direct photolysis of acyl azides, thermolysis of azidoformates and the α-elimination route to alkoxycarbonylnitrenes. This includes comparison of direct photolysis and photosensitized decomposition (22,43) and the study of the stereospecificity of addition to *cis*- and *trans*-2-butene (44) (see Chapter 6 for full discussion). The contribution from cyclization reactions is considerable. Smolinsky and Feuer (8b), in their work on azidoformate cyclization mentioned earlier, showed that at least partial retention of configuration resulted if the hydrogen involved was on an asymmetric carbon. Yamada and co-workers proved that the retention of configuration was complete (45). This is understandable if a singlet CH insertion (see **91**) was involved. Had the triplet nitrene been the intermediate, diradical **92** would have resulted on hydrogen abstraction. Sufficient time would probably elapse before spin inversion and bond formation for the asymmetry to be partially or wholly lost. Finally, in the study by Robinson and Hofer cited earlier, they showed by an independent synthesis that the γ-lactam **42** was formed with retention of configuration at C-20 indicating that acyl azides also operate by singlet CH insertion.

(*c*) In considering the ring size likely to result from acylnitrene insertion,

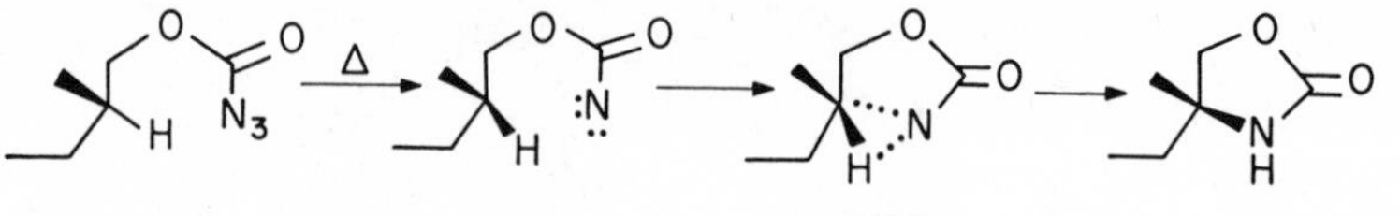

(91)

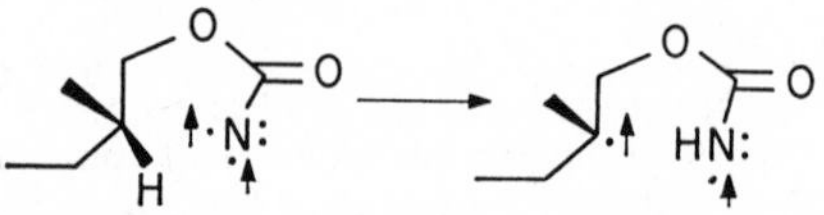

(92)

a statistical factor, the character (I° to III°) of accessible hydrogens, and proximity and steric effects must be considered. In a case of internal CH insertion in a conformationally rigid molecule both a methylene and a methyl group have a statistical factor of 1, the same as a tertiary hydrogen. (This assumes that an axial carbonylnitrene can only insert into an axial CH bond, while an equatorial one will show marked preference for an equatorial CH bond.) (*However, because of rotational flexibility the methyl hydrogens may have a somewhat better chance of adopting an ideal conformation for insertion.*) Ignoring steric effects, the statistical factor for each tertiary carbon, methylene, and methyl in an intermolecular insertion is 1, 2, and 3, respectively.

Both Breslow and colleagues (36) and Lwowski and colleagues (32) have determined the intrinsic (statistically corrected) insertion factor for the three types of hydrogen in both thermal and photochemical intermolecular reactions of azidoformates to be roughly 1:10:30 for primary, secondary, and tertiary CH bonds, respectively. The corresponding ratio for pivaloylnitrene in reaction with 2-methylbutane is 1:9:160 (46).

The work of Linke and Lwowski (23) and Brown and Edwards (22) described above shows that in acyl azide photolysis if equivalent methylenes are available, δ-lactam is favored over γ-lactam formation by a ratio of around 2:1. This was interpreted (22) as a consequence of transition state **93** for singlet insertion being favored over **94**. This ratio will be called the ring-size factor.

In the cases with rigid geometry related to podocarpic acid (**95**), the ratio of δ- to γ-lactam was roughly 4:1 despite the fact that δ-lactam formation required insertion into primary hydrogen. Taking into account

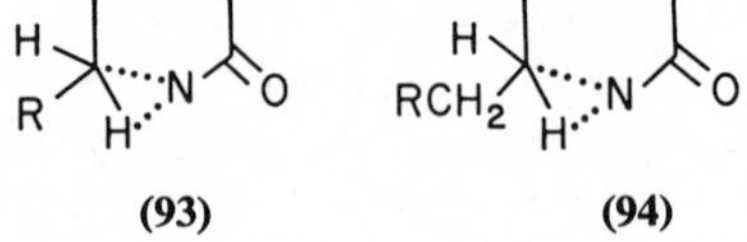

(93)   (94)

the intrinsic I° to II° insertion factor, the ring-size factor, and the statistical factor this ratio would be expected to be at least 1:10. Hence a dominant role was ascribed to the *proximity* of the axial azidocarbonyl and methyl groups (22).

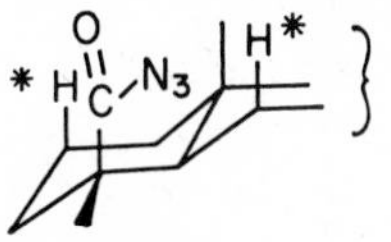

(95)

The importance of another factor was discerned by Meyer and Levinson (25). The acyl azide **96** gave a δ-lactam:γ-lactam ratio of 1:1.5, the reverse of the above examples. They attributed this to the steric hindrance to the approach to the starred hydrogens in **95** and **96**, whereas the 6-

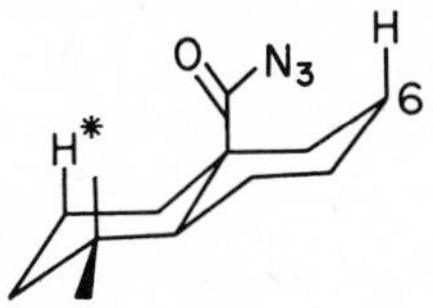

(96)

hydrogen in **96** was unencumbered. Hence the latter case, if the structure is correct, gives a truer measure of the magnitude of the proximity effect.

The nearly equivalent amounts of δ- and γ-lactams formed in Robinson and Hofer's work where insertion is into tertiary CH bonds in each case (see **42** and **43**), probably reflects greater steric impedance to attack on the one in C-17.

If the component not rigorously identified (see above) in the thermolysis studies of Breslow and colleagues is the tetrahydrooxazinone **73**, then again attack on equivalent methylenes gives more δ- (8%) than γ- (5%) lactam. However, the work of Kühling reported in Section IV-A leaves some doubt as to the favored transition state ring size for azidoformates. In the propyl azidoformate case, applying the 10:1 intrinsic factor for secondary vs. primary CH insertion, and the statistical factor, one expects a 7:1 ratio of γ- to δ-lactam. The ratio is actually 8:1, hence the smaller transition

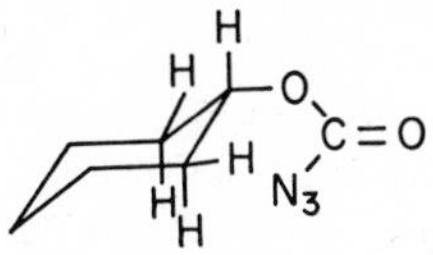

(97)

state is slightly favored. In addition, if one assumes that only the equatorial "ortho" CH bonds of hexahydrobenzyl azidoformate (see **97**) are competing with the tertiary CH bond, then a ratio of 1.5:1 γ- to δ-lactam is predicted in contrast to the 3:1 found. More examples are needed.

One of the most valuable features then, of acyl azide and azidoformate cyclizations is that *in contrast to the variety of radical-type substitutions on saturated carbon, substantial yields of six-membered nitrogen-containing rings can be formed.*

REFERENCES

1. T. Zincke and A. T. Lawson, *Ber.*, **20**, 1176 (1887); T. Zincke and H. Jaenke, *Ber.*, **21**, 546 (1888).
2. E. Noelting and K. Kohn, *Chem. Z.*, **18**, 1095 (1894).
3. P. A. S. Smith, B. B. Brown, R. K. Putney, and R. F. Reinisch, *J. Am. Chem. Soc.*, **75**, 6335 (1953).
4. J. H. Boyer and Ellzey, *J. Org. Chem.*, **26**, 4684 (1961).
5. P. J. Bunyan and J. I. Cadogan, *J. Chem. Soc.*, **1963**, 42.
6. P. A. S. Smith and B. B. Brown, *J. Am. Chem. Soc.*, **73**, 2435, 2438 (1951). P. A. S. Smith and J. H. Boyer, *J. Am. Chem. Soc.*, **73**, 2626 (1951).
7. K. H. Saunders, *J. Chem. Soc.*, **1955**, 3275.
8a. G. Smolinsky, *J. Am. Chem. Soc.*, **82**, 4717 (1960); **83**, 2489 (1961).
8b. G. Smolinsky and B. Feuer, *J. Am. Chem. Soc.*, **86**, 3085 (1964).
9. D. H. R. Barton and L. R. Morgan, *J. Chem. Soc.*, **1962**, 622.
10. D. H. R. Barton and A. N. Starratt, *J. Chem. Soc.*, **1965**, 2444.
11. G. Smolinsky, *J. Org. Chem.*, **27**, 3557 (1962).
12. L. Horner, A. Christmann and A. Gross, *Chem. Ber.*, **96**, 399 (1963).
13. A. Hassner and F. W. Fowler, *Tetrahedron Letters*, **1967**, 1545.
14. A. L. Logothetis, *J. Am. Chem. Soc.*, **87**, 749 (1965).
15. K. R. Brower, *J. Am. Chem. Soc.*, **83**, 4370 (1961); **85**, 1401 (1963). E. Whalley, *Advan. Phys. Org. Chem.*, **2**, 147 (1964).
16. Y. Yukawa and Y. Tsuno, *J. Am. Chem. Soc.*, **79**, 5530 (1957).
17. A. Fry and J. C. Wright, Dissertation of J. C. Wright, Univ. of Arkansas, Fayetteville, Ark., 1966.
18. S. Linke, G. T. Tisue, and W. Lwowski, *J. Am. Chem. Soc.*, **89**, 6308 (1967).
19. J. W. ApSimon and O. E. Edwards, *Proc. Chem. Soc.*, **1961**, 461; *Can. J. Chem.*, **40**, 896 (1962).
20. L. Horner and E. Spietschka, *Chem. Ber.*, **88**, 934 (1955); A. Schönberg, *Preparative Organische Photochemie*, Springer, Berlin, 1958, p. 190.
21. W. Antkowiak, O. E. Edwards, R. Howe, and J. W. ApSimon, *Can. J. Chem.*, **43**, 1257 (1965).
22. I. Brown and O. E. Edwards, *Can. J. Chem.*, **45**, 2599 (1967).
23. S. Linke and W. Lwowski, personal communication.
24. S. Masamune, *J. Am. Chem. Soc.*, **86**, 288–291 (1964).
25. W. L. Meyer and A. S. Levinson, *J. Org. Chem.*, **28**, 2859 (1963).
26. R. F. C. Brown, *Australian J. Chem.*, **17**, 47 (1965).
27. C. H. Robinson and P. Hofer, personal communication and 151st Meeting, American Chemical Society, Pittsburg, Pa., March 1966, *Abstr. Papers K-19.*

28. S. Huneck, *Chem. Ber.*, **98**, 2305 (1965).
29. G. Just and W. Zehetner, *Tetrahedron Letters*, **1967**, 3389.
30. I. Brown, O. E. Edwards, J. M. McIntosh, and D. Vovelle, *Can. J. Chem.*, in press.
31. N. C. Castellucci, M. Kato, H. Zendra, and S. Masamune, *Chem. Commun.*, **1967**, 473.
32. W. Lwowski and T. W. Mattingly, *Tetrahedron Letters*, **1962**, 277. See W. Lwowski, *Angew. Chem.*, **79**, 922 (1967); *Intl. Ed.*, **6**, 897 (1967).
33. R. Kreher and G. H. Bockhorn, *Angew. Chem.*, **76**, 681 (1964).
34. R. Puttner and K. Hafner, *Tetrahedron Letters*, **1964**, 3119.
35. W. V. Curran and R. B. Angier, *Chem. Commun.*, **1967**, 563.
36. D. S. Breslow, T. J. Prosser, A. F. Marcantonio, and C. A. Genge, *J. Am. Chem. Soc.*, **89**, 2384 (1967).
37. R. Huisgen and H. Blaschke, *Ann.*, **686**, 145 (1965).
38. R. Kreher and D. Kühling, Dissertation of D. Kühling, Univ. of Darmstadt, 1965. *Angew. Chem.*, **77**, 43 (1965).
39. I. Brown and O. E. Edwards, *Can. J. Chem.*, **43**, 1266 (1965).
40. A. C. Oehlschlager, P. Tillman, and L. H. Zalkow, *Chem. Commun.*, **1965**, 596.
41. O. E. Edwards, J. W. Elder, and M. Lesage, unpublished work.
42. P. Scheiner, *J. Org. Chem.*, **30**, 7 (1965).
43. W. Lwowski and T. W. Mattingly, *J. Am. Chem. Soc.*, **87**, 1947 (1965).
44. J. S. McConaghy and W. Lwowski, *J. Am. Chem. Soc.*, **89**, 2357, (1967).
45. S. Yamada, S. Terashima, and K. Achiwa, *Chem. Pharm. Bull. Japan*, **13**, 751 (1965).
46. G. T. Tisue, S. Linke, and W. Lwowski, *J. Am. Chem. Soc.*, **89**, 6303 (1967).

CHAPTER 8

Sulfonylnitrenes

DAVID S. BRESLOW

Hercules Incorporated,* Wilmington, Delaware

I. INTRODUCTION

Curtius and Lorenzen (1) reported the first synthesis of a sulfonyl azide in 1898. However, fifteen years passed before Curtius, in a lecture at Heidelberg (2,3), described, in work carried out with J. Rissom, the formation of toluidides from the thermolysis of arenesulfonyl azides in toluene, a reaction now known to involve the formation of a sulfonyl-nitrene intermediate. In the following year (4) he defined "starre" azides as those which can lose a molecule of nitrogen without suffering re-arrangement, and drew a picture of the residue (1) remarkably like one we would draw today to represent a sulfonylnitrene (2).

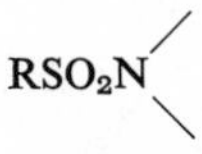

$$RSO_2N\diagup\diagdown \qquad\qquad RSO_2\ddot{N}\!\!\cdot\!\cdot$$

(1) 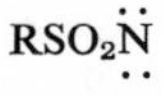 (2)

* Contribution No. 1474.

Although a few papers concerned with sulfonyl azides appeared in succeeding years, most of Curtius' work on sulfonyl azides was reported posthumously some fifteen years later (5,6).

Recent reviews (7–10) on sulfonyl azides, the predominant if not the sole precursors of sulfonylnitrenes, include only scattered references to work carried out between 1930 and 1960, but recently there has been a reawakening of interest in this interesting class of reactive intermediate.

II. SYNTHESIS AND PROPERTIES OF SULFONYL AZIDES

A considerable number of sulfonyl azides have been reported in the literature; monosulfonyl azides are listed in Table I and disulfonyl azides in Table II. Although the first sulfonyl azides were made by the nitrosation of the corresponding hydrazides (1) (method B in Tables I and II), the reaction of sulfonyl chlorides with sodium azide (method A in Tables I and II), the so-called Forster-Fierz procedure, has proved to be more popular.

$$RSO_2NHNH_2 + HNO_2 \longrightarrow RSO_2N_3 + 2H_2O$$

$$RSO_2Cl + NaN_3 \longrightarrow RSO_2N_3 + NaCl$$

Both methods generally give excellent yields; in fact, the yield variations shown in the Tables probably reflect differences in workup procedure, purity, etc., rather than differences in the actual course of the reactions. However, Cremlyn (30) has claimed that products obtained by nitrosation are more readily purified than those from sodium azide. Inasmuch as the hydrazides are usually prepared from the chlorides, method A is shorter than B, but since hydrazine is cheaper than sodium azide, there is probably little to choose economically. Aromatic sulfonyl chlorides are of course readily available by reaction of the aromatic compound with chlorosulfonic acid. Aliphatic sulfonyl chlorides are usually prepared from the corresponding sulfonic acids or mercaptans, but neither is widely available. However, if pure compounds are not required, sulfonyl chlorides can be made in one step by the Reed chlorosulfonation reaction, and Newburg (58) has described the synthesis of chlorinated alkanesulfonyl azides by this procedure.

The most frequently used procedure, introduced originally by Curtius (6), for reacting a sulfonyl chloride with sodium azide involves adding an ethanolic solution of a sulfonyl chloride to a concentrated aqueous sodium azide solution diluted with ethanol; the reaction is mildly exothermic and is over in a short time. Dilution with water then precipitates the azide. Dermer and Edmison (22) recommend keeping the sodium

TABLE I
Monosulfonyl Azides, RSO_2N_3

R	Method[a]	Yield, %	mp, °C	bp °C	bp mm	Other properties	Ref.
CH_3	A	61		56	0.5		11
	A	62	18	41–42	1.3	n_D^{22} 1.4695	12
	A			71	4.5		13
	A	55	18–20				14
C_2H_5	A	78	< −20	55–56	2.7	n_D^{22} 1.4619	12, 15
$(CH_3)_2CH$	A		−10				16
$n\text{-}C_4H_9$	A	78	−9 to −8.5				14, 17
$n\text{-}C_5H_{11}$	A						16
$C_2H_5CHCH_2CH_2$ | CH_3	A	95				Optically active (S)	18
$n\text{-}C_8H_{17}$	A	76	31–32			Oil	17
$n\text{-}C_{12}H_{25}$	A	84		52.3			17
CF_3	A	87	54		444		19
$C_6H_5CH_2$	A, B	90	54				20
	A	98	53.5–54.5				12
	B	90					14
C_6H_5	B	80–84	13–15	95–97	2	Oil	1
	B	92	13–14			Oil	21, 22
	A						23
	A		12.5–13.5				24, 25
	A		10.5–11.5				26
	B						27, 28
	B					n_D^{25} 1.5488	29

(continued)

247

TABLE I (continued)

R	Method[a]	Yield, %	mp, °C	bp °C	mm	Other properies	Ref.
p-CH₃C₆H₄	B		24–26				30
	B	96	22–23				23
	A	83	22	106–110	2		31–33, 12
	B	95	22				31
	A	90–92	21–22				34, 35
	A	100	20				36
	A	60, 86	19–20				25, 37
	A		18–19				26
o-C₆H₅C₆H₄			60–61				20a
o-ClC₆H₄	B	92	37–38				23
p-ClC₆H₄	A, B	97	39				38
	A	100	33				36
	A		31.5–32				26
p-BrC₆H₄	A	56	54.5–56				14
	A		54–55				26
	A	97	53–53.5				37
	A		50				39
o-IC₆H₄	A	63	30–32				25
p-CH₃OC₆H₄	A	96	55				12
	A		54–54.5				26
	A	95	51.5–52				14
	B	98	49–51				23
o-C₆H₅OC₆H₄	A	93	135 (dec.)			Dec. 135°	15
	A		79–80				37
	A						20a

Compound	Method	Yield (%)	M.p. (°C)	Ref.
o-C₆H₅SC₆H₄				20a
$o\text{-}C_6H_5SC_6H_4$				20a
$o\text{-}C_6H_5SO_2C_6H_4$				20a
$o\text{-}NO_2C_6H_4$	A	92	99.5–100	12
	A	78	74	37, 25
$m\text{-}NO_2C_6H_4$	A	92	71–72.5	14
	A	99	80.5–81	37
$p\text{-}NO_2C_6H_4$	A	93	78–80	12
	A	75	103	14
	A	88	102–102.5	26, 27
	B	95	101–102	23
$p\text{-}NH_2C_6H_4$		50	100–101	40
$p\text{-}CH_3CONHC_6H_4$	A	77–81	36	41
	B	62	113–114	30
			112–114 (dec.)	
$p\text{-}OCNC_6H_4$	A, B	93	107	40
$o\text{-}NCC_6H_4$	A	65–70		42
$p\text{-}[Cl^-\ {}^+N_2C_6H_4]$	A	40		43
$p\text{-}C_6H_5N{=}NC_6H_4$	B		117–118	40, 44
$4\text{-}(4\text{-}HOC_6H_4)N{=}NC_6H_4$	B		135–137	44
$4\text{-}[4\text{-}(CH_3)_2NC_6H_4]N{=}NC_6H_4$			158	40
$p\text{-}$ (ring)—NN$=$NC₆H₄	B			45
$4\text{-}(2\text{-}HO\text{-}5\text{-}ClC_6H_3)N{=}NC_6H_4$			135–137	44
$4\text{-}(3\text{-}CH_3\text{-}4\text{-}HOC_6H_3)N{=}NC_6H_4$			140	44
$4\text{-}(2\text{-}HO\text{-}1\text{-}naphthyl)\text{—}N{=}NC_6H_4$			161	40
$4(4\text{-}HO\text{-}1\text{-}naphthyl)\text{—}N{=}NC_6H_4$			173–175	44
			185–187	44

(continued)

TABLE I (continued)

R	Method[a]	Yield, %	mp, °C	bp °C	bp mm	Other Properties	Ref.
2,5-Cl$_2$C$_6$H$_3$	A	97	50–52.5				37
3,4-Cl$_2$C$_6$H$_3$	A	89	60–61				37
2,5-Br$_2$C$_6$H$_3$	A	92	65–67				37
2,3,5,6-(CH$_3$)$_4$C$_6$H	A, B	85	53.5–54				20a
1-Naphthyl	B		53				46
2-Naphthyl	A		44–46				1
	A		45				47
4-CH$_3$CONH-1-naphthyl	A	51	44–44.5				14
8-Quinolyl	B		154–156				30
2-Anthraquinonyl	A					Dec. 121°	15
F	A		153				48
Cl	A	65		38–39	215		19, 49
NH$_2$	A	90		116–117			49
n-C$_4$H$_9$NH	A	61	28–29				49
C$_6$H$_{12}$NH		28				Oil	49
C$_6$H$_5$NH		35				Oil	49
p-CH$_3$C$_6$H$_4$NH		50	48–49				49
p-ClC$_6$H$_4$NH		22	41–42				49
(C$_2$H$_5$)$_2$N		22	50–52				49
(n-C$_4$H$_9$)$_2$N	A	86				Oil	50
C$_6$H$_5$NCHO	A	70				Oil	49
	A	22	79–80				49
p-CH$_3$C$_6$H$_4$SO$_2$NC$_6$H$_5$	A	9	117–118				49

[a] Method A: sulfonyl halide plus sodium azide. Method B: sulfonyl hydrazide plus nitrous acid.

azide solution cold while adding the sulfonyl chloride. Boyer and co-workers (11), as well as Goerdeler and Ullmann (17), added dry sodium azide to methanolic solutions of sulfonyl chloride; the yields appear to be somewhat on the low side, however. Zalkow and Oehlschlager (59) recommended, in the synthesis of benzenesulfonyl azide, that the undiluted sulfonyl chloride be added to a cold sodium azide solution in aqueous alcohol, claiming the formation of up to 25% ethyl benzenesulfonate if the chloride were first dissolved in ethanol; in light of the very slow reaction of sulfonyl chlorides with alcohols in the absence of bases (60), this claim is difficult to rationalize. In any case, this problem is readily circumvented by running the reaction in aqueous acetone, which appears to be the solvent of choice at the present time. Shozda (61) has recommended sulfur dioxide as a solvent for the reaction, and in certain instances this might be advantageous.

Few other azide syntheses have been reported. Kobayashi and Yamamoto (62) attempted to prepare a sulfinyl azide by the reaction of a sulfinyl chloride with sodium azide, but they obtained a sulfonyl azide instead.

$$ArSOCl + NaN_3 \longrightarrow ArSO_2N_3 + ArSO_2SAr + ArSSAr$$

Maricich (62a) has shown that benzenesulfinyl azide can be prepared at $-35°$ (it showed none of the expected nitrene reactions on warming to $0°$, at which temperature it evolved nitrogen). Cremlyn (30) reported a 33% yield of p-acetamidobenzenesulfonyl azide (3) by nitration of the corresponding hydrazide; tosyl hydrazide reacted similarly.

$$p\text{-}CH_3CONHC_6H_4SO_2NHNH_2 + HNO_3 \xrightarrow[\substack{0° \\ 33\%}]{H_2SO_4} p\text{-}CH_3CONHC_6H_4SO_2N_3$$

$$(3)$$

A mechanism involving initial nitration was suggested:

$$ArSO_2NHNH_2 \longrightarrow \underset{\underset{NO_2}{|}}{ArSO_2NNH_2} \xrightarrow{-H_2O} \underset{\underset{NO}{|}}{ArSO_2\overset{\oplus}{N}{=}\overset{\ominus}{N}}$$

$$\xrightarrow{\text{reduction}} ArSO_2N_3$$

Since the final step involves reduction and rearrangement to the linear azide group, it might be more reasonable to postulate an initial reduction of the nitric acid to N_2O_3 (or a precursor) followed by normal nitrosation; the low yield of azide indicates considerable loss of hydrazide, presumably by oxidation.

As shown in Tables I and II, a wide variety of sulfonyl azides have been

TABLE II
Disulfonyl Azides

Structure	Method[a]	Yield, %	mp, °C	Other properties	Ref.
$N_3SO_2N_3$	A		Oil		51–53
$N_3O_2S(CH_2)_3SO_2N_3$	A	85	52		17
$N_3O_2S(CH_2)_4SO_2N_3$	A		85–86		16
$N_3O_2S(CH_2)_5SO_2N_3$	A	91	82		17
$N_3O_2S(CH_2)_6SO_2N_3$	A		88.5–91		16
$N_3O_2S(CH_2)_9SO_2N_3$	A		44.6–45.5		16
$N_3O_2S(CH_2)_{10}SO_2N_3$	A		89.5–91.5		16
Cyclopentane-1,3-$(SO_2N_3)_2$	A	93	39–43	Mixture of isomers	54
Cyclohexane-1,4-$(SO_2N_3)_2$	A	78	170–172	mp 175–176° Fischer block. *trans*-isomer?	54
$N_3O_2SCH_2$—(cyclohexane-1,4-diyl)—$CH_2SO_2N_3$	A		126–129.5	Mixture of isomers	16
m-$C_6H_4(CH_2SO_2N_3)_2$	A		129–131		16
p-$C_6H_4(CH_2SO_2N_3)_2$	A		164–165		16
m-$C_6H_4(SO_2N_3)_2$	A	70	82		55
$CH_3C_6H_3$-2,4-$(SO_2N_3)_2$	B	98	82–84		23, 55
1,2-$(CH_3)_2C_6H_2$-3,5-$(SO_2N_3)_2$	B		92–93		15
p,p'-$(—C_6H_4SO_2N_3)_2$	A		80		15
p,p'-$N_3O_2SC_6H_4N{=}NNHC_6H_4SO_2N_3$	A	63	133–134	Dec. 144–145°	56, 17
Naphthalene-1,5-$(SO_2N_3)_2$	A	90–96	177		40, 57
Naphthalene-2,6-$(SO_2N_3)_2$	A	66	152		17

[a] Method A: sulfonyl halide plus sodium azide. Method B: sulfonyl hydrazide plus nitrous acid.

252

reported. For example, Smolinsky and Feuer (18) prepared (*S*)-3-methyl-pentane-1-sulfonyl azide (**4**) in the following manner:

$$\underset{(+)\,(S)}{CH_3CH_2\overset{\underset{\displaystyle CH_3}{|}}{C}HCH_2CH_2OH} \xrightarrow[\underset{90\%}{}]{\begin{array}{c}1.\ TsCl\\2.\ KSCN\end{array}} CH_3CH_2\overset{\underset{\displaystyle CH_3}{|}}{C}HCH_2CH_2SCN \xrightarrow[22\%]{Cl_2-H_2O}$$

$$CH_3CH_2\overset{\underset{\displaystyle CH_3}{|}}{C}HCH_2CH_2SO_2Cl \xrightarrow[95\%]{\begin{array}{c}KN_3\\EtOH-H_2O\end{array}} CH_3CH_2\overset{\underset{\displaystyle CH_3}{|}}{C}HCH_2CH_2SO_2N_3$$
$$(\textbf{4})$$

Many arenesulfonyl azides with various functional groups on the aromatic ring have been prepared, either directly or indirectly. Thus, Curtius and Stoli (40) diazotized *p*-aminobenzenesulfonyl hydrazide (**5**) to *p*-azidosulfonylphenyldiazonium chloride (**6**), which underwent the normal reactions of a diazonium salt. The same compound was prepared by hydrolyzing *p*-acetamidobenzenesulfonyl azide (**7**) and diazotizing the resulting amine. Von Glahn and Rudner (45) reacted the diazonium salt

with piperidine to make the corresponding diazoamino derivative. Cremlyn (44) extended the diazo coupling reaction to a variety of phenols;

the parent compound, p-phenylazobenzenesulfonyl azide (8), was prepared from the hydrazide. Only in the attempted synthesis of 2-anthraquinone-

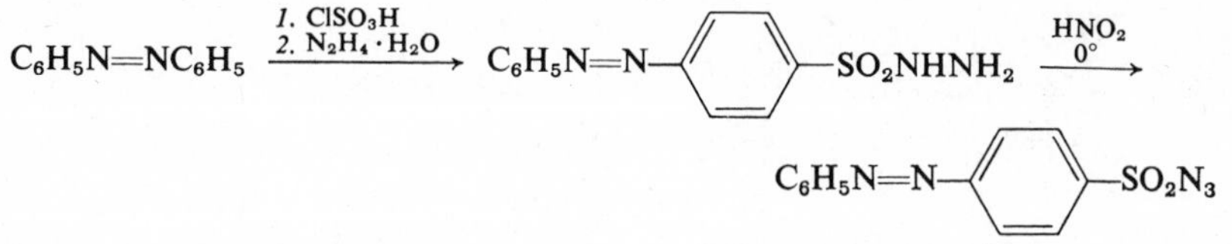

$$C_6H_5N{=}NC_6H_5 \xrightarrow[\text{2. } N_2H_4 \cdot H_2O]{\text{1. } ClSO_3H} C_6H_5N{=}N{-}\!\!\!\left\langle\right\rangle\!\!\!{-}SO_2NHNH_2 \xrightarrow[0°]{HNO_2}$$

$$C_6H_5N{=}N{-}\!\!\!\left\langle\right\rangle\!\!\!{-}SO_2N_3$$

(8)

sulfonyl azide (9) was the hydrazide nitrosation reported to fail; the sodium azide procedure was successful (48).

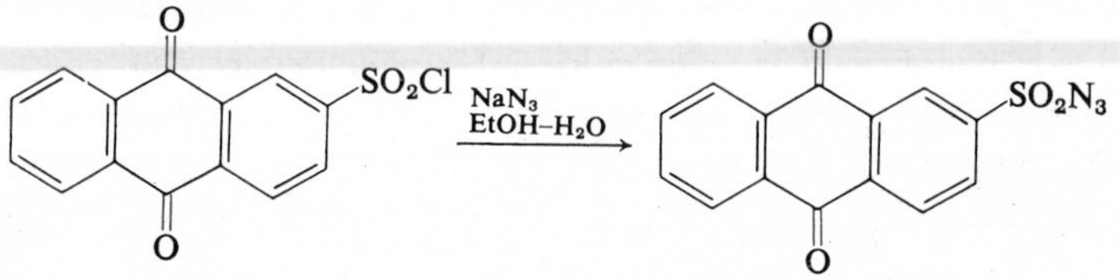

(9)

Danhaeuser and Pelz (42) prepared p-azidosulfonylphenyl isocyanate (11) via a Curtius rearrangement of the corresponding carbonyl azide (10).

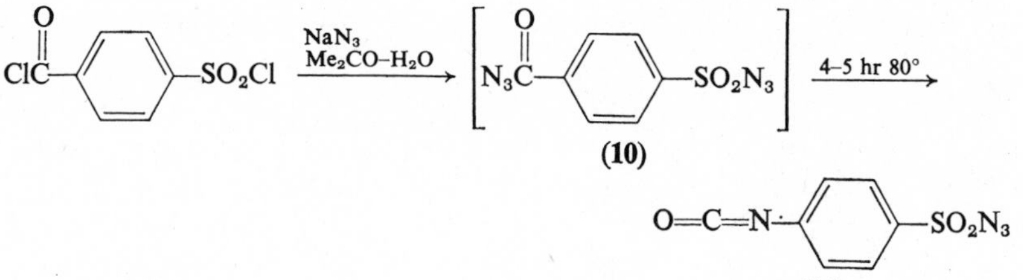

(10)

$$O{=}C{=}N{-}\!\!\!\left\langle\right\rangle\!\!\!{-}SO_2N_3$$

(11)

Even inorganic sulfonyl azides can be prepared from a sulfonyl halide and sodium azide. Thus, Ruff (19) prepared fluorosulfonyl azide (12) from pyrosulfuryl fluoride, while Shozda and Vernon (49) prepared it from sulfuryl fluoride and lithium azide.

$$S_2O_5F_2 + NaN_3 \xrightarrow[65\%]{\substack{CH_3NO_2 \\ 14 \text{ hr r.t.}}} FSO_2N_3 + [SO_2 + SO_2F_2]$$

(12)

$$SO_2F_2 + LiN_3 \xrightarrow[HCONMe_2]{Me_2O} FSO_2N_3$$

(12)

In a similar fashion the latter authors prepared chlorosulfonyl azide from chlorosulfonyl fluoride (or sulfuryl chloride). Hardy and Adams (50) patented the preparation of N,N-dialkylsulfamoyl azides (13) from the corresponding chlorides. Shozda and Vernon (49) extended the reaction to a variety of sulfamoyl chlorides, and showed that the same compounds could be made from the amine and chlorosulfonyl azide.

$$RR'NSO_2Cl + NaN_3 \xrightarrow{\text{MeCN}} RR'NSO_2N_3 + NaCl$$
(13)

$$RR'NH + ClSO_2N_3 \longrightarrow RR'NSO_2N_3 + RR'NH \cdot HCl$$
(13)

Although some of these compounds undergo reactions typical of organic sulfonyl azides, no nitrene reactions have been reported; presumably they have not been investigated.

Curtius and Schmidt (51–53) described the synthesis of sulfuryl azide (14) in a similar fashion.

$$SO_2Cl_2 + NaN_3 \xrightarrow[\text{r.t.}]{\text{24 hr}} N_3SO_2N_3$$
(14)

Because of its hazardous nature this compound has never been isolated and adequately identified. Although Schmidt (52) claimed in one instance to have made a chlorine-free preparation, this work must be somewhat suspect in view of the total lack of analysis or data on physical properties.

Aliphatic and simple aromatic sulfonyl azides are usually liquids or fairly low-melting, crystalline solids; disulfonyl azides are generally solids. The lower members of the series are distillable *in vacuo*; however, adequate shielding should be provided and it is recommended that only small samples be distilled. Sulfonyl azides are quite polar and are best recrystallized, as a rule, from methanol or ethanol. However, azide ion is such a good nucleophile that, by use of a reasonable excess, sulfonyl chlorides are converted into sulfonyl azides quantitatively, and frequently no purification is necessary.

The azide group in sulfonyl azides absorbs strongly at about 2130 cm.$^{-1}$ (11,28,44), and this absorption is relatively insensitive to substituents in p-substituted arenesulfonyl azides (62b). From their ultraviolet spectra Gal'perin and Balabanov (62b) concluded that the sulfonyl azide group on the benzene ring is electron-withdrawing.

All sulfonyl azides liberate nitrogen on heating above approximately 125°. Therefore, high-melting compounds melt with decomposition; this is often ignored in the literature, and there is occasional confusion between melting point and decomposition temperature (37). Aliphatic compounds are more stable than aromatics (63). According to Leffler and Tsuno (25) and Horner and Christmann (12), substituents on a benzene ring have only a small effect on the decomposition temperatures of the corresponding sulfonyl azides (see Section III-A). Curtius (47) reported that 2-naphthalenesulfonyl azide does not decompose in refluxing toluene, indicating it to be somewhat more stable than benzenesulfonyl azide.

Curtius (6) claimed that sulfonyl azides, with the exception of sulfuryl azide, are safe to handle, and according to Goerdeler and Ullmann (17) crystalline sulfonyl azides are not shock sensitive. However, in this laboratory shock sensitivity has been found to be a function of purity. For example, slightly impure methanesulfonyl azide has been found to be as sensitive to shock as nitroglycerine when tested in the standard drop test using a 2-kg weight (64); while pure azide is almost inert to impact, it is comparable to TNT in its explosive power. All sulfonyl azides decompose more or less vigorously and exothermally when strongly heated, and arenesulfonyl azides have been reported to detonate above 165° (65). However, sulfonyl azides containing sufficient organic matter (ca. 5 carbon atoms per azide group in aliphatic compounds) are no longer sensitive to shock, and even sensitive compounds can be handled safely in dilute solution.

III. THERMOLYSIS OF SULFONYL AZIDES

Although Curtius (4–6) classified sulfonyl azides as "starr" or non-rearranging, Reichle (68) isolated a 17.5% yield of azobenzene from the vapor-phase pyrolysis of benzenesulfonyl azide at 625°; Balabanov, Dergunov, and Gal'perin (66) isolated trace amounts from the decomposition in refluxing cyclohexanone. While Balabanov and co-workers suggested a sulfonylnitrene intermediate, it is just as reasonable to postulate, by analogy with the thermolysis of aliphatic sulfonyl azides (see below), initial elimination of sulfur dioxide and formation of phenyl azide, which is known to give azobenzene as one of its major decomposition products.

A. Kinetics

Although nitrenes had been suggested as intermediates in the decomposition of sulfonyl azides (2), no evidence for their actual existence appeared in the literature until quite recently. In 1963 Horner and Christmann (12) made the qualitative observation that tosyl azide evolved nitrogen at 130–135° at a rate apparently independent of the nature of the solvent. Leffler and Tsuno (24,25) reported that the decomposition of benzenesulfonyl azide followed first-order kinetics, with a rate constant of 1.48×10^{-3} min^{-1} at 126.8° in chlorobenzene, p-xylene, or nitrobenzene. The same rate was observed by gas evolution as by titration of unreacted azide with triphenylphosphine, indicating the absence of any stable, nitrogen-containing intermediate. The rate was unaffected by oxygen, quinone, hydroquinone, iodine, stilbene, diphenyl disulfide, or benzothiazolyl disulfide. In an independent study, Breslow and co-workers (16) investigated the thermolysis of p-toluenesulfonyl azide in a variety of solvents; the results are shown in Table III. The reactions, as

TABLE III

Decomposition of p-Toluenesulfonyl Azide in Various Solvents[a]

Solvent	Temp., °C	$k_1 \times 10^4$ sec^{-1}	Relative rate	Gas evolved, % of theory
Diphenyl ether	155	3.43	1	100
Tetradecane	155	3.80	1.11	—
Nitrobenzene	155	3.97	1.15	199
1-Octanol	155	3.63	1.06	96
n-Hexanoic acid	155	2.97	0.86	114
Dimethyl terephthalate	155	3.23	0.95	104
Diphenyl ether	145	1.44	1	—
1,4-Dichlorobutane	145	1.70	1.18	140

[a] Approximately 0.14M in azide.

measured by gas evolution, gave clean, first-order plots generally to 90% reaction. In spite of the wide diversity in the nature of the solvents and in the complexity of the reactions with the sulfonyl azide, as evidenced by the amount of gas evolved, there was surprisingly little variation in the rates of the reactions; the average half-life at 155° is 33 min. There can be little doubt from these results that the rate-determining step is the loss of nitrogen and the formation of an electron-deficient nitrene species.

$$RSO_2N_3 \xrightarrow{\Delta} RSO_2\ddot{N} + N_2$$

For the thermal decomposition of tosyl azide an activation energy of 36.0 kcal/mole was calculated. Balabanov and co-workers (66) reported the decomposition of benzenesulfonyl azide to be approximately twice as fast in cyclohexanone as neat, with $\Delta E^{\ddagger}$ of 33.0 kcal/mole and $\Delta S^{\ddagger}$ of 5.2 eu. Whereas Leffler and Tsuno (25) found a practically negligible substituent effect, Balabanov and co-workers (65) claimed that rate constants of p-substituted derivatives correlated well with a Hammett plot. The same authors attributed the higher activation energies (36.8–38.4 kcal/mole) to the absence of solvent. However, Takemoto, Fujita, and Imoto (26a) reported an activation energy of 36.4 kcal/mole for the decomposition of benzenesulfonyl azide in naphthalene, in excellent agreement with our results. The same authors obtained a $\rho = -0.1$ from a modified Hammett plot, in agreement with Leffler and Tsuno's claim for a very small substituent effect.

Leffler and Tsuno (24) found that t-butyl hydroperoxide catalyzes the decomposition of benzenesulfonyl azide and vice versa. Thus, at 126.7° in chlorobenzene a quantitative yield of nitrogen and a 90% yield of oxygen were obtained, based on the equations:

$$C_6H_5SO_2N_3 \longrightarrow C_6H_5SO_2\ddot{\ddot{N}} + N_2$$

$$(CH_3)_3C—OOH + R\cdot \longrightarrow RH + (CH_3)_3C—OO\cdot$$

$$2(CH_3)_3C—OO\cdot \longrightarrow O_2 + 2(CH_3)_3C—O\cdot$$

In the absence of azide, oxygen is formed at a negligible rate at this temperature. The induced decomposition of azide can be prevented completely by the addition of iodine to the chlorobenzene solution or by using p-xylene as solvent. Thus, there is little doubt that the azide decomposition involves radicals, but the kinetics are too complicated for a satisfactory mechanism to be written. The products formed from the azide were those to be expected from decomposition in an aromatic solvent (see Section III-B).

Breslow and co-workers (16) studied the decomposition of a number of aliphatic sulfonyl azides, both mono- and difunctional, in diphenyl ether. The rate was first-order in each case, and the calculated quantity of gas was evolved, within experimental error (Table IV). An activation energy

TABLE IV
Decomposition of Various Sulfonyl Azides in Diphenyl Ether[a]

Sulfonyl azide	Temp., °C	$k_1 \times 10^4$ sec^{-1}	Gas evolved, % of theory
1-Pentane-	166	4.46	102
1,4-Butanedi-	163	5.02	99
1,6-Hexanedi-	163	5.02	98
1,9-Nonanedi-	160	2.25	—
1,10-Decanedi-	163	4.45	100
1,4-Dimethylcyclohexane-α,α'-di-	163	4.82	98
m-Xylene-α,α'-di-	163	6.09	101
p-Xylene-α,α'-di-	163	5.78	96

[a] Approximately 0.1M in azide.

of 37.4 kcal/mole was found for 1,9-nonanedisulfonyl azide, only slightly higher than that for tosyl azide in spite of the considerably greater stability of the aliphatic derivative. The tetramethylene and hexamethylene derivatives appeared to be slightly less stable than 1-pentanesulfonyl azide, as did the *m*- and *p*-xylene derivatives, but the differences in all cases were very small.

In this laboratory (16,63) the decomposition of 1-pentanesulfonyl azide in an aliphatic solvent, mineral oil, was studied. Here a good first-order plot was not obtained by following the gas evolution. However, the points for the first half-life fell close to a straight line, and the rate constants in parentheses in Table V were calculated from the slope of this line. In spite of the fact that the calculated quantity of gas was evolved, mass spectroscopic analysis of the gas showed the presence of nitrogen, sulfur dioxide, and a small amount of *n*-pentane. In addition, the solution contained azide (absorption at 2100 cm^{-1}) even after being heated for 24 hr (about 60 half-lives). As shown in Table V, the amount of sulfur dioxide evolved was lowered only slightly by lowering the temperature; a somewhat greater effect was observed by lowering the concentration. 2-Propanesulfonyl azide gave off slightly more sulfur dioxide than the 1-pentane derivative, while *p*-toluenesulfonyl azide gave off very little and decomposed in a clean, first-order reaction. On the assumption that sulfur dioxide evolution might be a radical reaction, free radical traps, such as hydroquinone and sulfur, were added. Under these conditions the amount of sulfur dioxide was decreased to 3–4%, good first-order kinetics were observed, and the rate was reduced. It appears, therefore, that two simultaneous reactions were taking place: in one reaction, nitrogen was evolved, leaving a nitrene; in the other, sulfur dioxide was evolved in a radical chain decomposition. Inasmuch as the alkyl azide formed is stable under these conditions, each molecule gave off either nitrogen or sulfur dioxide, thereby accounting for the quantitative evolution of gas.

$$C_5H_{11}SO_2N_3 \longrightarrow C_5H_{11}SO_2\ddot{N} + N_2$$

$$R\cdot + C_5H_{11}SO_2N_3 \longrightarrow C_5H_{11}SO_2\cdot + RN_3$$

$$C_5H_{11}SO_2\cdot \longrightarrow C_5H_{11}\cdot + SO_2$$

$$C_5H_{11}\cdot + RH \longrightarrow C_5H_{12} + R\cdot$$

Because of several puzzling features in the reaction, the evolution of sulfur dioxide was investigated in greater detail. As the temperature was increased there was a small but significant increase in the amount of sulfur dioxide evolved. A secondary sulfonyl azide, 2-propanesulfonyl azide, evolved slightly more sulfur dioxide than the primary azide. *p*-Toluenesulfonyl azide under these conditions gave only traces of sulfur dioxide. Not only did the radical traps, such as hydroquinone, 2,6-di-*t*-butyl-*p*-cresol, and sulfur, decrease the sulfur dioxide evolution and convert

TABLE V

Thermal Decomposition of Sulfonyl Azides

Azide	Conc., M	Solvent	Additive	Conc., M	Temp., °C	$k_1 \times 10^4$ sec^{-1}	% SO_2
1-Pentane	0.27	Mineral oil			175.0	—	20.5
1-Pentane	0.27	Mineral oil			165.3	(8.0)[a]	19.2
1-Pentane	0.24	Mineral oil			155.8	—	18.0
1-Pentane	0.12	Mineral oil			163.0	—	16.5
2-Propane	0.28	Mineral oil			165.0	—	22.0
1-Pentane	0.25	Mineral oil	Hydroquinone	0.045	166.0	4.80	5.7
1-Pentane	0.25	Mineral oil	Hydroquinone	0.091	165.5	4.46	3.2
1-Pentane	0.25	Mineral oil	Sulfur	0.32	165.0	4.10	3.1
1-Pentane	0.24	Mineral oil	Air		165.7	(7.7)[a]	13.8
1-Pentane	0.26	Diphenyl ether			165.9	4.46	4.2
p-Toluene	0.26	Tetradecane			154.0	3.80	1.3

[a] Not first-order.

the decomposition to a clean, first-order reaction, but they also slowed down the rate. Thus at 165° in mineral oil, the straight line portion of the n-pentanesulfonyl azide decomposition gave k_1 of $8.00 \times 10^{-4} \sec^{-1}$. In the presence of $0.045M$ hydroquinone a k_1 of $4.80 \times 10^{-4} \sec^{-1}$ was obtained. The course of the reaction was followed by quenching a sulfonyl azide decomposition at various times and determining the extent of decomposition and the amount of sulfur dioxide formed. As shown in Table VI, the sulfur dioxide formation is not a first-order reaction and it is

TABLE VI

Rate of Sulfur Dioxide Formation from 1-Pentanesulfonyl Azide in Mineral Oil at 165°

	Azide reacted		SO_2 formed,
Time, min	mmoles	%	mmoles
3	0.34	13.1	0.155
8	0.79	30.4	0.318
15	1.23	47.4	0.445
25	1.66	64.0	0.440
∞	2.60	100	0.510

virtually complete by the time half the sulfonyl azide has decomposed. The puzzling question is the source of the initiator for the free-radical decomposition. One possibility would be impurities, presumably hydroperoxides, in the solvent. Most of the kinetic runs were carried out under nitrogen in a commercial mineral oil used without additional purification. Therefore, a comparable run was done in tetradecane which had been carefully purified by distillation and subsequently handled in a nitrogen atmosphere; the amount of sulfur dioxide was reduced from 19 to 16%, from which result it is difficult to draw any conclusion. Inasmuch as sulfur dioxide evolution ceased when only half the azide had been decomposed, it would appear that some if not all of the initiator would be consumed by the end of a run. To investigate this point, a normal run was carried out in mineral oil at 165°, the gas was vented at the end of the run, and a second azide sample was injected. Rate constants calculated from the slopes of the lines drawn through the points for the first 50% reaction were $7.63 \times 10^{-4} \sec^{-1}$ for the first run and $7.08 \times 10^{-4} \sec^{-1}$ for the second. Thus, although the rate did drop slightly in the second run, the change is very small and the reaction is still considerably faster than in the presence of an inhibitor. In another run, the air was not displaced by nitrogen. The rate behavior was unchanged, but the sulfur dioxide yield was lowered to

14%; if anything, air seems to inhibit the reaction. Furthermore, since sulfur dioxide is known to be a good decomposer for hydroperoxides it does not seem very likely that peroxidic impurities in the solvent are the initiators of the radical decomposition. A second possibility is impurities in the azides, but this does not appear to be very likely either. Although 1-pentanesulfonyl azide is a liquid and the sample used contained traces of chlorine, a sample purified by low-temperature recrystallization from n-hexane gave identical results. Furthermore, 1,10-decanedisulfonyl azide, which is a crystalline solid, and 2-propanesulfonyl azide both gave similar amounts of sulfur dioxide, and it does not appear to be very likely that all three azides would contain the same impurities at the same levels. The thermal cleavage of a sulfonyl azide to radicals is a third possibility.

$$RSO_2N_3 \longrightarrow RSO_2\cdot + N_3\cdot$$

This would be analogous to the cleavage of a sulfonyl chloride, which is considered to be a radical reaction. Although this possibility cannot be ruled out, it does not appear likely that the S—N bond would break as readily as the S—Cl bond. A more interesting possibility is that the nitrene triplet or a radical formed from it, catalyzes the radical decomposition, although other evidence indicates that little if any triplet is formed in sulfonylnitrene reactions.

In light of the finding that aliphatic sulfonyl azides undergo induced free-radical decomposition to give sulfur dioxide, one might question whether the nitrogen-forming reaction is also induced by free radicals, for example:

$$R\cdot + C_5H_{11}SO_2N_3 \longrightarrow C_5H_{11}SO_2\ddot{N}R + N_2$$

Extrapolation of the data in Table VI shows that at zero time 48% of the gas produced from 1-pentanesulfonyl azide is sulfur dioxide. As already mentioned, the reaction follows reasonably good first-order kinetics for the first half of the reaction in the absence of an inhibitor, and an average value of 8.0×10^{-4} sec^{-1} can be calculated for k_1. If we assume that the initial formation of both sulfur dioxide and nitrogen occur by first-order reactions, the relative amounts of the two products should be proportional to the relative magnitude of the individual rate constants for their formation. Applying this reasoning, a k_1 of 4.11×10^{-4} sec^{-1} for the nitrogen-forming reaction was calculated, which is in fairly good agreement with the actual value of 4.45×10^{-4} sec^{-1} found for the average of three inhibited runs in which the free-radical reaction had been almost completely suppressed. From this admittedly rough estimation it would appear that the evolution of nitrogen and the formation of a nitrene are not affected by the free radicals present in the system.

Recently, Leffler and Tsuno (24) reported the decomposition of benzenesulfonyl azide to be accelerated by *t*-butyl hydroperoxide and concluded that the decomposition to nitrene was induced by free radicals. In order to resolve this apparent disagreement, the decomposition of *p*-toluenesulfonyl azide was investigated. As mentioned previously, this compound decomposes in a strictly first-order fashion in tetradecane. The amount of sulfur dioxide was 0.75–1.3% of the evolved gas, and this was not changed by the addition of an equivalent amount of *t*-butyl hydroperoxide; neither was the rate. Since benzenesulfonyl radicals are reported (67) not to lose sulfur dioxide but to give other products, such as benzenesulfinic acid, *p*-toluenesulfonyl azide was decomposed in mineral oil at 130° in the presence of an equivalent amount of *t*-butyl hydroperoxide; calcium oxide was added to trap any sulfinic acid that might be formed. From the calcium salts present at the end of the reaction there was isolated an 11% yield of *p*-toluenesulfonic acid as its *p*-toluidine salt. No sulfinic acid was found, and no sulfonic acid was formed in the absence of hydroperoxide. Thus, the sulfur–nitrogen bond can be cleaved by the hydroperoxide; the expected sulfinic acid is undoubtedly oxidized to sulfonic acid before it can be trapped. Consequently, an aromatic sulfonyl azide will undergo free-radical decomposition if a source of radicals is provided, but not in its absence. It is conceivable that the lack of radical decomposition of an aromatic sulfonyl azide or of an aliphatic sulfonyl azide in an aromatic solvent is due to the formation of the corresponding sulfonanilide which, being an aniline derivative, is probably a fairly good radical trap.

B. Reactions with Hydrocarbons

Sloan, Breslow, and Renfrow (63) found that both alkane- and arenesulfonyl azides insert into the carbon–hydrogen bonds of saturated hydrocarbons. Thus, the thermolysis of 1-pentanesulfonyl azide in cyclohexane gave a 54% yield of *N*-cyclohexyl-1-pentanesulfonamide. No hydrogen-abstraction product, 1-pentanesulfonamide, could be isolated, although a 2% yield of a compound with a similar, but not identical, infrared spectrum was obtained (16).

$$C_5H_{11}SO_2N_3 + \left\langle \bigcirc \right\rangle \xrightarrow{\Delta} C_5H_{11}SO_2NH\text{—}\left\langle \bigcirc \right\rangle + N_2$$

Since about 20% of the sulfonyl azide would have been expected to decompose by loss of sulfur dioxide, about a fourth of the sulfonyl azide was not accounted for; the presence of cyclic sultams, formed by intramolecular carbon–hydrogen insertion, might have been expected, but no pure compounds could be isolated. Addition of sulfur as a radical trap to the reaction mixture increased the yield of insertion product to 72%, indicating that essentially all of the free-radical portion of the reaction had been inhibited.

2-Propanesulfonyl azide reacted with cyclohexane to give, by infrared analysis, a 60% yield of N-cyclohexyl-2-propanesulfonamide and a maximum of 3% of 2-propanesulfonamide. The latter was not isolated but a material with a similar spectrum appeared to be present in small quantities. p-Toluenesulfonyl azide gave a 58% yield of insertion product, but with this azide a 5% yield of the abstraction product was definitely isolated. Pritzkow and Timm (35) obtained quite different results. They reported a 17% yield of insertion product and a 40% yield of abstraction product by hydrolyzing the reaction mixture and determining the amounts of cyclohexylamine and ammonia. There is no apparent explanation for this low-insertion and high-abstraction yield.

In an attempt to determine the relative reactivity of primary, secondary, and tertiary carbon–hydrogen bonds with sulfonylnitrenes, Breslow and co-workers (16,69) analyzed the sulfonamides formed by the reaction of p-toluenesulfonyl azide with 2-methylbutane by gas chromatographic analysis. Although complete separations could not be made, vapor-phase chromatography showed a ratio of $(17 + 18):(15 + 16)$ of 1.53. More recent results (69a) indicate, however, that **18** is unstable under the reaction conditions, and that the ratio is much higher than this.

$$RSO_2N_3 + (CH_3)_2CHCH_2CH_3 \longrightarrow (CH_3)_2CHCH_2CH_2NHSO_2R +$$

(15)

$$\overset{\displaystyle CH_3}{\underset{\displaystyle |}{RSO_2NHCH_2CHCH_2CH_3}} + \underset{\displaystyle RSO_2NH}{(CH_3)_2CHCHCH_3} + \underset{\displaystyle RSO_2NH}{(CH_3)_2CCH_2CH_3}$$

(16)   **(17)**   **(18)**

$R = p\text{-}CH_3C_6H_4$

The classical reaction of sulfonylnitrenes, first reported by Curtius (2,3) in 1913, involves the insertion of a sulfonylnitrene into the C—H bond of an aromatic compound.

$$RSO_2N_3 \overset{\Delta}{\longrightarrow} RSO_2\ddot{N} + N_2 \qquad \text{rate-determining}$$

$$RSO_2\ddot{N} + ArH \longrightarrow RSO_2NHAr$$

Curtius (5,6) studied a number of sulfonyl azides with his co-workers— α-toluene- (20), benzene- (21), p-toluene- (31), p-chlorobenzene- (38), p-acetamidobenzene- (70), 1-naphthalene- (46), 2-naphthalene- (47), and β-anthraquinone- (48)—as well as two disulfonyl azides, m-benzene- (55) and 1,5-naphthalene- (57). These were reacted with a variety of aromatic compounds—benzene, toluene, p-xylene, naphthalene, aniline, N,N-dimethylaniline, etc.—generally by heating a dilute solution until the theoretical amount of nitrogen had been evolved. Representative examples described by Curtius and by later workers are shown in Table VII. Yields are probably determined more by ease of isolation than by the nature of the sulfonyl azide, at least in the absence of interfering groups.

Curtius was the first to investigate isomer distribution in the reaction and concluded that *ortho*-substitution predominated and *meta*-isomers were not formed (6). Thus, benzenesulfonyl azide was reported to react with toluene to give o- and p-benzenesulfonyltoluidides in a ratio of 2.4:1 (21). More recent work (Table VIII) has shown that by and large the amount of m-isomer formed is indeed quite small, but that in most experiments with *ortho–para* directing groups in the substrate considerable amounts of p-isomer are formed. The claim by Leffler and Tsuno (24,25) that the reaction of benzenesulfonyl azide with chlorobenzene gives a mixture of sulfonanilides containing 56% of the o-isomer, 25% of the m-isomer, and 19% of the p-isomer is in disagreement with the results of other authors (71,13).

Dermer and Edmison (22) were the first to attempt to determine quantitatively the orientation in the reaction of benzenesulfonyl azide with monosubstituted aromatics. Because their analytical procedure (acid hydrolysis and determination of the isomeric amines by ultraviolet spectroscopy) gave inconsistent results, the work was repeated by Heacock and Edmison (71), who isolated the mixtures of sulfonanilides and determined their composition by infrared. More recently, Abramovitch, Roy, and Uma (13) studied the decomposition of methanesulfonyl azide in monosubstituted aromatics and analyzed the products by vapor-phase chromatography. The two sets of results are compared in Table VIII. Both groups also ran competitive studies to determine relative reactivities as compared to benzene. These are reported in Table IX, as are the calculated apparent partial rate factors. From their results, Heacock and Edmison (71) drew the conclusion that benzenesulfonylnitrene reacts with an aromatic system as an electrophilic free radical, i.e., in the triplet state, $C_6H_5SO_2\ddot{N}\cdot$. However, Abramovitch and co-workers (13) pointed out several inconsistencies in their data. Thus, although homolytic substitution would be expected to show total rate ratios in competitive experiments with benzene close to unity, the greater reactivity of benzene as compared to anisole and phenol is not reasonable for an electrophilic radical. In this laboratory (16,69) we had found p-xylene to be twice as reactive as benzene toward tosyl azide, whereas the Heacock and Edmison toluene

TABLE VII

Reactions of Sulfonyl Azides with Aromatic Rings

Solvent	RSO_2N_3, R =	Product	Yield	Ref.
Benzene	CH_3	$RSO_2NHC_6H_5$	54.6	13
	C_6H_5	$RSO_2NHC_6H_5$	27	71
	$p\text{-}CH_3C_6H_4$	$RSO_2NHC_6H_5$	85 (crude)	31
Toluene	CH_3	$RSO_2NHC_6H_4CH_3$	72–77	13
	C_6H_5	$RSO_2NHC_6H_4CH_3$	50	21
		$RSO_2NHC_6H_4CH_3$	49	71
	$p\text{-}ClC_6H_4$	$RSO_2NHC_6H_4CH_3$	26 (crude)	38
o-Xylene	$p\text{-}CH_3CONHC_6H_4$	$RSO_2NHC_6H_3\text{-}3,4\text{-}(CH_3)_2$	25	30
p-Xylene	$C_6H_5CH_2$	$RSO_2NHC_6H_3\text{-}2,5\text{-}(CH_3)_2$		20
	C_6H_5	$RSO_2NHC_6H_3\text{-}2,5\text{-}(CH_3)_2$	67	25
	$p\text{-}CH_3C_6H_4$	$RSO_2NHC_6H_3\text{-}2,5\text{-}(CH_3)_2$	53 (crude)	31
	$p\text{-}ClC_6H_4$	$RSO_2NHC_6H_3\text{-}2,5\text{-}(CH_3)_2$	51 (crude)	38
	$o\text{-}IC_6H_4$	$RSO_2NHC_6H_3\text{-}2,5\text{-}(CH_3)_2$	Trace	25
	$o\text{-}NO_2C_6H_4$	$RSO_2NHC_6H_3\text{-}2,5\text{-}(CH_3)_2$	33	25
	$p\text{-}CH_3CONHC_6H_4$	$RSO_2NHC_6H_3\text{-}2,5\text{-}(CH_3)_2$	50	30
	$2\text{-}C_{10}H_7$	$RSO_2NHC_6H_3\text{-}2,5\text{-}(CH_3)_2$	12	47
	$\beta\text{-Anthraquinone}$	$RSO_2NHC_6H_3\text{-}2,5\text{-}(CH_3)_2$	11	48
Naphthalene	$C_6H_5CH_2$	$1\text{-}RSO_2NHC_{10}H_7$	76	20
	$p\text{-}ClC_6H_4$	$1\text{-}RSO_2NHC_{10}H_7$		38
	$2\text{-}C_{10}H_7$	$1\text{-}RSO_2NHC_{10}H_7$	15	47
Chlorobenzene	CH_3	$RSO_2NHC_6H_4Cl$	65	13
	C_6H_5	$RSO_2NHC_6H_4Cl$	34	71
			56	24, 25

Bromobenzene	C_6H_5	$RSO_2NHC_6H_4Br$	21	71
Anisole	CH_3	$RSO_2NHC_6H_4OCH_3$	67–81	13
	C_6H_5	$RSO_2NHC_6H_4OCH_3$	36	71
Methyl benzoate	C_6H_5	$RSO_2NHC_6H_4CO_2CH_3$	23	71
Aniline	C_6H_5	$RSO_2NHC_6H_4NH_2$	8	71
	p-ClC_6H_4	$RSO_2NHC_6H_4NH_2$	15	38
N-Methylaniline	C_6H_5	$RSO_2NHC_6H_4NHCH_3$	11	21
N,N-Dimethylaniline	$C_6H_5CH_2$	$RSO_2NHC_6H_4N(CH_3)_2$	17	20
	C_6H_5	$RSO_2NHC_6H_4N(CH_3)_2$	22	21
	p-ClC_6H_4	$RSO_2NHC_6H_4N(CH_3)_2$	15	38

TABLE VIII
Orientation in Aromatic Substitution by Sulfonyl Azides

| | Isomer ratios | | | | | |
| | $CH_3SO_2N_3$ | | | $C_6H_5SO_2N_3$ | | |
Substrate	o	m	p	o	m	p
$C_6H_5CH_3$	65.4	2.4	32.2	61	1	38
C_6H_5Cl	57.4	0.9	41.7	46	2	52
C_6H_5Br				50	5	45
$C_6H_5OCH_3$	55.5	1.2	43.3	71	2	27
$C_6H_5CO_2CH_3$				43	54	3
$C_6H_5NH_2$				40	7	53
C_6H_5OH				50	2	48
C_6H_5COCl				0	100	0

TABLE IX
Competitive Sulfonamidations of Aromatic Compounds

| | Total rate ratios ($C_6H_6 = 1.0$) | | Apparent partial rate factors | | | | | |
| | | | $CH_3SO_2N_3$ | | | $C_6H_5SO_2N_3$ | | |
Substrate	$CH_3SO_2N_3$	$C_6H_5SO_2N_3$	F_o	F_m	F_p	F_o	F_m	F_p
Toluene	1.86	1.00	3.65	0.13	3.59	1.8	0.03	2.3
Chlorobenzene	0.44	0.69	0.76	0.01	1.10	0.95	0.04	2.2
Bromobenzene		0.69				1.0	0.10	1.9
Anisole	2.54	0.96	4.23	0.09	6.6	2.0	0.06	1.6
Phenol		0.80				1.2	0.05	2.3
Methyl benzoate		0.38				0.49	0.62	0.07

results would have led to a prediction of comparable activity. The very low degree of *meta*-substitution (F_m) for toluene is also not in accord with a homolytic reaction. For these reasons Abramovitch, Roy, and Uma (13) reinvestigated aromatic substitution with methanesulfonyl azide. Inasmuch as aliphatic and aromatic sulfonyl azides decompose at somewhat different rates, it is difficult to make a direct comparison between the two results. However, the nitrenes would not be expected to differ greatly in their reactivities, and the Abramovitch results are internally more self-consistent.

Aromatic substitution can best be explained by consideration of a sulfonylnitrene as a highly reactive electrophilic reagent, in accord with toluene and anisole being more reactive and chlorobenzene less reactive than benzene toward methanesulfonyl azide in competitive experiments (Table IX). The very small rate differences indicate a radical substitution, and Smolinsky, Wasserman, and Yager (72) have shown by ESR measurements at 77°K that benzene- and p-toluenesulfonylnitrenes have triplet ground states. However, the spin conservation rule suggests that in the thermal decomposition of a sulfonyl azide the corresponding nitrene should be formed initially as a singlet, and evidence favors aromatic substitution being a singlet reaction. Thus, Tilney-Bassett (73) showed that anthracene reacts with benzenesulfonyl azide predominantly in the 1,2-position:

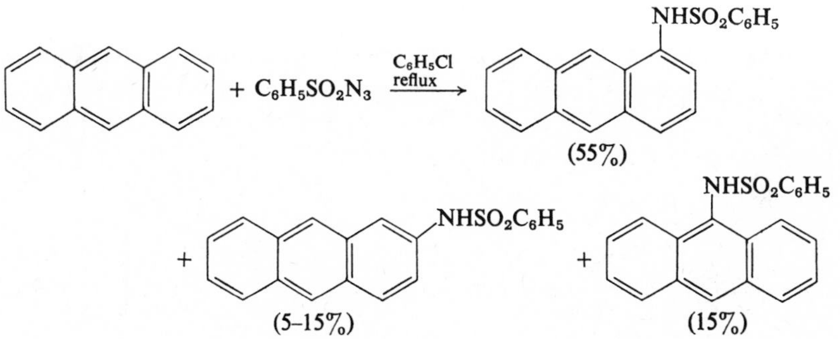

(55%)

(5–15%)

(15%)

Since anthracene is known to undergo free-radical substitution exclusively in the 9-position, it is apparent that this is not a radical reaction. Instead, the nitrene attacks the 1,2-position, which is known to have the highest electron density, and Tilney-Bassett suggested an aziridine intermediate or transition state (**19**). Abramovitch adopted the same intermediate to

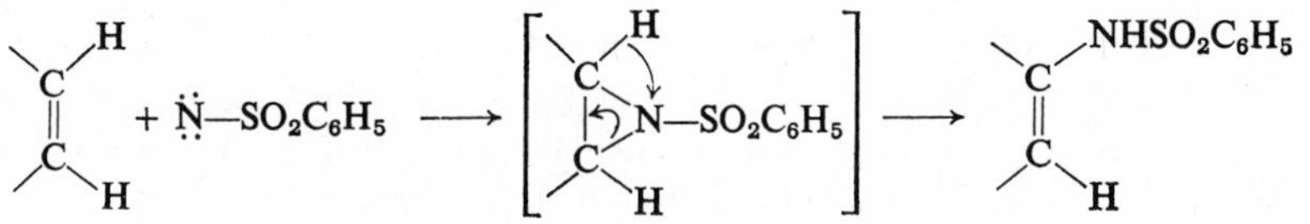

(19)

explain the nitrene being highly reactive (little difference in rate between benzene and anisole) but also highly selective (predominant *ortho–para* substitution with toluene and anisole). Thus, in the first step the nitrene reacts more or less indiscriminately, showing some preference for the molecule with a higher electron density, whereas in the second step the substituent determines the direction of ring opening.

By analogy with these results a sulfonylnitrene would be expected to attack preferentially the 1,2-position in naphthalene, since this is the area of highest electron density. It is rather surprising, however, that Curtius and his co-workers (6,20,38,47) claimed exclusive α-substitution with several sulfonyl azides; repetition of this work with more modern analytical tools would be desirable.

In order to determine the relative reactivities of aromatic and aliphatic hydrocarbons, tosyl azide was decomposed in an equimolar mixture of benzene and cyclohexane (16,69). Analysis of the aniline and cyclohexylamine formed by acid hydrolysis showed benzene to be about twice as reactive as cyclohexane. Thus, if the benzene reaction is considered to involve aziridine formation followed by rearrangement, a benzene "double bond" can be considered to be eight times more reactive with a sulfonyl nitrene than a carbon–hydrogen bond in cyclohexane.

Abramovitch and co-workers (20a) recently reported a number of intramolecular cyclizations of aromatic sulfonyl azides. Thus, biphenyl-2-sulfonyl azide yielded 6H-dibenzo[c,e][1,2]thiazine 5,5-dioxide (19a) on thermolysis, and 2-phenoxybenzenesulfonyl azide yielded the corresponding seven-membered ring compound, 6H-dibenz[1,4,5]oxathiazepine

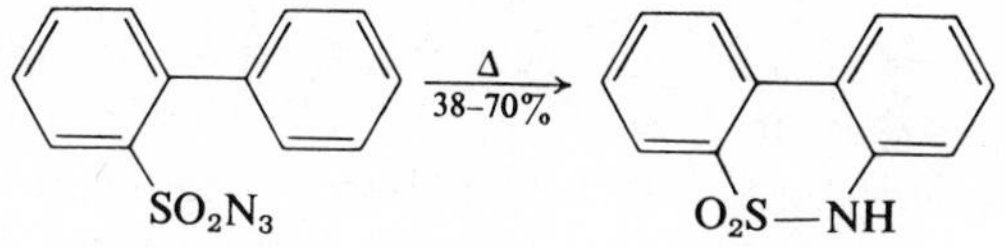

(19a)

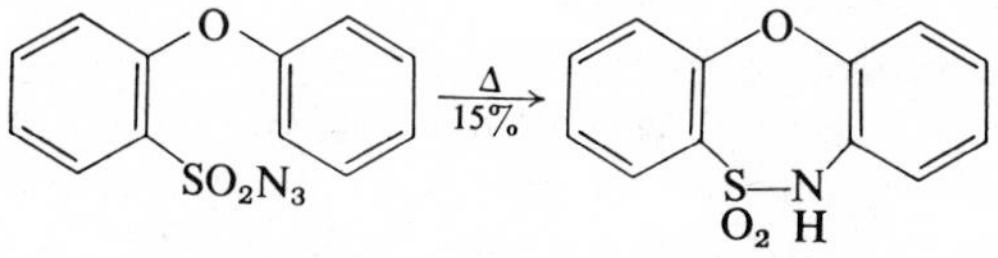

(19b)

5,5-dioxide (19b). The sulfur analog, diphenyl sulfide-2-sulfonyl azide, reacted with the sulfide sulfur (see Section III C), while the corresponding sulfone, diphenyl sulfone-2-sulfonyl azide, gave no cyclic products. Although α-toluenesulfonyl azide would not cyclize, a low yield of 4,5,7-trimethyl-2,3-dihydro-1,2-benzisothiazole 1,1-dioxide (19c), was obtained from durenesulfonyl azide. The failure of the α-toluene derivative to

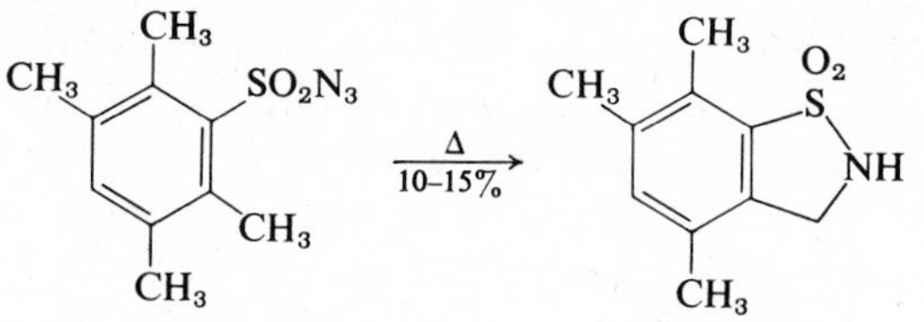

(19c)

cyclize is readily understandable if an aziridine intermediate must be involved in aromatic substitution.

The reaction of sulfonyl azides with aromatic compounds is not completely devoid of radical character, however. Thus, in this laboratory (69) a trace of dicumyl was isolated from the reaction of n-pentanesulfonyl azide with cumene, and Abramovitch and co-workers (13) isolated a trace of dibenzyl from the reaction of methanesulfonyl azide with toluene. However, it is impossible to decide at present whether these are reactions of the triplet nitrene or of residual radicals formed in the radical decomposition of aliphatic sulfonyl azides (16). A more interesting question is whether the substitution in the 9-position of anthracene is a measure of the amount of triplet character of the nitrene. Tilney-Bassett (73) carried out the reaction on a solution of anthracene in chlorobenzene, conditions which are conducive to singlet–triplet intersystem crossing; a study of the ratio of 1- to 9-substitution as a function of concentration would be highly desirable.

Dermer and Edmison (22) stated in their original paper in 1955 that benzenesulfonyl azide catalyzed the polymerization of methyl acrylate and acrylonitrile at 110°, and that the polymerizations were inhibited by hydroquinone; however, no experimental details were given. This work has often been cited (8–10,71) as evidence of the triplet character of sulfonylnitrenes. Dannley and Esayian (74) investigated the copolymerization of methyl methacrylate and styrene initiated by tosyl azide. The formation of a 1:1 copolymer is good evidence that a radical polymerization was involved, but the catalytic activity was extremely low, e.g., 3.05 ± 0.15% copolymer yield at 100° with catalyst and 1.5% without; 6.7 ± 0.6% with catalyst at 120° and 5.1% without. The fact that copolymer was formed at 60° (10% in 67 hr) casts some doubt on azide decomposition being the source of radicals. More recently, however, Imoto, Takemoto, and Fujita (26,26a) investigated the polymerization of acrylonitrile catalyzed by p-substituted arenesulfonyl azides, in which they claimed the control polymerized less than 0.1% in 8 hr at 135°. The polymerization rate was half-order in azide, as expected in a radical polymerization, and the polymer contained sulfur, presumably as an endgroup. However, here too the polymerization rate was very low, the azides being very poor catalysts. The authors interpreted their Hammett plot as indicating initiation by a radical with electronegative character, which does not seem reasonable in view of all the evidence for the electropositive nature of sulfonylnitrenes. The report by Ghosh and O'Driscoll (75) that sulfur dioxide is capable of initiating vinyl polymerization may be pertinent, in spite of their claim that acrylonitrile does not polymerize below 60°, inasmuch as aromatic sulfonyl azides do evolve some sulfur dioxide (16,66) at the elevated temperatures used by Imoto and his co-workers. Burleigh and Uraneck (76) recently patented the combination of cyclopentanesulfonyl azide and ferrous sulfate as a catalyst for the emulsion polymerization of butadiene. Actually, there is no valid reason for believing that a nitrene triplet should be a good polymerization catalyst—it might add to the activated double bond instead, for example.

A puzzling feature of the reaction of sulfonyl azides with aromatic compounds is the formation of some unsubstituted amide in most of the reactions.

$$\text{ArH} + \text{RSO}_2\text{N}_3 \xrightarrow[-\text{N}_2]{\Delta} \text{RSO}_2\text{NHAr} + \text{RSO}_2\text{NH}_2$$

Curtius (6) reported higher yields of sulfonamide as a result of "stormy" azide decomposition, and amide yields often surpassed those of anilide. In no case has the source of the hydrogen atoms been identified. Thus, even in the very careful work of Abramovitch and co-workers (13) the amount of 1,2-diphenylethane isolated (0.4%) from the reaction of toluene with methanesulfonyl azide was much too small to account for all the methanesulfonamide formed (24.3%). Several lines of evidence show quite conclusively that amide formation is not a simple radical reaction. Thus, little if any unsubstituted amide is formed in the reaction of aliphatic sulfonyl azides with aliphatic hydrocarbons (16,63), whereas 14% methane-sulfonamide was isolated from the reaction of the corresponding azide with benzene (13); radical abstraction of hydrogen atoms from benzene and not from an aliphatic hydrocarbon would be unprecedented. It is tempting to suggest that, since some tar is always formed in the aromatic reactions, the hydrogens are derived from reaction of product with nitrene, e.g.,

$$2ArNHSO_2R + RSO_2\ddot{\underset{\cdot\cdot}{N}} \longrightarrow RSO_2NH_2 + 2[Ar\dot{N}SO_2R]$$

$$\downarrow$$

$$\text{tar}$$

However, in the reaction of toluene with methanesulfonyl azide, Abramo-vitch, Roy, and Uma (13) accounted for all the azide introduced as either toluidide or unsubstituted amide. These authors suggested, therefore, the intermediate formation of benzyne, perhaps by decomposition of the initially formed aziridine.

An alternative explanation, that the nitrene abstracts two adjacent hydrogen atoms from the aromatic to give amide and a benzyne directly seems quite unlikely in view of the fact that a formylnitrene, probably as a triplet, abstracts adjacent hydrogen atoms from cyclohexane to yield cyclohexene and unsubstituted carbamate (77), whereas the unsubstituted sulfonamide yield is negligible under the same conditions (63). In any case, no evidence for the formation of a benzyne in this reaction has been forthcoming, although it would not be unreasonable for it to form tar under the reaction conditions.

Another puzzling question involves the difference between formyl-nitrenes and sulfonylnitrenes in their reactions with aromatic compounds. Why do the former give azepines and the latter only anilides, since an aziridine intermediate is postulated in both reactions? Do the two inter-mediates differ in kind or only in degree? Since it had been shown that the isomerization of *N*-carbethoxyazepine (**20**) to *N*-phenylurethan is catalyzed by acid (see Chapter 6, Section IVA2), it was considered likely that the sulfonamide always formed in the sulfonyl azide reaction is sufficiently acidic to cause the isomerization of an initially formed azepine (**21**) to the corresponding anilide, especially at the more elevated temperatures fre-quently used in sulfonyl azide reactions. Abramovitch and Uma (78) have recently demonstrated that **21** ($R = CH_3$) is actually formed in the reaction of methanesulfonyl azide with benzene. When the decomposition was

carried out at 120° in the presence of tetracyanoethylene, the TCNE adduct (**21a**, $R = CH_3$) was formed at the expense of the anilide. In fact, by carrying out the thermolysis at 80° for 100 hours (in the absence of TCNE) the presence of **21** ($R = CH_3$) and the absence of anilide could be demonstrated by thin layer chromatography. Thus, the two types of nitrene show parallel reactions.

Although bis(sulfonyl azides) show normal reactivity, sulfuryl azide seems to be in a class by itself. Thus, in 1915 Curtius (79) reported that heating sulfuryl azide in *p*-xylene eliminated sulfur dioxide and gave not a xylidine but a material with pyridine-like properties, which he formulated as an azepine (22) by analogy with the product formed from benzene and ethyl diazoacetate (80). In 1922 Curtius and Schmidt (51) reported the

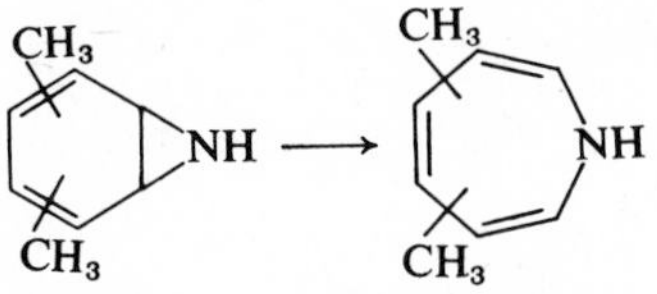

(22)

isolation of four compounds from the reaction after alkaline hydrolysis: Base **A**, $C_8H_{11}N$, which they called "pseudoxylidine," and which smelled like pyridine and made a variety of salts (hydrochloride, picrate, chloroplatinate); base **B**, C_8H_9N, with properties similar to **A**; base **C**, C_8H_9N, mp 112°, which was odorless but formed the usual salts; and **D**, C_8H_9N, mp 85°, which was neutral. Bertho (5) formulated **A** and **B** in the manner shown here. Schmidt (81,52) investigated the reaction with benzene and

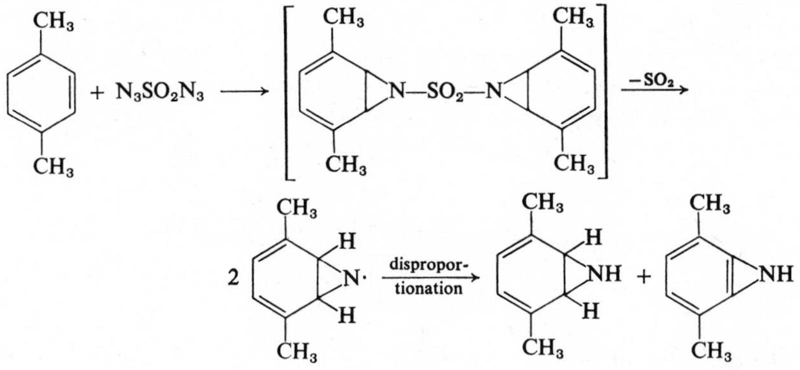

(A)   (B)

isolated, after alkaline hydrolysis, a 2.4% yield of pyridine, identified as its picrate, methiodide, and methylpyridinium chloroplatinate salts. To resolve this discrepancy Bertho, Curtius, and Schmidt (53) ran the reaction with *p*-cymene; the product isolated was assumed to be one of the isomeric methylisopropylpyridines. A repetition of this work would be highly desirable; as mentioned previously, even the synthesis of the starting material, sulfuryl azide, requires confirmation.

C. Reactions with Functional Groups

Curtius and his co-workers (6,20,21,31,38,47,55) treated a number of aromatic sulfonyl azides with aniline, *N*-methylaniline, and *N,N*-dimethylaniline. As shown in Table VII (see Section III-B) most azides gave modest yields (10–20%) of the *o*- and *p*-substituted sulfonanilides by addition of the sulfonylnitrenes to the aromatic rings; unsubstituted sulfonamides were also invariably formed.

$$ArSO_2N_3 + C_6H_5N\begin{smallmatrix}R(H)\\ \\R(H)\end{smallmatrix} \xrightarrow{\Delta} ArSO_2NHC_6H_4N\begin{smallmatrix}R(H)\\ \\R(H)\end{smallmatrix} + ArSO_2NH_2 + N_2$$

Curtius (6) noted that hydrazine derivatives were conspicuously absent. Many of the products can be attributed to non-nitrene reactions. Thus, aniline generally gave the corresponding anilide as the major product, formed by simple nucleophilic displacement, plus hydrazoic acid or its decomposition products.

$$ArSO_2N_3 + C_6H_5NH_2 \longrightarrow ArSO_2NHC_6H_5 + HN_3$$

Methylaniline and dimethylaniline with most azides gave 4,4'-methylene-bis(*N*-methyl- and *N,N*-dimethylanilines), respectively. Thus 15 g of benzenesulfonyl azide in excess dimethylaniline gave 9 g of 4,4'-methylene-bis(*N,N*-dimethylaniline) (21), although the yields were generally considerably lower. These products probably arise from the condensation of formaldehyde with the excess aniline used as solvent, the formaldehyde itself resulting from hydrolysis during work-up of the product formed by insertion of the nitrene into a C—H bond of the methyl group.

$$C_6H_5SO_2N_3 + (CH_3)_2NC_6H_5 \longrightarrow C_6H_5SO_2NHCH_2\overset{\overset{\displaystyle CH_3}{|}}{N}C_6H_5 \xrightarrow{H_2O}$$
$$C_6H_5SO_2NH_2 + HCHO + C_6H_5NHCH_3$$

$$(CH_3)_2NC_6H_5 + HCHO \longrightarrow (CH_3)_2N\!-\!\!\bigcirc\!\!-\!CH_2\!-\!\!\bigcirc\!\!-\!N(CH_3)_2$$

Other products are less easily explained. Thus, α-toluenesulfonyl azide gave the leuco base of malachite green (23) in its reaction with dimethylaniline (20). Since sulfur dioxide was also formed, perhaps a benzyl radical was air-oxidized to benzaldehyde, which then condensed with dimethylaniline.

$$C_6H_5CH_2SO_2N_3 \longrightarrow C_6H_5CH_2\cdot + SO_2 + N_3\cdot$$
$$C_6H_5CH_2\cdot + O_2 \longrightarrow C_6H_5CHO$$

$$C_6H_5CHO + C_6H_5N(CH_3)_2 \longrightarrow (CH_3)_2N\!-\!\!\bigcirc\!\!-\!\overset{\overset{\displaystyle }{|}}{\underset{\underset{\displaystyle C_6H_5}{|}}{CH}}\!-\!\!\bigcirc\!\!-\!N(CH_3)_2$$

(23)

Curtius and his co-workers (6,21,31,38,47) also reacted a number of sulfonyl azides with pyridine and isolated, often in good yields, compounds which they formulated as 2-, 3-, or 4-aminopyridine derivatives (24) by analogy with the reactions with benzene.

$$\text{ArSO}_2\text{N}_3 + \text{[pyridine]} \xrightarrow{\Delta} \text{[pyridine]}-\text{NHSO}_2\text{Ar} + \text{N}_2$$

(24)

Alamela and Ganapathi (82) reacted *p*-acetamidobenzenesulfonyl azide with pyridine and assumed, from Curtius' work, that the 3-aminopyridine derivative was formed; they also reacted it with thiazole, 2,4-dimethyl- and 2-hydroxy-4-methylthiazoles, and glyoxaline without forming "any new compounds." However, in 1947, both Datta (83) and Ashley, Buchanan, and Eason (41) proved that the product is actually *N*-(*p*-acetamidobenzenesulfonimido)pyridine (25), formed by addition of the nitrene to the pyridine nitrogen atom. The latter authors actually isolated four compounds from the reaction; one remains unidentified. 25 was identified by conversion to known products (41,83). 26 was identified by

$$\text{CH}_3\text{CONH}-\text{[C}_6\text{H}_4]-\text{SO}_2\text{N}_3 + \text{[pyridine]} \xrightarrow[-\text{N}_2]{\Delta} \text{CH}_3\text{CONH}-\text{[C}_6\text{H}_4]-\text{SO}_2\text{N}{-}\overset{\oplus}{\text{N}}\text{[pyridine]}$$

(29%)   (25)

$$+ \text{CH}_3\text{CONH}-\text{[C}_6\text{H}_4]-\text{SO}_2\text{NH}-\text{[C}_6\text{H}_4]-\text{NHCOCH}_3$$

(20%)   (26)

$$+ \text{CH}_3\text{CONH}-\text{[C}_6\text{H}_4]-\text{SO}_2\text{NH}_2 + \text{C}_{13}\text{H}_{15}\text{N}_3\text{O}_3\text{S}$$

(12%)

$$\text{CH}_3\text{CONH}-\text{[C}_6\text{H}_4]-\overset{\ominus}{\text{SO}_2\text{N}}{-}\overset{\oplus}{\text{N}}\text{[pyridine]} \xrightarrow[\text{Pd–C}]{\text{H}_2} $$

(25)

$$\text{CH}_3\text{CONH}-\text{[C}_6\text{H}_4]-\text{SO}_2\text{NH}_2 + \text{[pyridine]}$$

aq. HCl ↓

$$\text{NH}_2-\text{[C}_6\text{H}_4]-\text{SO}_3\text{H} + \text{[pyridinium]}\overset{\oplus}{\text{N}}{-}\text{NH}_2 \;\text{Cl}^{\ominus} \xrightarrow{\text{H}_2} \text{[piperidine]}\text{N}{-}\text{NH}_2$$

(27)

comparison with a known specimen; Ashley, Buchanan, and Eason suggested its formation by nucleophilic displacement of SO_2N_3 from the azide by sulfonamide ion. In light of the fact that the acetamido group should

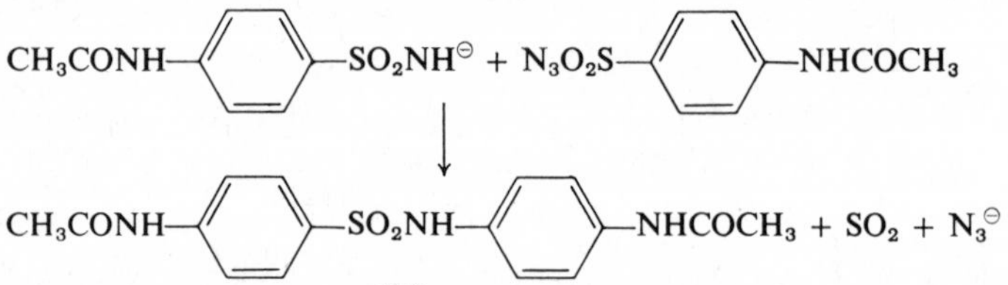

(26)

deactivate the aromatic ring toward nucleophilic substitution, this mechanism appears highly unlikely.

As further proof for the zwitterionic structure of **25** Datta (83) prepared it by condensing **27** with p-acetamidobenzenesulfonyl chloride; **27** was prepared by hydrolysis of the tosyl azide—pyridine condensate (**28**).

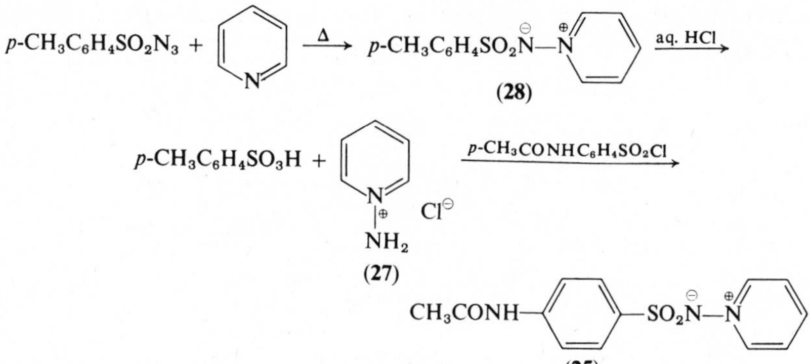

(28)

(27)

(25)

Finally, Datta synthesized **28** by an independent procedure from glutaconic dialdehyde monobenzoate (**29**).

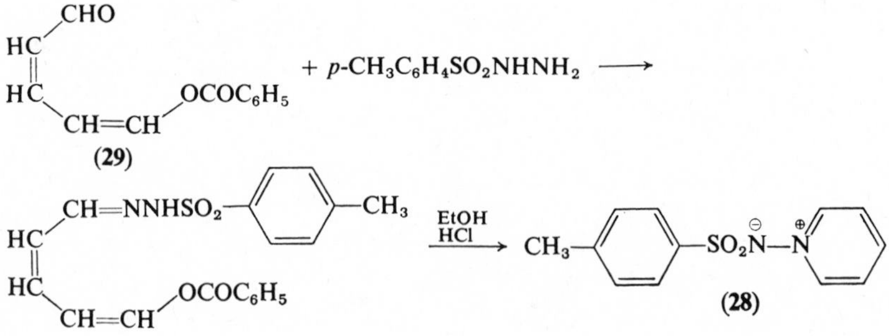

(29)

(28)

A number of 1-aminopyridinium derivatives have been prepared from the corresponding azides; they are listed in Table X. Attempts to prepare

TABLE X
Reaction of Sulfonyl Azides with Pyridine

$$RSO_2\overset{\ominus}{N}-\overset{\oplus}{N}$$

R	Yield, %	m.p., °C	Ref.
C_6H_5	47	152	21
	31	152	84
$p\text{-}CH_3C_6H_4$	46	210	31, 83
$p\text{-}ClC_6H_4$	22	191 (picrate)	38
	38	182	84
$p\text{-}CH_3CONHC_6H_4$	29	295–300 (dec.)	41
	14	284 (dec.)	83
	15	296–298	30
2-Naphthyl	7–10	193	47
	21	199–200 (picrate)	84

the corresponding 1-aminoquinolinium derivatives have led only to tars and unsubstituted amide (6,30). However, Cremlyn (30) reported the isolation of N-(p-acetamidobenzenesulfonimido)dimethylaniline **(30)** from the reaction of the corresponding azide, the only evidence for its structure, however, being the melting point difference between it and the p-substituted ring-addition product.

$$p\text{-}CH_3CONHC_6H_4SO_2N_3 + C_6H_5N(CH_3)_2 \xrightarrow[7\%]{\Delta} p\text{-}CH_3CONHC_6H_4SO_2\overset{\ominus}{N}-\overset{\overset{\displaystyle CH_3}{\overset{|}{\oplus}}}{\underset{\underset{\displaystyle CH_3}{|}}{N}}C_6H_5$$

(30)

Curtius (6) reported that benzaldehyde and nitrobenzene gave the unsubstituted sulfonamide as the only isolable product when heated with a sulfonyl azide. Dermer and Edmison (22) reported the evolution of a brown gas from the nitrobenzene reaction, presumably formed from initially liberated nitric oxide and air. Since the overall gas evolution followed first-order kinetics and the rate was independent of the solvent (16), there is little doubt that the rate-determining step is the formation of sulfonylnitrene. Leffler and Tsuno (25) investigated the decomposition of benzenesulfonyl azide in nitrobenzene in greater detail, as well as the decomposition of o-nitrobenzenesulfonyl azide. In both systems 0.5–0.6 mole of nitric oxide was formed per mole of nitrogen. In nitrobenzene as solvent NO evolution could not be inhibited by iodine, but dilution (1:10) with xylene eliminated it completely. o-Nitrobenzenesulfonyl azide, however, gave the same amount of NO in chlorobenzene as in p-xylene, a somewhat lower than normal yield of xylidide being isolated in the latter case (benzenesulfonyl azide in nitrobenzene gave only tar). Although the first-order rate constant for the decomposition of the o-nitro derivative appeared, by gas evolution, to be almost twice that of the unsubstituted derivative, the rate determined by azide disappearance (titration with triphenylphosphine) was essentially the same as the other azides. Thus, here too the rate-determining step is nitrene formation, followed in this case by an intramolecular reaction between nitrene and nitro group to give nitric oxide. o-Iodobenzenesulfonyl azide reacts in a similar fashion, i.e., the decomposition rate is normal, but the o-iodo group interferes with the usual nitrene reactions.

Balabanov, Dergunov, and Gal'perin (66) decomposed benzenesulfonyl azide in cyclohexanone. Surprisingly, no C—H insertion product was isolated, although only about a third of the azide added was accounted for. The cyclohexylidenecyclohexanone was probably formed by an aldol condensation catalyzed by sulfur dioxide, and the formation of a trace of azobenzene has already been discussed.

Horner and Christmann (12) found that the thermal decomposition of sulfonyl azides in dimethyl sulfoxide yield N-sulfonylsulfoximines (31). Although evidence is lacking that DMSO had no effect on the sulfonyl azide decomposition rate, it appears reasonable to assume that the reaction involves trapping of an initially formed nitrene.

Abramovitch and co-workers (20a) recently found a similar reaction in the cyclization of diphenyl sulfide-2-sulfonyl azide to 1-phenyl-1,3,2-benzodithiazole 3,3-dioxide (31a).

(31a)

It does not follow, of course, that all thermal reactions of sulfonyl azides go through a nitrene. For example, Shingaki and co-workers (85–88) reported that benzenethiol considerably accelerates the rate of decomposition of benzenesulfonyl azide to benzenesulfonamide. The effect has been shown to be caused by thiyl radicals, and undoubtedly a radical decomposition of azide is involved.

$$C_6H_5SO_2N_3 + C_6H_5SH \xrightarrow{\Delta} N_2 + C_6H_5SO_2NH_2 + C_6H_5SSC_6H_5$$

IV. PHOTOLYSIS OF SULFONYL AZIDES

There is no simple criterion for determining, in the photochemical decomposition of a sulfonyl azide, whether a sulfonylnitrene is involved in the reaction, since the kinetic analysis so useful in thermal decomposition (first-order rate independent of the nature of the solvent) cannot be used. Although there is little doubt, from the low temperature photolysis work of Smolinsky, Wassermann, and Yager (72), that sulfonylnitrenes can be formed photolytically, some uncertainty exists as to the actual intermediacy of a sulfonylnitrene in the photochemical reactions described below.

Horner and Christmann (12) were apparently the first to photolyze sulfonyl azides successfully, irradiation with a low-pressure mercury arc (2537 Å) in a sulfoxide solvent leading to N-sulfonylsulfoximines (32). Although, in light of the fact that the same reaction occurs by thermolysis (see Section III-C) it may seem reasonable to assume that the product is formed by condensation of sulfoxide with photolytically generated nitrene, the results obtained in the copper-catalyzed decomposition of sulfonyl azides (see Section V) may cast some doubt on this conclusion.

$$RSO_2N_3 \xrightarrow{h\nu} RSO_2\ddot{N} + N_2$$

$$RSO_2\ddot{N} + R_2'SO \xrightarrow{15-30\%} RSO_2N{=}\overset{\overset{O}{\|}}{S}R_2'$$

(32)

Better yields are obtained in sulfide solvents, the products being N-sulfonylsulfimines (33).

$$RSO_2N_3 + R_2'S \xrightarrow[48-55\%]{hv} RSO_2N{=}SR_2' + N_2$$
(33)

The reaction of sulfonyl azides with alcohols illustrates the difficulty in determining whether or not nitrenes are involved. In 1963 Horner and Christmann (12) reported the photolysis of tosyl azide in methanol to form N-(p-toluenesulfonyl)-O-methylhydroxylamine (34) as the major product, with minor amounts of ammonium tosylate, a volatile material (probably formaldehyde), and a solid tentatively formulated as 2-methoxy-4-methylbenzenesulfonamide, which the authors assumed might have been formed from 34 by methoxyl group migration.

$$p\text{-}CH_3C_6H_4SO_2N_3 + CH_3OH \xrightarrow{hv} N_2 + p\text{-}CH_3C_6H_4SO_2NHOCH_3$$
(44%) (34)

Although 34 could be formulated as arising from insertion of a sulfonyl nitrene into the H—O bond of methanol, the reaction is more complicated than this. Thus, Horner and Christmann found that the solution turned yellow upon irradiation and began to eliminate nitrogen only after about 45 min. If the solution was allowed to stand in the dark, however, it became acidic and liberated nitrogen as soon as irradiation was started, indicating perhaps the formation of a photolabile intermediate.

Lwowski and his co-workers (27,28) investigated the photolysis more thoroughly. In addition to the H—O insertion product (35), a compound identified as methyl N-phenylsulfamate (37) was isolated, formed presumably by reaction of the Curtius rearrangement product, N-sulfurylaniline (36), with methanol. The authors suggested that the azide reaction

$$C_6H_5SO_2N_3 + CH_3OH \xrightarrow{hv} N_2 + C_6H_5SO_2NHOCH_3 +$$
(15%)(35)

$$[C_6H_5NSO_2] + C_6H_5SO_2NH_2 + C_6H_5SO_3{}^-NH_4{}^+$$
(36)　　　(4.5%)　　　(8%)

$$\downarrow \text{MeOH}$$

$$C_6H_5NHSO_3CH_3$$
(23%)(37)

might involve a protonated species. It is tempting to speculate that the

$$C_6H_5\underset{\underset{O}{\|}}{\overset{\overset{O}{\|}}{S}}-\overset{\ominus}{N}-\overset{\oplus}{N}\equiv N \xrightarrow{CH_3OH} \underset{\underset{C_6H_5}{\overset{\delta+}{\diagup}}}{\overset{\overset{\delta-}{\overset{H-OCH_3}{\diagdown}}}{\underset{O}{\overset{O}{\diagup}}}}\overset{\delta+}{S} = = = \overset{}{N} - - - N_2 \xrightarrow{-N_2} \underset{}{\overset{H-OCH_3}{\overset{\diagup}{O}}} OS = NC_6H_5$$

$$\downarrow$$

$$C_6H_5NHSO_3CH_3$$

$$(37)$$

H—O insertion product (35) is derived from nitrene, but here too a protonated species may be involved.

The reaction of methanesulfonyl azide with isopropanol takes a totally different course. Reagan and Nickon (14) found that using either 2537 Å light directly or 3660 Å light with a benzophenone sensitizer led to a quantitative yield of amide and acetone.

$$CH_3SO_2N_3 + CH_3\underset{\underset{OH}{|}}{CH}CH_3 \xrightarrow{h\nu} N_2 + CH_3SO_2NH_2 + CH_3COCH_3$$

There is little doubt that this is a radical-chain reaction: the quantum yield in the acetone-sensitized reaction is 150, a mixture of ferrous chloride and hydrogen chloride at room temperature gives the same result, and Horner and Bauer (89) obtained tosyl amide and acetone from the thermal reaction (50–80°) of tosyl azide in isopropanol catalyzed by diethyl peroxydicarbonate. A free radical decomposition is also in accord with the photolysis of benzenesulfonyl azide in thiophenol reported by Shingaki (90).

$$C_6H_5SO_2N_3 + C_6H_5SH \xrightarrow{2537 Å} N_2 + C_6H_5SO_2NH_2 + C_6H_5SSC_6H_5$$

A reasonable mechanism for the isopropanol dehydrogenation has been proposed by Horner and Bauer:

$$(C_2H_5O\overset{\overset{O}{\|}}{C}O-)_2 \longrightarrow 2C_2H_5O\overset{\overset{O}{\|}}{C}O\cdot \longrightarrow 2C_2H_5O\cdot + 2CO_2$$

$$C_2H_5O\cdot + (CH_3)_2CHOH \longrightarrow (CH_3)_2\overset{\cdot}{C}OH + C_2H_5OH$$

$$RSO_2\overset{..}{N}{=}N{=}\overset{..}{N} + (CH_3)_2\overset{\cdot}{C}OH \longrightarrow RSO_2\overset{..}{N}{=}N{-}\overset{..}{N}{-}\underset{\underset{OH}{|}}{C}(CH_3)_2$$

$$\uparrow \quad \downarrow$$

$$RSO_2\overset{\cdot}{N}H + N_2 + (CH_3)_2CO \longleftarrow \underset{H\diagdown_{O}\diagup \overset{|}{C}(CH_3)_2}{RSO_2\overset{..}{N}\diagdown^{\overset{N}{\|}}{}N:}$$

$$RSO_2\overset{\cdot}{N}H + (CH_3)_2CHOH \longrightarrow RSO_2NH_2 + (CH_3)_2\overset{\cdot}{C}OH$$

What is surprising, however, is the statement by Reagan and Nickon that the reaction proceeds in wet isopropanol under reflux without irradiation. Since the azide is perfectly stable in isooctane at 100°, this cannot be a simple thermolysis, and the source of radicals causing the decomposition is unknown. This is reminiscent of the thermolysis results obtained in this laboratory (16), where sulfur dioxide elimination in a hydrocarbon solvent was also attributed to an unknown radical source (see Section III-A).

V. METAL-CATALYZED DECOMPOSITION OF SULFONYL AZIDES

There have been only two publications on the decomposition of sulfonyl azides catalyzed by metals. Kwart and Khan (91) reported in 1967 that copper metal catalyzes the decomposition of benzenesulfonyl azide in refluxing methanol; the major product was benzenesulfonamide, with minor amounts of methylenebis(benzenesulfonamide) (38) and 1,3,5-tris(benzenesulfonyl)hexahydro-s-triazine (39) resulting from condensation of the amide with formaldehyde formed in the reaction. Cuprous chloride

$$C_6H_5SO_2N_3 + CH_3OH \xrightarrow[\text{reflux}]{\text{Cu}} C_6H_5SO_2NH_2 + [HCHO] + N_2$$
$$(80\%)$$

$$C_6H_5SO_2NH_2 + HCHO \longrightarrow (C_6H_5SO_2NH)_2CH_2 + \underset{(38)}{}$$

was even more effective, but cuprous oxide was inert. Isopropanol gave an almost quantitative yield of amide plus about 15% acetone, while wet t-butanol gave amide and products expected from the t-butoxy radical. Green organosoluble copper complexes were formed in all the reactions. To account for the induced decomposition of azide, the authors postulated the initial formation of a copper–azide complex (40), which then might lose nitrogen to give a copper–nitrene complex (41). An amount of dimethyl sulfoxide about equivalent to the azide accelerated the decomposition even more; N-benzenesulfonyldimethylsulfoximine (42), the same product formed from azide and DMSO by thermolysis and photolysis, was the sole product of the reaction. In fact, even in the absence of copper 42 is formed slowly in refluxing methanol. By using optically active methyl

$$\underset{(40)}{\underset{\underset{SO_2R}{|}}{\overset{\overset{\ddot{N}}{\|}}{\underset{N}{Cu}}}\overset{N:}{}} \longrightarrow N_2 + [Cu \rightleftarrows NSO_2Ar] \longleftarrow$$

(41)

$$R_2CHOH$$

$$Cu° + ArSO_2NH_2 + R_2CO$$

$$R_2CHOH \longrightarrow R_2CHO—Cu \overset{\overset{N}{\|}}{\underset{HN}{}} \overset{N}{\underset{SO_2R}{|}} \longrightarrow R_2CHO—Cu—NHSO_2R + N_2$$

$$\downarrow H_2O$$

$$R_2CHOH + Cu^{2+} + RSO_2NH_2$$

p-tolyl sulfoxide Sabol, Davenport, and Andersen (92) showed that the reaction with tosyl azide to give the corresponding sulfoximine proceeds with retention of configuration, as would be expected of attack by nitrene.

$$\underset{(40)}{\underset{\underset{SO_2R}{|}}{\overset{\overset{\ddot{N}}{}}{\underset{\ddot{N}}{Cu}}}N:} \xrightarrow{\underset{DMSO}{MeOH}} \underset{(42 above)}{(CH_3)_2S} \overset{\overset{O}{\underset{\ddot{N}}{\|}}}{\underset{\underset{SO_2R}{|}}{\underset{N}{}}}N: \xleftarrow{MeOH} \begin{matrix} RSO_2\overset{\ominus}{\ddot{N}}—\overset{\oplus}{N}\equiv N: \\ (CH_3)_2\overset{}{S}—\overset{..}{\underset{..}{O}}: \end{matrix}$$

$$\downarrow -N_2$$

$$\underset{(42)}{(CH_3)_2\overset{\overset{O}{\|}}{S}=NSO_2R}$$

In a succeeding article Kwart and Khan (93) investigated the copper-catalyzed decomposition of benzenesulfonyl azide in cyclohexene. A bewildering variety of products were formed, chief among them being benzenesulfonamide, the aziridine from addition to the double bond (43), an enamine (44), and cyclohexanone:

$$C_6H_5SO_2N_3 + \hexagon \xrightarrow[reflux]{Cu}$$

$$\underset{(37\%)}{C_6H_5SO_2NH_2} + \underset{(15\%)\ (43)}{C_6H_5SO_2N\triangle} + \underset{(17\%)\ (44)}{C_6H_5SO_2NH} + \hexagon=O + SO_2 + N_2$$

Here, too, the authors invoke the formation of a copper–azide complex (40). However, in view of the results obtained in the photochemical

decomposition of sulfonyl azides in alcohols (see Section IV), it would seem just as reasonable to postulate a copper-catalyzed radical reaction in these decompositions, even though a trace of hydroquinone added to the cyclohexene reaction only eliminated one minor product. Here again the question of whether a nitrene reaction, or in this case perhaps a "nitrenoid" reaction, is involved is a very difficult one to answer with any degree of certainty.

VI. OTHER POSSIBLE SOURCES OF SULFONYLNITRENES

The most obvious source of a sulfonylnitrene other than azide decomposition would appear to be α-elimination of chloride ion from a compound such as Chloramine-T (45).

$$[p\text{-}CH_3C_6H_4SO_2NCl]^- Na^+ \longrightarrow p\text{-}CH_3C_6H_4SO_2\ddot{N} + NaCl$$

(45)

This possibility has been discussed (9) in relation, for example, to the synthesis of sulfimines from Chloramine-T and sulfides, since the same product is obtained by the photolysis of the corresponding sulfonyl azide (see Section IV). However, an alternative S_N2 displacement is just as plausible, and even less evidence is available to differentiate between the two than there is in the classical question of whether NH is involved in the Raschig synthesis of hydrazine (10).

Although Dermer and Edmison (22) reported that heating Chloramine-T in nitrobenzene results in the evolution of nitric oxide just as tosyl azide

$$p\text{-}CH_3C_6H_4SO_2\overset{..}{N} + R_2S$$

$$p\text{-}CH_3C_6H_4SO_2\overset{\ominus}{N}\text{—}Cl \quad -Cl^- $$

$$:\overset{..}{S}:$$
$$R \quad R$$

$$p\text{-}CH_3C_6H_4SO_2\overset{\ominus}{N}\text{—}\overset{\oplus}{S}R_2$$

does (see Section III-C), the fact that no other products could be isolated in either reaction is rather flimsy evidence for nitrene formation. Actually, in this laboratory (94) a number of salts of tosyl *N*-chloroamide were heated in cyclohexane with no evidence of any C—H insertion reaction. Unfortunately, the low solubility of these salts in the reaction medium renders this negative evidence also suspect; low yields in other nitrene reactions involving insoluble precursors have been reported (95). Quite recently, in fact, Carr, Seden, and Turner (95a) reported that heating Chloramine-B with dimethyl sulfoxide in the presence of copper gave an excellent yield of the corresponding *N*-sulfonylsulfoximine (**45a**); insertion into a C—H bond of dioxane to form 2-benzenesulfonamido-1,4-dioxane (**45b**) was cited as evidence for the intermediacy of a nitrene.

$$[C_6H_5SO_2NCl]^-\,Na^+ \; + \; Cu \xrightarrow[\;80\%\;]{\overset{Me_2SO}{\underset{3\,hr.\,80°}{}}} C_6H_5SO_2N\overset{\overset{O}{\|}}{=}S(CH_3)_2$$

$$70\% \left| \begin{matrix} dioxane \\ 40\,hr.\,25° \end{matrix}\right.$$

(**45a**)

C_6H_5SO_2NH

(**45b**)

An alternate synthesis of sulfonylnitrenes might involve abstraction of chlorine from Dichloramine-T (**46**) by a metal. Breslow and Sloan (94) found that treatment of Dichloramine-T in cyclohexane with metallic zinc yielded a small amount of the C—H insertion product, *N*-cyclohexyltosyl amide.

$$p\text{-}CH_3C_6H_4SO_2NCl_2 + \bigcirc + Zn \longrightarrow p\text{-}CH_3C_6H_4SO_2NH\text{—}\bigcirc + ZnCl_2$$

(**46**)

Since the C—H insertion reaction is a distinctive nitrene reaction, it is highly likely that a nitrene intermediate is involved, formed perhaps in a two-step reaction:

$$p\text{-}CH_3C_6H_4SO_2NCl_2 + Zn \longrightarrow p\text{-}CH_3C_6H_4SO_2N\underset{ZnCl}{\overset{Cl}{\diagdown}} \longrightarrow p\text{-}CH_3C_6H_4SO_2\ddot{N} + ZnCl_2$$

(47)

The success here might be attributed to some expected solubility of an intermediate covalent zinc complex (47) as opposed to the insolubility of Chloramine-T salts. Nevertheless, the fact that according to Abramovitch (95b) there is no formation of sulfonanilide in the reaction of benzene with Dichloramine-T and zinc under similar conditions throws some doubt on the formation of nitrene from these reagents.

A reasonable nitrene precursor would appear to be N-p-nitrobenzene-sulfonoxybenzenesulfonamide (48) investigated by Lwowski and Scheiffele (28). After several days at room temperature its salts (potassium or triethylammonium) in methanol or ethanol gave 65% yields of alkyl-N-phenylsulfamates (50), derived apparently from sulfurylaniline (49) formed by a Lossen-type rearrangement; the triethylammonium salt in aniline gave a 77% yield of N,N'-diphenylsulfamide (51). The rearrange-

$$C_6H_5SO_2NHOSO_2C_6H_4NO_2 \xrightarrow{\text{base}} C_6H_5SO_2\overset{\ominus}{N}OSO_2C_6H_4NO_2$$
$$\text{(48)} \qquad\qquad\qquad\qquad\qquad\qquad \Big\downarrow$$

$$\underset{\text{(50)}}{C_6H_5NHSO_2OR} \xleftarrow{\text{ROH}} \underset{\text{(49)}}{[C_6H_5NSO_2]} + p\text{-}NO_2C_6H_4SO_3^-$$

$$\Big\downarrow \text{PhNH}_2$$

$$C_6H_5NHSO_2NHC_6H_5$$
$$\text{(51)}$$

ment to 49 probably does not go through a nitrene intermediate. Thus, Lwowski and Scheiffele (28) found that decomposition of the soluble triethylammonium salt of 48 in toluene–methylene chloride gave no toluidides, the products expected if a nitrene intermediate were involved, even though hydrolysis of the reaction mixture yielded aniline, indicating that the Lossen rearrangement to N-sulfurylaniline had taken place. Similarly, the reaction in benzene gave no anilide. It is undoubtedly significant that Lwowski and Scheiffele did not report the presence of any of the H—O insertion product (35) in the methanolysis reaction. Since this is formed from sulfonyl azide but not from 48, it seems reasonable to postulate that 35 is formed from nitrene, and nitrene is not formed by α-elimination from 48.

$$C_6H_5SO_2N_3 \xrightarrow[-N_2]{h\nu} C_6H_5SO_2\overset{\cdot\cdot}{\underset{\cdot\cdot}{N}} \longleftarrow\!\!\!\!\times\!\!\!\!- C_6H_5SO_2NHOSO_2C_6H_4NO_2$$

$$\text{(48)}$$

with MeOH branch:

$$C_6H_5SO_2NHOCH_3$$
$$\text{(35)}$$

with base, MeOH branch:

$$C_6H_5NHSO_3CH_3$$

McFarland and Burkhardt (95c) investigated the reaction of benzene-sulfonyl isocyanate with phenylmagnesium bromide. Although *N*-benzoylbenzenesulfonamide, formed by addition of the Grignard reagent to the carbonyl group, was the major product, in one experiment they obtained a 13% yield of *N*-phenylbenzenesulfonamide, which they suggested might have been formed from a nitrene intermediate.

$$C_6H_5SO_2N\!\!=\!\!C\!\!=\!\!O \longrightarrow C_6H_5SO_2\overset{\cdot\cdot}{\underset{\cdot\cdot}{N}} + CO$$

$$\downarrow \text{PhMgBr}$$

$$C_6H_5SO_2N\!\!\begin{array}{c} \diagup C_6H_5 \\ \diagdown MgBr \end{array} \xrightarrow{H_2O} C_6H_5SO_2NHC_6H_5$$

Robson and Speakman (95d) have investigated the chemistry of alkane-sulfamidates. *N*-Trimethylammoniododecanesulfonamidate (**51a**) decomposed in dimethyl sulfoxide at 170° to give the expected sulfoximine (**51b**) and amide. In decalin in the presence of triphenylphosphine the phosphinimine (**51c**) was formed, as well as amide, the formation of the latter being suppressed entirely by the use of a large excess of phosphine. Although these reactions are most readily explained by postulating the formation of an intermediate sulfonylnitrene, it is surprising that no carbon–hydrogen insertion product was formed in the decalin reaction.

$$C_{12}H_{25}SO_2\overset{\ominus}{N}\!\!-\!\!\overset{\oplus}{N}(CH_3)_3 + (CH_3)_2SO \xrightarrow{170°} (CH_3)_3N +$$
$$\text{(51a)}$$

$$C_{12}H_{25}SO_2N\!\!=\!\!\underset{\underset{O}{\|}}{S}(CH_3)_2 + C_{12}H_{25}SO_2NH_2$$

$$\text{(51b)}$$

$$\textbf{51a} + P(C_6H_5)_3 \xrightarrow{\Delta} (CH_3)_3N + C_{12}H_{25}SO_2NH_2 + C_{12}H_{25}SO_2N\!\!=\!\!P(C_6H_5)_3$$
$$\text{(51c)}$$

VII. SULFONYL AZIDE REACTIONS NOT INVOLVING NITRENES

The preceding discussion describes reactions which undoubtedly do involve a nitrene intermediate or might conceivably involve a nitrene intermediate. However, there is a large body of literature on reactions of sulfonyl azides which quite definitely do not involve nitrenes. Evidence against nitrene intermediates comes from either the use of temperatures too low to involve sulfonyl azide thermolysis or from the nature of the products. Some of these are described briefly below.

A. Reactions with Active Hydrogen Compounds

In Curtius' early work (6) he described the reaction of malonic ester with a number of sulfonyl azides—benzene- (21,96), p-toluene- (32), α-toluene- (97), p-chlorobenzene- (38), p-acetamidobenzene- (40), 1-naphthalene- (46), 2-naphthalene- (47), and β-anthraquinone- (48)—in the presence of sodium ethoxide to give the sodium salt of a 5-hydroxytriazole (52). Acidification gave the corresponding 5-triazolone (53), which then isomerized to a diazo compound (54).

$$RSO_2N_3 + CH_2(CO_2C_2H_5)_2 \xrightarrow{\text{NaOEt}} RSO_2N \underset{\substack{| \\ C=C-CO_2C_2H_5 \\ | \\ ONa}}{\overset{N=N}{|}} \quad (52) \xrightarrow{\text{H}^+}$$

$$RSO_2N \underset{\substack{| \\ C-CH-CO_2C_2H_5 \\ \| \\ O}}{\overset{N=N}{|}} \quad (53) \longrightarrow N_2C\underset{CONHSO_2R}{\overset{CO_2C_2H_5}{<}} \quad (54)$$

This diazo synthesis lay dormant until Doering and DePuy (33) applied it to the preparation of diazocyclopentadiene (55).

$$p\text{-}CH_3C_6H_4SO_2N_3 + C_5H_5Li \xrightarrow{\text{Et}_2O} \langle \rangle = N_2 + p\text{-}CH_3C_6H_4SO_2NH_2$$
$$(35\%) \ (55) \qquad (29\%)$$

Since then this diazo transfer reaction has been applied to a variety of active-hydrogen compounds; the literature has been surveyed by Regitz (98). Regitz and Liedhegener (99) found that condensation of tosyl azide with diethyl malonate in acetonitrile using triethylamine as base gave an almost quantitative yield of diethyl diazomalonate; in many cases substitution of amine for alkoxide avoids side reactions. Regitz formulated the reactions shown in Scheme 1.

Many related syntheses have been reported. Thus, Fischer and Anselme (100) prepared aryl azides from anilinomagnesium halide and tosyl azide.

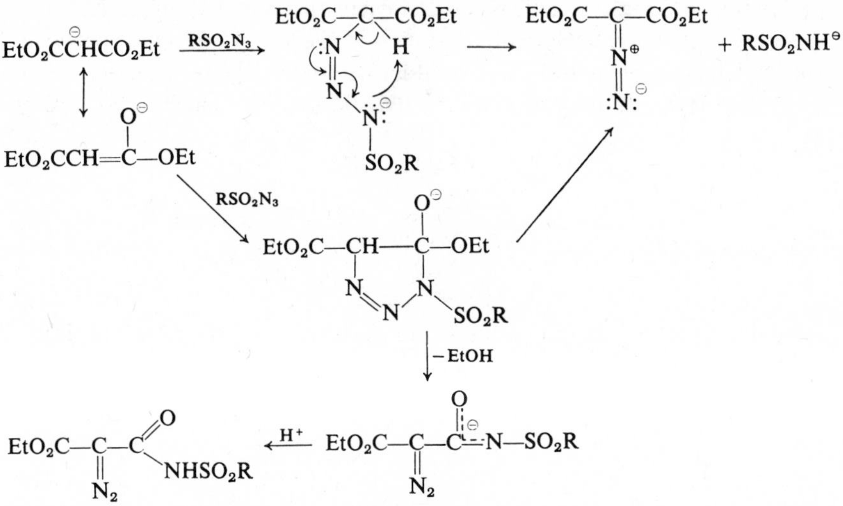

$$EtO_2CCHCO_2Et \xrightarrow{RSO_2N_3}$$

Scheme 1

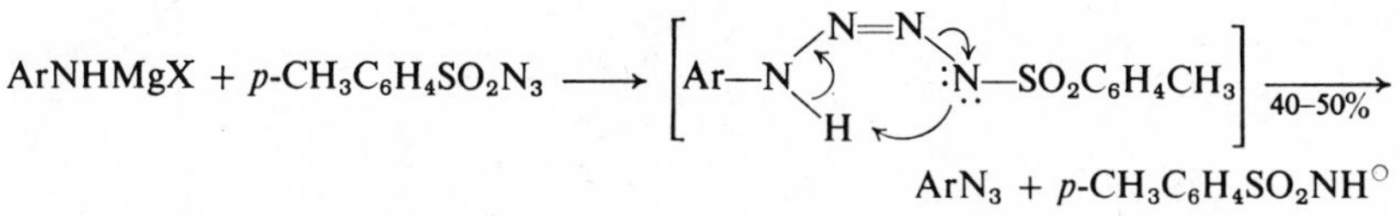

$$ArNHMgX + p\text{-}CH_3C_6H_4SO_2N_3 \longrightarrow \left[Ar-N \underset{H}{\overset{N=N}{\diagup}} \ddot{N}-SO_2C_6H_4CH_3 \right] \xrightarrow[40-50\%]{}$$

$$ArN_3 + p\text{-}CH_3C_6H_4SO_2NH^\circleddash$$

Ito (29) prepared phenyl azide from a phenyl Grignard:

$$C_6H_5MgBr + C_6H_5SO_2N_3 \xrightarrow{-15°,\ Et_2O} C_6H_5N=N-\underset{\underset{MgBr}{|}}{N}SO_2C_6H_5 \xrightarrow[82\%]{\overset{\Delta}{120-130°}}$$

$$C_6H_5N_3 + [C_6H_5SO_2MgBr]$$

Tedder and Webster (101) showed that even the anions of reactive phenols (resorcinol, phloroglucinol, α- and β-naphthol) would react with sulfonyl azides:

$$RSO_2N_3 + ArO^\circleddash \longrightarrow [RSO_2NH-N=N-Ar-O^\circleddash] \longrightarrow$$

$$RSO_2NH^\circleddash + {}^\circleddash O-Ar-\overset{\oplus}{N}{\equiv}N$$

B. Reactions with Unsaturated Compounds

One of the best known reactions of formylnitrenes, addition to a double bond to form an aziridine, has yet to be demonstrated with sulfonylnitrenes. However, in 1963 both Franz and Osuch (102) and Zalkow and Oehlschlager (59) reported the condensation of benzenesulfonyl azide with norbornene at 55–60°, a temperature so low that a nitrene could not have been involved. In fact, Franz and Osuch showed that styrene, cyclohexene, cyclopentene, maleic anhydride, N-phenylmaleimide, divinylsulfone, mesityl oxide, vinyl acetate, and p-quinone evolved no gas with benzenesulfonyl azide even at 100°. The product with norbornene has been shown (102,103,105) to be the *exo*-aziridine (**56**). Franz and Osuch (103) formu-

$$+ C_6H_5SO_2N_3 \xrightarrow[64\%]{\substack{\text{MeCN} \\ 55-60°}} \qquad NSO_2C_6H_5 + N_2$$

(**56**)

lated the reaction as analogous to an epoxidation. However, the intermediate formation of a triazoline (**57**) is probably just as reasonable, since with other azides it can sometimes be isolated. The same reaction

takes place with a variety of azides (102,49), and with a number of compounds with strained double bonds (34,104–109). With norbornadiene, however, the initial aziridine (**58**) is unstable and rearranges to **59**(104,106).

Curtius and Klavehn (110) had reported in 1930 that aromatic sulfonyl azides do not react with acetylenedicarboxylic acid esters, and Boyer, Mack et al. (11) found no reaction of tosyl azide with phenylacetylene. Huisgen and co-workers (111) demonstrated, however, that the reactions do take place, albeit slowly, to yield the corresponding 1,2,3-triazoles (**60**); the structure of the compound derived from phenylacetylene was not

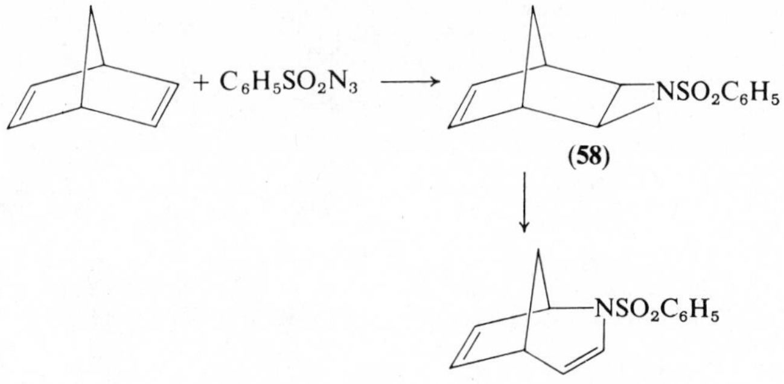

$$\text{(diagram)} + C_6H_5SO_2N_3 \longrightarrow \text{(diagram)} \text{NSO}_2C_6H_5$$

(58)

$$\downarrow$$

$$\text{(diagram)} \text{NSO}_2C_6H_5$$

(59)

$$p\text{-}CH_3C_6H_4SO_2N_3 + R\text{—}C\equiv C\text{—}R' \xrightarrow[50-75\%]{1-2 \text{ wk, } 70-80°}$$

R = CO₂Et, Ph

R' = CO₂Et, H

$$C_6H_5SO_2N \overset{N}{\underset{\underset{R}{C}=\underset{R'}{C}}{\diagup}} N$$

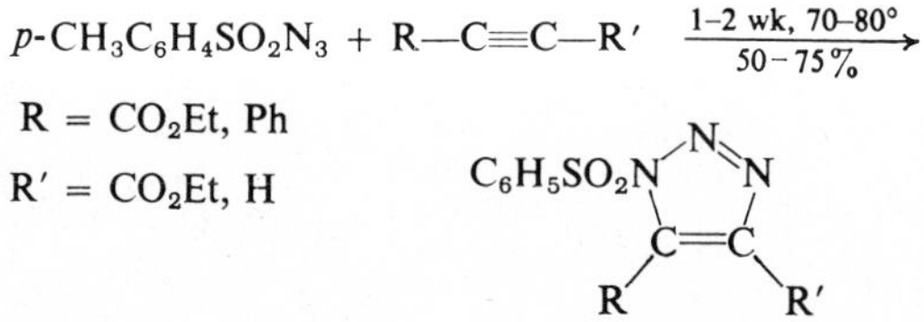

(60)

proved, however. According to Vita Finzi (112) the reaction with aryl-acetylenes yields either an amide (**61**) or a triazolium salt (**62**), depending on the substituents. Tosyl azide was purported to give a salt with phenyl-acetylene, but Vita Finzi did not obtain the same compound as Huisgen

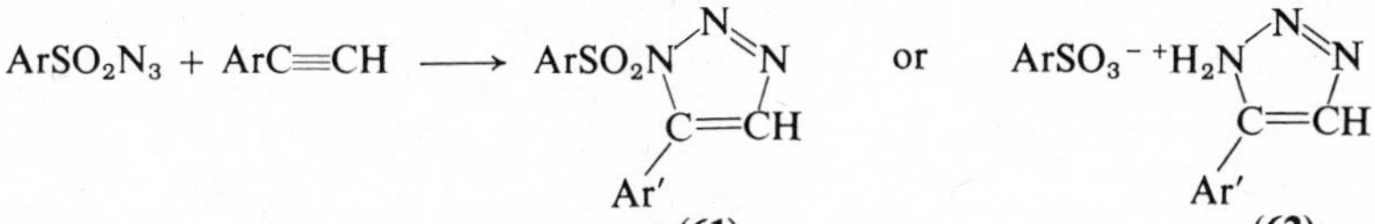

$$ArSO_2N_3 + ArC\equiv CH \longrightarrow ArSO_2N \overset{N}{\underset{\underset{Ar'}{C}=CH}{\diagup}} N \quad \text{or} \quad ArSO_3^- \, {}^+H_2N \overset{N}{\underset{\underset{Ar'}{C}=CH}{\diagup}} N$$

(61)   **(62)**

and co-workers (111). He did show, however, that the compound obtained by Boyer, Mack et al. (11) from tosyl azide and sodium phenylacetylide is the tosylate salt of 5-phenyltriazole (**62**, Ar = p-CH₃C₆H₄, Ar' = C₆H₅) and not the amide (**61**, Ar = p-CH₃C₆H₄, Ar' = C₆H₅) as reported by them. According to Robson, Tedder, and Webster (113) the lithium derivative of 1-hexyne reacts with tosyl azide to give an azotriazole derivative.

According to Fusco and co-workers (36,114) enamines react with sulfonyl azides to give amidines (**63** and **64**) with either rearrangement or elimination of diazomethane, depending on the nature of the enamine; triazoline intermediates were postulated. Ritchie and Rosenberger (115)

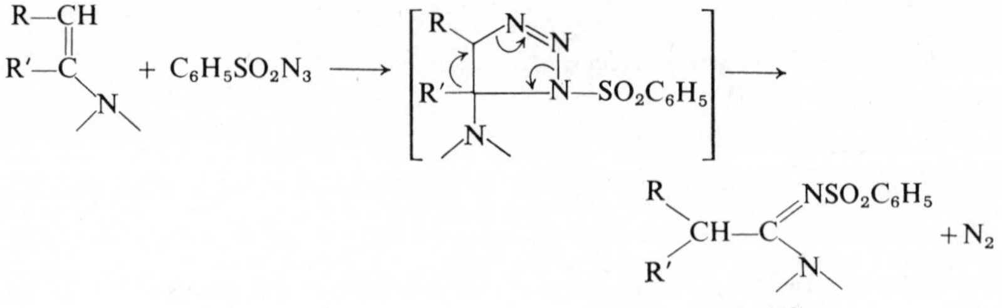

$$R-CH \\ R'-C \diagdown \diagup N \diagdown + C_6H_5SO_2N_3 \longrightarrow \left[\begin{array}{c} R \diagdown \quad N\diagdown N \\ R' \diagup \quad N-SO_2C_6H_5 \\ \diagup N \diagdown \end{array} \right] \longrightarrow$$

$$R \diagdown \quad \diagup NSO_2C_6H_5 \\ \quad CH-C \\ R' \diagup \quad \diagdown N \diagdown \quad + N_2$$

(63)

$$CH_2 \\ \parallel \\ C \\ R \diagup \diagdown N \diagdown + C_6H_5SO_2N_3 \longrightarrow \left[\begin{array}{c} N\diagdown N \\ R \quad N-SO_2C_6H_5 \\ \diagup N \diagdown \end{array} \right] \longrightarrow$$

$$\diagup NSO_2C_6H_5 \\ R-C \\ \diagdown N \diagdown \quad + CH_2N_2$$

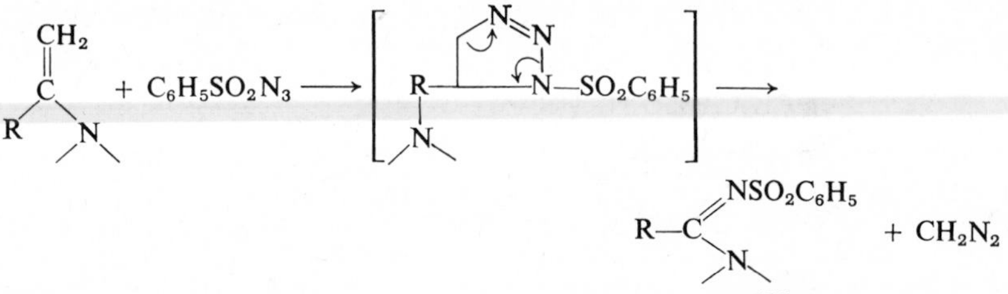

(64)

reported the same reaction with azomethines, the azomethine–enamine equilibrium being involved, and Kučera and Arnold (116) applied the reaction to the synthesis of the first aliphatic α-diazoaldehyde (65).

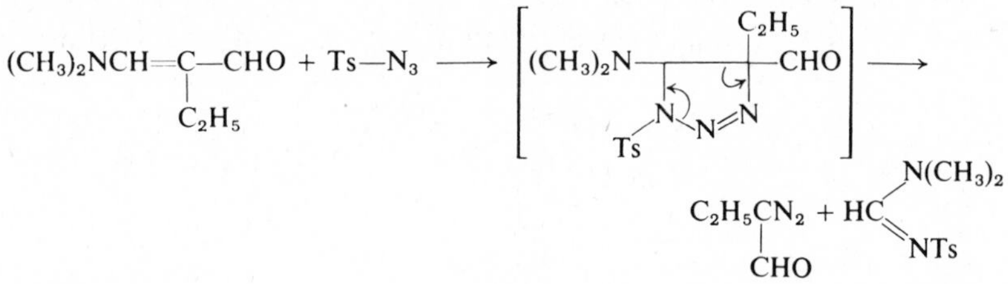

$$(CH_3)_2NCH=C-CHO + Ts-N_3 \longrightarrow \left[\begin{array}{c} C_2H_5 \\ (CH_3)_2N-\overset{|}{\underset{}{}}-\overset{|}{\underset{}{}}-CHO \\ \overset{|}{\underset{Ts}{N}}\diagdown N\diagup N \end{array} \right] \longrightarrow \\ \overset{|}{C_2H_5}$$

$$C_2H_5CN_2 + HC \diagup N(CH_3)_2 \\ \overset{|}{\underset{CHO}{}} \quad \diagdown NTs$$

(65)

There have been a number of reports on the reaction of sulfonyl azides with vinyl ethers (117–119,37), the products being the corresponding iminoesters (66 and 67) or polymer and dimer (68 and 69).

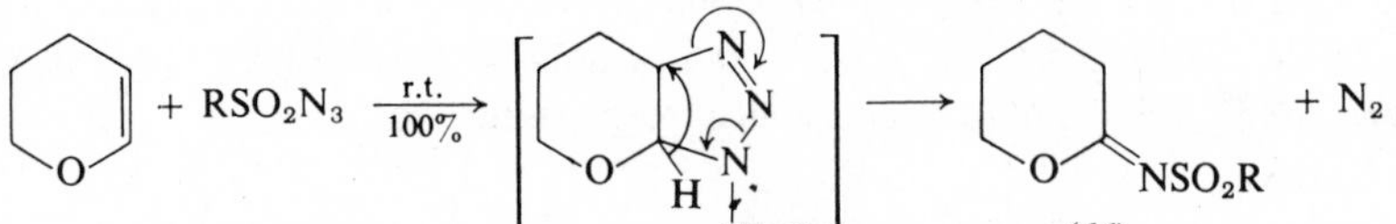

$$\left\langle \overset{}{\underset{O}{}} \right\rangle + RSO_2N_3 \xrightarrow[100\%]{r.t.} \left[\begin{array}{c} N \\ \overset{}{\underset{O}{}} \diagdown N \\ H \overset{|}{\underset{SO_2R}{}} N \end{array} \right] \longrightarrow \left\langle \overset{}{\underset{O}{}} \right\rangle \overset{}{\underset{NSO_2R}{}} + N_2$$

(66)

Grünanger and Vita Finzi (39,120) found that ethoxyacetylene reacts similarly to give a diazo iminoester (70).

$$CH_2=CH-OR' + RSO_2N_3 \longrightarrow \left[\begin{array}{c} OR' \\ \\ N \diagdown N \diagup N \\ SO_2R \end{array} \right] \xrightarrow{>0°} \begin{array}{c} NSO_2R \\ H-C \diagup \\ OR' \\ (67) \end{array} + CH_2N_2$$

$$\xrightarrow[\text{—}N_2 \quad 0°]{}$$

$$\left(\begin{array}{c} -CHCH_2N- \\ | \quad\quad | \\ OR' \quad SO_2R \end{array} \right)_n + \begin{array}{c} SO_2R \\ N \diagdown OR' \\ \\ R'O \diagup N \\ SO_2R \\ (69) \end{array}$$

$$(68)$$

$$C_2H_5O-C\equiv CH + RSO_2N_3 \xrightarrow{\text{r.t.}}$$

$$\left[\begin{array}{c} RSO_2-N \diagdown N \diagup N \\ C_2H_5O \end{array} \right] \longrightarrow \left[\begin{array}{c} N \\ \| \\ RSO_2N \diagup N^{\oplus} \\ | \\ C_2H_5O-C=CH \end{array} \right]^{\ominus} \longrightarrow \begin{array}{c} RSO_2N=C-OC_2H_5 \\ | \\ CHN_2 \\ (70) \end{array}$$

Bleiholder and Shechter (121) investigated the decomposition of benzene- and p-toluenesulfonyl azides in tetramethylallene. A rearrangement takes place to give N-(1,2,3-trimethyl-2-butenylidene)-arenesulfonamides (72) in modest yield. Although the temperature of the reaction was high enough for a nitrene to have been formed, the fact that the same type of product is obtained from p-nitrophenyl azide by heating the isolable intermediate triazoline makes it appear likely that a triazoline intermediate (71) is involved here too. The isolation of tosyl amide from the tosyl azide

$$\begin{array}{c} H_3C \diagdown \quad\quad CH_3 \\ C=C=C \\ H_3C \diagup \quad\quad CH_3 \end{array} + ArSO_2N_3 \xrightarrow[26\%]{\text{reflux}}$$

$$\left[\begin{array}{c} H_3C \diagdown \quad\quad\quad CH_3 \\ C=C \diagdown \quad C \\ H_3C \diagup \quad N\diagdown N \diagup N \diagup CH_3 \\ \quad\quad\quad SO_2Ar \end{array} \right] \xrightarrow{-N_2} \begin{array}{c} H_3C \diagdown \quad\quad CH_3 \\ C=C \diagdown CH_3 \\ H_3C \diagup \quad C \\ \| \\ NSO_2Ar \end{array}$$

$$(71) \quad\quad\quad\quad\quad\quad (72)$$

reaction might indicate the presence of a nitrene, but its involvement in the formation of the azomethine seems rather unlikely.

C. Reactions with Trivalent Phosphorus Derivatives

Although the reaction of simple azides with phosphines was reported by Staudinger and Meyer (122) in 1919, sulfonyl azides were apparently not investigated until 1961, when Goerdeler and Ullmann (17) reported that the condensation of a variety of sulfonyl azides with trialkyl phosphites and thiophosphites at room temperature or above takes place with the loss of nitrogen.

$$RSO_2N_3 + (R'O)_3P \longrightarrow RSO_2N{=}P(OR')_3 + N_2$$

Leffler and Tsuno (24) showed that the reaction with triphenylphosphine could be used as an analytical procedure for sulfonyl azides. However, they also showed (25) that the reaction is dependent on the solvent; in benzene at low temperatures a 1:1 complex, presumably the triazene (73), could be isolated, which broke down to the phosphinimine (74) in a clean first-order reaction on warming.

$$C_6H_5SO_2N_3 + (C_6H_5)_3P \longrightarrow C_6H_5SO_2N{=}N{-}N{=}P(C_6H_5)_3 \xrightarrow{\Delta}$$

(73)

$$C_6H_5SO_2N{=}P(C_6H_5)_3 + N_2$$

(74)

In chloroform or similar solvents, however, the reaction was much more complicated; the evolution of nitrogen was not quantitative, benzene-sulfonamide and triphenylphosphine oxide were isolated, and no phosphinimine was formed. They attributed these results to a radical nitrene decomposition. Franz and Osuch (123) investigated the reaction in more detail and isolated diphenyl sulfide and disulfide from the reaction in chloroform or acetonitrile; they postulated the initial formation of a sulfinyl azide.

$$(C_6H_5)_3P + C_6H_5SO_2N_3 \longrightarrow (C_6H_5)_3PO + C_6H_5SON_3 \longrightarrow$$

$$(C_6H_5)_3PO + C_6H_5SN_3 \longrightarrow 1/2\ C_6H_5SSC_6H_5 + 3/2\ N_2$$

$$C_6H_5SSC_6H_5 + (C_6H_5)_3P \longrightarrow C_6H_5SC_6H_5 + (C_6H_5)_3PS$$

The triazenes have been patented (124). Bock and Wiegräbe (125) discovered, apparently independently, that stable triazenes can be formed with sulfonyl azides and $(C_6H_5)_2PX$, where $X = C_6H_5$, C_6H_5NH, or $(n\text{-}C_3H_7)_2N$. Bock and Schnöller (125a) proved that addition is to the terminal nitrogen of the azide, to form 73, by ^{15}N labeling.

D. Miscellaneous Reactions

Fischer and Anselme (126) found that hydrazone anions are converted into diazo compounds on treatment with tosyl azide, presumably via a pentazene intermediate.

$$\underset{R}{\overset{R}{\diagdown}}C{=}N\overset{\ominus}{N}H + p\text{-}CH_3C_6H_4SO_2N_3 \xrightarrow{50\%} \underset{R}{\overset{R}{\diagdown}}C{=}N_2 + N_2 + p\text{-}CH_3C_6H_4SO_2NH^{\ominus}$$

A number of workers (127–129) have reacted ethylenes tetrasubstituted with electron-donating groups with tosyl azides. Thus, Hoffmann and Häuser (127) isolated **75** from tetramethoxyethylene.

$$\begin{array}{c} CH_3O \\ \\ CH_3O \end{array} C{=}C \begin{array}{c} OCH_3 \\ \\ OCH_3 \end{array} + \textit{p-}CH_3C_6H_4SO_2N_3 \xrightarrow{0°}$$

$$\textit{p-}CH_3C_6H_4SO_2N{=}C \begin{array}{c} OCH_3 \\ \\ OCH_3 \end{array} + N_2$$

$$(75)$$

Quast and Hünig (128) obtained a mixture of imine (**78**) and triazene (**79**) from the 2,2'-bis(benzothiazoline) (**76**), and claimed that **79** is derived from the intermediate carbene (**77**).

Winberg and Coffman (129) obtained a similar mixture from a tetra-aminoethylene (**80**).

Ried and Schön (130) showed that sulfonyl azides can be used as benzyne traps, benzotriazoles (81) being formed.

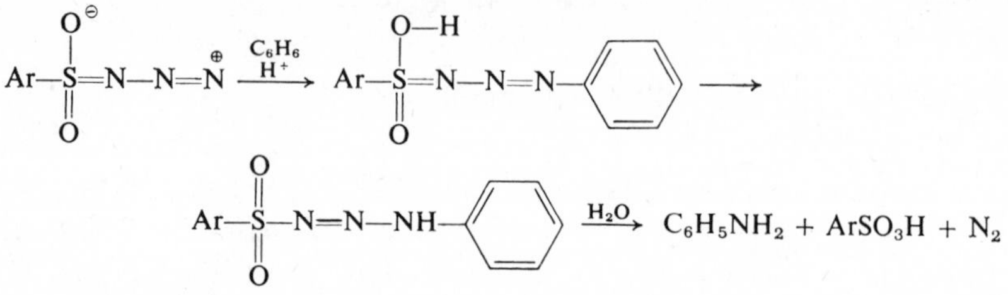

Kreher and Jäger (131) found that sulfonyl azides split out nitrogen or azide ion under Friedel-Crafts conditions to give amine and sulfone.

$$ArSO_2N_3 + C_6H_6 + AlCl_3 \longrightarrow C_6H_5NH_2 + C_6H_5SO_2Ar + N_2$$

Sidhu, Thyagarajan, and Bhalerao (132) found that benzenesulfonyl azide is an excellent aminating agent in the presence of benzenesulfonic acid; 50–60% yields of aniline and toluidines were obtained with benzene and toluene, the yield being somewhat lower with chlorobenzene. They formulated the reaction as follows:

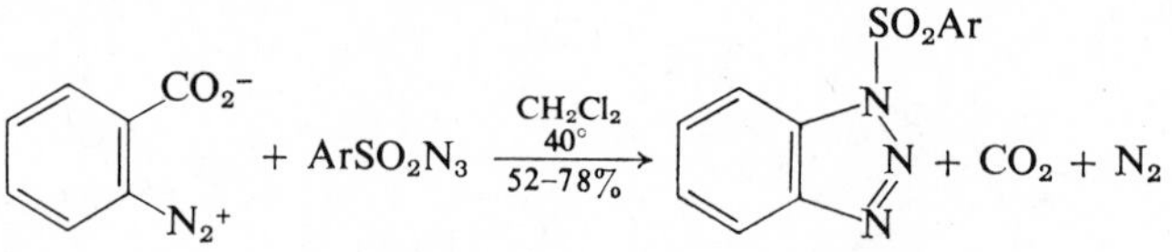

VIII. USES FOR SULFONYL AZIDES AND NITRENES

The first patents on sulfonyl azides involved the use of 4,4'-biphenyl-disulfonyl azide as a cross-linking and foaming agent for cellulose acetate (56), unsaturated alkyd resins (133), poly(vinyl acetal) resins (134), polystyrene (135), and polyethylene (136). Following this a number of mono- and disulfonyl azides were patented as blowing agents (15,50), including a diazoaminosulfonyl azide (82) prepared from sulfanilic acid (45).

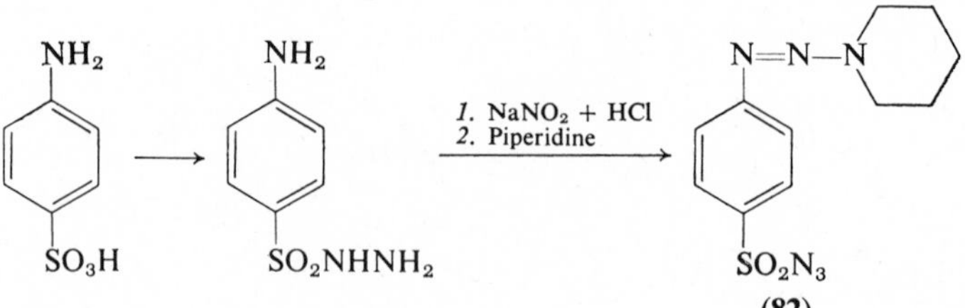

With the discovery of the ability of sulfonyl azides to insert into a carbon–hydrogen bond (63), a number of patents issued covering the use of mono- and poly(sulfonyl azides) for the modification or crosslinking of hydrocarbon polymers (137–145,54,58). In fact, it was the discovery that polypropylene and polyisobutylene could be crosslinked with disulfonyl azides (137) which led us to the conclusion that the C—H insertion reaction is a reaction of singlet nitrene, since these polymers are known to degrade with free radicals rather than crosslink. Sulfonyl azides are also useful for crosslinking poly(vinyl ethers) (146), poly(vinyl chloride) (147,148), and acrylic polymers (149). An isocyanatosulfonyl azide has been reacted with a hydroxyl-containing polymer through the isocyanate group and the polymer then insolubilized by irradiation (42).

REFERENCES

1. T. Curtius and F. Lorenzen, *J. Prakt. Chem.*, [2] **58**, 160 (1898).
2. T. Curtius, *Z. Angew. Chem.*, **26**, 134 (1913).
3. T. Curtius, *Chem. Ztg.*, **37**, 214 (1913); *Chem. Abstr.*, **8**, 670 (1914).
4. T. Curtius, *Z. Angew. Chem.*, **27**, 213 (1914).
5. A. Bertho, *J. Prakt. Chem.*, [2] **120**, 89 (1928).
6. T. Curtius, *J. Prakt. Chem.*, [2] **125**, 303 (1930) and succeeding papers.
7. F. Muth, in E. Müller, "Methoden der Organischen Chemie" (Houben-Weyl), Georg Thieme Verlag, Stuttgart, Germany, 4th Ed., Vol. IX, 1955, p. 653–4.
8. O. C. Dermer and M. T. Edmison, *Chem. Rev.*, **57**, 99 (1957).
9. L. Horner and A. Christmann, *Angew. Chem.*, **75**, 707 (1963); *Intern. Ed.*, **2**, 599 (1963).
10. R. A. Abramovitch and B. A. Davis, *Chem. Rev.*, **64**, 149 (1964).
11. J. H. Boyer, C. H. Mack, N. Goebel, and L. R. Morgan, *J. Org. Chem.*, **23**, 1051 (1958).
12. L. Horner and A. Christmann, *Chem. Ber.*, **96**, 388 (1963).
13. R. A. Abramovitch, J. Roy, and V. Uma, *Can. J. Chem.*, **43**, 3407 (1965).
14. M. T. Reagan and A. Nickon, *J. Am. Chem. Soc.*, **90**, 4096 (1968).
15. F. H. Adams, U.S. Pat. 2,830,029 (1958) (to American Cyanamid).
16. D. S. Breslow, M. F. Sloan, N. R. Newburg, and W. B. Renfrow, *J. Am. Chem. Soc.*, **91**, 2293 (1969).
17. J. Goerdeler and H. Ullmann, *Chem. Ber.*, **94**, 1067 (1961).
18. G. Smolinsky and B. I. Feuer, *J. Am. Chem. Soc.*, **86**, 3085 (1964).
19. J. K. Ruff, *Inorg. Chem.*, **4**, 567 (1965).
20. T. Curtius and F. W. Haas, *J. Prakt. Chem.*, [2] **102**, 85 (1921).
20a. R. A. Abramovitch, C. I. Azogu, and I. T. McMaster, *J. Am. Chem. Soc.*, **91**, 1219 (1969).
21. T. Curtius and J. Rissom, *J. Prakt. Chem.*, [2] **125**, 311 (1930).
22. O. C. Dermer and M. T. Edmison, *J. Am. Chem. Soc.*, **77**, 70 (1955).
23. N. A. Gol'dberg and G. P. Balabanov, *J. Org. Chem. USSR*, **1**, 1625 (1965).
24. J. E. Leffler and Y. Tsuno, *J. Org. Chem.*, **28**, 190 (1963).
25. J. E. Leffler and Y. Tsuno, *J. Org. Chem.*, **28**, 902 (1963).
26. M. Imoto, K. Takemoto, and R. Fujita, *Makromol. Chem.*, **100**, 249 (1967).

26a. K. Takemoto, R. Fujita, and M. Imoto, *Makromol. Chem.*, **112**, 116 (1968).
27. W. Lwowski, R. DeMauriac, T. W. Mattingly, Jr., and E. Scheiffele, *Tetrahedron Letters*, **1964**, 3285.
28. W. Lwowski and E. Scheiffele, *J. Am. Chem. Soc.*, **87**, 4359 (1965).
29. S. Ito, *Bull. Chem. Soc. Japan*, **39**, 635 (1966).
30. R. J. W. Cremlyn, *J. Chem. Soc.*, **1965**, 1132.
31. T. Curtius and G. Kraemer, *J. Prakt. Chem.* [2] **125**, 323 (1930).
32. T. Curtius and W. Klavehn, *J. Prakt. Chem.*, [2] **112**, 65 (1926).
33. W. v. E. Doering and C. H. DePuy, *J. Am. Chem. Soc.*, **75**, 5955 (1953).
34. R. Huisgen, L. Möbius, G. Müller, H. Stangl, G. Szeimies, and J. M. Vernon, *Chem. Ber.*, **98**, 3992 (1965).
35. W. Pritzkow and D. Timm, *J. Prakt. Chem.*, [4] **32**, 178 (1966).
36. R. Fusco, G. Bianchetti, and D. Pocar, *Gazz. Chim. Ital.*, **91**, 933 (1961); *Chem. Abstr.*, **56**, 14020 (1962).
37. D. L. Rector and R. E. Harmon, *J. Org. Chem.*, **31**, 2837 (1966).
38. T. Curtius and K. Vorbach, *J. Prakt. Chem.*, [2] **125**, 340 (1930).
39. P. Grünanger, P. Vita Finzi, and C. Scotti, *Chem. Ber.*, **98**, 623 (1965).
40. T. Curtius and W. Stoll, *J. Prakt. Chem.*, [2] **112**, 117 (1926).
41. J. N. Ashley, G. L. Buchanan, and A. P. T. Eason, *J. Chem. Soc.*, **1947**, 60.
42. J. Danhaeuser and W. Pelz, Belg. Pat. 665,429 (1965) (to Gevaert-Agfa), *Chem. Abstr.*, **64**, 12901 (1966). Agfa-Gevaert A.-G., French Pat. 1,470,623 (1967).
43. E. Schrader, *J. Prakt. Chem.*, [2] **96**, 180 (1918).
44. R. J. W. Cremlyn, *J. Chem. Soc.*, **1964**, 6235.
45. W. H. von Glahn and B. Rudner, U.S. Pat. 2,828,299 (1958) (to General Aniline and Film).
46. T. Curtius, H. Bottler, and G. Hasse, *J. Prakt. Chem.*, [2] **125**, 366 (1930).
47. T. Curtius, H. Bottler, and W. Raudenbusch, *J. Prakt. Chem.*, [2] **125**, 380 (1930).
48. T. Curtius and H. Derlon, *J. Prakt. Chem.*, [2] **125**, 420 (1930).
49. R. J. Shozda and J. A. Vernon, *J. Org. Chem.*, **32**, 2876 (1967). U.S. Pat. 3,418,088 (1968) (to DuPont).
50. W. B. Hardy and F. H. Adams, U.S. Pat. 2,863,866 (1958) (to American Cyanamid).
51. T. Curtius and F. Schmidt, *Ber.*, **55B**, 1571 (1922).
52. K. F. Schmidt, *Ber.*, **58B**, 2409 (1925).
53. A. Bertho, T. Curtius, and F. Schmidt, *Ber*, **60B**, 1717 (1927).
54. H. W. Bost and J. E. Mahan, U.S. Pat. 3,282,864 (1966) (to Phillips Petroleum).
55. T. Curtius and H. Meier, *J. Prakt. Chem.*, [2] **125**, 358 (1930).
56. J. B. Ott, U.S. Pat. 2,518,249 (1950) (to Monsanto).
57. T. Curtius and R. Tüxen, *J. Prakt. Chem.*, [2] **125**, 401 (1930).
58. N. R. Newburg, U.S. Pat. 3,287,376 (1966) (to Hercules).
59. L. H. Zalkow and A. C. Oehlschlager, *J. Org. Chem.*, **28**, 3303 (1963).
60. C. M. Suter, *The Organic Chemistry of Sulfur*, Wiley, New York, 1944, p. 507.
61. R. J. Shozda, *Inorg. Chem.*, **6**, 1919 (1967).
62. M. Kobayashi and A. Yamamoto, *Bull. Chem. Soc. Japan*, **39**, 2733 (1966).
62a. T. J. Maricich, *J. Am. Chem. Soc.*, **90**, 7179 (1968);
62b. V. A. Gal'perin and G. P. Balabanov, *J. Gen. Chem. USSR*, **38**, 889 (1968).
63. M. F. Sloan, D. S. Breslow, and W. B. Renfrow, *Tetrahedron Letters*, **1964**, 2905.
64. T. L. Davis, *The Chemistry of Powder and Explosives*, Vol. I, Wiley, New York, 1941, p. 21.

65. G. P. Balabanov, Y. I. Dergunov, and V. G. Golov, *Zh. Fiz. Khim.*, **40**, 2171 (1966); *Chem. Abstr.*, **65**, 19974 (1966).
66. G. P. Balabanov, Y. I. Dergunov, and V. A. Gal'perin, *J. Org. Chem. USSR*, **2**, 1797 (1966).
67. A. J. Rosenthal and C. G. Overberger, *J. Am. Chem. Soc.*, **82**, 108,117 (1960).
68. W. T. Reichle, *Inorg. Chem.*, **3**, 402 (1964).
69. M. F. Sloan, T. J. Prosser, N. R. Newburg, and D. S. Breslow, *Tetrahedron Letters*, **1964**, 2945.
69a. D. S. Breslow, E. C. Linsay, and H. Omura, unpublished observation.
70. T. Curtius and W. Stoll, *J. Prakt. Chem.*, [2] **125**, 364 (1930).
71. J. F. Heacock and M. T. Edmison, *J. Am. Chem. Soc.*, **82**, 3460 (1960).
72. G. Smolinsky, E. Wasserman, and W. A. Yager, *J. Am. Chem. Soc.*, **84**, 3220 (1962).
73. J. F. Tilney-Bassett, *J. Chem. Soc.*, **1962**, 2517.
74. R. L. Dannley and M. Esayian, *J. Polymer Sci.*, **45**, 105 (1960).
75. P. Ghosh and K. F. O'Driscoll, *J. Macromol. Sci.-Chem.*, **A1**, 1393 (1967).
76. J. E. Burleigh and C. A. Uraneck, U.S. Pat. 3,301,841 (1967) (to Phillips Petroleum).
77. D. S. Breslow, T. J. Prosser, A. F. Marcantonio, and C. A. Genge, *J. Am. Chem. Soc.*, **89**, 2384 (1967).
78. R. A. Abramovitch and V. Uma, *Chem. Commun.*, **1968**, 797.
79. T. Curtius, *Z. Angew. Chem.*, **28**, 5 (1915).
80. T. Curtius, *Z. Angew. Chem.*, **30**, 532 (1917).
81. F. Schmidt, *Ber.*, **55B**, 1581 (1922).
82. B. S. Alamela and K. Ganapathi, *Current Sci.*, **12**, 119 (1943); *Chem. Abstr.*, **38**, 5492 (1944).
83. P. K. Datta, *J. Indian Chem. Soc.*, **24**, 109 (1947).
84. G. L. Buchanan and R. M. Levine, *J. Chem. Soc.*, **1950**, 2248.
85. M. Takebayashi and T. Shingaki, *Kogyo Kagaku Zasshi*, **64**, 469 (1961); *Chem. Abstr.*, **57**, 2110 (1962).
86. M. Takebayashi, T. Shingaki, and T. Mitsuyama, *Sci. Rept. (Osaka Univ.)*, **10**, 35 (1961); *Chem. Abstr.*, **59**, 9493 (1963).
87. T. Shingaki, *Sci. Rept. Coll. Gen. Educ. Osaka Univ.*, **11**, 67 (1963); *Chem. Abstr.*, **60**, 6733 (1964).
88. T. Shingaki, *Sci. Rept. Coll. Gen. Educ., Osaka Univ.*, **11**, 81 (1963); *Chem. Abstr.*, **60**, 6734 (1964).
89. L. Horner and G. Bauer, *Tetrahedron Letters*, **1966**, 3573.
90. T. Shingaki, *Sci. Rept. Coll. Gen. Educ., Osaka Univ.*, **11**, 93 (1963); *Chem. Abstr.*, **60**, 6734 (1964).
91. H. Kwart and A. A. Khan, *J. Am. Chem. Soc.*, **89**, 1950 (1967).
92. M. A. Sabol, R. W. Davenport, and K. K. Andersen, *Tetrahedron Letters*, **1968**, 2159.
93. H. Kwart and A. A. Khan, *J. Am. Chem. Soc.*, **89**, 1951 (1967).
94. D. S. Breslow and M. F. Sloan, *Tetrahedron Letters*, **1968**, 5349.
95. W. Lwowski and T. J. Maricich, *J. Am. Chem. Soc.*, **87**, 3630 (1965).
95a. D. Carr, T. P. Seden, and R. W. Turner, *Tetrahedron Letters*, **1969**, 477.
95b. R. A. Abramovitch, private communication.
95c. J. W. McFarland and W. A. Burkhardt III, *J. Org. Chem.*, **31**, 1903 (1966).
95d. P. Robson and P. R. H. Speakman, *J. Chem. Soc.*, **1968B**, 463.
96. T. Curtius and G. Ehrhart, *J. Prakt. Chem.*, [2] **106**, 66 (1923).
97. T. Curtius and B. Jeremias, *J. Prakt. Chem.*, [2] **112**, 88 (1926).

98. M. Regitz, *Angew. Chem.*, **79**, 786 (1967); *Intern. Ed.*, **6**, 733 (1967).
99. M. Regitz and A. Liedhegener, *Chem. Ber.*, **99**, 3128 (1966).
100. W. Fischer and J. P. Anselme, *J. Am. Chem. Soc.*, **89**, 5284 (1967).
101. J. M. Tedder and B. Webster, *J. Chem. Soc.*, **1960**, 4417.
102. J. E. Franz and C. Osuch, *Tetrahedron Letters*, **1963**, 837.
103. J. E. Franz, C. Osuch, and M. W. Dietrich, *J. Org. Chem.*, **29**, 2922 (1964).
104. J. E. Franz and C. Osuch, *Chem. Ind. (London)*, **1964**, 2058.
105. L. H. Zalkow, A. C. Oehlschlager, G. A. Cabat, and R. L. Hale, *Chem. Ind. (London)*, **1964**, 1556.
106. A. C. Oehlschlager and L. H. Zalkow, *Chem. Commun.*, **1965**, 70.
107. A. C. Oehlschlager and L. H. Zalkow, *J. Org. Chem.*, **30**, 4205 (1965).
108. L. H. Zalkow and C. D. Kennedy, *J. Org. Chem.*, **28**, 3309 (1963).
109. A. C. Oehlschlager and L. H. Zalkow, *Can. J. Chem.*, **47**, 461 (1969).
110. T. Curtius and W. Klavehn, *J. Prakt. Chem.*, [2] **125**, 498 (1930).
111. R. Huisgen, R. Knorr, L. Möbius, and G. Szeimies, *Chem. Ber.*, **98**, 4014 (1965).
112. P. Vita Finzi, *Chim. Ind. (Milan)*, **47**, 1338 (1965).
113. E. Robson, J. M. Tedder, and B. Webster, *J. Chem. Soc.*, **1963**, 1863.
114. R. Fusco, G. Bianchetti, D. Pocar, and R. Ugo, *Chem. Ber.*, **96**, 802 (1963).
115. A. C. Ritchie and M. Rosenberger, *J. Chem. Soc.*, **1968C**, 227.
116. J. Kučera and Z. Arnold, *Tetrahedron Letters*, **1966**, 1109.
117. R. Huisgen, L. Möbius, and G. Szeimies, *Chem. Ber.*, **98**, 1138 (1965).
118. R. Harmon and D. Rector, *Chem. Ind. (London)*, **1965**, 1264.
119. J. E. Franz, M. W. Dietrich, A. Henshall, and C. Osuch, *J. Org. Chem.*, **31**, 2847 (1966).
120. P. Grünanger and P. Vita Finzi, *Tetrahedron Letters*, **1963**, 1839.
121. R. F. Bleiholder and H. Shechter, *J. Am. Chem. Soc.*, **90**, 2131 (1968).
122. H. Staudinger and J. Meyer, *Helv. Chim. Acta*, **2**, 635 (1919).
123. J. E. Franz and C. Osuch, *Tetrahedron Letters*, **1963**, 841.
124. J. E. Franz and C. Osuch, U.S. Pat. 3,282,895 (1966) (to Monsanto).
125. H. Bock and W. Wiegräbe, *Angew. Chem.*, **75**, 790 (1963); *Intern. Ed.*, **2**, 484 (1963).
125a. H. Bock and M. Schnöller, *Chem. Ber.*, **102**, 38 (1969).
126. W. Fischer and J. P. Anselme, *Tetrahedron Letters*, **1968**, 877.
127. R. W. Hoffmann and H. Häuser, *Tetrahedron Letters*, **1964**, 1365.
128. H. Quast and S. Hünig, *Angew. Chem.*, **76**, 989 (1964); *Intern. Ed.*, **3**, 800 (1964).
129. H. E. Winberg and D. D. Coffman, *J. Am. Chem. Soc.*, **87**, 2776 (1965).
130. W. Ried and M. Schön, *Chem. Ber.*, **98**, 3142 (1965).
131. R. Kreher and G. Jäger, *Angew. Chem., Intern. Ed.*, **4**, 706 (1965).
132. G. S. Sidhu, G. Thyagarajan, and U. T. Bhalerao, *Chem. Ind. (London)*, **1966**, 1301.
133. J. B. Ott, U.S. Pat. 2,532,240 (1950) (to Monsanto).
134. J. B. Ott, U.S. Pat. 2,532,241 (1950) (to Monsanto).
135. J. B. Ott, U.S. Pat. 2,532,242 (1950) (to Monsanto).
136. J. B. Ott, U.S. Pat. 2,532,243 (1950) (to Monsanto).
137. D. S. Breslow and H. M. Spurlin, U.S. Pat. 3,058,944 (1962) (to Hercules).
138. P. L. Johnstone, U.S. Pat. 3,137,745 (1964) (to Hercules).
139. D. S. Breslow and H. M. Spurlin, U.S. Pat. 3,203,936 (1965) (to Hercules).
140. D. S. Breslow and H. M. Spurlin, U.S. Pat. 3,203,937 (1965) (to Hercules).
141. D. S. Breslow, U.S. Pat. 3,220,985 (1965) (to Hercules).
142. G. B. Feild and P. H. Johnstone, U.S. Pat. 3,298,975 (1967) (to Hercules).

143. W. W. Cox, U.S. Pat. 3,336,268 (1967) (to Hercules).
144. G. B. Feild, U.S. Pat. 3,341,480 (1967) (to Hercules).
145. D. A. Palmer, U.S. Pat. 3,341,481 (1967) (to Hercules).
146. D. S. Breslow, U.S. Pat. 3,058,957 (1962) (to Hercules).
147. A. E. Robinson, U.S. Pat. 3,261,785 (1966) (to Hercules).
148. D. S. Breslow, U.S. Pat. 3,261,786 (1966) (to Hercules).
149. D. S. Breslow and F. E. Piech, U.S. Pat. 3,322,733 (1967) (to Hercules).

CHAPTER 9

Cyanonitrene

A. G. Anastassiou* (with Mrs. J. N. Shepelavy*)
Department of Chemistry, Syracuse University, Syracuse, New York

AND

H. E. Simmons and F. D. Marsh
Central Research Department, Experimental Station, E. I. du Pont de Nemours and Co., Inc., Wilmington, Delaware

* Work supported by the National Science Foundation under Grant GP 6618.

Cyanonitrene (NCN) is the youngest and most symmetrical member of the nitrene family. Nevertheless, its short history is a rich one from the viewpoint of both the spectroscopist and the chemist. The molecule was first reported by Herzberg and Travis (1) in 1964, who observed its generation in the photolysis of diazomethane, a rather unorthodox precursor. In the same year, the synthesis of cyanogen azide (N_3CN) (2) provided a more conventional progenitor for NCN, whose generation from the azide was soon realized both photolytically (3,4) and thermally (5). Subsequently, cyanogen azide has been employed as the exclusive source of NCN by both chemists and spectroscopists.

The goal of the present chapter is to present a unified account of the published studies concerned with NCN. To best achieve this end, we have divided our coverage into three major portions. Section I deals with theoretical aspects of the subject and provides the necessary background for better appreciation of the experimental work described later. Section II covers the spectroscopic studies designed to elucidate the structure of NCN and to determine the mechanism of both its formation and decay. Finally, Section III deals with chemical reactions of the molecule with various substrates.

I. THEORETICAL CONSIDERATIONS

Cyanonitrene is capable of only two symmetry modifications. It may be either linear $(D_{\infty h})$ (1) or bent (C_{2v}). The latter may, of course, assume either an acyclic (2) or a cyclic (3) structure depending on the magnitude of the bond angle and the extent to which the two nitrogen atoms interact.

$$\cdot N{=}C{=}N\cdot \qquad\qquad .N \overset{C}{\diagup\diagdown} N. \qquad\qquad N\overset{C}{\diagup\diagdown}N \quad \text{or} \quad N\overset{C}{\diagup\diagdown}N$$
$$\text{(1)} \qquad\qquad\qquad \text{(2)} \qquad\qquad\qquad \text{(3)} \qquad\qquad\qquad \text{(3)}$$

Although an *a priori* assignment of relative energies to these three structures is not possible without recourse to explicit calculation, strain-energy considerations suggest that, other factors being equal, 1 should represent the most stable and 3 the least stable of the three possible arrangements. Hence, we limit the following discussion of the ground and low-lying excited states of NCN to the acyclic structures 1 and 2. This preliminary choice is fully justified by a number of spectroscopic studies (Section II).

Cyanonitrene contains a total of twenty electrons. Of these, 14 are valence electrons and six are inner shell ($1s$) electrons. Valence theory suggests that in both 1 and 2 six valence electrons will occupy two orthogonal π orbitals, and eight will occupy σ-framework orbitals. The molecular orbitals (MO's) of NCN can be conveniently determined by allowing the various atomic orbitals (AO's) available to the central carbon to mix with the various possible two center MO's of a hypothetical N_2 species; i.e., the procedure employed may be depicted as $C + N{\equiv}N \rightarrow NCN$. The nature of the resulting three-center MO's will, of course, depend on the geometry of the system, i.e., whether NCN is assumed linear or bent. Low-energy, three-center MO's for a general, linear Y–X–Y arrangement were constructed in this fashion by Herzberg (6) (Fig. 1). In the right-hand side of this figure are shown the AO's of atom X and the two-center MO's of molecule Y_2. Interaction of these affords the three-center MO's shown in the left-hand side of the diagram. It should be noted in this context that

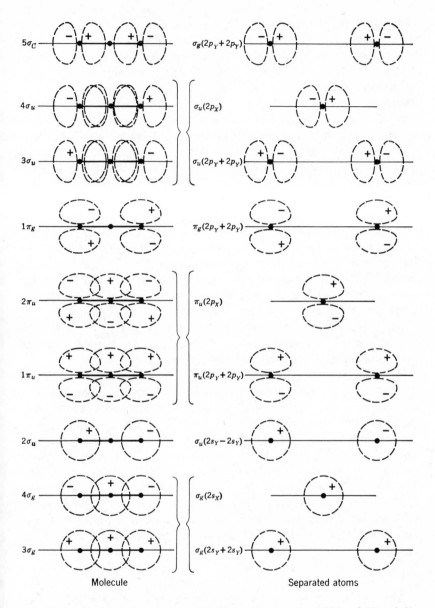

Molecule   Separated atoms

Fig. 1. Schematic representation of AO's and two-center MO's of X and Y_2, respectively (shown in the right) and of the three-center MO's of linear YXY (shown in the left).

307

only AO's and MO's of the same symmetry mix. For example, $\pi_u (2p_X)$ which is antisymmetric with respect to inversion mixes with $\pi_u (2p_Y + 2p_Y)$ also antisymmetric to inversion but does not interact with $\pi_g (2p_Y + 2p_Y)$ which is symmetric with respect to this operation.

Once the MO's are known, all one has to do in principle so as to obtain an appropriate configurational wavefunction of the ground state is to determine the occupancy of these levels. In practice this is not simple, since it requires knowledge of the energies of the various MO's which can be determined only through explicit calculation. Some information concerning the relative positions of the MO's may be obtained, however, through a commonly employed approximation. In this, the energies of molecular levels at small internuclear distances are approximated by those of atomic levels of an atom ("the united atom," which is isoelectronic with the molecule of interest) and the energies of molecular levels at large internuclear separations by those of the atomic levels of the individual atoms constituting the molecule ("the separated atoms"). The orbitals of the "united atom" are then correlated with those of the "separated atoms" within the restrictions imposed by both symmetry and the noncrossing rule.* This approximation, though convenient, is useful only when the internuclear distances in the molecule are known. A "united atom"–"separated atoms" energy correlation diagram for linear molecules of type Y–X–Y where Y is more electronegative than X (as required for NCN) has been given by Herzberg (6) (Fig. 2).

Simple inspection of this diagram indicates that at small internuclear distances the ground state is predicted to possess the open-shell configuration $\ldots 3\sigma_g^2 2\sigma_u^2 1\pi_u^4 4\sigma_g^2 3\sigma_u^2 2\pi_u^2$. At large internuclear distances the inner molecular shell is well defined as $\ldots 3\sigma_g^2 2\sigma_u^2 4\sigma_g^2 1\pi_u^4 3\sigma_u^2 \ldots$, but the choice of the highest occupied MO (frontier MO) is ambiguous since the $1\pi_g$ and $5\sigma_g$ levels are very close in energy. Therefore, the ground state here may be represented either by a $(\ldots 3\sigma_g^2 2\sigma_u^2 4\sigma_g^2 1\pi_u^4 3\sigma_u^2 5\sigma_g^2)$ closed-shell configuration or by a $(\ldots 3\sigma_g^2 2\sigma_u^2 4\sigma_g^2 1\pi_u^4 3\sigma_u^2 1\pi_g^2)$ open-shell arrangement. At intermediate C—N distances we are faced with the same problem of choice between $5\sigma_g$ and $1\pi_g$ as the frontier MO, since both these MO's appear to be equally responsive toward bond length changes. The two possible ground state configurations at intermediate bond lengths are thus a $(\ldots 3\sigma_g^2 2\sigma_u^2 1\pi_u^4 4\sigma_g^2 3\sigma_u^2 5\sigma_g^2)$ closed shell and a $(\ldots 3\sigma_g^2 2\sigma_u^2 1\pi_u^4 4\sigma_g^2 3\sigma_u^2 1\pi_g^2)$ open shell.

Although a definitive choice between the five different ground-state configurations suggested by the correlation diagram is not possible without explicit calculation, we can intuitively eliminate a few. For example, the arrangement predicted for short internuclear separations appears unlikely

* This states that levels of the same symmetry do not cross in energy.

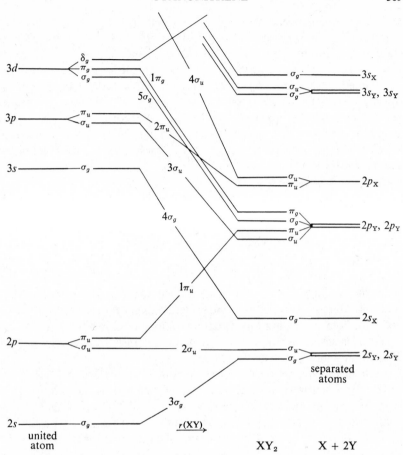

Fig. 2. United atom–separated atoms correlation diagram for the general case of linear YXY structures.

since in this a $2\pi_u$ MO is partially occupied while the normally lower energy $1\pi_g$ MO is left vacant. We can also tentatively eliminate the two configurations (one predicted for large and one for intermediate C—N separations) involving a total of five fully occupied σ MO's on the basis of simple valence considerations which suggest the presence of six π and eight σ valence electrons. Hence we are left with just the two open-shell arrangements in which the highest occupied level is π_g. Now, we can further limit our preference to one of these by considering the known energetics of carbon dioxide, a molecule similar to NCN in many respects. Self-consistent-field calculations (7) yield ($\ldots 3\sigma_g{}^2\, 2\sigma_u{}^2\, 4\sigma_g{}^2\, 1\pi_u{}^4\, 3\sigma_u{}^2\, 1\pi_g{}^4$) as the ground-state configuration of CO_2. Removal of two electrons from this arrangement would afford the open-shell arrangement ($\ldots 3\sigma_g{}^2\, 2\sigma_u{}^2$ $4\sigma_g{}^2\, 1\pi_u{}^4\, 3\sigma_u{}^2\, 1\pi_g{}^2$) for the ground state of NCN. We will now examine the ground state of NCN on the basis of this configuration. Simple multiplet theory requires that such an open-shell arrangement give rise to one $^3\Sigma_g{}^-$, one $^1\Sigma_g{}^+$ and two $^1\Delta_g$ states, shown in terms of the diagonals of the appropriate Slater determinants in equations 1–4.* Of these, the triplet should be least destabilized by electron repulsions and consequently ought to represent the ground state. This prediction based on strictly qualitative theory has been fully verified by both spectroscopy (Section II) and chemistry (Section III).

$$[|\ldots \pi_g(x)\bar{\pi}_g(y)| + |\ldots \bar{\pi}_g(x)\pi_g(y)|] \subset {}^3\Sigma_g{}^- \tag{1}$$

$$[|\ldots \pi_g(x)\bar{\pi}_g(y)| - |\ldots \bar{\pi}_g(x)\pi_g(y)|] \subset {}^1\Delta_g \tag{2}$$

$$[|\ldots \pi_g(x)\bar{\pi}_g(x)| - |\ldots \pi_g(y)\bar{\pi}_g(y)|] \subset {}^1\Delta_g \tag{3}$$

$$[|\ldots \pi_g(x)\bar{\pi}_g(x)| + |\ldots \pi_g(y)\bar{\pi}_g(y)|] \subset {}^1\Sigma_g{}^+ \tag{4}$$

In the present discussion we have centered our attention on linear NCN and in so doing concluded that the molecule ought to have a $^3\Sigma_g{}^-$ ground state, without providing any apparent rationale for our preference of a linear over a bent arrangement aside from possible strain energy factors. We will now briefly justify our choice.

In order to compare the energies and consequently the stabilities of two different geometric arrangements of a molecule one has to inquire as to what energy changes occur within each occupied MO upon changing its symmetry environment. This is conveniently accomplished by means of a symmetry–energy correlation diagram commonly known as a Walsh diagram, which is given in Figure 3 for Y–X–Y triatomics (6).

It is clear from this diagram that among occupied MO's, $3\sigma_g$ and $2\sigma_u$ are insensitive to distortion, $4\sigma_g$, $3\sigma_u$, $1\pi_g$, and one component of $1\pi_u(1\pi_u \to 5a_1)$ are destabilized on bending, and only one component of $1\pi_u(1\pi_u \to 1b_1)$ derives stabilization on going from $D_{\infty h}$ to C_{2v}. It is therefore clear that the molecule as a whole will be resistant to bending and that its most stable arrangement will be a linear one. It is interesting to note that the Walsh diagram also suggests that all excited states of NCN, except perhaps those involving occupancy of the $2\pi_u$ MO, should be linear as well.

So far we have discussed predictions concerning the structural details of NCN purely on a qualitative basis employing arguments based either on symmetry or comparison to carbon dioxide. We shall now examine the

---

* Here only the outer shell is depicted, as the inner shell is the same in each of these terms. Bars denote β spin while the absence of a bar indicates α spin.

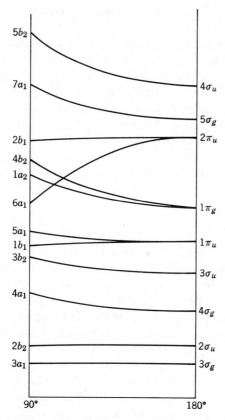

Fig. 3. Walsh diagram for YXY systems. $180° = D_{\infty h}$, $90° = C_{2v}$.

molecule from a more quantitative point of view. As a symmetrical tri-atomic, NCN is certainly amenable to description by the least approximate of theoretical methods. In addition, the large volume of structural infor-mation concerning this molecule gained recently by a variety of spectro-scopic and chemical studies, ought to provide an added incentive for carrying out such calculations, since comparison between calculated and observed properties should serve as a critical test of the various methods of computation. It is therefore surprising to find the literature devoid of any theoretical descriptions on NCN. In unpublished work (8), however, NCN was examined by the extended Hückel procedure recently developed by Hoffmann (9). We shall conclude the present section with a brief description of the results of this calculation.

The various quantities such as MO wavefunctions and energies, total energy, overlap populations and net charges, normally accessible by this type of calculation were determined at a variety of C—N bond distances ranging from 0.5 to 1.45 Å. Some salient features of the results are collected in Table I and Figure 4. In the table are shown the energies and types of the twelve MO's, constructed by appropriate mixing of the twelve outer-shell AO's available to the molecule, determined at the shortest and longest C—N bond lengths as well as at two intermediate internuclear separations which represent significant turning points of the results. Figure 4 is a diagrammatic representation of the energy response of the eight lowest MO's to bond alteration.

Examination of the results reveals several interesting points. For example, the total binding energy at the shortest internuclear separation studied is -248.95 eV and the ground-state configuration is predicted to be $\ldots \sigma_1^2 \sigma_2^2 \pi_1^4 \sigma_3^2 \pi_2^4$. This arrangement can only give rise to a singlet since all MO's involved are fully occupied. As the internuclear distance is increased the total binding energy rises until it reaches a maximum value of -266.14 eV at a C—N bond length of 0.92 Å. Lengthening the bonds from 0.50 to 0.92 Å also affects the energies of the various MO's. From Figure 4 we clearly see that among the low-lying MO's, σ_4 exhibits the greatest response to bond alteration, being depressed by as much as 22 eV. This large stabilization of σ_4 results in near degeneracy of this MO and π_2 at $r(\text{C—N}) = 0.92$ Å, so that it is difficult to choose between a $\ldots \sigma_1^2 \sigma_2^2 \pi_1^4 \sigma_3^2 \pi_2^4$ singlet and a $\ldots \sigma_1^2 \sigma_2^2 \pi_1^4 \sigma_3^2 \pi_2^3 \sigma_4^1$ triplet for the ground state. Increase of the bond length beyond 0.92 Å results in gradual decrease of the total binding energy of the molecule and a dampening of the response of σ_4 to bond lengthening. Nevertheless the stabilization of σ_4 proceeds faster than that of its immediate neighbor π_2 until finally when the C—N bonds are ca. 1.01 Å long the two MO levels cross, so that beyond this point the ground state of the molecule is best described by the open-shell configuration $\ldots \sigma_1^2 \sigma_2^2 \pi_1^2 \sigma_3^2 \sigma_4^2 \pi_2^2$. This arrangement, however, survives only within a very narrow range of internuclear separations ($\Delta r \sim 0.02$ Å), as a second energy crossing between a different set of σ and π MO's (σ_3 and π_1) occurs at ca. 1.03 Å. No additional energy crossings occur upon lengthening the bonds further to the studied limit, 1.45 Å, so that beyond 1.03 Å the ground state of NCN is predicted to possess the open-shell configuration $\ldots \sigma_1^2 \sigma_2^2 \sigma_3^2 \pi_1^4 \sigma_4^2 \pi_2^2$.

Finally, it would be interesting to examine the various quantities calculated by the extended Hückel method at the actual internuclear separation of ground-state NCN which as we shall see later, has been deduced spectroscopically at 1.23 Å. At this bond length the total binding energy is computed to be -264.58 eV, i.e., ca. 1.5 eV lower than the calculated maximum which as we have seen previously is predicted to occur at an internuclear separation of 0.92 Å. Hence the calculation proves incapable

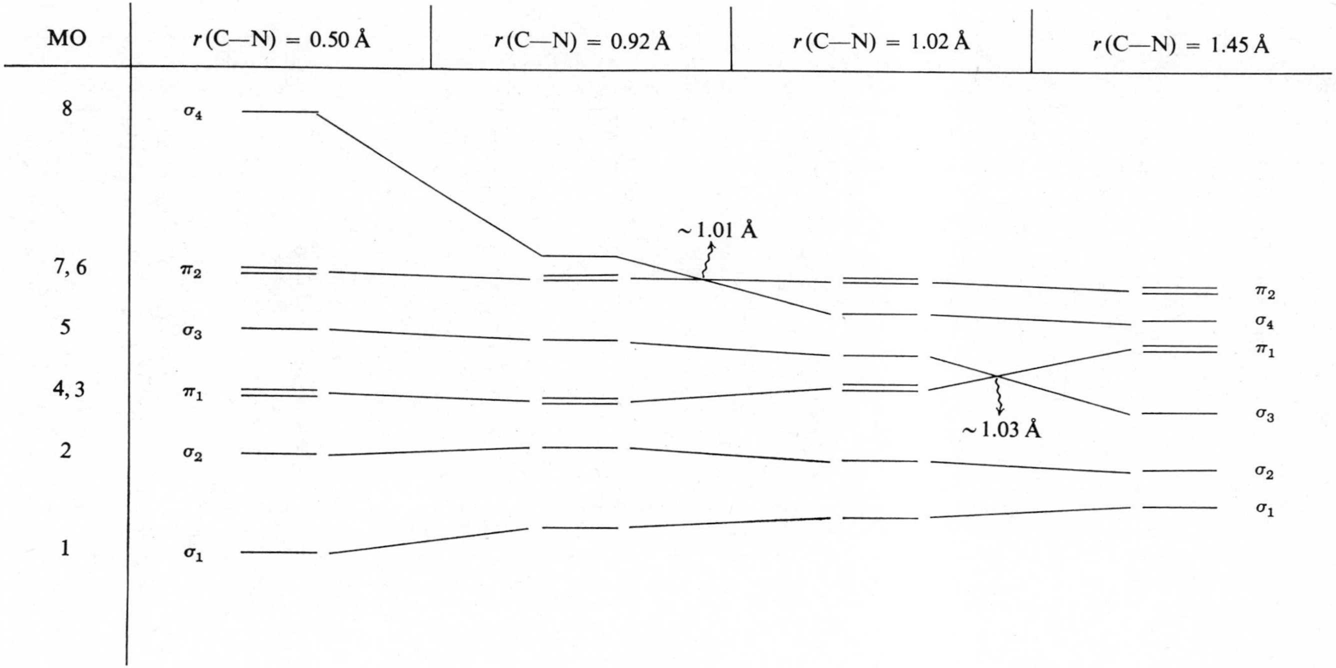

Fig. 4. Energy changes of the eight lowest-energy MO's of NCN as a function of C—N bond separation.

313

TABLE I
Results of Extended Hückel Calculations on NCN

MO	$r(C—N) = 0.50$ Å MO energy (eV)	Type MO	$r(C—N) = 0.92$ Å MO energy (eV)	Type MO	$r(C—N) = 1.02$ Å MO energy (eV)	Type MO	$r(C—N) = 1.45$ Å MO energy (eV)	Type MO
1	−34.685	σ_1	−32.143	σ_1	−31.550	σ_1	−29.125	σ_1
2	−22.131	σ_2	−25.478	σ_2	−25.838	σ_2	−26.124	σ_2
3	−18.292	π_1	−16.564	π_1	−16.186	π_1	−17.318	σ_3
4	−18.292	π_1	−16.564	π_1	−16.186	π_1	−14.846	π_1
5	−13.853	σ_3	−15.618	σ_3	−15.947	σ_3	−14.846	π_1
6	−8.611	π_2	−13.351	π_2	−13.626	σ_4	−14.768	σ_4
7	−8.611	π_2	−13.351	π_2	−13.591	π_2	−13.907	π_2
8	9.551	σ_4	−12.858	σ_4	−13.591	π_2	−13.907	π_2
9	81.034	π_3	1.494	π_3	−2.358	π_3	−9.912	π_3
10	81.034	π_3	1.494	π_3	−2.358	π_3	−9.912	π_3
11	742.764	σ_5	187.072	σ_5	104.933	σ_5	3.940	σ_5
12	3028.810	σ_6	356.321	σ_6	182.307	σ_6	15.555	σ_6

$$E_T(\cdots \sigma_1^2 \sigma_2^2 \pi_1^4 \sigma_3^2 \pi_2^4) = -248.949 \text{ eV}$$

$$E_T(\sigma_1^2 \sigma_2^2 \pi_1^4 \sigma_3^2 \pi_2^4) = -266.138 \text{ eV}$$

$$E_T(\cdots \sigma_1^2 \sigma_2^2 \pi_1^4 \sigma_3^2 \sigma_4^2 \pi_2^2) = -265.846 \text{ eV}$$

$$E_T(\cdots \sigma_1^2 \sigma_2^2 \sigma_3^2 \pi_1^4 \sigma_4^2 \pi_2^2) = -261.869 \text{ eV}$$

of predicting the C—N bond length of minimum energy. On the other hand, the configuration of the ground state predicted on the basis of the present calculation is exactly that deduced from both, experiment and the results of an SCF treatment on CO_2. In brief, the extended Hückel treatment of NCN correctly predicts the ground-state configuration but is unsuccessful in assessing the C—N internuclear separation of minimum energy.

II. SPECTROSCOPIC STUDIES

A. Identification and Structure

In 1960 Haggart and Winkler reported (10) that reaction of cyanogen, $(CN)_2$, with active nitrogen* at temperatures ranging from 80 to 400°C produces CN radicals. To rationalize the generation of CN the workers proposed a mechanism in which this species is generated synchronously with a fleeting fragment formulated as CN_2. The proposed scheme is reproduced in reaction sequence 5.

$$(CN)_2 + N \longrightarrow [(CN)_2 \cdot N] \longrightarrow CN + CN_2$$
$$CN_2 + N \longrightarrow CN + N_2 \tag{5}$$

Shortly thereafter Jennings and Linnett (11) obtained emission spectra of mixtures produced from reaction of active nitrogen with a variety of hydrocarbons, hydrochlorocarbons, and chlorocarbons and observed that in all cases the spectra consisted of emissions due to CN, NH, and CH fragments. In addition, the spectra of all reaction mixtures except those involving perchlorinated reactants, were characterized with a hitherto unobserved complex emission in the region of 3290 Å. Structural information concerning the fragment responsible for this emission was obtained from isotopic labeling experiments employing perdeuterocarbons rather than hydrocarbons as reactants. This change of substrate introduced no alterations in the array of bands in the 3290 Å region, thus strongly suggesting the absence of hydrogen in the fragment responsible for this emission. On this basis, and since oxygen was rigorously excluded from the reaction mixture, it was deduced that the species of interest must consist of only carbon and/or nitrogen. Furthermore, since none of the known emission characteristics of species such as N_2, C_2, and C_3 correspond to those of the 3290 Å emission, it was reasoned that the only likely candidates responsible for this emission are CN_2, C_2N, and excited $(CN)_2$.

---

* Active nitrogen is believed to be atomic nitrogen and is generated by passing molecular nitrogen through an electrical discharge at low pressures (11).

The 3290 Å band system was subsequently observed independently and under widely different conditions by McGrath and Morrow (12) and Herzberg and Travis (1,13). In the first of these studies* (12) it was found that flash photolysis of mixtures consisting either of cyanogen, ozone, and argon or cyanogen, nitrogen dioxide, and argon led to the formation of CN, NCO, and a third fragment displaying complex absorption in the region of 3250–3330 Å. Since cyanogen was found to be stable under the photolysis conditions employed, it was concluded that the oxygen source is essential for the production of the fragment absorbing at 3250–3330 Å and consequently that this species should contain oxygen. Furthermore since the same absorption was observed when a mixture of fulminic acid (HCNO) and argon was flash-photolyzed, the species in question was described as fulminate radical, CNO. Generation of this in mixtures containing either ozone or nitrogen dioxide, known to be sources of electronically excited (^1D) and ground-state (^3P) oxygen atoms, respectively, was interpreted to mean that atomic oxygen of either multiplicity can react with cyanogen to produce the fragment of interest. The proposed (12) mechanisms for the formation of CNO under both sets of conditions are depicted in sequences 6 and 7. It is to be noted, however, that atomic

$$O_3 \xrightarrow{h\nu} O_2 + O(^1D); \qquad O(^1D) + (CN)_2 \longrightarrow CNO + CN \qquad (6)$$

$$NO_2 \xrightarrow[h\nu]{} NO + O(^3P); \qquad O(^3P) + (CN)_2 \longrightarrow CNO + CN \qquad (7)$$

oxygen does not appear to be essential for the production of the species absorbing at 3250–3330 Å, since this is readily detected on flash photolysis of mixtures of $(CN)_2$ and O_2 with light possessing sufficient energy to fragment $(CN)_2$ but not O_2.

The assignment of the absorption at 3250–3330 Å to the CNO radical suffered one major drawback; it rested solely on an attempt to rationalize the fate of the reactants rather than on a detailed structural analysis of the fragment itself. In actual fact, it proved incorrect. Definitive identification of the fragment was provided by Herzberg and Travis (1,13) who observed that photolysis of diazomethane produces a species displaying an ultraviolet spectrum identical to that reported by McGrath and Morrow and by Jennings and Linnett. Detailed examination of various features of the spectrum led to NCN rather than CNO as the correct structure for the fragment (1,13). Though undoubtedly correct, this assignment cannot be readily rationalized mechanistically as it is difficult to envision the generation of NCN either from diazomethane or from fulminic acid and further to account for the requirement of an oxygen source in the experiments dealing with the photolysis of $(CN)_2$. In a careful reevaluation of the

* Both reports appeared concurrently.

results derived from these latter experiments, McGrath and Morrow (14,15) demonstrated in a kinetically consistent fashion that NCN is quite likely not a primary product of $(CN)_2$ and atomic oxygen, but rather arises from reaction between initially formed NCO and CN radicals, as depicted in sequence 8. Hence, the need of an oxygen source for the generation of NCN in these experiments. Additional work by Morrow and

$$O(^1D) + (CN)_2 \longrightarrow NCO + CN$$
$$NCO + CN \longrightarrow NCN + CO$$

(8)

McGrath (15) also established that NCO and consequently NCN is generated when $O(^1D)$ but not when $O(^3P)$ is employed. Furthermore (15), three new diffuse bands were detected at 4265, 3920, and 2807 Å, the kinetic behavior of which led Morrow and McGrath to tentatively assign them either to NCN or to NCNCO on the condition that this latter fragment rapidly fragments to NCN and CO.

We shall now elaborate on the work of Herzberg and Travis (1,13). As was stated before, these workers observed that diazomethane on flash photolysis produces among others a species exhibiting a group of absorption heads near 3290 Å. This absorption was found to be very weak when the flash photolysis was conducted on mixtures of diazomethane and an inert gas, but could be intensified considerably upon addition of cyanogen to the binary mixtures. The optimum composition of such mixtures containing N_2 as the inert diluent was determined at 1 part CH_2N_2, 5 parts $(CN)_2$, and 500 parts N_2. Under these conditions and at a pressure of 40 mm and a delay time* of 40 μsec, detailed analysis of the 3290 Å absorption was feasible. Revealing information concerning the elemental composition of the species of interest was obtained by employing isotopic variants of CH_2N_2. Thus, flash photolysis of CD_2N_2 produced a spectrum identical to that obtained under the same conditions from undeuterated materials, indicating the absence of any hydrogen in the molecule. On the other hand, flash photolysis of a mixture consisting of 60% ^{13}C-substituted diazomethane and 40% of the ^{12}C counterpart generated a spectrum in which every head due to the ordinary compound (^{12}C) is accompanied at longer wavelength by a slightly stronger head. This, strongly suggests that the molecule contains a single carbon atom. Furthermore, close inspection of the spectrum (from ordinary CH_2N_2) revealed, for reasons not to be discussed here, that the carbon in the molecule is undoubtedly situated between two *identical* first-row elements. Now, since molecules such as CB_2, CO_2, and CF_2 are expected (1) to display features different from those present in the 3290 Å band, it was concluded that the fragment

* This refers to the time lapse between the initial photolysis flash and activation of the analysis lamp.

giving rise to this band must be NCN. Once the elemental composition of the molecule was established, careful analysis of both its absorption and its emission spectra provided a number of significant structural details. For example, the nature and number of the observed band heads strongly suggest the 3290 Å band is due to a $^3\Sigma \rightarrow {}^3\Pi$ transition of a linear molecule. Furthermore, comparison of the rotational features of the spectrum with those of the spectrum of molecular nitrogen strongly indicates that the lower state of the transition ought to be either $^3\Sigma_g{}^-$ or $^3\Sigma_u{}^+$. In the first of these (shown in equation 1) the two highest energy electrons are evenly distributed among a doubly degenerate $1\pi_g$ MO, while in the $^3\Sigma_u{}^+$ configuration, shown in equation 9, two nonisoenergetic π MO's ($1\pi_u$ and $1\pi_g$) are partially occupied. Energy considerations thus point to a $^3\Sigma_g{}^-$ ground state. On this basis the upper state must be $3\Pi_u$, generated from the ground state through promotion of an electron from a $3\sigma_u$ MO to a $1\pi_g$ MO. The configuration giving rise to the $^3\Pi_u$ state is depicted in equation 10.

$$\ldots 3\sigma_g{}^2\, 2\sigma_u{}^2\, 4\sigma_g{}^2\, 1\pi_u{}^3\, 3\sigma_u{}^2\, 1\pi_g{}^3;\; {}^3\Sigma_u{}^-,\, {}^3\Sigma_u{}^+,\, {}^3\Delta_u,\, {}^1\Sigma_u{}^-,\, {}^1\Sigma_u{}^+,\, {}^1\Delta_u \quad (9)$$

$$\ldots 3\sigma_g{}^2\, 2\sigma_u{}^2\, 4\sigma_g{}^2\, 1\pi_u{}^4\, 3\sigma_u{}^1\, 1\pi_g{}^3;\; {}^3\Pi_u,\, {}^1\Pi_u \quad (10)$$

In their report, Herzberg and Travis indicated that even under optimum conditions photolysis of CH_2N_2 produces little NCN so that its absorption spectrum appears weak and detailed analysis is difficult. Nevertheless, upon careful study these workers were able to deduce values for bond lengths and bending frequencies for both the lower and upper state of the 3290 Å absorption. The C—N internuclear separation was estimated at 1.23 Å for both the lower ($^3\Sigma_g{}^-$) and upper ($^3\Pi_u$) state, and the bending frequencies were calculated to be 370 ± 50 and 460 ± 50 cm^{-1}, respectively. It was pointed out (1,16) that the "stiffer" nature of the upper state is quite likely due to the greater occupancy of the π_g degenerate MO which according to the Walsh diagram (Fig. 3) ought to be particularly resistant to bending.

The recent synthesis of cyanogen azide (N_3CN) (2) provided a unique source for direct and efficient generation of NCN. Thus, flash photolysis of N_3CN under conditions similar to those employed in the work with diazomethane produced a high concentration of NCN as evidenced by intense characteristic absorption in the 3290 Å region (3). In parallel experiments Wasserman, Barash, and Yager (4) photolyzed N_3CN in a variety of low-temperature matrices in order to generate NCN and record its electron paramagnetic resonance spectrum, in the event the molecule was a ground state triplet. An EPR signal was indeed produced, which upon careful examination was assigned to NCN. The EPR constants which led to this assignment were as follows: $E^* < 0.002$ cm^{-1} and

---

* This constant is a measure of the deviation of a molecule from linearity, i.e., $E = 0$ for a perfectly linear species.

$D^* = 1.544$ cm^{-1}, the latter being very similar to that deduced by Herzberg (1) ($2\lambda = D = 1.567$ cm^{-1}). Wasserman et al. also concluded, on the basis of a computation designed to match the experimentally determined value of D, that the nitrogen atoms of NCN possess a positive spin density, whereas the central carbon has a negative spin density. It should be noted in connection with the EPR experiments, that observation of a signal due to NCN requires that the decomposition of N_3CN be conducted through Pyrex ($\lambda > 3000$ Å), as unfiltered irradiation leads to rapid isomerization of the nitrene (4). We shall discuss this isomerization in detail in a subsequent section.

A large body of information with regard to the structure of NCN was obtained recently by Milligan et al. (17,18) on the basis of matrix isolation studies. In these, NCN was generated by filtered photolysis ($\lambda > 2800$ Å) of N_3CN in argon, nitrogen, carbon monoxide, and carbon dioxide matrices maintained either at 14 or 20°K. In each case the characteristic NCN absorption in the 3290 Å region was observed. In addition, a progression of bands appeared between 3000 and 2400 Å, which was also assigned to NCN on the basis of its rates of growth and disappearance which parallel those of the 3290 Å band. The high-energy progression was assigned to a fully allowed (parity, symmetry, and spin) $^3\Sigma_g^- \rightarrow {}^3\Sigma_u^-$ transition which, as can be readily seen from equations 1 and 5, involves promotion of an electron from a π_u to a π_g MO. Milligan et al. point out (18) that the splitting in the high-energy transition is incompatible with a bent upper state so that one can eliminate the possibility that the band is due to a $^3\Sigma_g^- \rightarrow \cdots 1\pi_g^1 2\pi_u^1$ transition on the basis of the occupancy of a $2\pi_u$† MO which, according to the Walsh diagram, ought to give rise to a bent upper state.‡ In the infrared, the mixture from filtered photolysis displays a medium band at 423 cm^{-1} as well as intense absorption at 1475 cm^{-1}. Both of these absorptions were assigned to NCN on account of their rates of growth and disappearance which parallel those of the 3290 Å band and the 3000–2400 Å progression. In addition, the shifts displayed by the infrared bands on employing $N_3{}^{13}CN$ and $N_3C^{15}N$ in the photolysis clearly indicate that the fragment responsible for these absorptions contains a single carbon and only one type of nitrogen. Now, energy considerations suggest that the 423 cm^{-1} band is due to a bending vibration (ν_2) and the 1475 cm^{-1} absorption to an antisymmetric stretching vibration (ν_3). Employing these values, Milligan et al. computed on the

* This constant is a measure of the magnetic dipole interaction of the unpaired electrons.

† $2\pi_u$ represents the highest-energy π MO of the system. See Figure 1.

‡ From the splitting of the progression Milligan et al. estimate a bending frequency of 520 cm^{-1} for a bent upper state. Therefore we have an unlikely situation in which a bent upper state has a higher bending frequency (by some 100 cm^{-1}) than a linear, generally more rigid, ground state ($^3\Sigma_g^-$).

basis of a simple valence potential, $k_{stretch} = 5.38$ mdyn/Å and $k_{bend} = 0.22$ mdyn/Å. The calculated stretching force constant is characteristic of C—N single bonds and consequently it is not consistent with the C—N bond length of 1.23 Å deduced by Herzberg and Travis which clearly indicates a carbon–nitrogen double bond. Milligan et al. pointed out that the source of this discrepancy could be a result of their neglect of a stretching-interaction constant in their calculation, which, if large and positive, may raise $k_{stretch}$ to a value generally associated with C=N bonds. Subsequent work showed that this was indeed the case (19). Thus, when NCN was produced in much higher concentration by decomposition of N_3CN with a 2288 Å cadmium line rather than a Pyrex filtered mercury arc, its infrared spectrum displayed a weak absorption at 2672 cm^{-1} in addition to the previously recorded bands. Employing $N_3{}^{13}CN$ and $N_3C^{15}N$ as sources of NCN, Milligan and Jacox were able to show this absorption to be a symmetric (ν_1)-antisymmetric (ν_3) stretching combination band. Thus $\nu_1 + \nu_3 = 2672$ cm^{-1}, and since $\nu_3 = 1475$ cm^{-1}, the infrared-inactive symmetric stretching frequency (ν_1) was estimated at 1197 cm^{-1}. On the basis of these values, the stretching interaction was computed to be 3.22 mdyn/Å and $k_{stretch}$ estimated at 8.60 mdyn/Å, a reasonable value for a C=N bond. It is interesting that a relatively large positive interaction constant is believed to be associated with delocalized

TABLE II

Thermodynamic Properties (cal/mole °K) of NCN (19)

$T(°K)$	$C_p°$	$(H° - H_0°)/T$	$-(F° - H_0°)/T$	$S°$
273.16	9.83	8.06	45.10	53.15
298.16	10.09	8.22	45.81	54.03
300	10.11	8.23	45.86	54.09
400	11.04	8.82	48.31	57.13
500	11.82	9.34	50.34	59.68
600	12.42	9.81	52.08	61.89
700	12.90	10.22	53.63	63.84
800	13.26	10.58	55.01	65.59
900	13.54	10.89	56.28	67.17
1000	13.76	11.16	57.44	68.61
1100	13.93	11.41	58.52	69.93
1200	14.06	11.63	59.52	71.15
1300	14.18	11.82	60.46	72.28
1400	14.27	11.99	61.34	73.33
1500	14.34	12.14	62.17	74.32
2000	14.58	12.72	65.74	78.47
3000	14.75	13.38	71.05	84.43

electrons (19,20). Milligan and Jacox employed their new data to estimate various thermodynamic properties of NCN over a wide range of temperatures. Their data are reproduced in Table II.

B. Mechanistic Details of the Formation of NCN from Photolysis of N_3CN

In the absence of external perturbation, the photodecomposition of N_3CN must occur with overall conservation of spin. Within this restriction, sequences 11–14 exhaust all possible low-energy spin combinations for the fragmentation. Of these, equations 12 and 13 are deemed unlikely since the light employed in the fragmentation of N_3CN does not possess the necessary energy to generate molecular nitrogen in an excited triplet

$$^1(N_3CN) \longrightarrow {}^1NCN + {}^1N_2 \tag{11}$$

$$^1(N_3CN) \longrightarrow {}^3NCN + {}^3N_2 \tag{12}$$

$$^3(N_3CN) \longrightarrow {}^1NCN + {}^3N_2 \tag{13}$$

$$^3(N_3CN) \longrightarrow {}^3NCN + {}^1N_2 \tag{14}$$

state. Hence, the electronic multiplicity of initially generated NCN must be the same as that of its azide precursor. Now we can further restrict our choice to one of the remaining two possibilities again on account of conservation of spin which requires that direct excitation of N_3CN lead to an excited singlet rather than an excited triplet state. On this basis, it is clear that direct photodecomposition of N_3CN ought to occur according to sequence 11 to generate NCN in an excited singlet state. This prediction has been substantiated both spectroscopically (21) and chemically (see Section III). In spectroscopic work, Kroto (21) was able to show that $^3\Sigma_g^-$ NCN is not a primary product but results from deactivation of initially generated singlet NCN. Thus, flash photolysis of neat gaseous N_3CN through quartz and a minimum delay time of 15 μsec produced weak absorption in the 3290 Å region ($^3\Sigma_g^- \to {}^3\Pi_u$) as well as new intense features close to 3327 Å. Rotational analysis of this new absorption led to its assignment as a $^1\Delta_g \to {}^1\Pi_u$ transition. Furthermore, by recording the spectrum at various delay times Kroto was able to clearly establish a time-dependent relation between the 3290 Å band and the 3327 Å array, whereupon the first of these increases in intensity as the second weakens. In the absence of an inert diluent the $^1\Delta_g \to {}^1\Pi_u$ array of peaks is weakly discernible up to 1000 μsec after the photolysis flash. In addition to the two NCN absorptions, bands due to CN, N_3, and CNC were also observed in these experiments.

The $^1\Delta_g \to {}^1\Pi_u$ transition of NCN was subsequently observed by Schoen (22) on flash photolysis of matrix-isolated N_3CN. Perhaps the most noteworthy feature of Schoen's results is an apparent wavelength dependence of the primary photochemical process in cyanogen azide. Thus, flash photolysis of the azide in a nitrogen matrix at 20.4°K through Pyrex ($\lambda > 2800$ Å) gave rise exclusively to the $^3\Sigma_g^- \rightarrow {}^3\Pi_u$ absorption of NCN, whereas illumination through Quartz ($\lambda > 2100$ Å) produced both the weak $^3\Sigma_g^- \rightarrow {}^3\Pi_u$ and intense $^1\Delta_g \rightarrow {}^1\Pi_u$ bands of NCN. To reconcile these unexpected results with the known ultraviolet spectrum of N_3CN (2), (λ_{max} 2200 Å (ϵ 2157), 2750 Å (ϵ 193)) Schoen proposed that though both the 2200 Å and the 2750 Å bands in the spectrum of N_3CN are due to spin allowed singlet–singlet transitions, fragmentation of the first occurs with overall conservation of spin to yield singlet NCN while collapse of the second occurs in violation to the conservation of spin requirement to produce triplet NCN. Alternatively, Milligan and Jacox (19) prefer viewing the purported direct formation of triplet NCN through a spin conserving process on the condition, of course, that the upper state of the 2750 Å band is a triplet. The various proposed decomposition sequences are shown in equations 15–17.

$$^1(N_3CN) \xrightarrow[\lambda > 2000 \text{ Å}]{h\nu} {}^1(N_3CN)^*(2200 \text{ Å}) \longrightarrow {}^1NCN + {}^1N_2 \qquad (15)$$

$$^1(N_3CN) \xrightarrow[\lambda > 2800 \text{ Å}]{h\nu} {}^1(N_3CN)^*(2750 \text{ Å}) \longrightarrow {}^3NCN + {}^1N_2 \qquad (16)$$

$$^1(N_3CN) \xrightarrow[\lambda > 2800 \text{ Å}]{h\nu} {}^3(N_3CN)^*(2750 \text{ Å}) \longrightarrow {}^3NCN + {}^1N_2 \qquad (17)$$

Now, although both explanations account for the direct formation of ^3NCN, they do not do so satisfactorily since it is not clear why spin is not conserved in the fragmentation of the azide from a low-energy excited state, or if the 2750 Å band is indeed due to a spin-forbidden transition, why should it be as intense as it is. In fact, we shall see in Section III that chemical work does not bear out the wavelength dependence of the primary photochemical process in cyanogen azide photolysis.

C. Isomerization of NCN

One major complication in studying the properties of photolytically generated NCN is introduced by the lability of this species under certain photolysis conditions. Thus, Wasserman et al. (4) observed in their EPR studies of NCN that, though this species is produced cleanly on irradiation of N_3CN through Pyrex ($\lambda > 3000$ Å), it rapidly isomerizes when subjected to illumination with $\lambda < 3000$ Å. The NCN-isomer displays an EPR signal with $D = 1.153 \text{ cm}^{-1}$ and $E < 0.002 \text{ cm}^{-1}$, indicating a linear ($C_{\infty v}$) triplet ground state. The linearity of this compound clearly limits the structural assignment to CNN (diazomethylene).*

---

* Features characteristic of unfiltered photolysis appear even when 2700 Å light (CCl₄ filter) is used in the fragmentation of N_3CN.

Ultraviolet absorption due to CNN was first reported by Robinson and McCarty (23) in 1960. These workers observed that photolysis of CH_2N_2 is solid krypton (4.2°K) with $\lambda < 3400$ Å produces, among others, a species absorbing at 4240 Å, which they tentatively identified as either CN_2 or HCN_2. The assignment of the 4240 Å band to CN_2 has since been firmly established by Milligan et al. (24) in their studies with N_3CN. These workers found that though irradiation of N_3CN in a matrix, with a mercury arc through Pyrex produces small amounts of NCN as the sole product, illumination with the full light of the mercury arc ($\lambda < 2100$ Å) produces a fragment with entirely different infrared and ultraviolet features* (17,18). These appear (24) at 393, 1241, and 2847 cm^{-1} and at 3964 and 4189 Å, respectively, and clearly belong to the same species, as they exhibit similar decay rates. Careful examination (24) of the spectra obtained from unfiltered photolysis of various isotopic variants of N_3CN ($N_3^{13}CN$, $N_3C^{15}N$) provided convincing evidence that the species responsible for these bands contains only one carbon and two nonequivalent nitrogen atoms. Hence, the isomer of NCN must be formulated as CNN. In addition, a study of the isotopic shifts of the various infrared features, served to identify the 393 cm^{-1} band as a bending mode and the 1241 and 2847 cm^{-1} as symmetric and antisymmetric stretching modes, respectively. On the basis of these assignments, Milligan and Jacox estimated a number of physical constants for CNN. Their results are reproduced in Tables III and IV.

TABLE III
Force Constants Calculated for CNN (24)

$k_{stretch}$ (CN)	19.5 mdyn/Å
$k_{stretch}$	14.7 mdyn/Å
k (CN—NN)	-3.9 mdyn/Å
k_{bend}	0.21 mdyn/Å

On the basis of the EPR data (4) and by analogy with cyanonitrene, CNN is predicted to possess a $^3\Sigma$ ground state. Furthermore, the Walsh diagram relating $C_{\infty v}$ and C_s geometries suggests that CNN ought to be linear in its various low-lying excited states as well. Assuming this to be true, Milligan and Jacox estimated from their UV data a value of 1355 cm^{-1} for an upper-state stretching vibration of CNN.

In principle, CNN could be formed either from NCN, or directly from an appropriate position isomer of N_3CN, such as **4** and/or **5** by loss of one molecule of nitrogen.

* See footnote on page 322.

TABLE IV

Thermodynamic Properties (cal/mole $°K$) of CNN ($^3\Sigma$), Calculated Assuming $r(CN) = 1.15$ Å and $r(NN) = 1.25$ Å (24)

$T(°K)$	$C_p°$	$(H° - H_0°)/T$	$[-(F° - H_0°)/T]$	$S°$
273.16	9.90	8.16	46.30	54.46
298.16	10.11	8.32	47.02	55.34
300	10.12	8.33	47.08	55.40
400	10.81	8.87	49.55	58.42
500	11.33	9.31	51.58	60.89
600	11.75	9.68	53.31	62.99
700	12.11	10.00	54.83	64.83
800	12.42	10.29	56.18	66.47
900	12.70	10.54	57.41	67.95
1000	12.94	10.77	58.53	69.30
1100	13.15	10.98	59.57	70.55
1200	13.33	11.16	60.53	71.69
1300	13.49	11.34	61.43	72.77
1400	13.63	11.50	62.27	73.76
1500	13.76	11.64	63.08	74.72
2000	14.18	12.23	66.52	78.76
3000	14.55	12.95	71.62	84.57

N_3NC

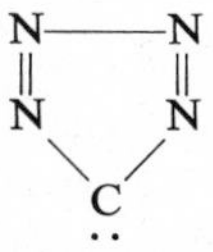

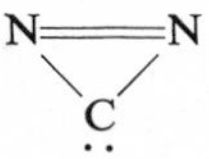

(4)   (5)   (6)

Originally (18), Milligan et al. proposed that high-energy photolysis of initially generated NCN leads exclusively to molecular nitrogen and atomic carbon as shown in sequence 18. This conclusion was arrived at on the basis of experiments involving unfiltered photolysis of NCN in CO and CO_2 matrices, which led to the formation of carbon suboxide (CCO) and CO, respectively. By elimination therefore, CNN was believed to form from N_3CN according to the sequence shown in equation 19 in which CNN is converted to N_2 and C.

$$NCN \xrightarrow[\lambda < 2800 \text{Å}]{h\nu} N_2 + C(^3P) \qquad (18)$$

$$N_3CN \xrightarrow[\lambda < 2800 \text{Å}]{h\nu} N_3 + CN \longrightarrow N_3NC \xrightarrow{-N_2}$$
$$CNN \longrightarrow N_2 + C \quad (19)$$

Accumulation of additional information, however, regarding chiefly the photolysis of isotopically labeled N_3CN, served to clearly exclude N_3NC

as a direct source of CNN. For example, unfiltered photolysis of $^{14}NC^{15}N$ (generated in high yields from photolysis of $N_3C^{15}N$ with the 2200 Å cadmium line) in an argon matrix generated equal amounts of $C^{14}N^{15}N$ and $C^{15}N^{14}N$ as well as some $C^{14}N^{14}N$. These observations are evidently inconsistent with either sequence 19, which ought to produce $C^{15}N^{14}N$ exclusively, or a path involving 5 as an intermediate, since this should lead to equal amounts of $C^{14}N^{14}N$ and $C^{15}N^{14}N$ and none of the observed $C^{14}N^{15}N$. The results are thus best rationalized (24) on the basis of decomposition of $^{14}NC^{15}N$ to $^{14}N\equiv^{15}N$ and atomic carbon, followed by reaction of these fragments to yield fully scrambled CNN. The formation of some unlabeled CNN is undoubtedly due to reaction of atomic carbon with $^{14}N\equiv^{14}N$ initially lost from the azide. The proposed mechanism is shown in sequence 20.

$$N_3C^{15}N \longrightarrow {}^{14}N\equiv^{14}N + {}^{14}NC^{15}N$$

$$^{14}NC^{15}N \longrightarrow {}^{14}N\equiv^{15}N + C(^3P) \longrightarrow C^{14}N^{15}N + C^{15}N^{14}N \qquad (20)$$

$$C(^3P) + {}^{14}N\equiv^{14}N \longrightarrow C^{14}N^{14}N$$

As indicated above, the results which led to sequence 20 were derived from photolysis of labeled N_3CN in an argon matrix. In contrast, unfiltered irradiation of $N_3C^{15}N$ in an unlabeled N_2 matrix, produced only fully unlabeled CNN. This result is not unexpected in view of the great excess of unlabeled N_2. In fact, failure to observe any labeled CNN under these latter conditions, serves to eliminate the possibility of an intramolecular isomerization of NCN to CNN by way of isodiazomethylene (6), since such a process ought to yield an equal proportion of $C^{14}N^{15}N$ and $C^{15}N^{14}N$ but not $C^{14}N^{14}N$, irrespective of the matrix environment. Isodiazomethylene may, however, be a precursor of atomic carbon.

Concerning the energy requirements for fragmentation of NCN to C and N_2, Milligan and Jacox (19) pointed out that since decomposition occurs readily with the full light of a mercury arc but not on illumination, either through Pyrex ($\lambda > 2800$ Å) or with the 2200 Å cadmium line, the absorption continuum, responsible for the fragmentation should possess a maximum close to 2537 Å. Furthermore, it was argued (18) that energy considerations require that the carbon be generated in its ^3P ground state by way of bent rather than linear NCN. It was also concluded, from results of experiments involving labeled N_3CN, that once formed, the atomic carbon reacts with N_2 to produce CNN but not NCN. This observation lends strong support to the notion that carbon here is generated exclusively in the ground state, as Moll and Thompson (25) have recently shown that vacuum ultraviolet irradiation ($\lambda \sim 1470$ Å) of carbon suboxide (C_3O_2) in a nitrogen matrix, which quite likely generates both

^1D and ^3P carbon (26), gives rise to both NCN and CNN. Thus, it appears that triplet carbon reacts with N_2 to produce only CNN while singlet carbon adds to N_2 to generate NCN and possibly CNN. The currently accepted (19,25) mechanism for reaction of atomic carbon with N_2 is depicted in sequence 21.

$$NCN(^3\Sigma_g{}^-) \xrightarrow[2200 < \lambda < 2800]{h\nu} {}^1N_2 + C(^3P) \longrightarrow CNN$$

$$OCCCO \xrightarrow[\lambda \sim 1470\ \text{Å}]{h\nu} 2CO + C(^3P, {}^1D) \qquad (21)$$

$$C(^1D) + N_2 \longrightarrow NCN + CNN\ (?)$$

The ready low-energy formation of active atomic carbon from NCN was recently utilized by Milligan and Jacox in the synthesis and subsequent study of a number of highly reactive species, not readily accessible otherwise. Three radicals produced in this manner are shown in sequences 22 (27), 23 (28), and 24 (29).

$$NCN \xrightarrow{h\nu} N_2 + C(^3P)$$
$$\xrightarrow{CO} :CCO \qquad (22)*$$
$$\xrightarrow{Cl_2} :CCl_2 \qquad (23)*$$
$$\xrightarrow{HCl} :CHCl \qquad (24)*$$

III. CHEMICAL INVESTIGATIONS

A. Reaction with Saturated Hydrocarbons

1. NCN from Thermolysis of N_3CN

Unlike most organic azides which require temperatures in excess of 100° for fragmentation to the corresponding nitrenes, cyanogen azide smoothly evolves nitrogen when heated to 40–50°C. The kinetics of the thermal decomposition of the azide have not been studied in detail but the rate of nitrogen evolution appears to be first order. In addition the rate of gas evolution is a function of the polarity of the solvent employed, being more rapid in nonpolar than in polar media (5,30). These facts are consistent with unimolecular fragmentation of the highly polar azide to two nonpolar species, N_2 and NCN. The suspected generation of the nitrene under these conditions received strong support from the formation alkyl cyanamides (7) on thermal decomposition of N_3CN in a hydrocarbon solvent (sequence 25) (5).

$$N_3CN + RH \xrightarrow{40-50°} N_2 + RNHCN \qquad (25)$$
$$(7)$$

* Photolyses of NCN were in each case conducted with the full light of a medium pressure mercury arc, in either a nitrogen or an argon matrix maintained at 14°K.

Definitive evidence that NCN rather than the starting azide is the immediate progenitor of **7** was furnished from experiments employing specifically N-15 labeled N_3CN. The results of these experiments shown in sequence 26 clearly establish that nitrogen atoms originally 3 and 4 in N_3CN lose their relative identity in the course of the reaction to produce cyclohexylcyanamide (**8**) in which the isotopic label is very nearly equally distributed among its two nitrogen atoms.* In this connection, it was also demonstrated that scrambling of the label does not occur prior to reaction. The experimental observations which led to this conclusion are depicted in sequence 27 (5,30).

The results shown in reactions 26 and 27 are thus uniquely consistent with the intermediacy of symmetrical NCN and clearly eliminate the possibility of direct C—H insertion by the azide, which could in principle

* It has been suggested (30) that the small excess label in the amines (28.7% ^{15}N instead of the theoretical 25%) is probably at least partly due to an isotopic bias of the insertion, i.e., a slight preference of the ^{15}N site over the ^{14}N site of NCN to undergo reaction.

occur as shown in sequence 28.*

$$
\begin{array}{ccc}
\overset{*\text{N}^{\ominus}}{\underset{\overset{\|}{\text{N}^{\oplus}}}{\|}} & \overset{*\text{N}^{\ominus}(\cdot)}{\underset{\overset{|}{\text{N}}}{\|}} & \overset{*\text{N}}{\underset{\overset{|}{\text{N}}}{\||}} \\
\overset{\|}{\underset{|}{*\text{N}\,\text{H}\!-\!\text{R}}} & \overset{}{\underset{|}{*\text{N}\!-\!\text{H}\ \text{R}^{\oplus}(\cdot)}} & \overset{*}{\underset{|}{\text{H}\!-\!\text{N}\!-\!\text{R}}} \\
\text{CN} & \text{CN} & \text{CN}
\end{array}
\qquad (28)
$$

The mild conditions employed for its thermal generation, require that the nitrene be produced in a low-lying electronic state (5). Furthermore, since spin-allowed radiationless decay is usually very rapid compared to diffusion it was deemed reasonable to assume that reacting NCN will be in the lowest electronic state of a certain multiplicity. Therefore, only the lowest-energy configuration $(\ldots 3\sigma_g{}^2\,2\sigma_u{}^2\,4\sigma_g{}^2\,1\pi_u{}^4\,3\sigma_u{}^2\,1\pi_g{}^2)$ need be seriously considered for reacting NCN. This arrangement gives rise to the three chemically distinct electronic distributions shown in equations 29–31, for which the three distinct modes of reaction shown in equations 32–34 are formally possible (5,30). Of these, reactions 32 and 33 represent the commonly accepted two-step and one-step mechanisms for triplet and singlet species,† while reaction 34 becomes a distinct possibility only with highly electrophilic species such as NCN which possesses a vacant low-energy MO, e.g., π_g of S_2.

$$T_1 = \cdots \pi_g(x)\pi_g(y) \qquad (29)$$

$$S_1 = \cdots \pi_g(x)\pi_g(y) \qquad (30)$$

$$S_2 = \cdots \pi_g{}^2(x) \quad \text{or} \quad \cdots \pi_g{}^2(y) \qquad (31)$$

$$\text{RH} + \text{NCN}(T_1) \longrightarrow \text{R}\ldots\text{H}\ldots\text{NCN} \longrightarrow \overset{\cdot}{\text{R}} + \underset{\text{H}}{\overset{\cdot}{\text{N}}\text{CN}} \longrightarrow \text{RNHCN} \quad (32)$$

$$\text{RH} + \text{NCN}(S_1) \longrightarrow \underset{\text{H}^{\delta-}}{\overset{\delta^+\text{R}}{\vdots}}\!\!\diagdown\!\!\text{NCN} \longrightarrow \text{RNHCN} \quad (33)$$

$$\text{RH} + \text{NCN}(S_2) \longrightarrow \underset{\delta^+}{\text{R}}\ldots\text{H}\ldots\underset{\delta^-}{\text{NCN}} \longrightarrow \overset{\oplus}{\text{R}} + \underset{\text{H}}{\overset{\ominus}{\text{N}}\text{CN}} \longrightarrow \text{RNHCN} \quad (34)$$

In order to distinguish between the three possible processes and consequently assess the configuration of reacting NCN, two sets of experiments were conducted. The first of these involved the determination of the

---

* Here the dipolar structure is to be associated with singlet N_3CN and the diradical species with triplet N_3CN.

† It has been pointed out that S_1 of NCN has the appropriate symmetry for direct interaction with a donor C—H bond (32).

relative reactivity of various types of C—H links toward NCN. The results of these experiments are reproduced in sequences 35–37 and Table V. It is seen thus that NCN displays considerable bias in choosing among its C—H partners, exhibiting relative affinities of 67:9–14.8:1 toward tertiary, secondary, and primary C—H bonds, respectively. The extent of the discrimination was deemed to be consistent with either path 32 or path 33 but insufficient to account for sequence 34 which involves the generation of a carbonium ion. These intramolecular competitions, thus served to tentatively eliminate sequence 34 as a possibility.

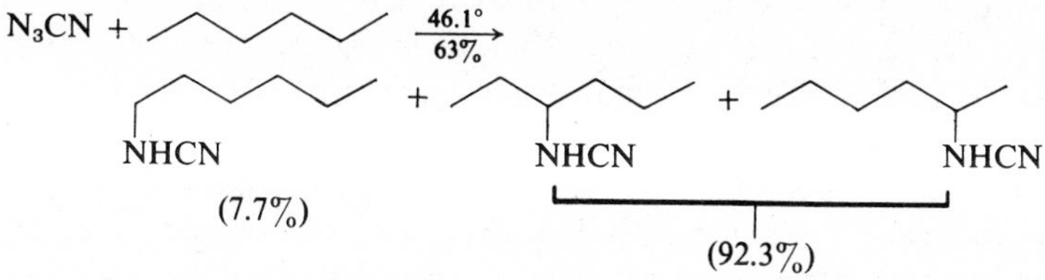

$$N_3CN + \quad \xrightarrow[63\%]{46.1°} \quad$$

NHCN   +   NHCN   +   NHCN   (35)

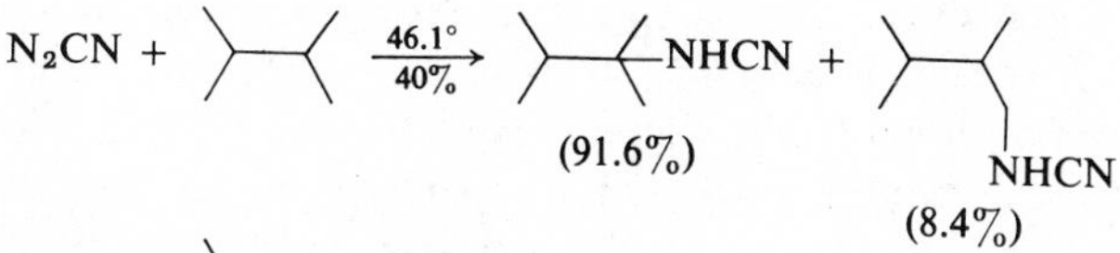

(7.7%)   (92.3%)

$$N_2CN + \quad \xrightarrow[40\%]{46.1°} \quad \text{—NHCN} + \quad \text{NHCN}$$ (36)

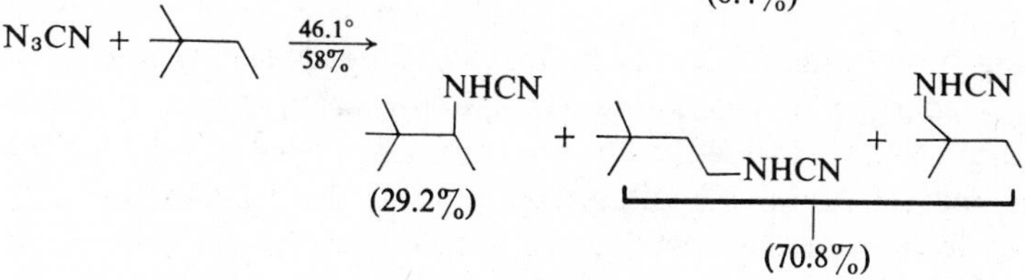

(91.6%)   (8.4%)

$$N_3CN + \quad \xrightarrow[58\%]{46.1°} \quad$$

NHCN   NHCN

+   —NHCN   +   (37)

(29.2%)   (70.8%)

TABLE V[a]

Reactivity of Cyanonitrene with C—H Bonds of Saturated Hydrocarbons[b]

A			B		
HC	H	Affinity	HC	H	Affinity
2,3-Dimethyl-	Primary	1.0	n-Hexane	Secondary (α)	1.62
butane	Tertiary	67.0		Secondary (β)	1.30
n-Hexane	Primary	1.0	Cyclo-	Secondary	1.00
	Secondary	9.0	hexane		
			Cyclo-	Secondary	1.21
			heptane		
2,2-Dimethyl-	Primary	1.0	Cyclo-	Secondary	1.2 ± 0.1
butane	Secondary	14.8	octane		

[a] Values corrected for statistical factors.
[b] Reaction temperature: 46.1°.

The second set of experiments was designed to distinguish between the remaining possibilities 32 and 33 and involved the determination of the stereochemical details of the insertion reaction, employing neat *cis*- and *trans*-1,2-dimethylcyclohexanes as substrates. Insertion of NCN into the tertiary C—H links of either stereoisomer was found to be fully stereospecific. This result clearly favors the one-step process 33 over the two-step sequences 32 and 34 both of which involve intermediates which are not likely to preserve the asymmetry of the starting hydrocarbon (30). This preliminary choice in favor of a concerted insertion was fully substantiated by the results of subsequent experiments designed to ascertain the stereochemistry of the insertion reaction in the presence of various inert diluents (31,32). The reasoning which prompted the conception of these experiments can be briefly summarized as follows: if in the absence of a solvent NCN reacts with tertiary C—H bonds, exclusively as a singlet, one ought to be able to artificially induce the deactivation of this species to the ground-state triplet prior to reaction, by providing a suitable inert medium capable of effectively dissipating the energy released in the $^1NCN \rightarrow {}^3NCN(^3\Sigma_g{}^-)$ decay. Reacting NCN under these conditions would then be partly or wholly in its triplet ground state and should react with C—H functions in the manner indicated in sequence 32 to produce product with partial or total loss of stereochemistry. Inert diluents incorporating heavy atoms such as the heavy halogens (Br, I), should be particularly effective in promoting the spin-forbidden singlet to triplet conversion of NCN*. Consequently, the stereochemistry of the insertion was examined by employing dilute solutions of the 1,2-dimethylcyclohexanes in dichloromethane and dibromomethane. As anticipated, the insertion of NCN into the tertiary C—H bonds of the hydrocarbons was partly stereospecific in CH_2Cl_2 and fully stereorandom in the heavier solvent CH_2Br_2. A thermochemical "heavy atom" effect was thus established and the mechanism of the insertion of thermally generated NCN into tertiary C—H bonds, was demonstrated to occur by sequence 33 in the absence of inert diluent, by path 32 in the presence of CH_2Br_2 and by a combination of 32 and 33 in CH_2Cl_2 (31). The same reaction was also examined with acetonitrile and ethyl acetate as inert diluents. The first of these proved to be fully effective in preserving the original asymmetry while the second was quite ineffective in this respect (32). The combined results dealing with the stereochemistry of insertion are reproduced in Table VI. Sequence 36 was advanced to rationalize the various findings (32). Within this scheme k_1 is much larger than k_2 in the absence of a solvent, comparable to k_2 in CH_2Cl_2 and much smaller than k_2 in CH_2Br_2

---

* Heavy atoms are known to enhance the probability of spin-forbidden transitions through coupling of spin and orbital angular momenta.

TABLE VI

Stereochemistry of the Insertion of Thermally Generated Cyanonitrene into the Tertiary C—H Bonds of cis- and trans-1,2-Dimethyl-cyclohexane as a Function of Solvent

Reaction	Solvent	Concn.,[a] %	Hydrocarbon	Temp.,[b] °C	% cis- RNHCN (9)	% trans- RNHCN (10)
1	None	100	cis	43.5	>98	<2
2	None	100	trans	43.5	<2	>98
3	None	100	74% cis 26% trans	43.5	83	17
4	CH_2Cl_2	10	cis	41.0	75	25
5	CH_2Cl_2	10	trans	41.0	36	64
6	CH_2Cl_2	10	74% cis 26% trans	41.0	70	30
7	CH_2Cl_2	2	cis	41.0	62	38
8	CH_2Cl_2	2	trans	41.0	39	61
9	CH_2Br_2	10	cis	43.5	52	48
10	CH_2Br_2	10	trans	43.5	52	48
11	CH_2Br_2	10	74% cis 26% trans	43.5	54	46
12	CH_2Br_2	10	trans	53.0	50	50
13	CH_3CN	10	cis	53.0	>98	<2
14	CH_3CN	10	trans	53.0	<2	>98
15	$CH_3COOC_2H_5$	10	cis	53.0	53	47
16	$CH_3COOC_2H_5$	10	trans	53.0	44	56
17	$CH_3COOC_2H_5$	2	cis	53.0	52	48
18	$CH_3COOC_2H_5$	2	trans	53.0	50	50

[a] Volume percent of hydrocarbon in solvent.
[b] Maintained to within ±0.1°C.

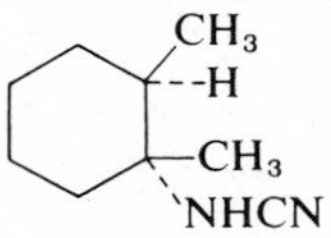

(9)

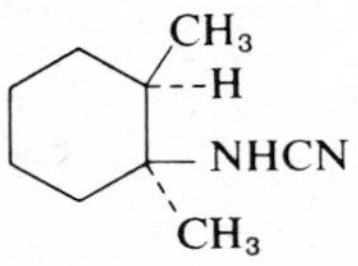

(10)

and possibly $CH_3COOC_2H_5$. In CH_3CN, k_5 is probably large compared with k_3, leading to results identical with those obtained in the case where $k_1 \gg k_2$. In $CH_3COOC_2H_5$, a large k_6 may also be partly responsible for the similarity of the results to those obtained when $k_2 \gg k_1$.

$$
\begin{array}{c}
\text{SNCN} \\
\uparrow \\
\text{solvent } (S) \;\Big|\; k_6 \\
\\
N_3CN \xrightarrow[-N_2]{\Delta} \;{}^1NCN \xrightarrow[k_1]{RH} RNHCN \\
\Big\downarrow k_2 \qquad\qquad\qquad \Big\uparrow k_4 \\
{}^3NCN \xrightarrow[k_5]{RH} R \cdot\cdot NHCN \\
\text{solvent } (S) \;\Big|\; k_5 \\
\downarrow \\
\text{SNCN}
\end{array}
\qquad (38)
$$

The effective control of the multiplicity of reacting NCN provided by the reaction medium led to an examination of the relative affinity of ground state NCN for tertiary and secondary C—H bonds. Trans-1,2-dimethylcyclohexane and cyclopentane were employed as substrates in these intermolecular competition experiments. From the results tabulated in Table VII one can readily estimate that triplet NCN (reaction 22) discriminates ca. 2.1 times better between the tertiary C—H bond of the *trans*-dimethylcyclohexane and the secondary C—H links of cyclopentane

TABLE VII
Relative Affinity of NCN for the Tertiary C—H Bonds of *cis*- and *trans*-1,2-Dimethylcyclohexane and the C—H Bonds of Cyclopentane and Cyclohexane at $43.5 \pm 0.1°$

Reaction	Solvent	Hydrocarbon[a]	C—H bond	Affinity[b,c]
19	None	A	3°	4.2
		C	2°	1.0
20	None	B	3°	2.4
		C	2°	1.0
21	None	B	3°	3.5
		D	2°	1.0
22	CH_2Br_2	B	3°	5.0
		C	2°	1.0

[a] A, *cis*-1,2-dimethylcyclohexane; B, *trans*-1,2-dimethylcyclohexane; C, cyclopentane; D, cyclohexane.

[b] In each reaction, the secondary C—H bonds are assigned the arbitrary value of 1.0.

[c] Values are corrected for statistical factors.

than does singlet NCN (reaction 20). Reactions 19 and 20, show that singlet NCN reacts some 1.7 times faster with the tertiary C—H bond of the *cis*-dimethylcyclohexane than with the corresponding function of the *trans* isomer. Interestingly, a similar factor can be estimated from the results of reaction 3 and those of 4 and 5 (in conjunction with those of 9 and 10) shown in Table VI. This small but reproducible difference in reactivity may reflect the steric demands imposed on the reaction by singlet NCN, since the tertiary hydrogens of the *trans* compound, being *cis* to the methyl groups, should be less accessible sterically.

The stereospecificity of C—H insertion requires that NCN react by way of S_1 rather than S_2. This is not unexpected as simple energy considerations would suggest that one-step insertion should be more beneficial energetically than hydride abstraction, since C—H rupture in the former process occurs concurrently with the formation of two new bonds whereas in the latter process only one new bond is formed. The two sequences are depicted diagrammatically in Figure 5. It is interesting to note in this connection that one should in principle be able to lower the first transition state in the reaction of S_2 NCN by choosing a hydrocarbon capable of generating a stable carbonium ion on hydride abstraction. In fact, it is not inconceivable that proper choice of a substrate may reverse the energies shown in Figure 5* and thus induce NCN to react through a hydride abstraction–recombination process by the S_2 arrangement† (33).

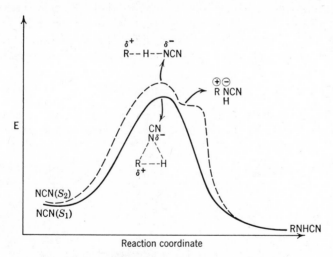

Fig. 5. Potential energy diagrams depicting C—H insertion by S_1 (—) and S_2 (– –) NCN.

* For the sake of simplicity S_1 and S_2 are depicted as being isoenergetic.
† A. G. Anastassiou assumes full responsibility for this statement.

2. NCN from-Photolysis of N_3CN

As already indicated in Section III, NCN is readily generated on photolysis of N_3CN either in the gas phase or in a solid matrix. The azide also fragments to NCN and nitrogen upon illumination in a liquid solution (34). While the matrix experiments require the multiplicity of initially generated NCN to be a function of the light energy employed to fragment N_3CN, the results of liquid-phase photolysis of the azide indicate no such wavelength dependence. Thus, irradiation of solutions of N_3CN in *cis*-1,2-dimethylcyclohexane, through Vycor (absolute cutoff point ~ 2130 Å), Corex (absolute cutoff point ~ 2580 Å) and Pyrex (absolute cutoff point ~ 2800 Å) generated the cyanamides from insertion into the tertiary C—H bond in a totally stereospecific fashion (reactions 23, 24, and 25, Table VIII). This result indicates that photofragmentation of the azide with light ranging from 2100 to ca. 3000 Å leads exclusively to singlet NCN and by implication that both the 2200 Å and 2750 Å bands in the spectrum of N_3CN arise from singlet to singlet transitions. These chemical results are in keeping with theory. Furthermore, the rather unlikely possibility that stereospecific insertion in the liquid state photolysis is due to some excited NCN triplet (perhaps $^3\Pi_u$) was eliminated on the basis of experiments involving sensitized fragmentation of N_3CN in neat *cis*-1,2-dimethylcyclohexane(35). C—H insertion under these conditions was found not to be stereospecific, the loss of stereochemistry paralleling the efficacy of the sensitizer in promoting the decomposition of the azide (reactions 27 and 28, Table VIII). These results clearly point to triplet NCN as the species responsible for nonstereospecific C—H insertion and in addition establish a close similarity between the triplet nitrene generated from

TABLE VIII[a]

Stereochemistry of the Insertion of Photolytically Generated Cyanonitrene into the Tertiary C—H Bonds of *cis*-1,2-Dimethylcyclohexane[b]

Reaction	Wavelength, Å	Relative rate	Sensitizer	% cis-RNHCN	% trans-RNHCN
23	> 2120		None	> 98	< 2
24	> 2580		None	> 98	< 2
25	> 2800		None	> 98	< 2
26	3100–4250	1.0	None	> 98	< 2
27	3100–4250	1.5	Fluorenone	87	13
28	3100–4250	2.9	Benzophenone	56	44

[a] Temperature: 15–18°C.
[b] Neat.

triplet N_3CN and that produced from deactivation of singlet NCN in their reactions with C—H bonds. The failure of reaction 27 and, to a lesser extent, 28 to occur by the totally stereorandom course followed by reactions 9 and 10 (Table VI) is undoubtedly due to the presence, in the former two, of some singlet NCN, formed from direct photolysis of the azide (reaction 26 Table VIII). In fact, on the basis of the relative rates of azide fragmentation shown in Table VIII it is evident that competitive direct photolysis actually predominates when fluorene is employed as the sensitizer (reaction 27). On the other hand, in the presence of benzophenone, sensitized photofragmentation is the predominant source of the nitrene.

As stated in Section II, there exists a large body of well documented spectroscopic information concerning the energetics of few low-lying NCN states of the same multiplicity. In great contrast, very little is known about the relative positions of states differing in total spin, except that the lowest of these must be a triplet. A reasonable energy level diagram, which satisfies the commonly encountered energy alternation between singlets and triplets, is depicted in Figure 6. Schoen (22) has recently advanced an alternate arrangement in which the $^3\Pi_u$ rather than the $^1\Delta_g$ state is depicted as the first electronically excited states, i.e., $E(^3\Sigma_g{}^-) < E(^3\Pi_u) < E(^1\Delta_g)$. This conclusion was arrived at on the basis of the already discussed spectroscopic observation that irradiation of N_3CN through Pyrex leads exclusively to triplet NCN, coupled with the assumption that the states of N_3CN should correlate in energy with those of NCN. It was reasoned that since decomposition of N_3CN from the upper state of the 2750 Å transition fails to generate singlet NCN, the lowest-energy NCN singlet ($^1\Delta_g$) must be higher than the ground state by at least 104 kcal/mole (2750 Å) minus the energy required to rupture the N—N_2 bond, estimated at ca. 9 kcal/mole. $E(^1\Delta_g)-E(^3\Sigma_g{}^-)$ was thus estimated to be at least $104 - 9 = 95$ kcal/mole which is higher than the $^3\Pi_u{}^{-3}\Sigma_g{}^-$ separation, observed at 86 kcal/mole (3300 Å). The basic assumption, however, that the various states of NCN should correspond in energy with those of N_3CN is clearly not consistent with the chemical evidence that N_3CN fragments under mild

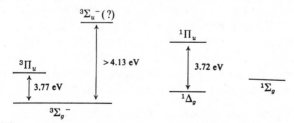

Fig. 6. Observed spectroscopic states of cyanonitrene (32).

thermal treatment with overall conservation of spin to produce singlet NCN. Hence, all available chemical evidence is consistent with the originally advanced (32), theoretically sound, energy diagram shown in Figure 6.

B. Reaction with Olefinic Hydrocarbons

Unlike most azides, cyanogen azide reacts readily with both strained and relatively strain-free olefinic functions at or below ambient temperature. The products generated in the process are usually alkylidene cyanamides (11) and/or *N*-cyanoaziridines (12), as shown in sequence 39 (2,36).

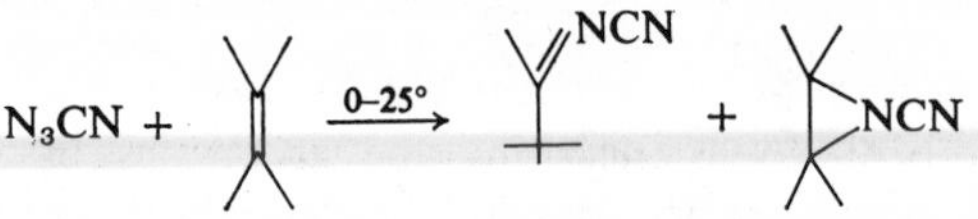

$$N_3CN + \quad \overset{0\text{--}25°}{\longrightarrow} \quad \overset{NCN}{\underset{(11)}{\bigvee}} \quad + \quad \underset{(12)}{\overset{}{\bigtriangleup}}NCN \qquad (39)$$

The pronounced affinity of N_3CN for double bonds introduces serious complications in studying the reactions of NCN with olefins, since the products generated by both the nitrene and its azide precursor would, in the absence of isotopic labeling, be indistinguishable. It is not surprising therefore that the only NCN-olefin reaction studied in detail to date employed cyclooctatetraene which reacts only sluggishly with the azide at room temperature. Reaction between the two substrates at this temperature required several hours, and led exclusively to alkylidene cyanamide (13). On the other hand, treatment of the azide with a dilute, boiling (78°), solution of cyclooctatetraene in ethyl acetate produced 13 and a new product identified as the 1,4 NCN–COT adduct (14) (37,38). Both these products formed independently as they were shown not to interconvert when subjected to the reaction conditions. This piece of information coupled with the known fragmentation of the azide to NCN and nitrogen at temperatures appreciably lower than that employed in the reaction, led to the conclusion that 14 is a product not of N_3CN but of NCN. In addition, since 14 represents the first *bona fide* example of 1,4 addition in either the carbene or the nitrene field it was deemed necessary to obtain definitive information concerning its source, i.e., to determine whether it arises directly from NCN and COT, or is perhaps a product of isomerization of an unstable primary product such as the 1,2 adduct 15. Upon interrupting the high-temperature reaction at some early stage, 15 was indeed detected. However, though exceedingly unstable, 15 is not the source of 14 as the two are not interconvertible under the reaction conditions.*

---

* Though 14 and 15 do not interconvert thermally under the reaction conditions, 14 does isomerize to 15 photochemically (39).

Spatial requirements do suggest that whereas **15** may form in one-step by way of singlet NCN, **14** must be generated in a stepwise fashion conceivably through triplet NCN (37). Support for such a scheme was obtained from experiments employing various types of diluents. (Table IX (38)).

TABLE IX

Relative Amounts of **14** and **15** from Reaction of NCN with Cyclooctatetraene in Various Solvents

Solvent[a]	NCN–COT product, %		[14]/[15]
	14	**15**	
CH_2Br_2	76	24	3.3
$CH_3COOC_2H_5$	56	44	1.3
CH_3CN	17	83	0.2

[a] ca. 1% solution of COT in solvent.

Simple inspection of the data reveals that the ratio of **14** to **15** is greatest in CH_2Br_2 (a solvent favoring reaction by ^{3}NCN), smallest in CH_3CN (a solvent favoring reaction by ^{1}NCN), and assumes an intermediate value in ethyl acetate. In addition, these experiments served to determine the predominant source of alkylidene cyanamide **13** in the high-temperature reaction. The proportion of this material was found to invariably decrease relative to **14** and **15** with increasing dilution. This concentration dependence was interpreted to mean that even at 78° **13** is predominantly, if not solely, produced from N_3CN (bimolecular process) rather than NCN (unimolecular decomposition).

The following mechanism accounts for all the data obtained in connection with the reaction of N_3CN and NCN with COT.*

* In sequence 41, full arrows represent predominant paths, while dashed arrows depict uncertain or minor processes.

N_3CN $\xrightarrow{\text{COT}}$ (13)

$-N_2$ | Δ

^1NCN $\xleftarrow{\text{COT}}$ (15)   (41)

collisional
deactivation

^3NCN $\xleftarrow{\text{COT}}$ (14)

Finally, it would be instructive to briefly consider the question of stereochemistry of the addition reaction. Here, as in the insertion reaction, the stereochemical course will undoubtedly depend on both electronic multiplicity and distribution of frontier electrons. On this basis, S_1 NCN must add in one step to produce aziridines in a stereospecific fashion,* while both $T_1^\cdot$ and S_2 NCN ought to lead to three-membered products by way of distinct intermediates (32).† The addition in this latter case will be stereospecific only in the unlikely event that ring closure of the initial intermediate is appreciably faster than rotation around the newly formed carbon–carbon single bond. Hence, T_1 and S_2 NCN would lead to aziridines stereoselectively at best. The various proposed mechanisms for addition to C=C bonds by the low-lying states of NCN are depicted in sequences 42–44. In connection with paths 42 and 44, it is interesting to note that in these, the presence of distinct reactive intermediates makes possible the generation of acyclic products as well. For reasons akin to those presented for the insertion reaction, singlet NCN would be expected to add to olefins preferentially through path 43, but the reaction may follow path 44 if the olefin substrate is capable of generating a highly stabilized dipolar intermediate (33).

* It was shown that S_1 has the appropriate symmetry for direct interaction with a donor C=C bond (32).

† It should be noted in this context that one can reach the exact reverse conclusion concerning the mode of addition open to each of the two NCN singlets, in terms of the correlation procedure employed recently to describe the reaction of CH_2 with double bonds (40,41). Thus, it can be readily shown that of the two combinations, $NCN(S_1)$ + olefin and $NCN(S_2)$ + olefin, only the second correlates properly with the resulting aziridine. On this basis, one would expect S_2 rather than S_1 NCN to add stereospecifically to double bonds.

$$C=C + NCN(T_1) \longrightarrow \overset{\cdot NCN}{\underset{C-\dot{C}}{|}} \longrightarrow \overset{CN}{\underset{C-C}{\overset{N}{\triangle}}} + \overset{NCN}{\underset{C-C}{\|}} \qquad (42)$$

$$C=C + NCN(S_1) \longrightarrow \overset{CN}{\underset{\dot{C}=\dot{C}}{\overset{N}{\triangle}}} \longrightarrow \overset{CN}{\underset{C-C}{\overset{N}{\triangle}}} \qquad (43)$$

$$C=C + NCN(S_2) \longrightarrow \overset{\overset{\ominus}{N}CN}{\underset{C-C^{\oplus}}{|}} \longrightarrow \overset{CN}{\underset{C-C}{\overset{N}{\triangle}}} + \overset{NCN}{\underset{C-C}{\|}} \qquad (44)$$

C. Reaction with Aromatic Hydrocarbons

Cyanonitrene reacts readily with aromatic substrates (42). Thus, at 45–60°C, solutions of cyanogen azide in aromatic hydrocarbons smoothly evolve nitrogen to produce N-cyanoazepines (16). These products, though isolable upon careful workup at or below ambient temperature, readily dimerize on mild heating and isomerize to arylcyanamides (17) in the presence of traces of acid. Acid hydrolysis of the arylcyanamides leads first to arylureas (18) and eventually to the corresponding anilines (19).

The intermediacy of NCN in the generation of the azepines was demonstrated (42) through reaction of appropriately labeled N_3CN with p-xylene as shown in sequence 46.

Pertinent information with regard to the multiplicity of azepine-forming NCN was obtained from competition experiments between cyclohexane and benzene, toluene and trifluoromethylbenzene. As can be readily seen from the results (43) of these competition runs, tabulated in Table X, the C—H bonds of cyclohexane are appreciably less reactive toward NCN than the double bonds of either of the three aromatic substrates examined.

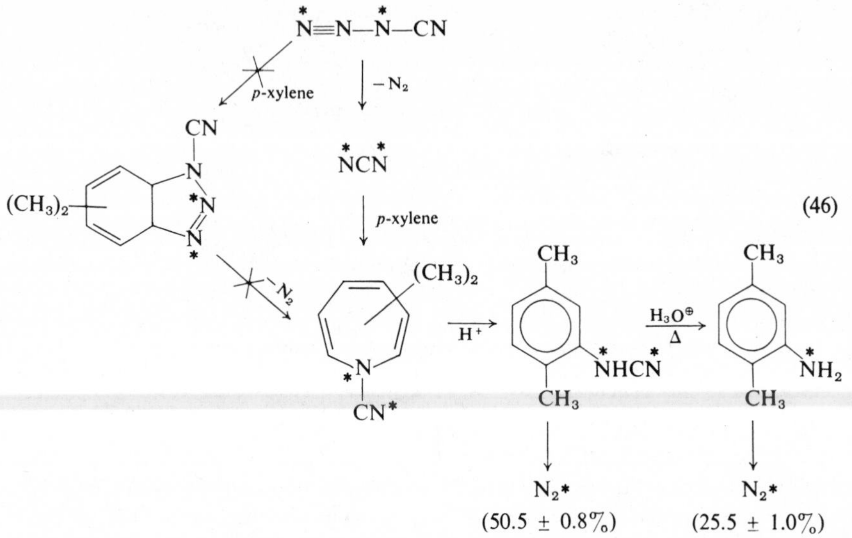

(46)

Furthermore, comparing the relative reactivities collected in Table X with those shown in Table VII we readily see that the aromatic double bonds examined are also more reactive toward NCN than the tertiary C—H bonds of the 1,2-dimethylcyclohexanes which were shown to react with singlet NCN prior to deactivation to its triplet ground state. Hence, we conclude that in the absence of an inert diluent azepines are formed strictly by way of singlet NCN.*

Reaction of singlet NCN with aromatic substrates may occur in one of two ways, depending on the frontier electronic distribution of the nitrene. The two distinct ways by which S_1 and S_2 NCN ought to interact with monosubstituted benzenes are depicted in sequence 47. In this, S_1 NCN is shown to react by concerted addition while S_2 NCN is depicted to act strictly as an electrophile. There exist appreciable experimental data favoring direct addition (path a) over electrophilic substitution (path b). In the first place, in all the reactions studied azepines appear to be the exclusive primary products. This fact is clearly consistent with path a but not with path b in which azepines and aryl cyanamides would be expected to form concurrently. Secondly, the inability of NCN to substantially discriminate between double bonds of aromatics bearing electron-releasing substituents (toluene) and those of aromatics bearing electron-withdrawing groups (trifluoromethylbenzene) strongly suggests a transition state with

---

* This conclusion rests of course on the assumption that aromatic and aliphatic substrates display comparable efficiency in collisionally deactivating singlet NCN to its triplet ground state. There is no reason to believe this not to be the case.

TABLE X

Relative Affinity of NCN for the C—H Bonds of Cyclohexane and the C=C Bonds of Selected Aromatics,
at 55°C (43)

Competition Mixture[a]	% Reaction	Affinity[b,c]
Benzene	53	23
Cyclohexane		1
Toluene	61	21
Cyclohexane		1
Toluene	100	20
Cyclohexane		1
Trifluoromethylbenzene	62	14
Cyclohexane		1

[a] Neat.

[b] In each reaction, the secondary C—H bonds are assigned
the arbitrary value of 1.0.

[c] Values are corrected for statistical factors.

very little charge separation. Clearly, of the two possible mechanisms,
only that involving direct addition may be viewed to occur through a
relatively nonpolar transition state. Thirdly, reaction of NCN with the
various monosubstituted benzenes appears to produce in each case all
three possible *N*-cyanoazepines shown in sequence 48. Furthermore in
two carefully examined cases, X = F and X = CF_3, the three isomeric
azepines were found to form in comparable proportions. Now, the formation of all three azepines in the case of fluorobenzene does not militate
against path *b* since this substrate, being *ortho–para* directing, should,
within the frame of this mechanism, generate both an "*ortho*" (**21**, X *ortho*
to NCN) and a "*para*" (**21**, X *para* to NCN) "*σ*" complex. Of these, the

$(X = CH_3, Cl, F, CF_3)$ (22)   (23)   (24)

first could cyclize to the two N-cyanoazanorcaradiene precursors of 22 and 23, while the second σ complex should lead exclusively to 24. The situation with trifluoromethylbenzene is quite different since this substrate, being a meta director, will give rise to a single σ complex (21, X *meta* to NCN) which upon cyclization should give 23 and 24, but not 22. Thus, unless the azanorcaradiene precursors of the azepines undergo rapid (perhaps circulambulatory*) isomerization the formation of all three azepines from a *meta* directing substituted benzene certainly tends to discredit path *b*.

In conclusion, it appears that, although more information is required before issuing any definitive statements concerning the mechanistic details of the reaction of NCN with aromatic substrates, all the presently available data in connection with monosubstituted benzenes point strongly to fairly indiscriminate double bond addition by NCN. It should be stressed that the alternate process, i.e., electrophilic attack by S_2 NCN (path *b*) may become the predominant path of reaction, with benzenes bearing highly electron-donating substituents in strongly polar media.

D. Cyanonitrene Products from Reaction of N_3CN with Lewis Bases (45)

Oxygen bases such as alcohols and alkyl ethers react with N_3CN on mild heating to produce alkoxycyanamides and α-alkoxyalkylcyanamides respectively, as shown in sequences 49 and 50 for t-butyl alcohol and diethyl ether. Formally, alkoxycyanamides (e.g., 25) and α-alkoxyalkyl-cyanamides (e.g., 26) are products of overall insertion of NCN into alcoholic O—H bonds and ethereal α-C—H links, respectively. Nevertheless, it is doubtful whether the nitrene is at all involved in these transfor-

$$\left. \right|{\!\!\!-} OH + N_3CN \xrightarrow[30-40°]{} N_2 + \left. \right|{\!\!\!-} ONHCN \qquad (49)$$

(25)

(26)

---

* The term was coined to describe the isomerization of tropilidines (44).

mations, as N_3CN fragmentation in both alcoholic and ethereal solvents occurs far more readily than in media of comparable polarity but lacking a basic oxygen function.* Consequently, products such as **25** and **26** are believed to form by way of N_3CN rather than cyanonitrene. Reasonable mechanistic possibilities which account for the observed rate enhancements are shown in sequences 51 and 52.

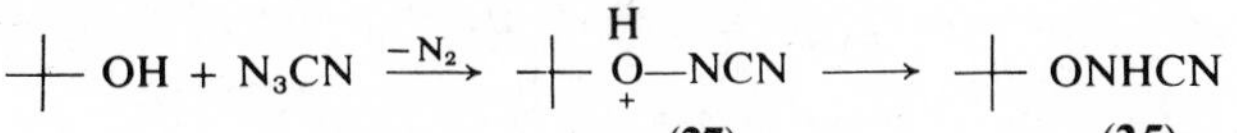

(51)

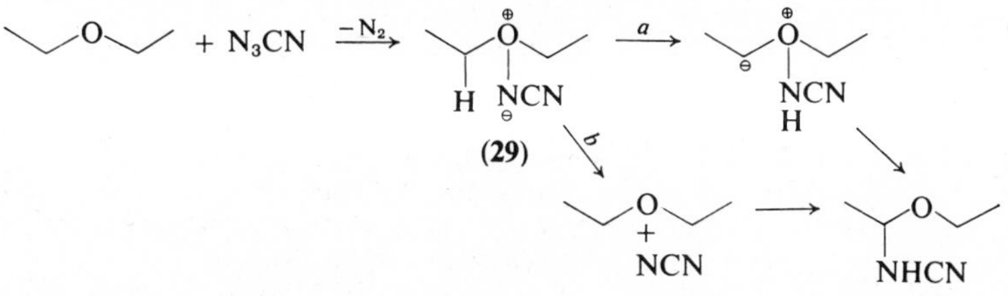

(52)

It is interesting to note in connection with sequence 52 that collapse of oxonium ion (**29**) to product (**26**) may in principle occur either by dissociation to ether and NCN followed by selective insertion of the nitrene into the α-C—H bonds of the ether or by transfer of an acidic α proton (45) to the NCN moiety of **29** followed by energetically beneficial 1,2 migration of NHCN. Both paths are mechanistically reasonable and ought to be distinguishable if appropriately labeled N_3CN is employed.

Finally reaction of cyanogen azide with stronger Lewis bases such as dialkyl sulfides, tertiary amines and tertiary phosphines occurs exothermically at 0–20°C to afford stable "onium" salts, quantitatively. Three examples are shown in sequences 53–55. In all three of these reactions N_3CN rather than NCN is clearly the reactive intermediate.

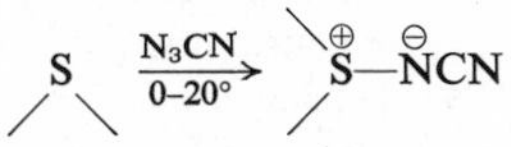

(53)

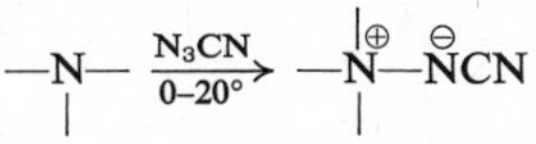

(54)

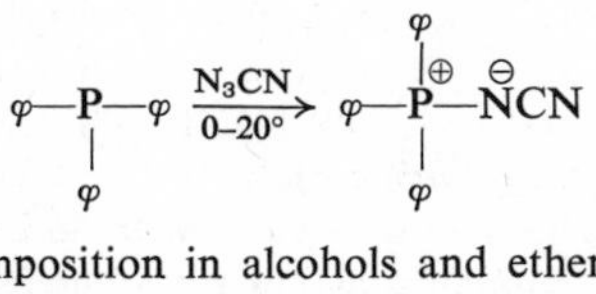

(55)

* In fact, N_3CN decomposition in alcohols and ethers occurs *exothermically* at ca. 30–40°C. In this temperature range, the rate of unassisted N_3CN thermolysis is virtually nil.

REFERENCES

1. G. Herzberg and D. N. Travis, *Can. J. Phys.*, **42**, 1658 (1964).
2. F. D. Marsh and M. E. Hermes, *J. Am. Chem. Soc.*, **86**, 4506 (1964).
3. G. J. Pontrelli and A. G. Anastassiou, *J. Chem. Phys.*, **42** 3735 (1965).
4. E. Wasserman, L. Barash, and W. A. Yager, *J. Am. Chem. Soc.*, **87**, 2075 (1965).
5. A. G. Anastassiou, H. E. Simmons, and F. D. Marsh, *J. Am. Chem. Soc.*, **87**, 2296 (1965).
6. G. Herzberg, *Electronic Spectra of Polyatomic Molecules*, D. Van Nostrand, Princeton, N.J., 1966.
7. J. F. Mulligan, *J. Chem. Phys.*, **19**, 347 (1951).
8. H. E. Simmons, unpublished calculations.
9. R. Hoffmann, *J. Chem. Phys.*, **39**, 1397 (1963).
10. C. Haggart and C. A. Winkler, *Can. J. Chem.*, **38**, 329 (1960).
11. K. R. Jennings and J. W. Linnett, *Trans. Faraday Soc.*, **46**, 1737 (1960).
12. W. D. McGrath and T. Morrow, *Nature*, **203**, 619 (1964).
13. G. Herzberg and D. N. Travis, *Nature*, **204**, 988 (1964).
14. W. D. McGrath and T. Morrow, *Nature*, **204**, 988 (1964).
15. T. Morrow and W. D. McGrath, *Trans. Faraday Soc.*, **62**, 642 (1966).
16. Wm. H. Smith and G. E. Leroi, *J. Chem. Phys.*, **45**, 1784 (1966).
17. D. E. Milligan, M. E. Jacox, J. J. Comeford, and D. E. Mann, *J. Chem. Phys.*, **43**, 756 (1965).
18. D. E. Milligan, M. E. Jacox, and A. M. Bass, *J. Chem. Phys.*, **43**, 3149 (1965).
19. D. E. Milligan and M. E. Jacox, *J. Chem. Phys.*, **45**, 1387 (1966).
20. J. W. Linnett and M. F. Hoare, *Trans. Faraday Soc.*, **45**, 844 (1949).
21. H. W. Kroto, *J. Chem. Phys.*, **44**, 831 (1966).
22. L. J. Schoen, *J. Chem. Phys.*, **45**, 2773 (1966).
23. G. W. Robinson and M. McCarty, *J. Am. Chem. Soc.*, **82**, 1859 (1960).
24. D. E. Milligan and M. E. Jacox, *J. Chem. Phys.*, **44**, 2850 (1966).
25. N. G. Moll and W. E. Thompson, *J. Chem. Phys.*, **44**, 2684 (1966).
26. L. J. Stief and V. J. DeCarlo, *J. Chem. Phys.*, **43**, 2552 (1965).
27. M. E. Jacox and D. E. Milligan, *J. Chem. Phys.*, **43**, 3734 (1965).
28. D. E. Milligan and M. E. Jacox, *J. Chem. Phys.*, **47**, 703 (1967).
29. M. E. Jacox and D. E. Milligan, *J. Chem. Phys.*, **47**, 1626 (1967).
30. A. G. Anastassiou and H. E. Simmons, *J. Am. Chem. Soc.*, **89**, 3177 (1967).
31. A. G. Anastassiou, *J. Am. Chem. Soc.*, **88**, 2322 (1966).
32. A. G. Anastassiou, *J. Am. Chem. Soc.*, **89**, 3184 (1967).
33. Experiments to test this point are currently under way at Syracuse University.
34. A. G. Anastassiou and J. N. Shepelavy, *J. Am. Chem. Soc.*, **90**, 492 (1968).
35. A. G. Anastassiou, R. B. Hammer, and J. N. Shepelavy, manuscript in preparation.
36. A. G. Anastassiou, *J. Org. Chem.*, **31**, 1131 (1966).
37. A. G. Anastassiou, *J. Am. Chem. Soc.*, **87**, 5512 (1965).
38. A. G. Anastassiou, *J. Am. Chem. Soc.*, **90**, 1527 (1968).
39. A. G. Anastassiou and R. P. Cellura, *Chem. Commun.*, **1967**, 762.
40. R. Hoffmann, *J. Am. Chem. Soc.*, **90**, 1475 (1968).
41. A. G. Anastassiou, *Chem. Commun.*, 1968.
42. F. D. Marsh and H. E. Simmons, *J. Am. Chem. Soc.*, **87**, 3529 (1965).
43. F. D. Marsh and H. E. Simmons, manuscript in preparation.
44. J. A. Berson and M. R. Willcott, *Rec. Chem. Progr.*, **27**, 139 (1966).
45. F. D. Marsh, unpublished observations.

CHAPTER 10

Aminonitrenes (1,1-Diazenes)

DAVID M. LEMAL

Department of Chemistry, Dartmouth College, Hanover,
New Hampshire, 03755

I. INTRODUCTION

Earlier chapters of this monograph have led one to conceive of the typical nitrene as a high-energy, hyper-reactive beast with a dual nature, at once electrophilic and radical-like in its triplet ground state (1–4). When the substituent attached to (formally) monovalent nitrogen is a basic amino function, however, a dramatic change in properties results (5). Dialkylaminonitrenes (1) fail to manifest free radical behavior and, strikingly, they accept a proton on the "nitrene" nitrogen to give rather

stable cations (2). The ease of formation and apparently the thermodynamic stability of aminonitrenes increase sensitively with the basicity of
the amino substituent.* These facts indicate that 1 has a singlet ground
state which can be represented by the resonance forms shown below. They

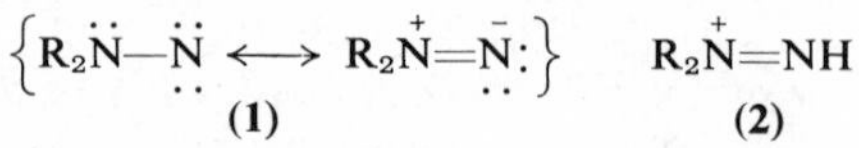

$$\left\{ R_2\ddot{N}-\ddot{N} \longleftrightarrow R_2\overset{+}{N}=\overset{-}{N}\!: \right\} \qquad R_2\overset{+}{N}=NH$$

(1)   (2)

further suggest that the dipolar structure contributes very importantly to
the hybrid, no doubt because this structure alone obeys the octet rule. A
close parallel exists between aminonitrenes and their carbon analogs,
diaminocarbenes (3). In contrast to conventional carbenes, the latter are
excellent nucleophiles, better represented as ylids than as divalent carbon
derivatives (6). The existence and properties of aminonitrenes have been

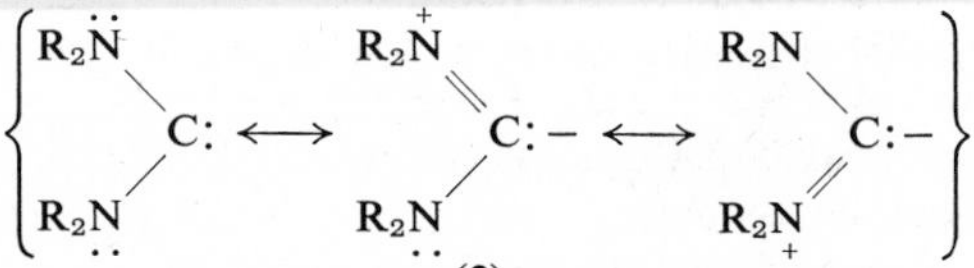

$$\left\{ \begin{array}{c} R_2\ddot{N} \\ \diagdown \\ C\!: \\ \diagup \\ R_2\ddot{N} \end{array} \longleftrightarrow \begin{array}{c} R_2\overset{+}{N} \\ \diagdown\!\!\diagdown \\ C\!:\!- \\ \diagup \\ R_2N \end{array} \longleftrightarrow \begin{array}{c} R_2N \\ \diagdown \\ C\!:\!- \\ \diagup\!\!\diagup \\ R_2\overset{+}{N} \end{array} \right\}$$

(3)

deduced from a large body of chemical information; no direct physical
evidence for these species has appeared in the literature. Such evidence
does exist for their conjugate acids (2), however.

Aminonitrenes are known by numerous aliases, notably: azamine,
aminoimido intermediate, N-imene and diazene (2). If the name "aminonitrene" is generally misleading with regard to structure and reactivity,
the corresponding term for the conjugate acid (2), "aminonitrenium
ion," is *really* a misnomer. There simply is no cationic divalent nitrogen
present. Hence the concise terms "diazene" and, for the conjugate acid,
"diazenium ions" introduced by McBride (7) will be used in this chapter.
The term "diazene" is sometimes used in naming derivatives of diimide
($HN=NH$); in the present context unsymmetrical attachment to the N_2
group will be understood throughout.

The azomethinimine tautomer (4) of a diazene will be seen to play a
significant role in diazene chemistry. Although methods for producing all
three species will be considered together, for clarity's sake separate

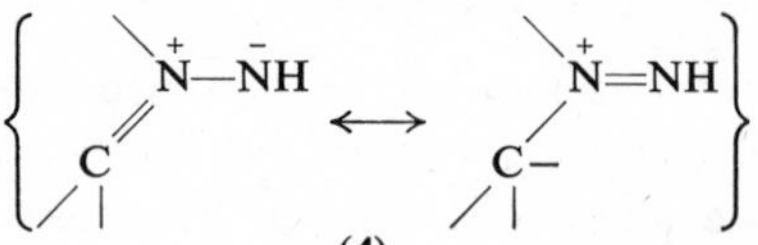

$$\left\{ \begin{array}{c} \diagdown \\ \overset{+}{N}-\overset{-}{N}H \\ \diagup \\ C \\ \diagup\,| \end{array} \longleftrightarrow \begin{array}{c} \diagdown \\ \overset{+}{N}=NH \\ \diagup \\ \overset{-}{C} \\ \diagup\,| \end{array} \right\}$$

(4)

* Among aminonitrenes, those with dialkyl substituents clearly provide the sharpest
contrast with "typical" nitrenes. Powerfully electron-attracting substituents render
even an aminonitrene distinctly electrophilic, and it is quite conceivable that triplet
ground states will be found in extreme cases.

sections of this article will be devoted to the reactivity characteristic of diazenes, diazenium ions, and the related azomethinimines.

Finally, the relationship of diazenes to their symmetrically substituted counterparts, azo compounds, deserves attention. As unstable isomers of azo compounds, diazenes might be expected to lose molecular nitrogen very readily, thus providing, as azo compounds do, an important method for generating carbon–carbon bonds (8,9,10a,11a). This expectation has

$$R—N{=}N—R \xrightarrow[-N_2]{\Delta \text{ or } h\nu} R—R, \text{ etc.}$$

$$\left[\begin{array}{c} R \\ \diagdown \\ {\overset{+}{N}}{=}{\overset{-}{N}} \\ \diagup \\ R \end{array} \right] \xrightarrow[-N_2]{?} R—R$$

been realized only in part, for their very instability makes diazenes susceptible to modes of reaction other than nitrogen loss. Nonetheless, the fragmentation of diazenes is a particularly interesting and potentially the most useful facet of their chemistry.

II. GENERATION OF DIAZENES, DIAZENIUM IONS, AND THE CORRESPONDING AZOMETHINIMINES

This chapter will be concerned solely with 1,1-disubstituted diazenes and derived species. Reactions designed to produce monosubstituted diazenes (5) or diazenium ions (6) generally result in deamination (10b), since the

$$\left. \begin{array}{c} [R\overset{+}{N}H{=}\overset{-}{N}] \\ \textbf{(5)} \\ [R\overset{+}{N}H{=}NH] \\ \textbf{(6)} \end{array} \right\} \longrightarrow [R—N{=}NH] \longrightarrow R—H + N_2$$
$$\hspace{6.5cm} \textbf{(7)}$$

proton at N-1 is readily lost to give a monosubstituted diimide (7) either directly or via 5 or 6. Correspondingly, unsubstituted diazene is known only as its symmetrical tautomer diimide.

Certain of the methods for generating diazenes, as well as their fragmentation, are reminiscent of diazo chemistry (9a,11). This is surely appropriate, for diazo compounds may be thought of as "alkylidene-diazenes."

A. Oxidation of 1,1-Disubstituted Hydrazines

This is the most venerable and perhaps the most versatile method for producing diazenes. Long before the turn of the century it was known that tetrasubstituted tetrazenes (8) result from the two-electron oxidation of unsymmetrically disubstituted hydrazines (12). A wide variety of

$$R_2NNH_2 \xrightarrow{-4H} R_2N-N=N-NR_2$$
(8)

oxidizing agents effects this transformation; among the most commonly used are mercuric oxide, lead tetraacetate, and halogens. The observation made by Busch and Weiss (13) in 1900 that mercuric oxide oxidation of 1,1-dibenzylhydrazine yields bibenzyl and molecular nitrogen (eq. 1) lends

$$(C_6H_5CH_2)_2NNH_2 \xrightarrow{HgO} C_6H_5CH_2CH_2C_6H_5 + N_2 \tag{1}$$
(100%)

credence to the notion that diazenes are formed in such oxidations, and the studies of McBride and his co-workers a decade ago provided elegant confirmation (7,14,15). Oxidizing 1,1-dialkylhydrazines with powerfully acidic halogen solutions, this group prepared rather stable solutions of diazenium salts and convincingly characterized them (Section III-A). Neutralization of these solutions freed the diazenes, giving tetrazenes in nearly quantitative yield; in effect, then, diazenes had been "trapped" by protonation in the oxidation experiments.

Though not directly related to diazenes, alkoxydiazenium ions merit brief comment. Prepared by alkylation of nitrosamines, which is accomplished with agents as mild as dimethyl sulfate, the salts (9) are isolable and quite stable provided that X^- is a poor nucleophile (16,17). Hünig and coworkers have used 1-aryl-1-alkyl-2-alkoxydiazenium salts in an ingenious

$$R_2\overset{+}{N}=N \quad X^-$$
$$\diagdown$$
$$OR'$$
(9)

synthesis of α-carbonyl azo compounds (18). Like alkylation, protonation of nitrosamines also occurs on oxygen, giving hydroxydiazenium ions (16b).

B. Decomposition of 1,1-Disubstituted 2-Sulfonylhydrazine Salts

1,1-Disubstituted hydrazines can be prepared by alkylation of hydrazine or a monosubstituted hydrazine, but amination of the appropriate secondary amine is a more general method (10b). This can be carried out directly using hydroxylamine-O-sulfonic acid or chloramine, but the indirect approach of nitrosating and then reducing with lithium aluminum hydride, zinc in acetic acid, aluminum amalgam or other reagents is often to be preferred. Sulfonylation, typically with tosyl chloride, yields crystalline derivatives in the proper oxidation state to give diazenes directly (eq. 2).

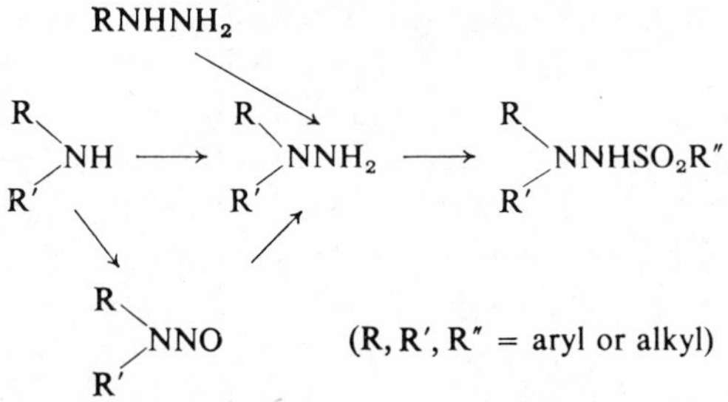

RNHNH$_2$

$$\begin{array}{ccc}
\overset{R}{\underset{R'}{\diagdown}}\text{NH} & \longrightarrow & \overset{R}{\underset{R'}{\diagdown}}\text{NNH}_2 & \longrightarrow & \overset{R}{\underset{R'}{\diagdown}}\text{NNHSO}_2\text{R}'' \qquad (2)
\end{array}$$

$$\overset{R}{\underset{R'}{\diagdown}}\text{NNO} \qquad (R, R', R'' = \text{aryl or alkyl})$$

Carpino discovered in 1957 (19,20) that 1,1-dibenzyl-2-benzenesulfonylhydrazine is readily degraded by aqueous base (eq. 3) to bibenzyl, nitrogen, and benzenesulfinate in analogy to equation 1. He had taken advantage here of the considerable "leaving ability" of the benzenesulfinate group, just as Bamford and Stevens had done earlier in developing their valuable route to diazo compounds (21). The timing of the deprotonation relative to the N—S cleavage will be discussed in Section III-C.

$$\overset{C_6H_5CH_2}{\underset{C_6H_5CH_2}{\diagdown}}\text{N}{-}\text{NHSO}_2\text{C}_6\text{H}_5 \xrightarrow[\Delta]{20\% \text{ NaOH}}$$

$$C_6H_5CH_2{-}CH_2C_6H_5 + N_2 + NaO_2SC_6H_5 \quad (3)$$
$$(85\%)$$

The Carpino method for generating diazenes is more sensitive to structural variations than is hydrazine oxidation. In cases where the basicity of N-1, the amino nitrogen, is very low, violent conditions are required to induce the sulfinate group to leave. This fact supports the view that a sharp increase in energy content of a diazene accompanies a decrease in electron availability at the substituted nitrogen (22,23). Sulfonylhydrazine salt **10**, which fragments to carbazole (eq. 4), is a case in point.

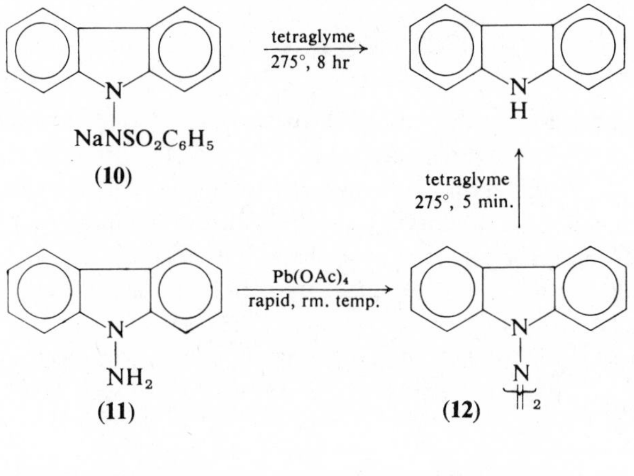

(4)

(5)

The available evidence bearing on the reaction has been interpreted in terms of diazene and tetrazene (12) intermediates (23). Consistent with this conclusion is the efficient conversion of authentic tetrazene (12) to carbazole under the reaction conditions, presumably via fragmentation to carbazolyl radicals which abstract hydrogen from the solvent. Whatever the mechanism for equation 4, the extreme reluctance of sulfonylhydrazine salt 10 to lose benzenesulfinate ion is a dramatic consequence of the low basicity of the carbazole nitrogen. The gentle conditions which suffice for transforming N-aminocarbazole (11) into tetrazene (12) (eq. 5) demonstrate that the hydrazine oxidation method for generating diazenes is not subject to the structural limitations of the Carpino method (23,24).

On the other hand, the Carpino method can work rapidly and efficiently at room temperature, as it does, for example, with 1,1-diethyl-2-benzene-sulfonylhydrazine in hydroxylic media (Section V-B) (25). It has distinct advantages for mechanistic studies of diazene behavior. In particular, it is chemically simple and unequivocal, it leads to a single inert by-product (the metal sulfinate), and it is adaptable for use in a wide variety of media. Finally, sulfonylhydrazine salts which stoutly resist thermolysis are not all hopeless: certain types fragment smoothly when irradiated (Section IV-C-3).

Diazenium ions* as well as diazenes can be generated from many sulfonylhydrazines and studied in neutral and basic media (27), despite the observation of McBride and his co-workers that neutralization of diazenium ion solutions produces tetrazene even at very low pH (down to pH 0 for concentrated diazenium ion solutions!) (7). The key difference between these experiments is that equilibrium constants for ionization of sulfonylhydrazines to diazenium and sulfinate ions (eq. 6) are very small even in highly polar media. Hence dimer formation is avoided, yet the diazenium ions can be trapped by appropriate reagents (Section III-C).

$$R_2NNHSO_2R' \rightleftharpoons R_2\overset{+}{N}=NH \quad {}^-O_2SR' \qquad (6)$$

Tautomerization of a diazene to the corresponding azomethinimine probably occurs via the diazenium ion, which can deprotonate at carbon as well as at nitrogen (eq. 7). Hence any method for generating diazenes in a protic medium is potentially a route to azomethinimines as well, and for this reason product compositions often depend greatly upon the character of the solvent (25). Decomposition of dry sulfonylhydrazine

---

* Wawzonek and McKillip treated a number of tosylhydrazines with concentrated hydrochloric acid at elevated temperatures and interpreted the complex product mixtures which resulted in terms of diazenium ion intermediates (26).

$$\begin{array}{c} RCH_2 \\ \diagdown \\ \overset{+}{N}=\overset{-}{N} \\ \diagup \\ RCH_2 \end{array} \underset{-H^+}{\overset{H^+}{\rightleftharpoons}} \begin{array}{c} RCH_2 \\ \diagdown \\ \overset{+}{N}=NH \\ \diagup \\ RCH_2 \end{array} \overset{-H^+}{\longrightarrow}$$

$$\left\{ \begin{array}{c} RCH_2 \\ \diagdown \\ \overset{+}{N}=NH \\ \diagup \\ RCH- \end{array} \longleftrightarrow \begin{array}{c} RCH_2 \\ \diagdown \\ \overset{+}{N}-\overset{-}{NH} \\ \diagdown\diagdown \\ RCH \end{array} \right\} \quad (7)$$

salts in aprotic media have not yielded products traceable to azomethinin-
imine intermediates, but even small amounts of a hydroxylic cosolvent
can make such products predominate. Indeed, azomethinimine formation
has been known to occur even when the medium was entirely aprotic
(triethylamine), provided that the sulfonylhydrazine was present with its
proton, not as a metal salt (28,29).

C. Reaction of Difluoramine with Secondary Amines

Several years ago Bumgardner and his co-workers discovered that
difluoramine (13) transforms secondary amines into products which appear
to derive from diazene intermediates (eq. 8) (30,31). They postulated

$$R_2NH + HNF_2 \xrightarrow{-2HF} [R_2\overset{+}{N}=\overset{-}{N}] \longrightarrow products \qquad (8)$$
$$(13)$$

base-promoted elimination of hydrogen fluoride from difluoramine to
yield fluoronitrene ($:\ddot{N}—F$), which was then intercepted by a second
molecule of amine. Direct displacement by amine on difluoramine is also
a possible pathway, though a recent study of difluoramine hydrolysis in
aqueous base at high pressures supports the view that fluoronitrene is
formed (32). In any event, elimination of another hydrogen fluoride
molecule would yield the diazene; intermediate diazenium ions would also
explain some of the results of secondary amine–difluoramine reactions.

Since the reagent can be prepared only inefficiently by direct fluorination of ammonia (33), the indirect route shown in equation 9 has found

$$NH_3 \xrightarrow[F_2]{excess} NF_3 \xrightarrow[375°]{Cu} F_2N—NF_2 \xrightarrow{C_6H_5SH} HNF_2 \qquad (9)$$
$$(13)$$

considerable use (34,35). Difluoramine, a gas at room temperature (bp
−23°), is normally handled in a vacuum system; the compound is worthy
of respect, as it tends to detonate when frozen or thawed (mp −116°) (35).
Nevertheless, the difluoramine method for generating diazenes boasts the
considerable advantage of utilizing secondary amines, which are more
easily available than hydrazines or sulfonylhydrazines. Incidentally,

difluoramine also makes possible the deamination of primary amines in a single step (30,31).

Klopotek and Hobrock have found that isopropyl N,N-difluoro-carbamate reacts with secondary amines to give a number of products, but that in at least one case a diazene or a diazenium ion intermediate is clearly implicated (36). The urethan is apparently functioning as a source of difluoramine and/or fluoronitrene here.

D. Reaction of Angeli's Salt and Piloty's Acid with Secondary Amines

At the turn of the century Angeli (37) made the intriguing observation that tetrazenes, together with other products, are formed from secondary amine salts by the action of sodium nitrohydroxamate, now known as Angeli's salt (14) (eq. 10). Rave has recently provided strong support for

$$R_2\overset{+}{N}H_2Cl^- + Na_2N_2O_3 \xrightarrow{H_2O} R_2N\!-\!N\!=\!N\!-\!NR_2 \qquad (10)$$
$$(14)$$

Angeli's view that diazenes are intermediates here (38). Choosing sulfonyl-hydrazines which yield products of diazene fragmentation, rearrangement, and tautomerization, he found that the same products resulted from the corresponding Angeli's salt–secondary amine reactions. Angeli interpreted the varied chemistry of sodium nitrohydroxamate in terms of decomposition to nitroxyl (15), which could occur as shown in equation 11; he assumed that nitroxyl then attacked the substrate under study. The formation of nitroxyl has been confirmed by others (39,40).

$$\longrightarrow NO_2^- + [HN\!=\!O] \quad (11)$$
$$(15)$$

N-Benzenesulfonylhydroxylamine, called Piloty's acid (16) (41), was also considered by Angeli to be a nitroxyl source (42), and in fact he observed strong similarities in the chemistry of Angeli's salt and Piloty's acid. Though yields are poorer, Rave has found that Piloty's acid transforms secondary amines into some of the same products obtained using Angeli's salt (43). Whereas the salt requires a proton source to give nitroxyl, Piloty's acid decomposes rapidly in aqueous alkali, probably via equation 12.

$$HONHSO_2C_6H_5 \xrightarrow{^-OH} \left[^-O\!-\!\overset{H}{N}\!-\!SO_2C_6H_5 \right] \longrightarrow [HN\!=\!O] + C_6H_5SO_2^- \quad (12)$$
$$(16) \hspace{9.5cm} (15)$$

Smith and Hein (40) have shown that Angeli was incorrect in assuming the intermediacy of nitroxyl in certain reactions of 14 and 16, notably the Angeli-Rimini reaction (eq. 13)* and the halide–oxime transformation

---

* This is used as a test for aldehydes because the resulting hydroxamic acids (17) form highly colored complexes with ferric ion.

$$\underset{(16)}{\overset{\overset{\displaystyle O}{\|}}{RCH}} + HONHSO_2C_6H_5 \xrightarrow[\text{alkali}]{\text{aqueous}} \underset{(17)}{\overset{\overset{\displaystyle O}{\|}}{RCNHOH}} \qquad (13)$$

$$\overset{\overset{\displaystyle O}{\|}}{RCH} + HONSO_2C_6H_5 \longrightarrow \left[\overset{\overset{\displaystyle O^-}{|}}{\underset{\underset{\displaystyle SO_2C_6H_5}{|}}{RCH-NOH}} \right] \longrightarrow \overset{\overset{\displaystyle O}{\|}}{RC-NHOH} \qquad (13a)$$

(eq. 14); their mechanisms for these reactions are presented in equations 13a and 14a, respectively. Aldehydes and halides are electrophilic re-

$$\underset{(14)}{RCH_2X} + Na_2N_2O_3 \longrightarrow RCH{=}NOH \qquad (14)$$

$$RCH_2{-}X + \overset{Na^+{}^-O}{\underset{O_-}{N{=}N}}{\overset{O^-}{\diagup}} \; Na^+ \xrightarrow{-NaX} \left[RCH_2{-}N\overset{O^-\,Na^+}{\underset{NO_2}{\diagdown}} \right] \xrightarrow{-NaNO_2}$$

$$[RCH_2{-}N{=}O] \longrightarrow RCH{=}NOH \qquad (14a)$$

agents, however, in contrast to amines; attack by Angeli's salt, even in protonated form, on a secondary amine seems unlikely, but nitroxyl should find a good nucleophile irresistible. At present, then, Angeli's original view of the reaction with secondary amines, expressed in modern terms in equation 5, appears to be correct.

$$R_2\overset{..}{N}H + H\overset{..}{N}{=}O \longrightarrow \left[R_2\overset{..}{N}{-}N\overset{OH}{\underset{H}{\diagup}} \right] \longrightarrow [R_2\overset{+}{N}{=}NH] \rightleftharpoons$$

$$[R_2\overset{+}{N}{=}\overset{-}{N}] \longrightarrow R_2N{-}N{=}N{-}NR_2 + \text{other products} \qquad (15)$$

Like difluoramine, Angeli's salt produces a diazene from a secondary amine in a single step, and this reagent has the additional advantage of being easily prepared on a large scale, stored, and handled. Angeli's salt is made by nitration of hydroxylamine with ethyl nitrate in sodium methoxide solution (44). A serious limitation of this approach to diazenes is that aqueous solvents seem to be required for satisfactory results. Preliminary attempts to use Angeli's salt for deaminating primary amines have not been encouraging, but the method deserves further experimentation (43).

Hammick and his co-workers found that t-alkyl nitroso compounds are readily destroyed by red light, yielding alkenes, hyponitrous acid, and decomposition products of the latter (eq. 16) (45). These results were most

$$\underset{\underset{NO}{|}}{(CH_3)_2CCH_2R} \xrightarrow[\text{light}]{\text{red}} (CH_3)_2C{=}CHR + HON{=}NOH + HNO_3 + HNO_2 + N_2 \qquad (16)$$

simply explained in terms of photolytic C—N cleavage to nitric oxide and a t-alkyl radical followed by disproportionation to alkene plus nitroxyl, which dimerized to the labile hyponitrous acid (eq. 17). With the hope of

$$(CH_3)_2C-CH_2R \xrightarrow{h\nu} \left[(CH_3)_2C \cdots CH \cdots H \cdots N=O\right] \longrightarrow$$
$$\underset{\textstyle N=O}{}$$

$$(CH_3)_2C=CHR + [HNO] \qquad (17)$$
$$(15)$$

generating diazenes via nitroxyl but in nonpolar solvents, McGregor photolyzed the beautiful blue nitrosoisobutane in the presence of secondary amines (46). The interesting array of products which was isolated included none that were traceable to diazenes, however. Possibly nitroxyl was intercepted by the amines as planned, but in nonaqueous media the resulting adduct found other reactions more attractive than dehydration to the diazene (eq. 15).

E. Reduction of Nitrosamines

Overberger and his co-workers (47b,c) observed that sodium dithionite in aqueous alkali or lithium in liquid ammonia brings about fragmentation of certain nitrosamines (8), as illustrated in equation 18. Product composi-

$$\underset{\underset{\textstyle NO}{\textstyle N}}{\overset{\textstyle}{H_5C_6 \quad C_6H_5}} \xrightarrow[^-OH]{Na_2S_2O_4} \quad \underset{H_5C_6 \quad C_6H_5}{} + C_6H_5(CH_2)_3CH=CHC_6H_5 \qquad (18)$$

tions here and in other examples are very similar to those obtained by mercuric oxide oxidation of the corresponding hydrazines (47a,c), so it is believed that diazene intermediates are involved once again. Unfortunately, the attractive idea of deoxygenating nitrosamines to diazenes with triphenylphosphine is reported to fail (48a)*.

When fragmentation of a diazene is the desired reaction, dithionite reduction is a convenient route to choose; its use for the study of diazenes is sharply limited, however, because the normal course of the reduction includes a second two-electron step leading to the hydrazine (eq. 19) (47b).

$$R_2N-NO \xrightarrow{2e} [R_2\overset{+}{N}=\overset{-}{N}] \xrightarrow{2e} R_2NNH_2 \qquad (19)$$

* Note added in Proof. A very recent communication claims success with the reagent ethyl diphenylphosphinite. The low basicity of the amine (benzotriazole) from which the nitroso compound was derived may be responsible for success here, so the generality of this deoxygenation reaction is a question for the future (J. I. G. Cadogan and J. B. Thomson, *Chem. Commun.*, 770 (1969)).

Special structural features (Section IV-C) are required if loss of molecular nitrogen is to compete with further reduction. Yet a third kind of reaction, reductive N—N cleavage, is observed with certain substrates such as diphenylnitrosamine (47b). Even nitrosamines which do yield diazene fragmentation products when treated with alkaline dithionite are reduced to hydrazines with zinc in acetic acid, lithium aluminum hydride (but note ref. 47c), or aluminum amalgam (8,48b).

F. Cleavage of N-Aminosulfoximines

Very recently Rees' group has intercepted several acyl-substituted diazenes by dimethyl sulfoxide to give crystalline dimethylsulfoximine derivatives. Photolysis[49a] or thermolysis[49b] of the sulfoximines regenerated the diazenes, as indicated by formation of the same products obtained from oxidation of the corresponding hydrazine.

III. CHEMISTRY OF DIAZENIUM IONS

A. Characterization

As noted earlier, McBride's group prepared solutions of 1,1-dialkyl-diazenium salts by halogen or bromic acid oxidation of the corresponding hydrazines in strongly acidic media (14). Treatment with stannous chloride reduced the dialkyldiazenium ions quantitatively to the starting hydrazines (eq. 20). Neutralized by dropwise addition of sodium hydroxide solution,

$$R_2N—NH_2 \qquad R_2\overset{+}{N}{=}NH \ Br^- \qquad (20)$$

diazenium salt solutions gave nearly theoretical yields of tetraalkyltetrazenes. These dimeric products were not present in protonated form prior to neutralization, for the characteristic ultraviolet absorption of tetra-zenium salts was absent, and these salts were found to resist reduction by stannous chloride. Interestingly, neutralization of diazenium salt solu-tions by addition to aqueous base gave decreased yields of tetrazenes, and in some cases practically none (7). Probably tautomerization of the dia-zenes to azomethinimines occurred under these conditions and the latter led to rearrangement and hydrolysis products. Generated by the inverse mode of addition, a diazene finds itself surrounded by powerfully electro-philic diazenium ions and its capture by one of these competes effectively against tautomerization (eq. 21). Statistical mixtures of all possible tetrazenes result from neutralization of solutions containing a number of different diazenium ions (7).

$$\overset{+}{R_2N}=NH \xrightarrow{\ ^-OH\ } [\overset{+}{R_2N}=\overset{-}{N}] \xrightarrow{\overset{+}{R_2N}=NH}$$

$$\overset{+}{R_2N}=N-\underset{H}{N}-NR_2 \rightleftharpoons \overset{+}{R_2N}-N=N-NR_2 \quad (21)$$

1,1-Dimethyldiazenium perchlorate was isolated, albeit not in pure form, from the reaction of iodine, silver perchlorate, and 1,1-dimethylhydrazine in anhydrous ether (14). Direct spectral evidence for this cation is available as well. Urry and his co-workers discovered that during the bromine oxidation of 1,1-dimethylhydrazinium bromide (in deuterium oxide at 0°) disappearance of the methyl singlet initially present in the NMR spectrum is accompanied by growth of a new singlet 1.4 ppm farther downfield (δ 4.78), attributable to the diazenium ion methyls (cf. tetramethyl tetrazenium bromide, δ 2.86) (50). Ultraviolet spectra of dimethyldiazenium salt solutions exhibit a maximum at $\approx$280 mμ, $\epsilon \approx$150 in $4M$ perchloric acid (7). The absorption is virtually independent of hydrogen ion con-centration but is enhanced considerably if chloride ion is present in high concentration. This phenomenon has been ascribed to formation of an ion pair or of the covalent adduct of chloride to the cation ($(CH_3)_2N-NHCl$).

B. Attack on Unsaturated Systems

Diazenium ions are capable of functioning as dienophiles in Diels-Alder-like reactions. Urry's group obtained good yields of tetrahydro-pyridazinium salts (18) from the reaction of dimethyldiazenium bromide with butadiene (15) and with isoprene (eq. 22) (50a). Evidence for the structures included hydrogenolysis to the putrescine derivatives (19).

Equally interesting is their observation that this diazenium bromide adds efficiently to styrene (50a), giving a bromohydrazine as its hydro-bromide (20). Hydrogenolysis of 20 yielded 2-phenethylamine (76%),

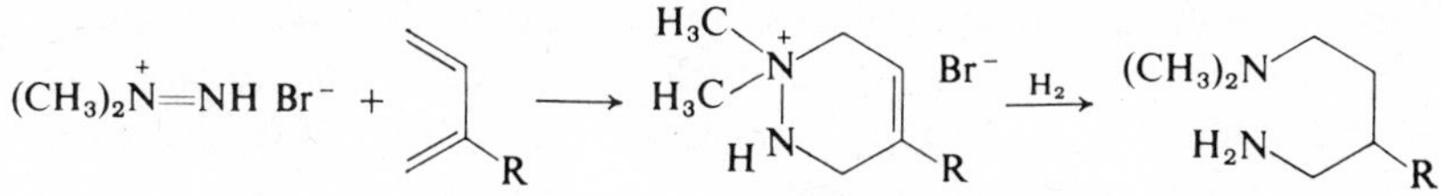

$$(CH_3)_2\overset{+}{N}{=}NH\ Br^- \ +$$

(22)

(18)   **(19)**

$(\sim 75\%)$ $(R = H, CH_3)$

dimethylamine and 1,1-dimethyl-2-(2-phenethyl)-hydrazine (15%). Whereas equation 22 may represent a concerted transformation, the addition to

$$(CH_3)_2\overset{+}{N}{=}N\,HBr^- \ + \ C_6H_5CH{=}CH_2 \ \xrightarrow[0°]{HBr} \ (CH_3)_2\overset{+}{N}{-}NHCH_2\overset{|}{C}HC_6H_5$$
$$Br^- \qquad\qquad Br$$

(20)
(76%)

styrene more likely entails electrophilic attack on the hydrocarbon followed by bromide ion addition. Treatment of **20** with aqueous sodium carbonate resulted in ring closure to the azetidinium ion **21**, isolated as its perchlorate. Hydrogenolysis of the perchlorate gave the known 2-dimethylamino-2-phenethylamine. In aqueous sodium hydroxide the ultimate product from **20** was the hydrazinoalcohol **22**, but an NMR experiment confirmed that **21** was an intermediate in this transformation.

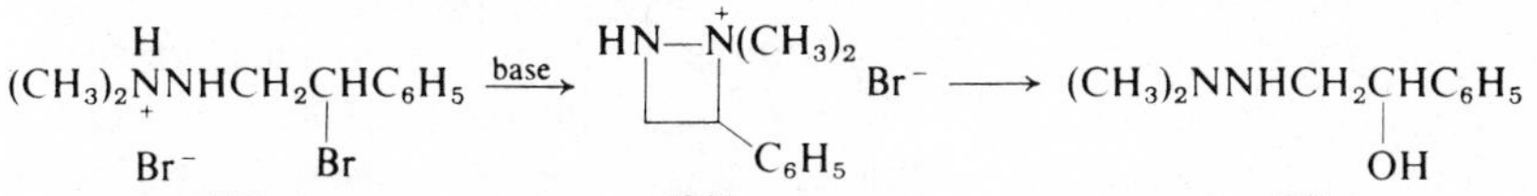

$$(CH_3)_2\underset{Br^-}{\overset{+}{N}}NHCH_2\overset{|}{\underset{Br}{C}}HC_6H_5 \ \xrightarrow{base} \ HN{-}\overset{+}{N}(CH_3)_2 \quad Br^- \ \longrightarrow \ (CH_3)_2NNHCH_2\overset{|}{\underset{OH}{C}}HC_6H_5$$

(20)   **(21)**   **(22)**

Attack of dimethyldiazenium ion on norbornadiene is reported to yield, among other products, a hydrazinium bromide which is transformed on basic workup into an aminoaziridine (eq. 23). Surprisingly, the configuration of the latter is believed to be *endo*, whereas an *exo*-aminoaziridine is obtained starting with norbornene (50b).

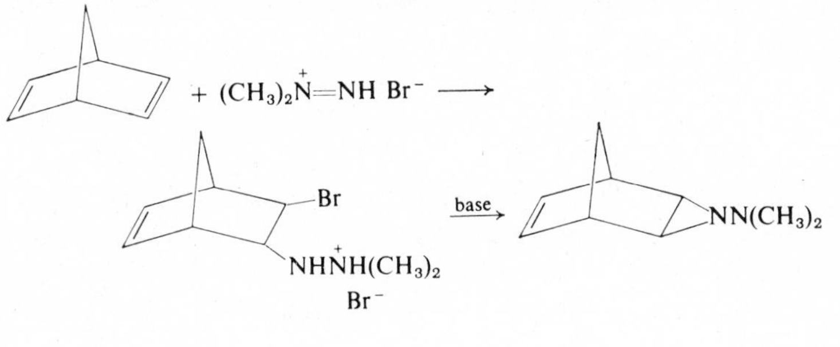

$$+ \ (CH_3)_2\overset{+}{N}{=}NH\ Br^- \ \longrightarrow$$

$$\xrightarrow{base} \ \overset{}{\diagup}NN(CH_3)_2 \qquad (23)$$

Br

$\overset{+}{NHNH}(CH_3)_2$

Br^-

Hünig and his co-workers found that oxidation of 1,1-diarylhydrazines in the presence of aromatic amines and phenols yielded coupling products such as **23**, which are dyes (51). Oxidative coupling was successful even

(23)

with the weak nucleophile 1-phenylpyrazoline in place of the amine or phenol, but no coupling products were isolated when 1-alkyl-1-aryl-hydrazines or dialkylhydrazines were substituted for their diaryl counterparts. Evidence was adduced that dye formation followed this pathway: generation of the diaryldiazenium ion, electrophilic substitution of this ion on the ring, and dehydrogenation to the fully conjugated system.

C. Exchange Equilibria

As explained in Section II-B, neutralization of diazenium salt solutions produces tetrazenes while the medium is still strongly acidic, but it is nonetheless possible to generate and study diazenium ions in neutral and basic media without formation of tetrazenes. Underbrink and Rave discovered that many 1,1-disubstituted sulfonylhydrazines ionize to diazenium and sulfinate ions readily and reversibly in polar media (27). Ion concentrations are low enough to have gone undetected by NMR spectroscopy, so the stability of such solutions against tetrazene formation is understandable.

The evidence for ionization derives from exchange experiments, a dramatic example of which occurs when aqueous solutions of 1,1-dimethyl-2-methanesulfonylhydrazine (**24**) and sodium p-toluenesulfinate are mixed. The insoluble 1,1-dimethyl-2-p-toluenesulfonylhydrazine (**25**) crystallizes immediately in near-quantitative yield, thus testifying to the rapidity of the ionization. When the reactants are so chosen that all components of the

$$(CH_3)_2NNHSO_2CH_3 + Na^{+-}O_2SC_7H_7(p) \longrightarrow$$
(**24**)

$$Na^{+-}O_2SCH_3 + (CH_3)_2NNHSO_2C_7H_7(p)$$
(**25**)

exchange equilibrium are soluble, the rate of approach to and position of equilibrium are conveniently studied by NMR spectroscopy. Support for the ionization mechanism for exchange, represented in general form in equation 24, is drawn from several aspects of the NMR studies. First,

$$
\begin{array}{c}
R_1 \\
\diagdown \\
N\!-\!NHSO_2R_3 \rightleftharpoons \\
\diagup \\
R_2
\end{array}
\left[
\begin{array}{c}
R_1 \\
\diagdown \\
\overset{+}{N}\!=\!NH \\
\diagup \\
R_2
\end{array}
\right]
+ R_3SO_2^{\,-}
$$

$$
\left[
\begin{array}{c}
R_1 \\
\diagdown \\
\overset{+}{N}\!=\!NH \\
\diagup \\
R_2
\end{array}
\right]
+ R_4SO_2^{\,-} \rightleftharpoons
\begin{array}{c}
R_1 \\
\diagdown \\
N\!-\!NHSO_2R_4 \\
\diagup \\
R_2
\end{array}
\quad (24)
$$

exchange rates are rapid at room temperature only in highly polar solvents, and only for 1,1-dialkyl-2-sulfonylhydrazines. Diminishing the basicity of N-1 by substituting a phenyl for an alkyl group reduces exchange rates several hundredfold and 1,1-diphenyl-2-sulfonylhydrazines are quite inert. Initial exchange rates are independent of metal sulfinate concentration, thus ruling out a bimolecular event as or preceding the rate-determining step. Increasing electron-withdrawing character in R_3 both accelerates exchange and drives the equilibria of equation 24 toward the right as expected. Finally, exchange is sharply inhibited for **24** by high concentrations of hydroxide ion, so it is the free sulfonylhydrazine, not its salt, which loses sulfinate ion (pK_a's of typical sulfonylhydrazines are ~ 11) (52). The N—S bond must be stronger in the salt than in the free sulfonylhydrazine, to be sure, but cleavage of this bond in the salt requires no charge separation. Hence the strong preference for forming dimethyldiazenium ion over dimethyldiazene is very striking.

The above considerations reveal an "energetically cheap" way to generate diazenes from sulfonylhydrazines. Since diazenium ions deprotonate very readily, the sequence ionization–deprotonation can produce diazenes rapidly at room temperature from sulfonylhydrazines whose dry salts decompose only at elevated temperatures. The point is forcefully illustrated by Underbrink's observation that 1,1-dimethyl-2-methanesulfonylhydrazine (**24**) is stable in water and stable as its salt in strong aqueous solid hydroxide, but is destroyed in minutes by dilute base (27). The rearrangement product formaldehyde methylhydrazone (Section V-5) was found, but the mixture has not been fully characterized.

$$
(CH_3)_2NNHSO_2CH_3 \!-\!
\begin{cases}
\xrightarrow{\;H_2O\;} \text{no reaction} \\[4pt]
\xrightarrow{\;dil.\ NaOH\;} \text{rapid decomposition} \\[4pt]
\xrightarrow{\;conc.\ NaOH\;} \text{no reaction}
\end{cases}
$$
(**24**)

The NMR spectrum of **24** in aqueous solution is not perturbed by addition of such excellent nucleophiles as azide, thiosulfate, or thiocyanate ion despite the rapid exchange which occurs in the presence of sulfinate ions. Very likely these ions do intercept dimethyldiazenium ions, but the resulting adducts reionize with equilibrium constants considerably larger than those of sulfonylhydrazines. Hence no net reaction is observed. The N—S bond of sulfonamides and sulfonylhydrazines owes part of its strength to overlap of the lone-pair orbital on N-2 with vacant d-orbitals on sulfur. One might expect that nucleophiles other than sulfinate which are capable of forming (partial) double bonds to a diazene could displace sulfinate from a sulfonylhydrazine. Bisulfite can do this, in fact, and so can phosphines as illustrated in equation 25. The structure of adduct **26**

$$(CH_3)_2NNHSO_2C_7H_7(p\text{-}) + Bu_3P \xrightarrow[\text{fast}]{CH_3OH}$$
$$\textbf{(25)}$$

$$\left\{ (CH_3)_2NNH\overset{+}{-}PBu_3 \longleftrightarrow (CH_3)_2N\overset{+}{N}H=PBu_3 \right\} \quad (25)$$
$$-O_2SC_7H_7(p\text{-})$$
$$\textbf{(26)}$$

was corroborated by its facile hydrolysis to 1,1-dimethylhydrazine in aqueous base (27). Zimmer has prepared similar hydrazinophosphonium salts from hydrazines and triphenylphosphine dibromide (53a).

As a final example of diazenium ion-trapping by nucleophiles, the oxidation of guanidine derivative **27** is of interest (14). Although the intermediacy of diazenium ion **28** has not been established, the cyclization step is probably addition of primary amino group to diazenium function in this species.

IV. CHEMISTRY OF DIAZENES

A. Tetrazene Formation

Oxidation of 1,1-disubstituted hydrazines is the best general method for preparing symmetrically tetrasubstituted tetrazenes, and yields are often high (53b). Numerous tetrazenes have also been obtained by decomposition of sulfonylhydrazine salts (generally in aprotic media) (23) and by the reaction of secondary amines with Angeli's salt (38). If the intermediate diazene is so constituted that fragmentation with loss of molecular nitrogen is facile, however, no tetrazene may result regardless of the method chosen. Certain 1-aryl-1-benzylhydrazines have been found to suffer rearrangement so readily during oxidation that again little or no tetrazene is produced. Both of these reaction types will be considered (Sections IV-C and V-B), but first the "normal" course deserves further comment.

Mechanisms for tetrazene formation remain in large part a matter for conjecture. Probably simple head-to-head coupling of diazenes occurs under certain circumstances, for example in the thermal decomposition of 1,1-dialkyl-2-sulfonylhydrazine salts in aprotic media. Tetrazene formation by neutralization of 1,1-dialkyldiazenium salt solutions almost certainly involves interception of the nucleophilic diazene by a diazenium ion, however, as discussed in Section III-A. Evidence presented below indicates that diazenes with electron-withdrawing substituents are actually quite electrophilic. Interception of such a diazene by a nucleophilic precursor (e.g., hydrazine or sulfonylhydrazine anion) leading ultimately to tetrazene, might well occur to the exclusion of diazene dimerization (eq. 26) (54). Such a pathway is much less appealing for dialkyldiazenes,

$$(C_6H_5)_2N-NH_2 \xrightarrow{[O]} \left[(C_6H_5)_2\overset{+}{N}=\overset{-}{N}\right] \xrightarrow{(C_6H_5)_2NNH_2}$$

$$\left[(C_6H_5)_2N-\overset{-}{N}-\overset{+}{\underset{H_2}{N}}-N(C_6H_5)_2\right] \longrightarrow \left[(C_6H_5)_2N-N-N-N(C_6H_5)_2\right] \xrightarrow{[O]}$$
$$ \underset{H \ \ H}{}$$

$$(C_6H_5)_2N-N=N-N(C_6H_5)_2 \quad (26)$$

however. In short, tetrazenes may be produced via any of a number of mechanisms, and the dominant pathway may be expected to depend upon the nature of the diazene, the diazene-forming reaction, and reaction conditions.

Forgione and collaborators made the interesting discovery that mercuric oxide oxidation of 3-amino-2-oxazolidinone gave stereospecifically the *cis* tetrazene, whereas other methods yielded the stabler *trans* form (55). Formation of the *cis* isomer was attributed to oxidation and coupling of amino groups coordinated to neighboring metal atoms in the mercuric oxide crystal lattice.

B. Attack on Unsaturated Systems

Certain diazenes are capable of adding to olefins to yield *N*-aminoaziridines. Dornfeldt (56a) found that lead tetraacetate oxidation of *N*-amino-carbazole (11) in cyclohexene gave in addition to much tetrazene a small amount ($\sim 5\%$) of the adduct 29. Structure determination included ring opening with aqueous perchloric acid to *trans*-hydrazinoalcohol (30), an authentic sample of which was prepared from the lithium salt of 11 and cyclohexene oxide.

Atkinson and Rees (56b) have obtained very much better yields of aminoaziridines by lead tetraacetate oxidation of 3-aminobenzoxazolinone

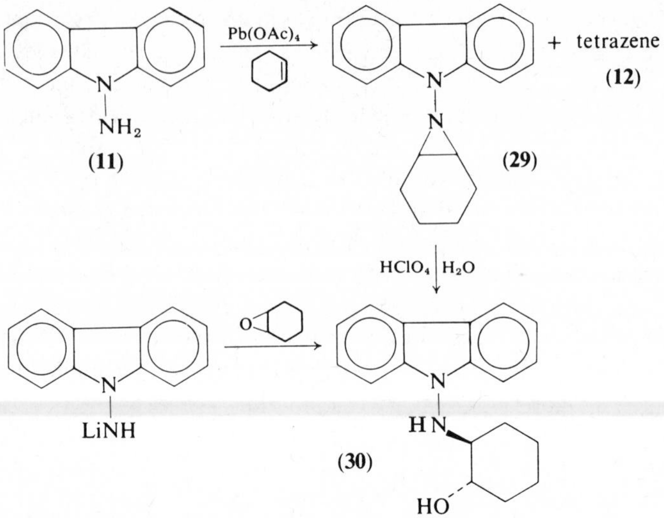

(12)

(11) (29)

(30)

31) in the presence of olefins and dienes.* Interestingly, addition to *cis*- and *trans*-2-butene is stereospecific within the limits of detection ($<5\%$) even at low olefin concentrations (eq. 27). Thus it is a good presumption

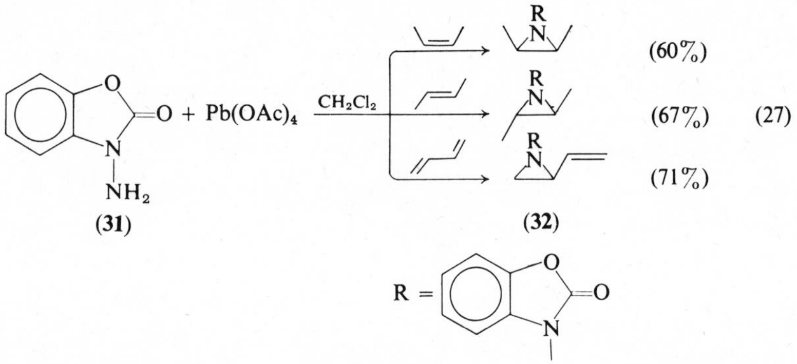

(60%)

(67%) (27)

(71%)

(31) (32)

that the diazene reacts as a singlet (57), and the results at low olefin concentration further suggest that the ground state is singlet. In contrast to the 1,4-additions of dimethyldiazenium ions to dienes noted earlier, this diazene gave exclusively 1,2-addition products with 1,3-butadiene (eq. 27), isoprene, 2,3-dimethyl-1,3-butadiene and 2,5-dimethylhexa-2,4-diene. Proof of structure for the butadiene adduct (**32**) included catalytic hydrogenolysis to 3-butylaminobenzoxazolinone (**33**), identical with an authentic sample, and acetolysis to yield **34**. 3-Pyrrolines, the products of net 1,4-addition of diazene to diene, were obtainable indirectly, for the

---

* Oxidation of the closely related *N*-aminoöxindole led only to rearrangement (Section V-B).

RNH(CH$_2$)$_3$CH$_3$ (33)

$$\overset{R}{\underset{(32)}{\triangle N}} \diagdown \diagup \quad \xrightarrow[HOAc]{H_2} \quad RNHCH_2\underset{\underset{R}{|}}{C}HCH=CH_2 \quad (34)$$

RNCH$_2$CHCH=CH$_2$ (34)
OAc

(35)

1,2-adducts were found to isomerize thermally to the corresponding pyrrolines (e.g., **32** → **35**) (56c). Much lower temperatures sufficed than those required for the analogous vinylcyclopropane–cyclopentene rearrangement.

Very recent work from Rees' group has expanded the range of both diazenes and olefins from which aziridines have been derived (49a). All of the diazenes to date are substituted with electron-withdrawing groups, and attempts to add dialkyldiazenes have so far met with failure (52). Strikingly, negatively-substituted olefins are very effective traps even for the diacyldiazene phthaloyldiazene (49a). The simplest interpretation of these facts is that the negative substituents on a diazene greatly increase its energy content and reactivity without depriving it completely of its nucleophilic character. Stated differently, the contribution from the dipolar form of a typical diazene is so overwhelming that it can be diminished sharply and still leave a firm imprint on the species' reactivity. Can one design a diazene with powerful enough electron-withdrawing groups that the ground state is triplet? Perhaps a pair of carbonyl substituents (**36**) will suffice (the phthaloyldiazene-olefin reaction need not have involved the ground state), but this question has yet to be answered. The first success in producing 1*H*-azirines by nitrene addition to acetylenes has just been reported by the same group at Leicester (56d). In harmony with their expected antiaromaticity, these substances suffer a remarkable spontaneous rearrangement to 2*H* isomers, the products actually isolated and characterized.

$$R_2NNH_2 \xrightarrow[EtC\equiv CEt]{Pb(OAc)_4} \left[R_2N-N\diagup\diagdown\overset{Et}{\underset{Et}{|}} \right] \longrightarrow N\diagup\diagdown\overset{\overset{NR_2}{|}\ Et}{\underset{Et}{}}$$

$$R_2N^- =$$

Finally, the capture of negatively-substituted diazenes by another unsaturated function, the sulfoxide group, is a new development (Section II-F), though familiar for years with other nitrenes (49). It remains to be determined whether or not sulfoximine derivatives can be prepared from the better-stabilized dialkyldiazenes.

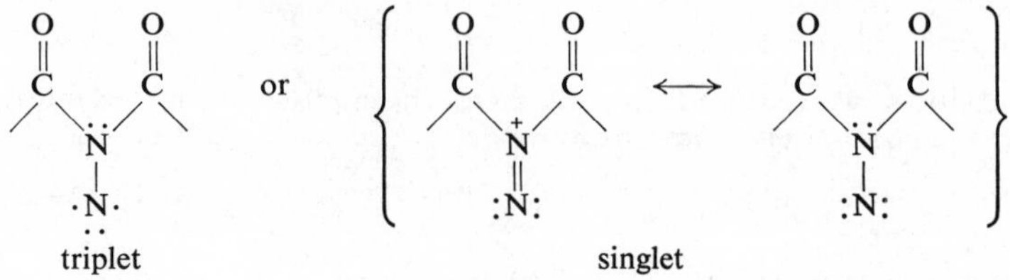

(36)

A reaction which might be interpreted in terms of intramolecular diazene attack upon an olefin was discovered by Bumgardner and Freeman (eq. 28) (31). Other pathways to pyrazoline (37) appear more likely,

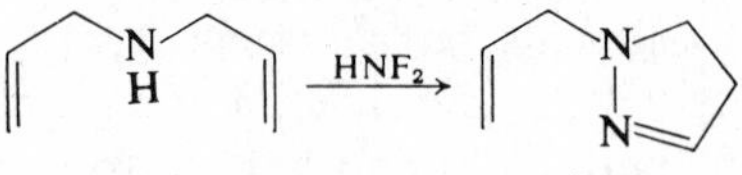

(28)

however; in particular, attack by the much-more-electrophilic diazenium ion (eq. 29), or rearrangement to hydrazone (38) (Section V-B) followed by cyclization (5) are viable alternatives.

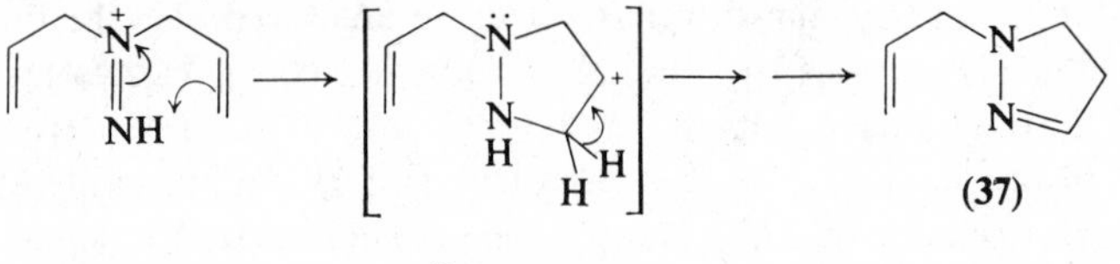

(29)

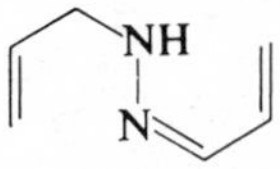

(38)

C. Fragmentation

This is the most useful, perhaps the most intriguing, and, in any event, the most studied reaction of diazenes. Although the use of high dilution and high temperatures may expand the range of diazenes which are capable of expelling molecular nitrogen, to date every diazene which has been reported to fragment in appreciable yield possesses one or more of these features: (*1*) substituents which can stabilize the fragmentation transition state, (2) a small ring structure, or (3) a concerted pathway for fragmentation. For convenience fragmentations will be discussed in terms of discrete diazene intermediates, but it should be borne in mind that in some cases carbon–nitrogen bond cleavage may occur during the act of diazene formation.

1. *Substituent-Assisted Fragmentation*

Since 1955 numerous papers have appeared dealing with diazenes bearing aryl or cyano groups on the α-carbons. These studies were an outgrowth of the early observation by Busch and Weiss (13) that mercuric oxide oxidation of 1,1-dibenzylhydrazine gave bibenzyl, and from Kenner and Knight's interpretation of this fragmentation in terms of a diazene intermediate (58). Mechanistic details of this disarmingly simple-looking reaction remain in doubt, but the considerable body of evidence related to fragmentations of this type is summarized here.

First of all, it is agreed that diazene intermediates are generated in these reactions, for bibenzyls are obtained when 1,1-dibenzylhydrazines are oxidized with any of a wide variety of oxidizing agents (e.g., mercuric oxide or acetate, potassium permanganate, bromine, *t*-butyl hypochlorite, Fehling's solution, *p*-benzoquinone, or air) or when any of the other methods in Sections II-B-E for preparing diazenes are employed (8,19,30, 38,59). By analogy to azo chemistry, it is reasonable to hypothesize that cleavage of the two carbon–nitrogen bonds of a diazene is concerted* except in highly unsymmetrical cases.† Since most of the examples to be discussed are rather symmetrical, only the two mechanistic alternatives presented in equations 30 and 31 will be considered here. The latter is actually orbital symmetry-forbidden, but strictly speaking only so if the transition state is symmetrical (see Section IV-C-2).

$$\left[\begin{array}{c}\text{ArCH}_2 \quad \text{CH}_2\text{Ar}' \\ \text{N} \\ \text{N} \end{array}\right] \xrightarrow{-\text{N}_2} [\text{ArCH}_2\cdot \quad \cdot\text{CH}_2\text{Ar}'] \longrightarrow \text{ArCH}_2\text{CH}_2\text{Ar}' \quad (30)$$

$$\left[\begin{array}{c}\text{ArCH}_2 \quad \text{CH}_2\text{Ar}' \\ \text{N} \\ \text{N} \end{array}\right] \longrightarrow \left[\begin{array}{c}\text{ArCH}_2\text{-----CH}_2\text{Ar}' \\ \text{N} \\ \text{N} \end{array}\right] \xrightarrow{-\text{N}_2} \text{ArCH}_2\text{CH}_2\text{Ar}' \quad (31)$$

* Two events which are concerted overlap in time, of course, but at no time need they be advanced to the same degree.

† Evidence for concerted fragmentation has been presented by S. Seltzer, *J. Am. Chem. Soc.*, **83**, 2625 (1961), in refs. 97b and 98b, and elsewhere. For stepwise loss of nitrogen, see W. A. Pryor and K. Smith, *ibid.*, **89**, 1741 (1967) and references contained therein.

In support of the latter representation, Hinman and Hamm (59a) discovered that oxidation of several 1-benzyl-1-(p-substituted benzyl)-hydrazines under widely varying conditions led to a single bibenzyl, the monosubstituted one, and that alkaline decomposition of corresponding sulfonylhydrazines produced the same result (eq. 32). Had free benzylic radicals been generated, one might have expected substantial quantities of the two symmetrical bibenzyls.

$$X-\langle\bigcirc\rangle-CH_2 \\ \quad\quad\quad\quad\quad NNH_2 \xrightarrow[-N_2]{[O]} X-\langle\bigcirc\rangle-CH_2CH_2-\langle\bigcirc\rangle \quad (32) \\ \langle\bigcirc\rangle-CH_2$$

Overberger's group (60) prepared the R,R optical isomer of N-nitrosobis(α-methylbenzyl)amine (39) and reduced it with alkaline sodium dithionite. A mixture (63%) of the three possible diastereomers of the bibenzyl resulted: $\frac{1}{2}l$(R, R), $\frac{1}{3}meso$(R, S), and $\frac{1}{6}d$(S, S) (eq. 33). Thus a great

$$\begin{array}{c} H_3C \quad H \quad H \quad CH_3 \\ \diagdown \quad \diagup \diagdown \quad \diagup \\ C \quad\quad C \quad\quad \xrightarrow[-OH]{Na_2S_2O_4} C_6H_5CH-CHC_6H_5 \quad (33) \\ \diagup \diagdown \quad N \quad \diagup \diagdown \\ C_6H_5 \quad\quad\quad C_6H_5 \quad\quad CH_3 \\ | \\ NO \\ (39) \quad\quad\quad\quad (d, l, meso) \end{array}$$

deal of inversion occurs prior to formation of the new carbon–carbon bond, belying equation 31. Mercuric oxide oxidation of the hydrazine corresponding to 39 gave a similar mixture of bibenzyls (76%) but one containing even more of the *meso* and d isomers.

The possibility that configuration is lost via reversible deprotonation at the α-positions in the intermediate diazene was considered by the authors. The appropriate experiment, viz., generation of the intermediate in a deuterated medium and search for deuterium in the bibenzyl, has not been performed, but somewhat related deuterium exchange experiments of Underbrink (52) hint at the outcome. Working in aqueous and methanolic base, he found tautomerization of diethyldiazene to the corresponding azomethinimine to be irreversible; the same may be expected for bis(α-methylbenzyl)diazene. If so, α-deprotonation cannot be a preliminary to bibenzyl formation.

Another way of bringing the Hinman-Hamm and Overberger results into harmony was to postulate that the α-methyls present in the latter's experiments brought about a change in mechanism (eq. 31 → eq. 30), a suggestion defensible on both steric and electronic grounds. Accordingly, Overberger and Marullo (61) synthesized the *m*-amino analog of **39**, viz., **40**, and decomposed it with sodium dithionite in base. Only the mixed bibenzyl (72%) was obtained, however, a result exactly like those of Hinman and Hamm.

(40)

The hypothesis that free radicals are generated seems to account best for the presently available facts underlying this dilemma. Crossover experiments make a convincing case against free radicals only if it be assumed that those which escape cage recombination are successful at finding one another in the reaction medium. It is by no means clear that escaped benzylic radicals could elude attack by oxygen, mercuric oxide, sodium dithionite, etc., depending upon the choice of reaction conditions; indeed, yields of bibenzyls are often low.

Somewhat embarrassing to the cage-recombination scheme is the fact that high yields of bibenzyls (without crossing over) have been secured in a few instances, most impressively in the aqueous alkaline decomposition of 1-benzyl-1-*p*-methoxybenzyl-2-benzenesulfonylhydrazine to *p*-methoxybibenzyl, 93% (59a). Cage recombination of benzylic radicals generated from azo compounds is not very efficient, as shown by Nelsen and Bartlett (62) for azocumene (27% in toluene at 40°) and by Greene (63) for α,α'-diphenylazoethane (32% in benzene at 105°). These decompositions are imperfect models for sulfonylhydrazine salt fragmentations in at least two respects, however. First, the geminate radicals from a (*trans*) azo compound are generated relatively far apart and insulated from one another by a nitrogen molecule, in contrast to the nearest-neighbor radical pair which a diazene would yield. Recent work by Koenig and Deinzer (64) on the cage recombination of *t*-butoxy radicals suggests that cage recombination efficiencies should reflect this difference dramatically. For sulfonylhydrazine decompositions conducted in aqueous base a second factor should be considered. Large geminate hydrocarbon radicals might escape from one another less readily in water than in the nonpolar media required to dissolve the corresponding azo compounds. In brief, the notion that bibenzyls are formed from diazenes via coupling of free radicals is at least defensible as a working hypothesis, and it will prove useful in explaining the chemistry of ring diazenes described below.

Cyclic counterparts of the above dibenzylnitrosamines and hydrazines were also investigated by Overberger and his co-workers. The *cis* and *trans* isomers of *N*-amino-2,6-diphenylpiperidine (**41**) were synthesized, then subjected to mercuric oxide treatment to give 1,2-diphenylcyclo-pentane (**42**) and 1,5-diphenyl-1-pentene (**43**) (47a). From *cis* **41**, *cis* **42**

was obtained (65%) together with 25% of **43**; *trans* **41** yielded primarily *trans* **42** (59%) plus *cis* **42** (12%) and **43** (14%). Very similar results were obtained in the alkaline dithionite reduction of the *cis* and *trans* nitros-amines corresponding to **41**.

Although this view is not favored by the authors (9), the present reviewer chooses to interpret these data in terms of benzylic biradical formation followed by coupling to give cyclopentane **42** and disproportionation to give olefin **43**. That only the *cis* isomer of **42** was detected among the products from *cis* **41** argues against this simple picture, but not con-vincingly, for the alternative of a fully concerted pathway to cyclopentane (*1*) fails to explain the incomplete stereospecificity observed in the degra-dation of *trans* **41** and (*2*) requires that some independent route to olefin exist. It is known that ring closure of 1,4-biradicals is typically only an order of magnitude slower than rotation about single bonds with conse-quent loss of configuration (65). The energetically preferable ring closure of 1,5-biradicals generated in a favorable conformation for cyclization may well compete effectively with loss of configuration. Comparison of the products from *cis* and *trans* **41** with those from thermal decomposition of the azo compounds *cis* and *trans* **44** should be most informative, but at present only a fragmentary comparison is possible. A single isomer of **44**,

configuration unknown, has been synthesized and pyrolyzed to yield a mixture (65%) of *cis* and *trans* **42**, composition unreported, plus some **43** (66). Incidentally, ionic cleavage of the diazenes derived from **41** was seriously considered, but further experiments by Overberger and Herin

(67), supported by the fact that no products incorporating solvent were found in the degradation of *cis* and *trans* **41**, excluded this possibility.

Oxidation by bromine in ethanol of a single stereoisomer each of hydrazines **45** and **46** gave much olefin accompanied by a mixture of both 1,2-dimethyl-1,2-dicyanocyclopentanes from **45** and a single cyclopentane from **46** (68). Again the results are in accord with biradical intermediates,

(**45**) R = H
(**46**) R = CH$_3$

particularly in light of thermolysis results for a related azo compound. Decomposition of **47** (69), which certainly proceeds via a biradical, yields 2-cyanopropene (**48**) plus a single cyclobutane, presumably *trans* (**49**).* Oxidation of the 2,6-dicyano- and 2,6-dicarboxamido derivatives of

(**47**)　　　　(**48**)　　　(**49**)
　　　　　　　~30%　　　~50%

1-aminopiperidine with mercuric oxide yielded fragmentation products (70), but similar treatment of the 2,6-dimethyl compound gave only the tetrazene (71).

Overberger, Valentine, and Anselme (47c) have found that *cis*- and *trans*-1-amino-2,5-diphenylpyrrolidine (**50**) give good yields of 1,2-diphenylcyclobutanes (**51**) accompanied by some styrene when treated with mercuric oxide. The *trans*-amine yields only *trans*-hydrocarbon (85%), but

* An important study of the thermal, photolytic, and photosensitized decomposition of *meso*- and *dl*-i. has recently been reported (P. D. Bartlett and N. A. Porter, *J. Am. Chem. Soc.*, **90**, 5317 (1968)). Both thermal and direct photochemical decomposition are highly stereoselective processes, like thermolysis of **47**.

(**i.**)

the *cis*-amine gives *cis* (66%) and *trans* (6.5%). The authors believe that a fully concerted process (eq. 31) is responsible for the high degree of stereoretention; once again this reviewer feels that a 1,4-biradical can account for all of the results. Similar results were found in the alkaline dithionite reduction of the related nitrosamines, though isomerization of starting material under the reaction conditions obscured the stereochemical outcome.

$$+ C_6H_5CH{=}CH_2$$

(50)   **(51)**

An *N*-aminopyrrolidine derivative unadorned by phenyl groups fragments to ethylene, but no cyclobutane, and there is strong evidence that this process is concerted (Section IV-C). It is difficult indeed to explain why a bond between α-carbons should be forming in the transition state for nitrogen loss from **50** but not from its unsubstituted counterpart; on the other hand, it makes sense that α-phenyl substituents permit **50** to yield a biradical, a luxury which the naked parent system can ill afford. Cohen and co-workers prepared and pyrolyzed the azo compound **52** having the carbon skeleton of **50**; it yielded the isomeric hydrazone, nitrogen, and some styrene, but much of the hydrocarbon product remains unidentified (72). Presumably the same 1,4-biradical is involved here.

(52)

The examples of fragmenting diazenes considered so far have all carried radical-stabilizing substituents on both α-carbons. Carpino (48b) found, and others (22,54) confirmed, that a single phenyl would suffice, as illustrated in equation 34 (54). Dropwise addition of the hydrazine

$$+ CH_2{=}CH(CH_2)_3C_6H_5$$
$$(\sim 26\%)$$

$$+ CH_3(CH_2)_2CH{=}CHC_6H_5 + N_2 \qquad (34)$$
$$(\sim 15\%)$$

produced the results shown, but rapid addition led to tetrazene (58%) accompanied by only 17% of hydrocarbons and $\sim 32\%$ nitrogen. Hence the lifetime of the diazene is long enough for a bimolecular process to be made to compete effectively with the unimolecular fragmentation. Interestingly, dibenzylmercury was the major product from mercuric oxide degradation of hydrazine **53** (48b).

$$C_6H_5CH_2NC(CH_3)_3 \xrightarrow[CH_2Cl_2]{HgO} (C_6H_5CH_2)_2Hg + C_6H_5C(CH_3)_3$$

$$\underset{NH_2}{|}$$

(63)   (47% crude)   (24%)

A special case of "substituent assisted" fragmentation is particularly useful. Reactions which can produce diazenes of the type **54** yield fragmentation products because of the driving force of generating *two* molecules of nitrogen. When the precursor is a 1,2,3-triazole or triazoline (**55**),

(54)   **(55)**

the availability of a fully concerted pathway leading to three stable molecules undoubtedly contributes to the ease of ring splitting. An acetylene synthesis based on this idea will be discussed with related processes in Section IV-C-3. The organic product need not be a low-energy species, however. Campbell and Rees discovered that oxidation of 1-aminobenztriazole (**56**) with lead tetraacetate generates benzyne almost quantitatively (73–75). In the absence of a trapping agent, the major product

is biphenylene (83%); with tetracyclone present, 1,2,3,4-tetraphenylnaphthalene is obtained in 95% yield. The combination of low temperature and good yields recommend this procedure for wider usage than it has enjoyed to date. The approach to arynes via diazenes is a powerful one, as illustrated by generation of the extremely reactive 1,8-dehydronaphthalene (*peri*-naphthalyne, **58**) (76–78a). Oxidation of aminotriazine **57** in benzene gave fluoranthene, in the 1,2-dichloroethylenes gave *cis*- and *trans*-dichloroacenaphthenes with virtually complete stereospecificity, and in the hydrogen donor solvent cyclohexane gave naphthalene (76). It was concluded that the reacting species is singlet, but the possibility that its ground state is triplet was not ruled out rigorously.

Benzocyclopropenone (**79a**) and benzynequinone (**79b**) have now been generated from precursors analogous to **56** and **57**.

2. Small Ring Fragmentation

Bumgardner, Martin, and Freeman (30) made the remarkable observation that both aziridine and azetidine suffer deamination when treated with difluoramine, yielding ethylene (80%), and cyclopropane (40%), respectively (equations 35 and 36). The view that diazenes are intermediates in these reactions gains support from the fact that alkaline

$$\text{HNF}_2 \longrightarrow CH_2{=}CH_2 + N_2 \qquad (35)$$

$$\text{HNF}_2 \longrightarrow \triangle + N_2 \qquad (36)$$

dithionite reduction of N-nitrosoazetidine also gave cyclopropane. The corresponding experiment with N-nitrosoaziridine was not feasible, as this substance fragments to nitrous oxide and ethylene below room temperature (80).

Eschenmoser and his co-workers (81) have developed a most ingenious as well as synthetically useful reaction related, formally at least, to decomposition of the 3-ring diazene (ethylenediazene (59)). Condensation of an

(59)

N-aminoaziridine, e.g., the 2-phenyl derivative, with an α,β-epoxyketone yields the hydrazone; when heated or photolyzed this suffers a spectacular fragmentation which generates a new carbonyl, an acetylenic linkage, an olefin, and nitrogen (eq. 37). Although the ethylenediazene is surely not a

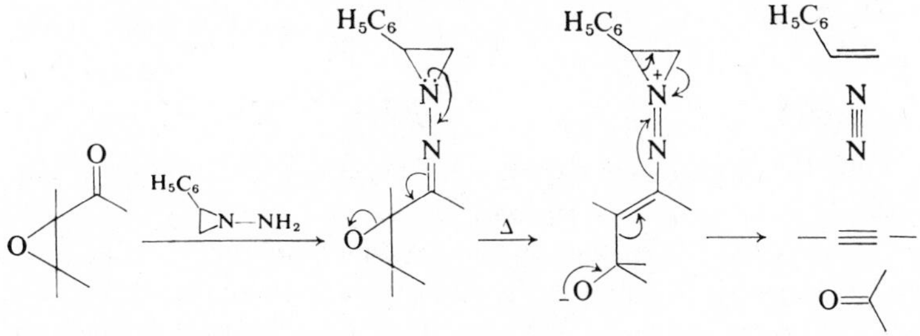

(37)

discrete intermediate in this transformation, it may form and collapse simultaneously as suggested in equation 37. Alternatively, the hydrazone may fragment into olefin plus diazo compound, which subsequently disintegrates. This route to acetylenic ketones was an outgrowth of the Swiss workers' discovery that α,β-epoxyketones are similarly cleaved by tosylhydrazine (82–85).

Freeman and Graham recently demonstrated that the deamination with difluoramine of cis- and trans-2,3-dimethylaziridine proceeds stereospecifically ($\geq 96\%$ retention of configuration), as shown in equations 38 and 39 (86). This result takes on unusual interest in light of Hoffmann's

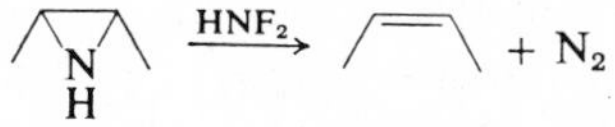

(38)

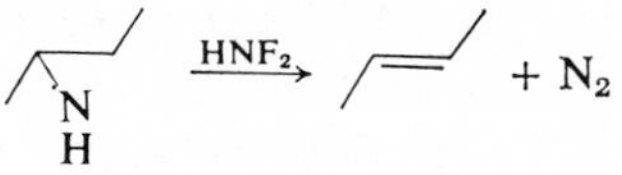

(39)

prediction on orbital symmetry grounds that fragmentation of ethylene-diazene (**59**) should be a nonconcerted process (87). A closely related problem exists in the fragmentation of episulfones, e.g., **60**, to olefin plus sulfur dioxide, where again stepwise bond cleavage was predicted yet products are formed stereospecifically (88). A pronounced solvent effect

on thermolysis rates for episulfones creates serious doubt that this reaction is concerted (89), however, so the initial prediction may yet prove to be correct. The fragmentation of *N*-nitrosoaziridines alluded to above is also a stereospecific transformation (90), which would be anticipated on orbital symmetry grounds for pyramidal hybridization at the ring nitrogen (**61**), but not for trigonal (**62**). In general, nitrosamines are trigonally hybridized

(91), but the 3-ring geometry opposes this hybridization so strongly that nitrosoaziridines may be exceptions. Pyramidal geometry for 3-ring diazenes is much less likely than for the corresponding nitrosamines.

Fragmentation of 3-membered heterocycles is intimately related to 3-ring formation by carbene or nitrene addition to olefins, and the same mechanistic questions are very much alive in this field as well. A large body of experimental evidence supports the "Skell criterion," viz., that stereospecific addition to a double bond is characteristic of a singlet carbene (or nitrene), and nonstereospecific addition is characteristic of triplets (57). Addition of a singlet carbene via the expected symmetrical transition state **63** is forbidden on orbital symmetry grounds, however. Extended Hückel

calculations by Hoffmann (92) for the methylene–ethylene reaction indicate that the transition state is unsymmetrical (**64**), but that the addition is nonetheless concerted. The situation is very likely analogous for fragmentation of 3-ring diazenes, nitrosoaziridines, and sulfones (93b), but a firm conclusion would be premature.

The decomposition of Δ^1-1,2-diazetine (**65**) and its derivatives assumes special interest because **65** is isomeric with ethylenediazene (**59**). Nothing

is known about the fragmentation stereochemistry of Δ^1-1,2-diazetine for the only known representatives of the ring system (94, 95) lack stereochemical "labels." It is significant, though, that **66–68** are quite stable thermally despite ideal geometry and large negative ΔH values for fragmentation to olefin and nitrogen (95). Their stability is understandable only in orbital symmetry terms: concerted fragmentation is forbidden (93).

The stereochemistry of azetidine deamination by difluoramine, currently under investigation (96), will provide a most intriguing comparison with the thermolysis stereochemistry of 1-pyrazolines (azo isomers of 4-ring diazenes, cf. **69** and **70**). Crawford and his co-workers (97,98) have

adduced powerful evidence that 1-pyrazolines lose nitrogen to give singlet trimethylene with the geometry shown in **71**. This species is further

believed to possess an antisymmetric electronic state as indicated in **72**, not only on the basis of Hoffmann's extended Hückel calculations* but also because *conrotatory* ring closure of the intermediate is required to explain the strikingly novel stereoselectivities (eq. 40) found by the groups of Crawford and McGreer (99) (disrotatory cyclization would be expected for the symmetric counterpart of **72**). Moreover, **72** can be formed from 1-pyrazoline with preservation of orbital symmetry by way of transition state **73** (92).

---

* That this species is calculated to be the *ground state* of trimethylene at large CCC angles is at first surprising; the antisymmetric nonbonding MO lies below the symmetric one because the latter is destabilized by interaction with the C—H σ-orbitals of the central methylene group (92).

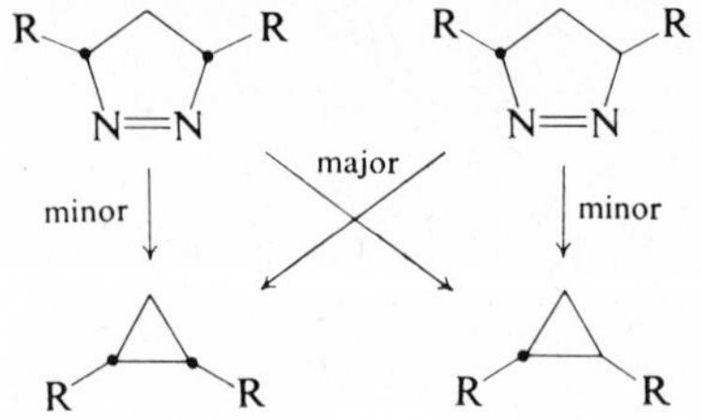

$$(40)$$

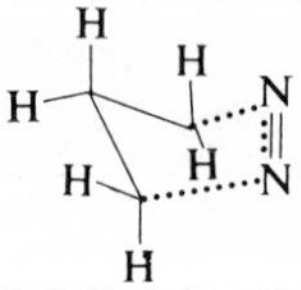

(73) 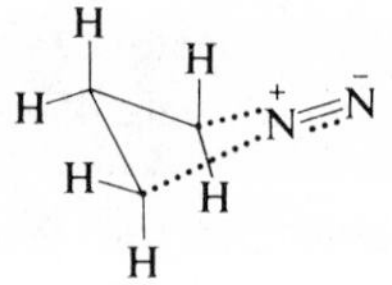   (74)

The corresponding transition state **74** from diazene **69** suffers from two disadvantages relative to **73**, namely, (*1*) a more severe H—H repulsion exists on the "underside," and (*2*) the energy of the antisymmetric "nonbonding" orbital of the product trimethylene is calculated to be higher for smaller values of the CCC bond angle. On the other hand, the planar, unsymmetrical transition state **75** should lead *concertedly* to cyclopropane and nitrogen, for the incipient trimethylene is generated on a portion of its potential energy surface (assuming Hoffmann's calculations to be correct) from which ring closure is all downhill. Expulsion of

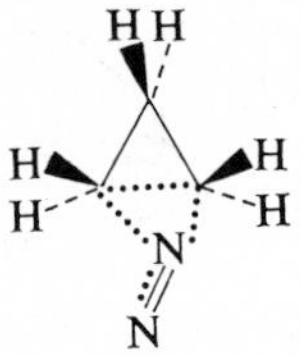

(75) 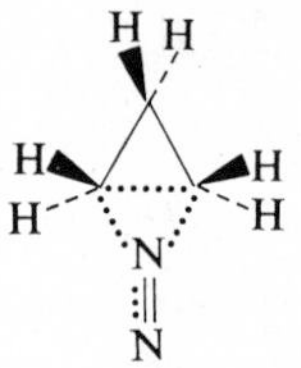   (76)

nitrogen "with a twist" as shown in **75** is postulated because the transition state which is both planar and symmetrical **76** can yield ground state fragments only by non-conservation of orbital symmetry. If diazenes derived from azetidines do indeed fragment concertedly via transition states of the type **75**, cyclopropanes must be formed stereospecifically in the sense *opposite* to that characteristic of pyrazoline thermolyses. As an example, the isomers **77** and **78**, which belong to the same point group, would both fragment to nitrogen and 1,2-dimethylcyclopropane, but **77** would yield the *cis* isomer whereas **78** is known to give primarily *trans*

(97b)! At present the dichotomy of mechanism is mere speculation,*† but this contrast is perhaps significant: olefins accompany cyclopropanes in the products of pyrazoline decomposition, yet no propene is found in the deamination of azetidine (30).

It is now apparent that classification of aziridine and azetidine deamination in a category distinct from "Concerted fragmentation···" may be misleading; this seems appropriate nevertheless until further information regarding mechanism is brought to light. Internal strain is undoubtedly an important cause of fragmentation of small-ring diazenes, but the availability of concerted decomposition pathways may be at least as significant.

3. Concerted Fragmentation to Stable Products

When sulfonylhydrazine salt **79** is heated in diglyme, vigorous gas evolution occurs and yields of nitrogen and ethylene up to about 80% are obtained (23). The absence of cyclobutane in the product is evidence that

fragmentation of tetramethylenediazene (**80**) is concerted, i.e., that it does not proceed via a 1,4-biradical intermediate. Confirming this conclusion, the homologous salt **81** decomposes without nitrogen loss; no low energy,

* Recent work on the thermal isomerization of cyclopropanes is germane to the arguments developed here (J. A. Berson and J. M. Balquist, *J. Am. Chem. Soc.*, **90**, 7343 (1968); W. L. Carter and R. G. Bergman, *ibid.*, **90**, 7344 (1968); R. J. Crawford and T. R. Lynch, *Can. J. Chem.*, **46**, 1457 (1968)).

† Note Added in Proof. The decomposition of S,S-dialkylthietanes displays in even greater degree the crossed-over stereoselectivities found by Crawford and McGreer (B. M. Trost, W. L. Schinski and I. B. Mantz, *J. Am. Chem. Soc.*, **91**, 4320 (1969)).

concerted pathway is available to pentamethylenediazene. In the orbital correlation diagram for fragmentation of **80** all occupied orbitals of the products derive from occupied orbitals of the diazene, so the synchronous pathway bears the blessing of Woodward and Hoffmann (93).

The pyrrole derivative **82** provides an amusing contrast with the labile **79**. Instead of yielding acetylene and nitrogen, this salt survived (86% recovery) an hour at 225° and decomposed only slowly at 275° in tetraglyme (23). Pyrrole ($\sim 10\%$) was identified in the complex product mix-

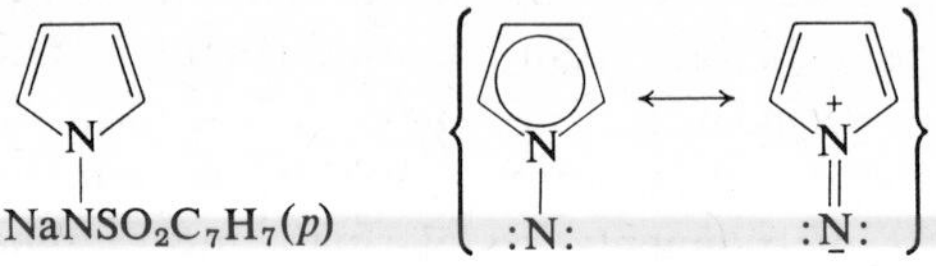

NaNSO₂C₇H₇(p)

(82)   (83)

ture; its formation may well have paralleled that of carbazole from the sulfonylhydrazine salt **10**. In analogy to the carbazole series, the reluctance of **82** to lose sodium sulfinate is traceable to the aromaticity of the pyrrole nucleus: delocalization of the pyrrole nitrogen's lone pair electrons makes them less available to the terminal nitrogen of diazene **83**. Carpino has made similar observations on phenylated aminopyrrole derivatives (100).

To return to systems which do fragment, pyrazoline **84** yields nitrogen, acetonitrile, and isobutylene when treated with difluoramine (30). Willey

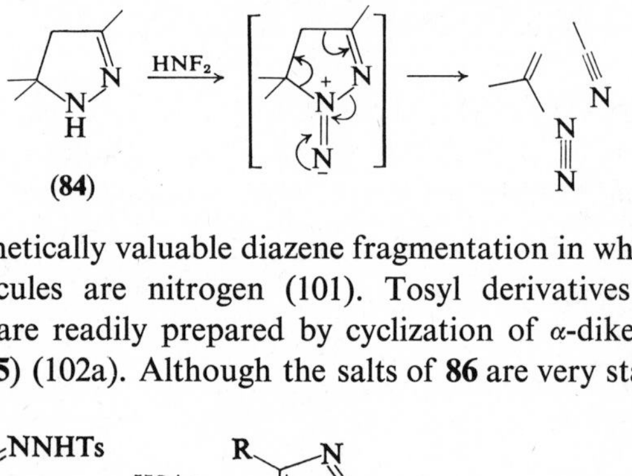

(84)

devised a synthetically valuable diazene fragmentation in which two of the product molecules are nitrogen (101). Tosyl derivatives of 1-amino-triazoles (**86**) are readily prepared by cyclization of α-diketone bistosylhydrazones (**85**) (102a). Although the salts of **86** are very stable thermally

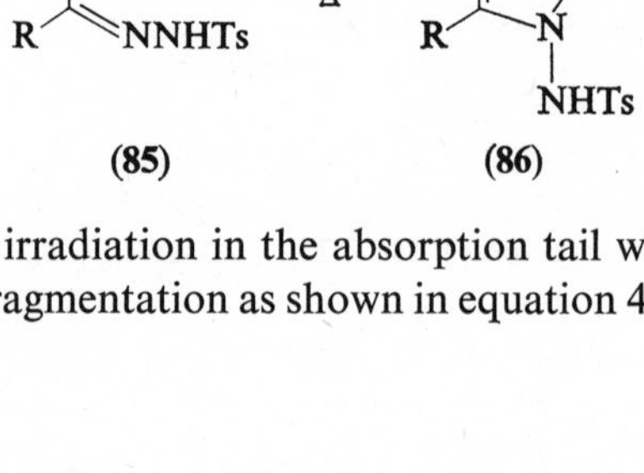

$$\xrightarrow[\Delta]{\text{HOAc}}$$

(—Ts = —SO₂C₇H₇(p))

(85)   (86)

(cf. **82**), irradiation in the absorption tail which extends to about 300 mμ effects fragmentation as shown in equation 41. One may imagine formation

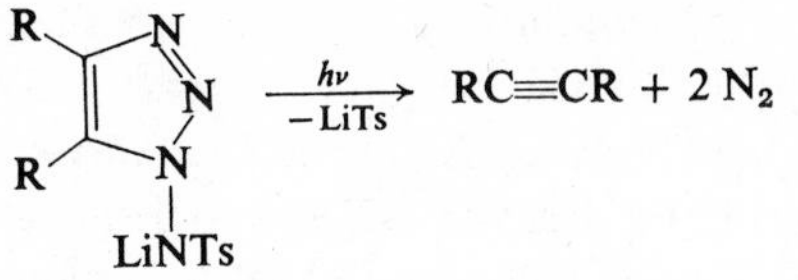

$$\text{(structure)} \xrightarrow[-\text{LiTs}]{h\nu} RC{\equiv}CR + 2 N_2 \qquad (41)$$

of a bisdiazo compound (78c) from the diazene (eq. 42) in analogy to bishydrazone oxidations (102a), but synchronous fragmentation of the diazene into three pieces is orbital symmetry-allowed. The power of this

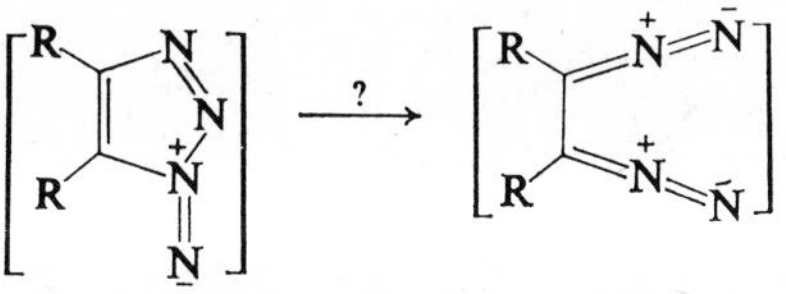

$$\left[\text{structure}\right] \xrightarrow{?} \left[\text{structure}\right] \qquad (42)$$

general acetylene synthesis is revealed in equation 43: it is an efficient method for preparing such highly strained representatives as cyclohexyne (101) (cf. the related aryne synthesis, Section IV-C-1). The aromatic products (87) from trapping the cycloalkynes with tetracyclone are formed in 54, 56, and 77% yield for $n = 4, 5, 6$, respectively. In the $n = 6$

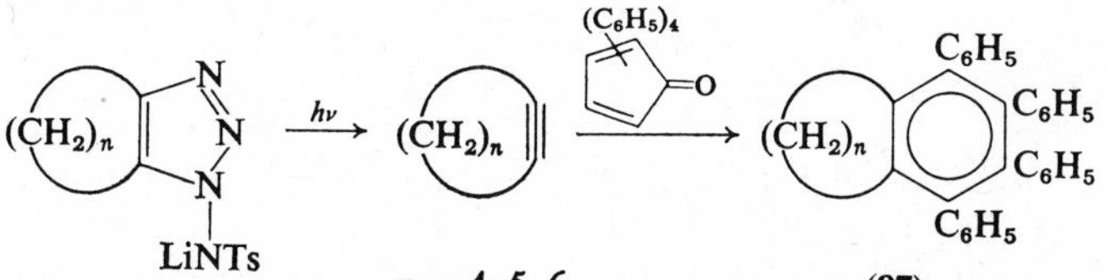

$$(\text{CH}_2)_n \text{(structure)} \xrightarrow{h\nu} (\text{CH}_2)_n \xrightarrow{(\text{C}_6\text{H}_5)_4} (\text{CH}_2)_n \text{(structure)} \qquad (43)$$

$$n = 4, 5, 6 \qquad (87)$$

case, the tetracyclone was added after irradiation was stopped; this was successful even for $n = 5$ (not for $n = 4$), but the yield was only 26% instead of the 56% obtained with trapping agent present during irradiation. Oxidation of 1-aminotriazoles is another practical route to cycloalkynes (78c, 102b).

Campbell and Rees discovered that lead tetraacetate oxidation of 2-aminobenztriazole (88) also leads to fragmentation, with the formation of cis, cis-mucononitrile (90) (74). An intermediate dinitrene was postulated, but the fragmentation is depicted here as concerted (89) on orbital

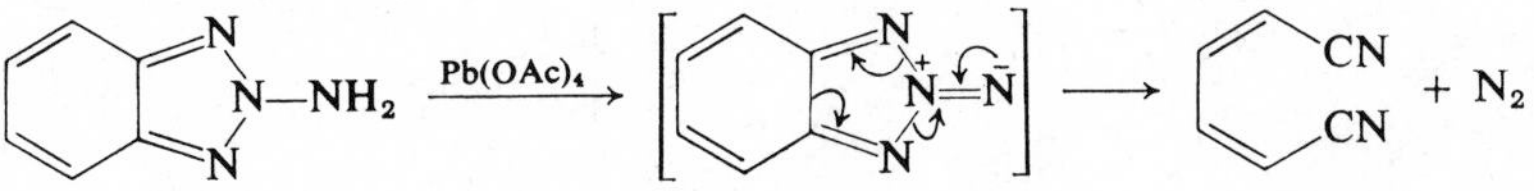

$$\text{(88)} \xrightarrow{\text{Pb(OAc)}_4} \left[\text{(89)}\right] \longrightarrow \text{(90)} + N_2$$

(88)   (89)   (90)
                                               64%

symmetry grounds. Mucononitrile is also obtained by pyrolysis of
o-diazidobenzene, probably via the same diazene (103).

Stereochemistry has been clear-cut in the various 5-membered-ring
fragmentations considered thus far, but stereochemical choice comes into
play when the C—N bonds to be broken lie in a plane perpendicular (or
roughly so) to the π-system being generated. Such a case is 3-pyrroline.
McGregor discovered that this substance is degraded readily to butadiene
and nitrogen when heated in aqueous solution with (monoprotonated)
Angeli's salt (eq. 44) (104). Two stereochemical possibilities exist, **91** and

$$\text{(eq. 44)}$$

92, if one assumes synchronous collapse of an intermediate diazene.
Orbital symmetry considerations lead to the conclusion that the disrotatory pathway **(91)** is allowed, the conrotatory **(92)** forbidden. The

matter was settled experimentally by synthesis and degradation with
Angeli's salt of *cis*- and *trans*-2,5-dimethyl-3-pyrroline (**93** and **94**). The
cis-pyrroline yielded only a single 2,4-hexadiene, the *trans,trans* isomer,
and the *trans*-pyrroline gave only the *cis,trans*-diene. Therefore the fragmentation is indeed disrotatory. In principle, **93** could yield *cis,cis*- as

well as *trans,trans*-2,4-hexadiene by way of a disrotatory process **(95)**, but the serious methyl–methyl repulsion which would develop along this pathway precludes its competing with the alternative course. McGregor found identical stereochemistry in the thermal fragmentation of *cis*- and *trans*-2,5-dimethyl-3-sulfolene **(96)** to diene plus sulfur dioxide (105).

(95)   **(96)**

Fragmentation of benzo derivatives of 3-pyrrolines, i.e., isoindolines, has been investigated by Baker's group (106) and by Carpino (107). Sulfonylhydrazine **97** was degraded by aqueous alkali, presumably via *o*-xylylene **(98)**, to a mixture of *o*-xylylene dimer **(99)**, benzocyclobutene **(100)**, *s*-dibenzocyclooctadiene **(101)** and a little *o*-xylene. The labile

(97)   **(98)**

(99)   **(100)**   **(101)**

dimer **99** was isolated as its dibromide **(102)** (108) in a closely related experiment, mercuric oxide oxidation of *N*-aminoisoindoline (107). The stable products **100**, **101**, and *o*-xylene were also obtained in the pyrolysis of molten sulfone **103** at 280–300° (109).

(102)   **(103)**

Diazene-generating experiments with 1,3-diphenylisoindoline derivatives (106,107) gave rise to some stereochemical confusion which has recently been resolved by Carpino (110). He found that *trans*-hydrazine hydrochloride **104** gave *cis*-1,2-diphenylbenzocyclobutene (**105**) (27%) when oxidized with manganese dioxide, whereas the *cis*-hydrazine (**106**) was transformed into the *trans*-benzocyclobutene (**107**) (81%) by mercuric

(104)   (105)

(106)   (107)

oxide. Gentle alkaline degradation of the tosyl derivative of **106** also occurs stereospecifically, but the tosyl derivative of the *trans*-hydrazine gives a 3:1 mixture of **105** and **107**. The proportion of **107** increases with the vigor of the reaction conditions, and it is believed to arise by isomerization of the labile *cis* isomer (**105**) (or less likely via isomerization of the *o*-xylylene precursor of **105**).

Like the stereochemical results with simple pyrrolines, these data are neatly in accord with the theory of concerted reactions (93,111). Collapse of *trans*-diazene **108** in the predicted disrotatory fashion would give the reactive *o*-quinodimethane (**109**); cyclization of the latter, essentially an 8 π-electron system, should occur in the conrotatory sense to give the observed hydrocarbon (**105**). Similarly, the *cis*-diazene should, according

(108)   (109)   (105)

to theory, yield **106** via σ-quinodimethane **110**. The groups of Huisgen (112) and Quinkert (113) have studied the stereochemistry of ring opening of **105** and **107**, which requires only gentle heating, by intercepting the **109** and **110** with dienophiles. Strictly conrotatory opening is observed.

(110)   (111)   (112)

Incidentally, pyrolysis of the *cis* and *trans* sulfones (111) requires elevated temperatures and yields not benzocyclobutenes but 9-phenyl-9,10-dihydroanthracene (112) (114). The latter is also formed when benzocyclobutenes 105 and 107 are pyrolyzed.

Brief warming with aqueous alkali transformed tosylhydrazine 113 nearly quantitatively into 9, 10-dihydrophenanthrene (114) (19,20). The di-

$$\xrightarrow{\text{base}}$$

(114)

NHTs

95%

(113)

hydroazepine derivative 113 is a benzolog of the isoindoline system just discussed; but the stereochemistry predicted for the 113 → 114 transformation is the same; i.e., *cis* substituents at the α-carbons of 113 should lead to a *trans*-substituted dihydrophenanthrene and *trans* groups on 113 should give a *cis*-substituted product. The individual steps in the sequence from diazene to dihydrophenanthrene should occur in the *opposite* stereochemical sense, however, as illustrated in equation 45*. With regard to orbital symmetry considerations benzo and vinyl groups are interchangeable; hence stereochemical predictions for both monocyclic unsaturated diazenes and their benzo derivatives can be summarized in terms of 115:

* Note Added in Proof. Thermal decomposition of the azo isomer of this *trans*-diazene (R=CH₃) has been found to yield little dihydrophenanthrene, and that which is formed has the "wrong" (*trans*) stereochemistry (D. M. Lemal and G. Guhn, unpublished results)! Models indicate that the intermediate quinodimethane is a rigid, lock-washer shaped species which cannot readily assume a conformation suitable for the allowed disrotatory cyclization. Hence another symmetry allowed process, an antarafacial [1,7] sigmatropic hydrogen shift, dominates even though the molecular geometry is unfavorable for this event as well. The resulting 2-ethyl-2'-vinylbiphenyl constitutes ∼90% of the product; the remainder is *trans*-9,10-dimethyl-9,10-dihydrophenanthrene, arising via symmetry forbidden but geometrically ideal conrotatory cyclization.

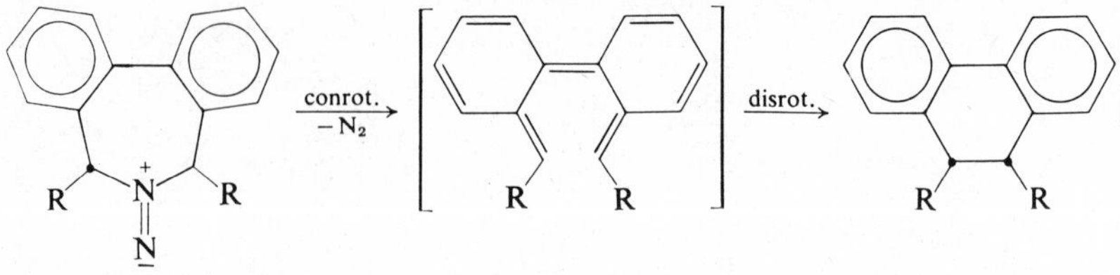

$$(45)$$

ring opening should be conrotatory for n even, disrotatory for n odd. Although an attempt to test the predictions for $n = 2$ is underway (115), the experimental answer is not yet forthcoming.

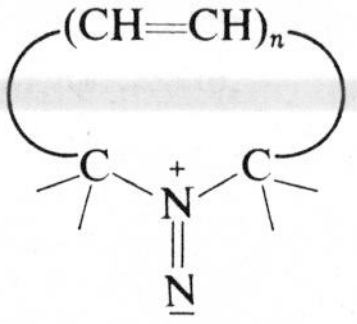

(115)

In contrast to **113**, tosylhydrazine **116** (R = Ts) is degraded to a nitrogen-containing compound, not acenaphthene, by hot aqueous base (116). Similarly, oxidation of **116** (R = H) with several different reagents gives nothing tractable or low yields of a substance presumed to be the tetrazene. To be sure, no quinoidal structure can be written for the frag-

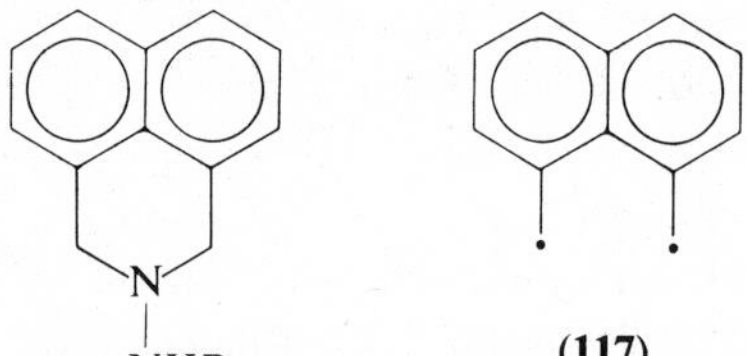

(116) R = H, Ts

(117)

mentation intermediate **117**, but it will be recalled that even dibenzyldiazene readily decomposes. Formation of biradical **117** from the diazene is not orbital symmetry forbidden, so the most likely explanation is simply steric inhibition of resonance in **117** (116). Nonbonded repulsion between methylene hydrogens in the *peri* biradical prevents the molecule from achieving a planar or even near-planar conformation. Acenaphthene (59%) can be obtained by neat pyrolysis of the azo compound **118**, but the latter loses nitrogen less readily than the labile **119**, in harmony with the behavior of the isomeric diazenes.

(118)   (119)

Another point regarding the alkaline degradation of **116** (R = Ts) is noteworthy. The rate of destruction of this sulfonylhydrazine is markedly lower than for **113** or **97** under the same conditions. This observation is difficult to explain if discrete diazene intermediates are generated in each case; it strongly suggests that the diazenes from **113** and **97** are fragmenting as they form.

Consistent with **116**, hydrazine **120** yields no pyracene when treated with activated manganese dioxide (117). Oxidation of **121** does give 1,2-dihydropyracyclenes (low yields), presumably because the quinoidal intermediate **122** is considerably lower in energy than a *peri* biradical even though both would undoubtedly be twisted.

(120)   (121)   (122)

V. CHEMISTRY OF DIAZENE TAUTOMERS

A. Dimer Formation

Rave found that decompositon of *N*-(benzenesulfonamido)piperidine and *N*-(tosylamido)-morpholine sodium salt (**123** and **124**) in hydroxylic solvents yielded the hexahydro-*s*-tetrazines **125** and **126**, respectively, accompanied by little if any tetrazene (38). Mercuric oxide transformed the tricyclic compounds into dehydro and bisdehydro derivatives (**127** and **128**), from which the saturated systems could be regenerated by lithium aluminum hydride reduction. Though the hexahydro-*s*-tetrazines stoutly resisted hydrolysis, canary yellow **128** (Y = 0) was cleaved almost immediately by dilute hydrochloric acid at room temperature. With benzaldehyde present, the benzal derivative of *N*-aminomorpholin-3-one

(123) Y = CH₂, Ar = C₆H₅   (125) Y = CH₂
(124) Y = O, Ar = C₇H₇(p-)   (126) Y = O

(127) Y = CH₂, O   (128) Y = CH₂, O   (129)

(129) was obtained (55%) from both halves of the molecule, thus confirming the original structure assignment for 123 (43).

The formation of 125 and 126 is understandable in the following terms: the diazenium ions 130 may be formed directly by ionization of the sulfonylhydrazines or by protonation of initially formed diazenes; deprotonation of these ions at an α-carbon gives the diazene tautomers (131) which can simply dimerize head-to-tail (132). The possibility has not been

(130)   (131)   (132)
Y = CH₂, O

excluded, however, that azomethinimine 131 reacts with another species, such a starting material, to reach the hexahydro-s-tetrazine by a more circuitous route. Consistent with mechanisms requiring the 1,3-dipolar species is the fact that 125 and 126 are not formed at all when dry sodium salts of the tosylhydrazine precursors are decomposed in hot diglyme. Under these conditions the tetrazenes 133 and their pyrolysis products are obtained. As explained in Section II-B, hydroxylic solvents greatly

(133)   Y = CH$_2$, O

facilitate formation of azomethinimines from sulfonylhydrazines. Even small amounts may suffice, as the best yield (84%) of **125** was obtained when the salt **123** was decomposed in diglyme containing traces of methanol. An alumina surface can also catalyze azomethinimine formation, for chromatography of **123** with chloroform gave **125** directly (70%) (43). Finally, **125** and **126** were obtained in low yield together with considerable tetrazene (**133**) and small quantities of tetrahydro-s-tetrazine (**127**) when piperidine and morpholine were treated with Angeli's salt (monoprotonated) in aqueous solution (38).

In earlier work, Höft and Rieche had discovered that oxidation by a number of methods of N-aminotetrahydroisoquinoline (**134**) yielded the hexahydro-s-tetrazine **136**, presumably via the diazenium ion and the 1,3-dipolar tautomer **135** (118). Formation of **136** instead of tetrazene

regardless of reaction conditions is undoubtedly attributable in large part to activation of a pair of α-hydrogens by the aryl substituent. Schmitz and Ohme obtained **136** simply by deprotonating iminoimmonium salt **138** prepared in turn by acid hydrolysis of the diaziridine **137** (119).

Being 1,3-dipolar species, azomethinimines are often trapped as adducts with dipolarophiles such as carbon disulfide or dimethyl acetylenedicarboxylate. Since this subject has been well reviewed by Huisgen (120, 121), a single example (one closely related to the chemistry just discussed) will suffice. Huisgen's group found that N-aminoquinolinium ion (**139**) is transformed via **140** into hexahydro-s-tetrazine **141** by potassium carbonate in dimethylformamide (122). When carbon disulfide was

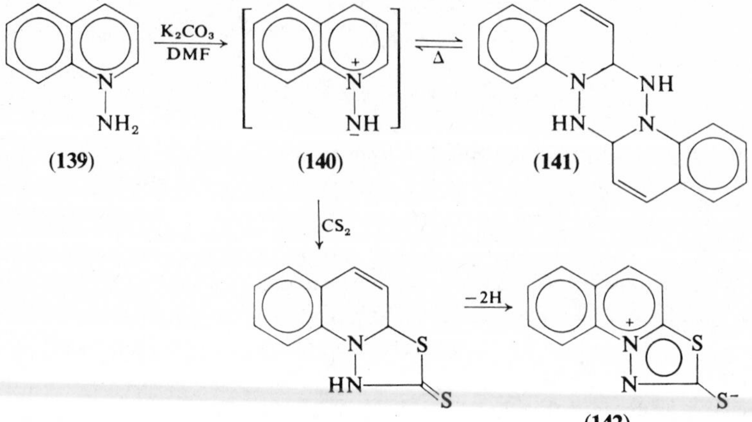

(139)   (140)   (141)

(142)

present, the intermediate dipolar species was intercepted efficiently and the resulting adduct was isolated as its air-oxidation product, the dithio-sydnone **142**. In solution compound **141** exists in equilibrium with its dipolar halves. The aromaticity of the heterocyclic ring of **140** provides a driving force for the facile reversion, for neither the saturated hexahydro-s-tetrazines **125** and **126** nor **136** dissociates readily. Pyrolysis of the latter will be considered in the section to follow.

B. Rearrangement

In 1936 Busch and Lang discovered that mercuric oxide oxidation of 1-aryl-1-benzylhydrazines gives either of two products, or both: the expected tetrazene and a rearranged compound, the arylhydrazone of a benzaldehyde (eq. 46) (123). Twenty-five years later Carter and Stevens

$$\begin{matrix} Ar \\ \diagdown \\ \diagup \\ Ar'CH_2 \end{matrix} N{-}NH_2 \xrightarrow{HgO} \begin{matrix} Ar \quad\quad CH_2Ar' \\ \diagdown \quad\quad \diagup \\ N{-}N{=}N{-}N \\ \diagup \quad\quad\quad \diagdown \\ Ar'CH_2 \quad\quad Ar \end{matrix} + ArNHN{=}CHAr'$$

(46)

followed up this work with a study of the decomposition of 1-aryl-1-benzyl-2-sulfonylhydrazines in aqueous alkali (22). Here hydrazones are the major products, accompanied by products of radical fragmentation of the intermediate diazenes (Section IV-C-1), but not by tetrazenes (eq. 47).

$$\begin{matrix} Ar \\ \diagdown \\ \diagup \\ Ar'CH_2 \end{matrix} N{-}NHSO_2R \xrightarrow[\Delta]{aq.\ NaOH} ArNH{-}N{=}CHAr' + NaO_2SR$$

(143)

(47)

They found that decomposition rate is first-order in the anion of **143**, and that it was affected only slightly by choice of R and Ar'. Ar was not varied widely, but it was clear that electron-donating substituents here accelerated and withdrawing substituents decelerated the reaction. The English authors further noted that decomposition of a 1,1-dibenzyl-2-sulfonylhydrazine is conspicuously faster than that of its benzylphenyl counterpart and that the diphenyl analog is quite inert. These observations led them to conclude that the rate-determining step of the rearrangement is expulsion of sulfinate ion from the anion of **143** to give the diazene, whose stability is greater, the greater the electron availability at the substituted nitrogen.

A crossover experiment was performed in which a mixture of sulfonyl-hydrazine **144** and **145**, whose decomposition rates are nearly identical, was heated with aqueous alkali (eq. 48). No "crossed" hydrazones were

$$p\text{-ClC}_6\text{H}_4 \diagdown \quad\quad\quad\quad \text{C}_6\text{H}_5 \diagdown$$
$$\phantom{p\text{-ClC}_6\text{H}_4}\text{N—NHTs} + \quad\quad \phantom{\text{C}_6\text{H}_5}\text{NNHTs} \xrightarrow[\Delta]{\text{aq. NaOH}}$$
$$\text{C}_6\text{H}_5\text{CH}_2 \diagup \quad \textbf{(144)} \quad p\text{-ClC}_6\text{H}_4\text{CH}_2 \diagup \quad \textbf{(145)}$$

$$p\text{-ClC}_6\text{H}_4\text{NHN}{=}\text{CHC}_6\text{H}_5 + \text{C}_6\text{H}_5\text{NHN}{=}\text{CHC}_6\text{H}_4\text{Cl}(p\text{-}) \quad (48)$$

detected by paper chromotography under conditions where benzaldehyde phenylhydrazone would show up clearly, so the rearrangement is intra-molecular.

The key question, namely, how the migration to the terminal nitrogen occurs in the "diazene-hydrazone rearrangement," remained a mystery. Menger pursued this question after discovering that even simple 1,1-di-alkyl-2-sulfonylhydrazine salts yielded hydrazones when decomposed in hydroxylic media (25). The most obvious hypothesis was that an alkyl group of the diazene underwent a 1,2-shift to give an azo compound, which then tautomerized to the hydrazone (eq. 49). To test this idea, the

$$\overset{+}{\text{R}_2\text{N}}{=}\overset{-}{\text{N}} \xrightarrow{\text{~R}} \text{R—N}{=}\text{N—R} \xrightarrow{\text{~αH}} \text{hydrazone} \quad\quad (49)$$

decomposition of 1,1-diethyl-2-benzenesulfonylhydrazine **(146)** to acet-aldehyde ethylhydrazone (eq. 50) was carried out with azoethane present. Since 1,1-dialkyl-2-sulfonylhydrazines ionize readily in polar solvents, it was possible to perform the rearrangement rapidly at room temperature (Section III-C). Under these gentle conditions the added azo compound was recovered unscathed, thereby demonstrating that it is not a reaction intermediate.*

$$\text{Et}_2\text{NNHSO}_2\text{C}_6\text{H}_5 \xrightarrow[\text{DEG Na salt}]{\text{DEG}} \text{EtNHN}{=}\text{CHCH}_3 \quad\quad (50)$$
$$\textbf{(146)} \quad\quad\quad\quad\quad\quad\quad\quad (70\text{–}80\%)$$

* Identifying several minor products as well, Urry and Ikoku (28) found that the *di-* as well as monopropylhydrazone of propionaldehyde is a significant product from reactions generating 1,1-dipropyldiazene, and indeed acetaldehyde *di*ethylhydrazone

$$(RCH_2)_2\overset{+}{N}{=}NH + RCH_2NHN{=}CHR \longrightarrow$$
$$RCH{=}N{-}N{=}CHR + (RCH_2)_2NNH_2 \longrightarrow \qquad \text{(i.)}$$
$$RCH{=}NNH_2 + (RCH_2)_2N{-}N{=}CHR$$

This result had been foreshadowed by the observation that 1,1-diphenyl-2-benzenesulfonylhydrazine sodium salt gave on pyrolysis diphenylamine (25,124) (in analogy to its carbazole-derived counterpart, Section II-B) but none of the azobenzene to be expected on the basis of equation 49. The claim that certain 1-allyl-1-arylhydrazines yield azo compounds when treated with mercuric oxide (125,126) is apparently incorrect; this matter is presently being reinvestigated (127). On the other hand, Baumgarten's group discovered that lead tetraacetate oxidation of N-aminooxindole (**147**) gives 3-cinnolinol (**148**), formally the product of 1,2-acyl or aryl migration in an intermediate diazene (128). An N-aminopyridone-pyridazine rearrangement newly discovered by Rees and Yelland seems to involve a similar step (49b). These processes are apparently unrelated to the major pathway of the "diazene–hydrazone rearrangement."

In aprotic media, the sodium salts of 1,1-dialkyl-2-sulfonylhydrazines such as **146** decompose to give tetraalkyltetrazenes in good yield, but no hydrazones (eq. 51) (25). Higher temperatures are required than in

$$R_2N{-}NSO_2C_6H_5 \xrightarrow[\text{110–120°}]{\text{tetraglyme}} R_2N{-}N{=}N{-}NR_2 \qquad (51)$$
$$\text{Na}$$

$$R = CH_3, 74\% \qquad R = Et, 80\% \qquad R = \text{i-Pr}, 72\%$$

hydroxylic media since ionization of free sulfonylhydrazine is obviously not feasible here. Even very small amounts of methanol in the tetraglyme have been known to change the product from tetrazene to hydrazone. This dramatic effect of solvent on product composition, reminiscent of that found in the decomposition of arenesulfonamidopiperidine and -morpholine salts (Section V-A), argues once again for the formation of azomethinimine intermediates. As noted earlier, Underbrink showed that sulfonylhydrazine **143** does not undergo deuterium exchange at the

---

is the major byproduct of equation 50 (29,52). The appearance of dialkylhydrazone has been explained in terms of equation i (50b). Alternate possibilities have recently been suggested by Koga and Anselme (139).

α-carbons during rearrangement, so azomethinimine formation is apparently irreversible (at least under his conditions) (52).

The 1,3-dipolar species **149** derived from **146** could be imagined to cyclize to a diaziridine (**150**), cleavage of which might yield the observed hydrazone. Therefore, diaziridine **150** (129) was synthesized and com-

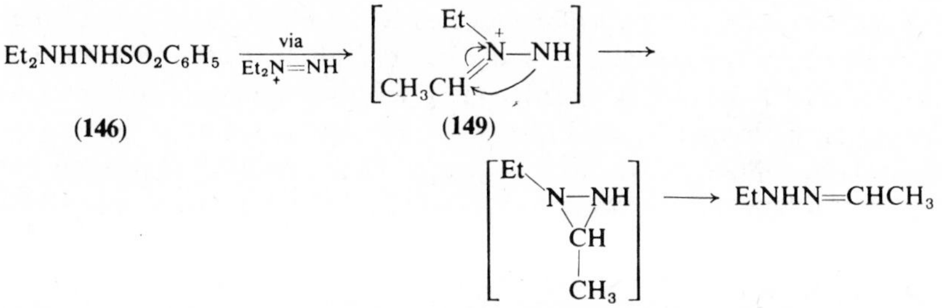

$$Et_2NHNHSO_2C_6H_5 \xrightarrow[\overset{via}{Et_2\overset{+}{N}=NH}]{} \left[\begin{array}{c} \overset{Et}{\underset{CH_3CH}{\diagdown}} \overset{+}{\underset{\diagup}{C}} N-NH \end{array} \right] \longrightarrow$$

(**146**)   (**149**)

$$\left[\begin{array}{c} \overset{Et}{\diagdown} \\ N-NH \\ \underset{\diagup}{CH} \\ | \\ CH_3 \end{array} \right] \longrightarrow EtNHN=CHCH_3$$

(**150**)

bined with **146** in a rearrangement experiment (25). Diaziridines are quite base-stable, so it is no surprise that **150** was recovered unchanged, thus ruling out the above scheme.

At this stage the hypothesis expressed in equation 52 was deemed most

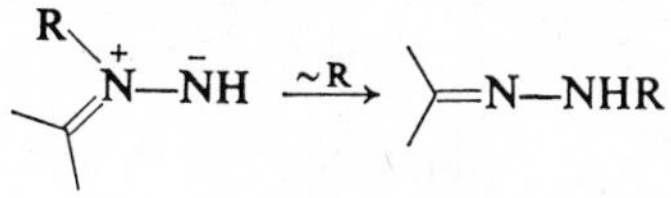

$$\overset{R}{\diagdown} \overset{+}{N}-\overset{-}{N}H \xrightarrow{\sim R} \hspace{0.3cm} \diagup =N-NHR \hspace{1cm} (52)$$

$$\overset{C_6H_5}{\underset{C_6H_5CH}{\diagdown}} \overset{+}{N}-\overset{-}{N}H$$

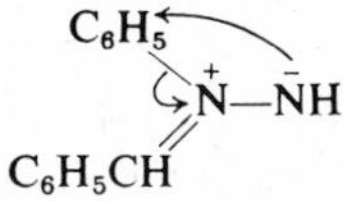

(**151**)

likely, despite certain drawbacks. 1-Phenyl-1-benzyl-2-benzenesulfonylhydrazine was chosen to test this idea, since equation 52 makes a clear commitment in favor of phenyl migration (**151**). Lichter prepared ^{15}N-labeled material by nitrosation of phenylbenzylamine with labeled nitrous acid, followed by lithium aluminum hydride reduction to the hydrazine and sulfonylation (29). Decomposition in dry ethylene glycol/sodium ethylene glycolate gave the hydrazone, which was reductively cleaved to aniline and benzylamine by sodium in liquid ammonia (eq. 53). Diazotization

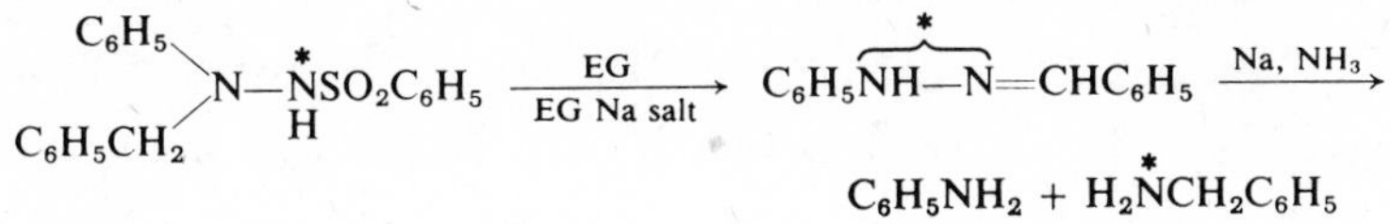

$$\overset{C_6H_5}{\underset{C_6H_5CH_2}{\diagdown}} N-\overset{*}{\underset{H}{N}}SO_2C_6H_5 \xrightarrow[\text{EG Na salt}]{\text{EG}} C_6H_5\overset{\frown}{NH}-\overset{*}{N}=CHC_6H_5 \xrightarrow{Na,\ NH_3}$$

$$C_6H_5NH_2 + H_2\overset{*}{N}CH_2C_6H_5 \hspace{1cm} (53)$$

of the aniline, then hydrolysis, gave molecular nitrogen which was virtually all mass 28, corresponding to $\geq 98.7\%$ net *benzyl* migration!

This violent contradiction of the prediction based on equation 52 underlined the need for another, somewhat subtler labeling experiment, for the possibility still existed that dialkylsulfonylhydrazines (the testing ground for earlier mechanistic hypotheses) rearranged via equation 52 even though arylbenzylsulfonylhydrazines clearly do not. Accordingly, Lichter synthesized ^{15}N-labeled 1,1-diethyl-2-benzenesulfonylhydrazine (**146**) and decomposed it in very dry diethylene glycol containing its sodium salt. Since the hydrocarbon substituents of **146** are identical, a hydrazone degradation different from that of equation 53 was required to preserve the distinction between migrating and stationary groups. Hence the acetaldehyde ethylhydrazone was gently quaternized and cleaved to acetonitrile and ethyldimethylamine in one step, using methyl iodide and silver oxide in dimethylformamide (eq. 54) (130). Mass spectral analysis of the purified

$$\text{Et}_2\overset{*}{\text{N}}\text{NHSO}_2\text{C}_6\text{H}_5 \xrightarrow[\text{DEG Na salt}]{\text{dry DEG}} \text{Et}\overset{*}{\overbrace{\text{NHN}}}=\text{CHCH}_3 \xrightarrow[\text{DMF}]{\text{CH}_3\text{I, Ag}_2\text{O}}$$

$$\left[\begin{array}{c} \text{CH}_3 \\ | \quad * \\ \text{EtN}-\text{N}=\text{CHCH}_3 \\ | \\ \text{CH}_3 \quad \text{H} \end{array}\right] \longrightarrow \text{EtN(CH}_3)_2 + \overset{*}{\text{N}}\equiv\text{CCH}_3 \qquad (54)$$

acetonitrile showed it to contain 98.3% of the label, meaning that N-2 of the starting material becomes the doubly bonded nitrogen of the product (29). This result is completely in harmony with the previous labeling experiment, and it points to the remarkable conclusion that the azomethinimine substituent which migrates is the partially double-bonded one. Note, incidentally, that the present experiment confirms the deduction that azoethane is not an intermediate in the decomposition of 1,1-diethyl-2-benzenesulfonylhydrazine (**146**).

Rigorously anhydrous conditions were employed in the decomposition of labeled **146** to exclude the possibility that azomethinimine **149** was hydrolyzed to aldehyde and hydrazine, which recondensed to yield properly labeled hydrazone (eq. 55). Though insignificant here, this hydrolysis–recondensation mechanism may be important in the decomposition of certain sulfonylhydrazines in aqueous solution.

$$\left[\begin{array}{c} \text{Et} \\ \diagdown \quad + \quad * \\ \text{N}-\text{NH} \\ \diagup\diagup \\ \text{CH}_3\text{CH} \end{array}\right] \xrightarrow{\text{H}_2\text{O}} [\text{EtNH}\overset{*}{\text{N}}\text{H}_2 + \text{CH}_3\text{CHO}] \xrightarrow{-\text{H}_2\text{O}} \text{EtNH}\overset{*}{\text{N}}=\text{CHCH}_3$$
$$(\mathbf{149}) \qquad\qquad\qquad\qquad\qquad\qquad\qquad\qquad\qquad (55)$$

Another scheme which would accomplish the necessary net migration of the "ethylidene" group in **149** involves head-to-tail dimerization of this 1,3-dipole followed by fragmentation to two molecules of hydrazone (eq. 56). Precedents for the dimerization have already been discussed, and this pathway is all the more appealing because (unsubstituted) hydrazones of many aldehydes are known to dimerize reversibly to hexahydro-s-tetrazines closely related to **152** (131,132).

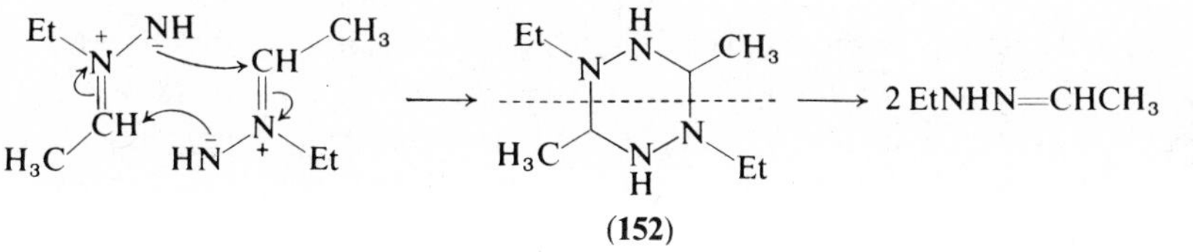

$$\longrightarrow \quad 2\,EtNHN{=}CHCH_3 \quad (56)$$

(**152**)

On the other hand, *N*-tosylamidopyrrolidine (**153**) decomposes with rearrangement to the ring-expanded hydrazone 2,3,4,5-tetrahydropyridazine (**154**) in dry methanol/sodium methoxide (29) as well as in aqueous

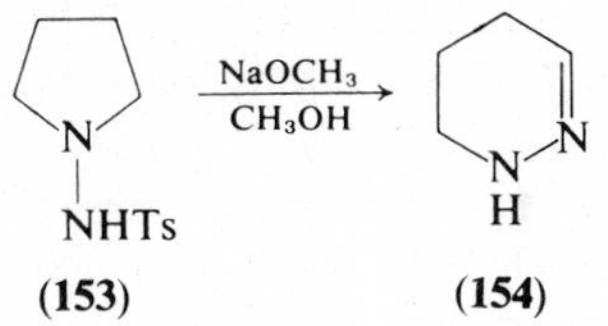

$$\xrightarrow[\text{CH}_3\text{OH}]{\text{NaOCH}_3}$$

(**153**)   (**154**)

base (58% yield); analogy to equation 56 would require formation of a 12-membered ring bishydrazone **155**. Angeli's salt transforms pyrrolidine itself into (**154**) (68%) (38), as does isopropyl *N,N*-difluorocarbamate

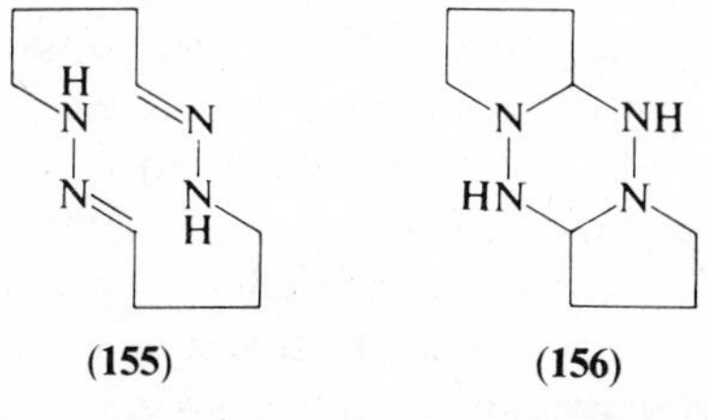

(**155**)   (**156**)

(Section II-C) (36). Since decomposition of *N*-arenesulfonamidopiperidine and -morpholine salts in hydroxylic media gives stable hexahydro-s-tetrazines (Section V-A), it is remarkable that in the 5-ring counterpart **153** rearrangement takes precedence over formation of **156** (none of which has been found). In this connection, Schmitz and Ohme discovered that hexahydro-s-tetrazine **136** fragments at high temperatures to yield the ring-expanded hydrazone **157**, among other products (119), perhaps via the azomethinimine halves of **136**.

(136) $\xrightarrow{250°}$ **(157)** + + NH$_3$

(10–30%)   (34%)

The behavior of N-arenesulfonamidopyrrolidine salts provides another striking illustration of the influence of solvents upon the course of sulfonylhydrazine decompositions: instead of rearrangement, fragmentation to ethylene and nitrogen occurs in the aprotic solvent diglyme, as has been noted (Section IV-C-3). In triethylamine it appears that the two pathways for destruction of **153** exist in competition, consistent with the fact that triethylammonium ions are available as a proton source even though at low concentration (29). Finally, oxidation of N-aminopyrrolidine with mercuric oxide in ether at 0° follows neither route, but yields the tetrazene instead (23)! These conditions are unfavorable for rearrangement, and the low temperature apparently decelerates fragmentation sufficiently for the diazene to react bimolecularly.

The **153** → **154** transformation dampens one's enthusiasm regarding the dimerization-cleavage mechanism for rearrangement expressed in equation 56, but Carter and Stevens' crossover experiment (eq. 48) crushes it altogether. Because this experiment is critically important, it was repeated by Guhn and Lichter, who analyzed the product hydrazone mixtures mass-spectrometrically. They confirmed the English workers' results and set an upper limit of 1% on the yield of crossed hydrazones. The absence of crossover also proves that the hydrolysis–recondensation mechanism (eq. 55) was not operative here even though the solvent was water.

In short, none of the mechanistic hypotheses considered above has withstood the test of experiment. The rearranging species cannot be a diazene or a diazenium ion (dialkyldiazenium ions are too stable, Section III-A), nor can it be a sulfonylhydrazine anion since rearrangement is observed in hydrazine oxidation, in the Angeli's salt-secondary amine reaction, etc. All evidence is consistent with formation of an azomethinimine intermediate,* but schemes requiring cleavage or dimerization of this 1,3-dipolar species are demolished by proof that the rearrangement is

---

* "Diazene-hydrazone rearrangement" (25) still seems to be an appropriate designation. The first intermediate common to the various types of reaction in which rearrangement has been observed must be either diazene or diazenium ion. Moreover, these two species are probably in rapid equilibrium in hydroxylic media and at least interconverting under most rearrangement conditions.

intramolecular. Cyclization to a diaziridine followed by a reopening has also been clearly ruled out.

The only hypothesis which appears to be in harmony with all the experimental data is a simple, but theoretically dubious one, namely, that the "alkylidene group" of an azomethinimime intermediate migrates directly to the terminal nitrogen (eq. 57). Such a migration, requiring as it

$$\underset{\underset{\displaystyle -C}{\big\|}}{N}{-}\bar{N}H \longrightarrow \underset{\underset{\displaystyle /C\diagdown}{}}{\bar{N}}{-}\overset{+}{N}H \longrightarrow NH{-}N\diagup \qquad (57)$$

does simultaneous rearrangement of both the π and the σ framework of the azomethinimine, would be without precedent. Whatever detailed model one constructs for the triangular transition state for migration, there is an embarrassingly large number of electrons to be accommodated in the available low-lying orbitals (29). Since rearrangement occurs very rapidly at room temperature in favorable cases, one can ill afford the luxury of postulating a high-energy transition state. Thus there is at present an impasse between theory and experiment, and "alkylidene migration" must be regarded as merely a working hypothesis.

There is no compelling reason to believe that the imine proton of the postulated 1,3-dipolar intermediate is essential for the rearrangement per se, though its ability to leave is required for hydrazone formation. In other words, rearrangement may not be restricted to those azomethinimines which are tautomers of diazenes: the "diazene–hydrazone rearrangement" may be representative of a more general "azomethinimine rearrangement." With appropriate substituents at the imine nitrogen, reversible or even degenerate rearrangement might be demonstrable, even though intensive investigations of azomethinimine chemistry over the past several years have not revealed such a process (120,121). To test the idea, Lichter synthesized the bicyclic tosylhydrazine **158** as a precursor for azomethinimine **159**, a candidate for degenerate rearrangement (eq. 58).

(158)   (159)

$$(58)$$

The dipolar intermediate was generated readily from **158** (probably via the diazenium ion) in polar solvents containing a tertiary amine, as indicated by the formation of trapping products (**160–162**) and aligomers (e.g., **163**). The bistosyl compound (**162**) was actually formed during tosylation of the hydrazine precursor of **158**, but almost certainly by way of **158** and **159**.

If **158** were available in optically active form, optical rotary dispersion measurements on trapping products could serve as a criterion for rearrangement. Since the azomethinimines of equation 58 are enantiomers, racemic products would have to result if rearrangement of **159** were fast with respect to its capture rate. Optically pure **158** was obtained by fractional crystallization of the two diastereomeric *d*-camphor-10-sulfonylhydrazines **164**, an interesting reductive cleavage of **164** with lithium aluminum hydride and tosylation of the resulting hydrazine enantiomers (**165**) (29). Decomposition of *d*-**158** in dimethyl sulfoxide containing

triethylamine and carbon disulfide yielded adduct **160** with the rotation $[\phi]_D = +682°$ (CH$_3$OH), clear testimony that rearrangement had failed (eq. 59) (133). Geometric restrictions imposed by the rigid bicyclic skeleton

$$(59)$$

(158)   **(162)**

(absolute configurations unknown)

of **159** may have precluded rearrangement, or trapping may simply have been rapid enough to compete successfully with rearrangement. To test the latter possibility, experiments in which the concentration of trapping agent is varied are currently underway.

Whether or not dipolar species **159** be capable of rearrangement, the conclusion that azomethinimines are intermediates in the diazene–hydrazone rearrangement is difficult to avoid. Nonetheless, a complete understanding of this deceptively simple-looking transformation must await further cogitation and experiment.

VI. A FORWARD GLANCE

To presume to chart the destiny of any field of chemistry, one must be blind to the enormous influence wielded by Serendipity and, antithetically, by Murphy's Law over the course of scientific research! This bittersweet thought notwithstanding, it is appropriate here to reiterate some of the still-unanswered questions already encountered, to pose others, and to call attention to potentialities not yet realized in diazene chemistry.

Consider first the fragmentation of diazenes. Does "substituent-assisted" fragmentation generally proceed via free radicals, as this review has suggested, or do concerted and stepwise pathways exist in close competition? Will fragmentation of the theoretically interesting diazenes derived from azetidines and 2,7-dihydroazepines proceed with the anticipated stereochemistry? Can diazenes which lack all of the structural features conducive to fragmentation be compelled to lose nitrogen by generating them at elevated temperatures and high dilution? If so, this could be a valuable method for making carbon–carbon bonds. Synthetic chemists should find intriguing the stereospecific fragmentation of various cyclic diazenes, for ring degradation has become an important aspect of organic synthesis. Particularly in the construction of molecules studded with asymmetric centers, rings are incorporated as "stereochemical scaffolding," and must be dismantled before the finished molecule can emerge (134). Finally, diazenes incorporating two potential nitrogen molecules can yield high-energy species on fragmentation, as noted earlier. This capability will undoubtedly find notable new applications in the study of short-lived intermediates.

Regarding the "diazene–hydrazone" transformation, what is the true nature of the elusive rearrangement step(s)? Is there a more general azomethinimine rearrangement?

Further probing into the nature of the diazene function itself should be rewarding. Even if they are inert to simple olefins, are the nucleophilic dialkyldiazenes capable of forming aziridines by addition to *negatively* substituted olefins, in analogy to Rees' work (49a) and to the apparent behavior of certain nucleophilic carbenes (135)? Can diazenes, like other nitrenes (136) and carbenes (137), be trapped in the form of transition metal complexes? Is it possible to prepare diazenes whose ground states are triplet? If so, ESR examination of these species and determination of the relative energy of lowest singlet and triplet states as a function of structure should be most interesting.

As pointed out in the Introduction, all our present knowledge of diazenes is chemical in origin: there is no direct physical evidence for their existence. Using flow techniques, one should be able to examine diazenes spectroscopically in solution. Hydrazine oxidation or, more simply, diazenium ion deprotonation should be readily adaptable for this purpose. A method for generating diazenes in the vapor phase or in a rigid matrix would be a great boon to those wishing to delve into their physical properties. Photolysis or thermolysis of N-aminosulfoximines (49) or azidamines $(R_2N—N{=}\overset{+}{N}{=}\overset{-}{N})$ might be such a method. The simplest representative (and a very dangerous one) of the latter class of compounds, N-azidodimethylamine, has been reported (138,139), Carpino has now discovered that N-amino-7-azanorbornadiene derivatives lose the nitrogen bridge thermally, probably as the diazene, owing to the driving force of aromatizing the carbon skeleton (140). These and related approaches to diazenes promise a wealth of presently unattainable information on these fascinating species.

ACKNOWLEDGMENTS

The author is greatly indebted to the many members of his research group who have contributed over the past several years to his understanding of diazene chemistry. Foremost among them are Drs. R. L. Lichter, S. D. McGregor, F. Menger, T. W. Rave, and Mr. C. D. Underbrink. He wishes to thank Drs. J.-P. Anselme, L. A. Carpino, J. P. Freeman, F. D. Greene, F. D. Marsh, C. G. Overberger, and W. H. Urry for furnishing unpublished information and valuable counsel. Generous financial support during the preparation of the manuscript was provided by the National Science Foundation and the U.S. Army Research Office (Durham).

REFERENCES

1. W. Lwowski, *Angew. Chem. Intern. Ed. Engl.*, **6**, 897 (1967).
2. R. A. Abramovitch and B. A. Davis, *Chem. Rev.*, **64**, 149 (1964).
3. L. Horner and A. Christmann, *Angew. Chem. Intern. Ed. Engl.*, **2**, 599 (1963).
4. W. Kirmse, *Angew. Chem.*, **71**, 537 (1959).
5. J. A. Bonham, "Diazene Intermediates," Univ. of Illinois Seminar, October 12, 1964.
6. D. M. Lemal, in *The Chemistry of the Amino Group*, S. Patai, Ed., Interscience, New York, 1968, Chapter 12, and references contained therein.
7. W. R. McBride and E. M. Bens, *J. Am. Chem. Soc.*, **81**, 5546 (1959).
8. C. G. Overberger, *Rec. Chem. Progr.*, **21**, 21 (1960).
9. C. G. Overberger, J. P. Anselme, and J. G. Lombardino, *Organic Compounds with Nitrogen–Nitrogen Bonds*, Ronald, New York, 1966; (a) Chapter 5.
10. P. A. S. Smith, *Open-Chain Nitrogen Compounds*, Vol. 2, Benjamin, New York, 1966; (a) Chapter 11; (b) Chapter 9.
11. H. Zollinger, *Azo and Diazo Chemistry*, Interscience, New York, 1961; (a) Chapter 12.
12. See, for example, E. Fischer and H. Troschke, *Ann.*, **A199**, 294 (1879); E. Renouf, *Chem. Ber.*, **13**, 143 (1880).
13. M. Busch and B. Weiss, *Chem. Ber.*, **33**, 2701 (1900).
14. W. McBride and H. W. Kruse, *J. Am. Chem. Soc.*, **79**, 572 (1957).
15. W. H. Urry, H. W. Kruse, and W. R. McBride, *J. Am. Chem. Soc.*, **79**, 6568 (1957).
16. (a) A. Schmidpeter, *Tetrahedron Letters*, **1963**, 1421; (b) *Angew. Chem. Intern. Ed. Engl.*, **3**, 151 (1964). For complexation of nitrosamines with Lewis acids, see A. Schmidpeter, *Chem. Ber.*, **96**, 3275 (1963).
17. S. Hünig, L. Geldern and E. Lücke, *Angew. Chem. Intern. Ed. Engl.*, **2**, 327 (1963); S. Hünig, G. Büttner, J. Cramer, L. Geldern, H. Hansen, and E. Lücke, *Chem. Ber.*, **102**, 2093 (1969); S. Hünig and H. Hansen, *Chem. Ber.*, **102**, 2109 (1969).
18. Th. Eicher, S. Hünig, and H. Hansen, *Angew. Chem. Intern. Ed. Engl.*, **6**, 699 (1967). See also Th. Eicher, S. Hünig and P. Nikolaus, *Angew. Chem. Intern. Ed. Engl.*, **6**, 699 (1967); S. Hünig and G. Büttner, *Angew. Chem. Intern. Ed. Engl.*, **8**, 451 (1969).
19. L. A. Carpino, *Chem. Ind.*, **1957**, 172.
20. L. A. Carpino, *J. Am. Chem. Soc.*, **79**, 4427 (1957).
21. W. R. Bamford and T. S. Stevens, *J. Chem. Soc.*, **1952**, 4735.
22. P. Carter and T. S. Stevens, *J. Chem. Soc.*, **1961**, 1743.
23. D. M. Lemal, T. W. Rave, and S. D. McGregor, *J. Am. Chem. Soc.*, **85**, 1944 (1963).
24. H. Wieland and A. Susser, *Ann.*, **392**, 169 (1912).
25. D. M. Lemal, F. Menger, and E. A. Coats, *J. Am. Chem. Soc.*, **86**, 2395 (1964).
26. S. Wawzonek and W. McKillip, *J. Org. Chem.*, **27**, 3946 (1962).
27. D. M. Lemal, C. D. Underbrink, and T. W. Rave, *Tetrahedron Letters*, **1964**, 1955.
28. W. H. Urry and C. Ikoku, *Abstracts, 146th National Meeting Am. Chem. Soc.*, Denver, Colo., Jan. 1964, p. 25c.

29. R. L. Lichter, Ph.D. Dissertation, Univ. of Wisconsin, Madison, Wis., 1967.
30. C. L. Bumgardner, K. J. Martin, and J. P. Freeman, *J. Am. Chem. Soc.*, **85**, 97 (1963).
31. C. L. Bumgardner and J. P. Freeman, *J. Am. Chem. Soc.*, **86**, 2233 (1964).
32. W. le Noble and D. Skulnik, *Tetrahedron Letters*, **1967**, 5217.
33. S. I. Morrow, D. D. Perry, M. S. Cohen, and C. Schoenfelder, *J. Am. Chem. Soc.*, **82**, 5301 (1960).
34. O. Ruff and E. Hanke, *Z. Anorg. Allgem. Chem.*, **197**, 394 (1931); C. B. Colburn and A. Kennedy, *J. Am. Chem. Soc.*, **80**, 5004 (1958); J. P. Freeman, A. Kennedy, C. B. Colburn, *J. Am. Chem. Soc.*, **82**, 5304 (1960).
35. C. B. Colburn, in *Advances in Fluorine Chemistry*, Vol. 3, M. Stacey, J. C. Tatlow, and A. G. Sharpe, Eds., Butterworths, Washington, 1963, p. 92.
36. D. L. Klopotek and B. G. Hobrock, *Abstracts, 153rd Natl. Mtg. Am. Chem. Soc., Miami Beach, Fla., April, 1967*, p. O 200; private communication with D. L. Klopotek.
37. A. Angeli, *Chem. Zentr.*, **71**, II, 857 (1900); **72**, I, 889 (1901).
38. D. M. Lemal and T. W. Rave, *J. Am. Chem. Soc.*, **87**, 393 (1965).
39. J. Veprek-Siska, V. Pliska, F. Smirous, and F. Vesely, *Collection Czech. Chem. Commun.*, **24**, 687 (1959).
40. P. A. S. Smith and G. E. Hein, *J. Am. Chem. Soc.*, **82**, 5731 (1960), and references contained therein.
41. O. Piloty, *Chem. Ber.*, **29**, 2559 (1896).
42. A. Angeli, *Chem. Zentr.*, **73**, II, 691 (1902).
43. T. W. Rave, Ph.D. Dissertation, Univ. of Wisconsin, Madison, Wis., 1965. T. W. Rave and D. M. Lemal, manuscript in preparation.
44. A. Angeli, *Chem. Zentr.*, **67**, I, 799 (1896); H. R. Hunt, Jr., J. R. Cox, Jr., and J. D. Ray, *Inorg. Chem.*, **1**, 938 (1962).
45. K. D. Anderson, C. F. Crumpler, and D. L. Hammick, *J. Chem. Soc.*, **1935**, 1679; D. L. Hammick and M. W. Lister, *ibid*, **1937**, 489.
46. S. M. McGregor, Ph.D. Dissertation, Univ. of Wisconsin, Madison, Wis., 1966.
47. (a) C. G. Overberger, J. G. Lombardino, and R. G. Hiskey, *J. Am. Chem. Soc.*, **79**, 6430 (1957); (b) *J. Am. Chem. Soc.*, **80**, 3009 (1958); (c) C. G. Overberger, M. Valentine, and J.-P. Anselme, *J Am. Chem. Soc.*, **91**, 687 (1969).
48. (a) L. Horner and H. Hoffmann, *Angew. Chem.*, **68**, 473 (1956); (b) L. A. Carpino, A. A. Santilli, and R. W. Murray, *J. Am. Chem. Soc.*, **82**, 2728 (1960).
49. D. J. Anderson, T. L. Gilchrist, D. C. Horwell, and C. W. Rees, *Chem. Commun.*, **146** (1969); (b) C. W. Rees and M. Yelland, *ibid.*, **377** (1969).
50. (a) W. H. Urry, P. Szecsi, C. Ikoku, and D. W. Moore, *J. Am. Chem. Soc.*, **86**, 2224 (1964); (b) W. H. Urry, private communication.
51. S. Hünig, H. Balli, E. Breither, F. Brühne, H. Geiger, E. Grigat, F. Müller, and H. Quast, *Angew. Chem. Intern. Ed. Engl.*, **1**, 640 (1962).
52. C. D. Underbrink and D. M. Lemal, unpublished results.
53. (a) H. Zimmer and G. Singh, *J. Org. Chem.*, **29**, 1579 (1964); (b) I. Bhatnagar and M. V. George, *ibid.*, **33**, 2407 (1968), and references contained therein.
54. C. G. Overberger and L. P. Herin, *J. Org. Chem.*, **27**, 417 (1962).
55. P. S. Forgione, G. S. Sprague, and H. F. Troffkin, *J. Am. Chem. Soc.*, **88**, 1080 (1966).
56. (a) W. Dornfeldt and D. M. Lemal, unpublished results, 1966; (b) R. S. Atkinson and C. W. Rees, *Chem. Commun.*, 1230 (1967); *J. Chem. Soc.*, (C), 772 (1969); (c) *Chem. Commun.*, 1232 (1967); *ibid.*, 631 (1968); *J. Chem. Soc.*, (C), 778

(1969); (d) D. J. Anderson, T. L. Gilchrist, and C. W. Rees, *Chem. Commun.*, 147 (1969).

57. For a good discussion of the "Skell criterion," see P. P. Gaspar and G. S. Hammond in W. Kirmse, "Carbene Chemistry," Academic Press, New York, 1964, Chapter 12.

58. J. Kenner and E. C. Knight, *Chem. Ber.*, **69**, 341 (1936).

59. (a) R. L. Hinman and K. L. Hamm, *J. Am. Chem. Soc.*, **81**, 3294 (1959); (b) Tetrazenes and other identifiable products are also obtained in certain circumstances See, for example, ref. 53b and G. Koga and J.-P. Anselme, *J. Am. Chem. Soc* **91**, 4323 (1969).

60. C. G. Overberger, N. P. Marullo, and R. G. Hiskey, *J. Am. Chem. Soc.*, **83**, 1374 (1961).

61. C. G. Overberger and N. P. Marullo, *J. Am. Chem. Soc.*, **83**, 1378 (1961).

62. S. F. Nelson and P. D. Bartlett, *J. Am. Chem. Soc.*, **88**, 143 (1966).

63. F. D. Greene, private communication.

64. T. Koenig and M. Deinzer, *J. Am. Chem. Soc.*, **90**, 7014 (1968).

65. L. K. Montgomery, K. Schueller, and P. D. Bartlett, *J. Am. Chem. Soc.*, **86**, 622 (1964).

66. C. G. Overberger and J. G. Lombardino, *J. Am. Chem. Soc.*, **80**, 2317 (1958).

67. C. G. Overberger and L. P. Herin, *J. Org. Chem.*, **27**, 2423 (1962).

68. C. G. Overberger and B. S. Marks, *J. Am. Chem. Soc.*, **77**, 4104 (1955).

69. C. G. Overberger, G. Kesslin, and N. R. Byrd, *J. Org. Chem.*, **27**, 1568 (1962).

70. C. G. Overberger and S. Altscher, *J. Org. Chem.*, **31**, 1728 (1966).

71. C. G. Overberger, L. C. Palmer, B. S. Marks, and N. R. Byrd, *J. Am. Chem. Soc.*, **77**, 4100 (1955).

72. C. H. Wang, S. Hsiao, E. Saklad, and S. G. Cohen, *J. Am. Chem. Soc.*, **79**, 2661 (1957); S. G. Cohen, S. Hsiao, E. Saklad, and C. H. Wang, *J. Am. Chem. Soc.*, **79**, 4400 (1957).

73. C. D. Campbell and C. W. Rees, *Proc. Chem. Soc.*, 296 (1964); *J. Chem. Soc.*, (C), 748, 752 (1969). R. W. Atkin and C. W. Rees, *Chem. Commun.*, 152 (1969).

74. C. D. Campbell and C. W. Rees, *Chem. Commun.*, 192 (1965); *J. Chem. Soc.*, (C), 742 (1969). C. W. Rees and R. C. Storr, *Chem. Commun.*, 1305 (1968).

75. For related attempts to prepare benzyne, see ref. 79b.

76. C. W. Rees and R. C. Storr, *Chem. Commun.*, 193 (1965); *J. Chem. Soc.*, (C), 756, 760, 765 (1969).

77. R. W. Hoffmann, G. Guhn, M. Preis and B. Dittrich, *J. Chem. Soc.*, (C), 769 (1969).

78. R. W. Hoffmann, "Dehydrobenzenes and Cycloalkynes," Academic Press New York, 1967. (a) pp. 311–313; (b) pp. 80, 81; (c) pp. 332–336.

79. (a) M. W. Ao, E. M. Burgess, A. Schauer, and E. A. Taylor, *Chem. Commun.*, 220 (1969); J. Adamson, D. L. Forster, T. L. Gilchrist, and C. W. Rees, *ibid.*, 221 (1969); (b) C. W. Rees and D. E. West, *ibid.*, 647 (1969).

80. C. L. Bumgardner, K. S. McCallum, and J. P. Freeman, *J. Am. Chem. Soc.*, **83**, 4417 (1961).

81. D. Felix, J. Schreiber, K. Piers, U. Horn, and A. Eschenmoser, *Helv. Chim. Acta*, **51**, 1461 (1968).

82. A. Eschenmoser, D. Felix, and G. Ohloff, *Helv. Chim. Acta*, **50**, 708 (1967).

83. J. Schreiber, D. Felix, A. Eschenmoser; M. Winter, F. Gautschi, K. H. Schulte-Elte, E. Sundt, and G. Ohloff; J. Kalvoda, H. Kaufmann, P. Wieland, and G. Anner, *Helv. Chim. Acta*, **50**, 2102 (1967).

84. M. Tanabe, D. F. Crowe, R. L. Dehn, and G. Detre, *Tetrahedron Letters*,

1967, 3739; M. Tanabe, D. F. Crowe, and R. L. Dehn, *Tetrahedron Letters,* **1967,** 3943.

85. For a closely related fragmentation reaction, see P. Wieland, H. Kaufmann, and A. Eschenmoser, *Helv. Chim. Acta,* **50,** 2108 (1967).

86. J. P. Freeman and W. H. Graham, *J. Am. Chem. Soc.,* **89,** 1762 (1967).

87. R. Hoffmann and R. B. Woodward, *Abstracts, 150th Natl. Mtg. Am. Chem. Soc., Atlantic City, N.J., Sept., 1965,* No. 17S.

88. N. Tokura, T. Nagai, and S. Matsumura, *J. Org. Chem.,* **31,** 349 (1966); N. P. Neureiter, *J. Am. Chem. Soc.,* **88,** 558 (1966).

89. F. G. Bordwell, J. M. Williams, Jr., E. B. Hoyt, Jr., and B. B. Jarvis, *J. Am. Chem. Soc.,* **90,** 429 (1968).

90. R. D. Clark and G. K. Helmkamp, *J. Org. Chem.,* **29,** 1316 (1964). For decomposition stereochemistry in other hetero 3-ring systems see G. E. Hartzell and J. N. Paige, *ibid.,* **32,** 459 (1967); B. M. Trost and S. Ziman, *Chem. Commun.,* 181 (1969).

91. G. J. Karabatsos and R. A. Taller, *J. Am. Chem. Soc.,* **86,** 4373 (1964); C. E. Looney, W. D. Phillips, and E. L. Reilly, *J. Am. Chem. Soc.,* **79,** 6136 (1957).

92. R. Hoffmann, *Abstracts, 151st Natl. Mtg. Am. Chem. Soc., Pittsburgh, Pa., April, 1966,* K109; R. Hoffmann, *J. Am. Chem. Soc.,* **90,** 1475 (1968).

93. (a) R. Hoffmann and R. B. Woodward, *Accounts of Chemical Research,* **1,** 17 (1968); *J. Am. Chem. Soc.,* **87,** 2046 (1965); (b) R. B. Woodward and R. Hoffman, *Angew. Chem.,* in press.

94. D. E. Applequist, M. A. Lintner and R. Searle, *J. Org. Chem.,* **33,** 254 (1968).

95. H. J. Eméleus and G. L. Hurst, *J. Chem. Soc.,* 3276 (1962); E. A. V. Ebsworth and G. L. Hurst, *ibid.,* 4840 (1962). N. Rieber, J. Alberts, J. A. Lipsky, and D. M. Lemal, *J. Amer. Chem. Soc.,* in press.

96. J. P. Freeman, private communication.

97. (a) R. J. Crawford, R. J Dummel, and A. Mishra, *J. Am. Chem. Soc..,* **87,** 3023 (1965); (b) R. J. Crawford and A. Mishra, *J. Am. Chem. Soc..,* **87,** 3768 (1965); *J. Am. Chem. Soc.,* **88,** 3963 (1966).

98. (a) R. J. Crawford and G. L. Erikson, *J. Am. Chem. Soc.,* **89,** 3907 (1967); R. J. Crawford and L. H. Ali, *J. Am. Chem. Soc.,* **89,** 3908 (1967); (b) B. H. Al-Sader and R. J. Crawford, *Can. J. Chem.,* **46,** 3301 (1968); (c) A. Mishra and R. J. Crawford, *ibid.,* **47,** 1515 (1969).

99. D. E. McGreer, N. W. K. Chiu, M. G. Vinje, and K. C. K. Wong, *Can. J. Chem.,* **43,** 1407 (1965).

100. L. A. Carpino, *J. Org. Chem.,* **30,** 736 (1965).

101. F. G. Willey, *Angew. Chem. Intern. Edit. Engl.,* **2,** 138 (1964).

102. (a) G. Wittig and A. Krebs, *Chem, Ber.,* **94,** 3260 (1961), and references contained therein; (b) G. Wittig and H. L. Dorsch, *Ann. Chem.,* **711,** 46 (1968). G. Wittig and J. Meske-Schueller, *ibid.,* **711,** 65 (1968).

103. J. H. Hall, *J. Am. Chem. Soc.,* **87,** 1147 (1965); J. H. Hall and E. Patterson, *J. Am. Chem. Soc.,* **89,** 5856 (1967).

104. D. M. Lemal and S. D. McGregor, *J. Am. Chem. Soc.,* **88,** 1335 (1966).

105. S. D. McGregor and D. M. Lemal, *J. Am. Chem. Soc.,* **88,** 2858 (1966).

106. W. Baker, J. F. W. McOmie, and D. R. Preston, *Chem. Ind.,* **1960,** 1305; *J. Chem. Soc.,* **1961,** 2971.

107. L. A. Carpino, *J. Am. Chem. Soc.,* **84,** 2196 (1962).

108. L. A. Errede, *J. Am. Chem. Soc.,* **83,** 949 (1961).

109. M. P. Cava and A. A. Deana, *J. Am. Chem. Soc.,* **81,** 4266 (1959).

110. L. A. Carpino, *Chem. Commun.*, **1966**, 494. For related work, see L. A. Carpino, *J. Org. Chem.*, **34**, 461 (1969).
111. R. B. Woodward and R. Hoffmann, *J. Am. Chem. Soc.*, **87**, 395 (1965); H. C. Longuet-Higgins and E. W. Abrahamson, *J. Am. Chem. Soc.*, **87**, 2045 (1965).
112. R. Huisgen and H. Seidl, *Tetrahedron Letters*, **1964**, 3381.
113. G. Quinkert, K. Opitz, W. W. Wiersdorff, and M. Finke, *Tetrahedron Letters*, **1965**, 3009.
114. M. P. Cava, M. J. Mitchell, and A. A. Deana, *J. Org. Chem.*, **25**, 1481 (1960).
115. D. M. Lemal, G. Guhn, and M. Kinsky, unpublished results.
116. L. A. Carpino, *J. Am. Chem. Soc.*, **85**, 2144 (1963).
117. L. A. Carpino and S. Göwecke, *J. Org. Chem.*, **29**, 2824 (1964); L. A. Carpino, J. Ferrari, S. Göwecke, and S. Herliczek, *J. Org. Chem.*, **34**, 2009 (1969).
118. E. Höft and A. Rieche, *Angew. Chem.*, **73**, 907 (1961).
119. E. Schmitz and R. Ohme, *Chem. Ber.*, **95**, 2012 (1962).
120. R. Huisgen, R. Grashey, and J. Sauer, in *The Chemistry of the Alkenes*, S. Patai, Ed., Interscience, New York, 1964, p. 739.
121. R. Huisgen, *Angew. Chem. Intern. Ed. Engl.*, **2**, 565 (1963).
122. R. Huisgen, R. Grashey, and R. Krischke, *Tetrahedron Letters*, **1962**, 387.
123. M. Busch and K. Lang, *J. Prakt. Chem.*, **144**, 291 (1936).
124. L. A. Carpino, *Abstracts, 130th Natl. Mtg. Am. Chem. Soc., Atlantic City, N.J., Sept., 1956*, p. 18-0.
125. A. Michaelis and C. Claessen, *Chem. Ber.*, **22**, 2233 (1889).
126. A. Michaelis and K. Luxembourg, *Chem. Ber.*, **26**, 2174 (1893).
127. D. M. Lemal and R. S. Feld, unpublished results.
128. H. E. Baumgarten, P. L. Creger, and R. L. Zey, *J. Am. Chem. Soc.*, **82**, 3977 (1960).
129. E. Schmitz and D. Habisch, *Chem. Ber.*, **95**, 680 (1962).
130. For a closely related method of hydrazone degradation, see R. F. Smith and L. E. Walker, *J. Org. Chem.*, **27**, 4372 (1962).
131. T. Kauffmann, G. Ruckelshauss, and J. Schulz, *Angew. Chem. Intern. Ed. Engl.*, **3**, 63 (1964).
132. W. Skorianetz and E. sz. Kovats, *Tetrahedron Letters*, **1966**, 5067.
133. D. M. Lemal, R. L. Lichter, and G. Guhn, unpublished results.
134. The nearly completed synthesis of vitamin B_{12} by R. B. Woodward, A. Eschenmoser, and their co-workers is an example *par excellence* of this approach (R. B. Woodward, "Recent Advances in the Chemistry of Natural Products," IUPAC 5th International Symposium on the Chemistry of Natural Products, London, England, July 8–13, 1968).
135. W. M. Jones, M. E. Stowe, E. E. Wells, Jr., and E. W. Lester, *J. Am. Chem. Soc.*, **90**, 1849 (1968).
136. M. Dekker and G. R. Knox, *Chem. Commun.*, 1243 (1967); J. Ashley-Smith, M. Green, N. Mayne, and F. G. A. Stone, *ibid.*, 409 (1969).
137. E. O. Fischer and R. Aumann, *Chem. Ber.*, **102**, 1495 (1969), and earlier papers.
138. H. Bock and K.-L. Kompa, *Angew. Chem. Intern. Ed. Engl.*, **1**, 264 (1962); *Z. Anorg. Allgem. Chem.*, **333**, 238 (1964). Dr. F. D. Marsh (du Pont) brought this work to our attention.
139. A manuscript by G. Koga and J.-P. Anselme describes the complex reaction of tosyl azide with the anion of 1,1-dibenzylhydrazine. Evidence is presented that *N*-azidodibenzylamine as well as dibenzyldiazene are formed as intermediates.
140. L. A. Carpino, private communication.

CHAPTER 11

Nitrenium Cations

P. T. LANSBURY

Department of Chemistry, State University of New York at Buffalo, Buffalo, New York, 14214

Numerous incisive and well-conceived experimental studies of univalent, neutral nitrogen species, nitrenes, have permeated the chemical literature in recent years and this subject constitutes the major portion of this book. By contrast, *cationic* univalent nitrogen intermediates have received little attention. Several types of these can be envisaged, e.g.,

$$
\begin{array}{ccc}
& \underset{|}{\overset{R}{\underset{|}{C}}} & \overset{R}{\diagdown}\,\overset{R}{\diagup} \\
\overset{R}{\underset{|}{\oplus}}\text{N:} & \overset{\|}{\underset{\cdot\,\oplus}{N}} & \underset{\oplus}{N:} \\
\overset{\cdot}{} & & \\
\textbf{(A)} & \textbf{(B)} & \textbf{(C)}
\end{array}
$$

Structure **A**, a radical-cation, is isoelectronic with monovalent carbon, the first representative of which has just been reported by Gunning et al. (1). Structure **B** is the molecular ion from electron impact upon nitriles (2), which is usually of very low intensity, even when compared with $M - 1$ and $M + 1$ peaks, the latter corresponding to H atom abstraction from another species. We will restrict ourselves to type **C** intermediates, which need not have any radical character, in the singlet state at least. Ions of the latter type have been referred to as azonium, iminium, and methyleniminium ions among other terms but we will use the term *nitrenium* ions since they are closely related to nitrenes. Such nitrenium ions were considered in early studies of the well-known Beckmann rearrangement of oximes (3), but were soon discarded in view of the stereospecific *trans* nature of the reaction shown in Scheme 1. Nitrenium ions supposedly would lead to nonstereospecific reactions, the amide ratio being independent of configuration in the original oxime.

Before discussing investigations of such nitrogen cations which have been pursued in our laboratories (4,5), it seems worthwhile to compare the relationship of such intermediates with nitrenes, especially since predictions regarding the former based on the more extensive studies of the latter may point the way to significant future experiments.

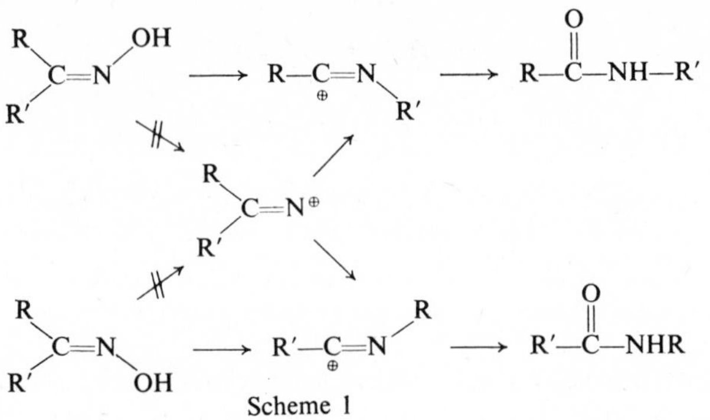

Scheme 1

Inspection of the canonical forms which can be formulated for various unsaturated nitrenes reveals that we may consider such species as having

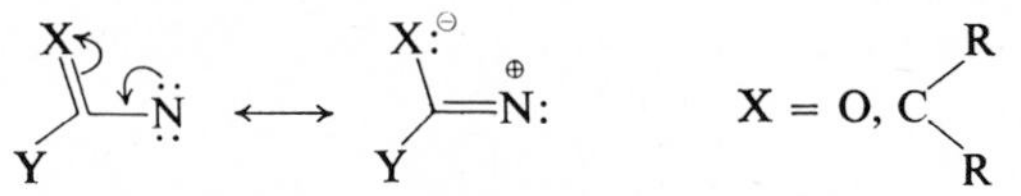

partial nitrenium ion character. Contributions of structure **B** are expected to be minimal, however, since these are charge-separated; furthermore, nitrenes are usually generated in nonpolar media where dipolar resonance character is least important. Protonation of unsaturated nitrenes (or more accurately, their generation from conjugate acids of nitrene precursors) may produce the related univalent nitrenium ions, which possess a sextet of electrons around nitrogen and may exist in singlet and triplet states like their conjugate bases. In the case where X—H is —C—H, **D** represents

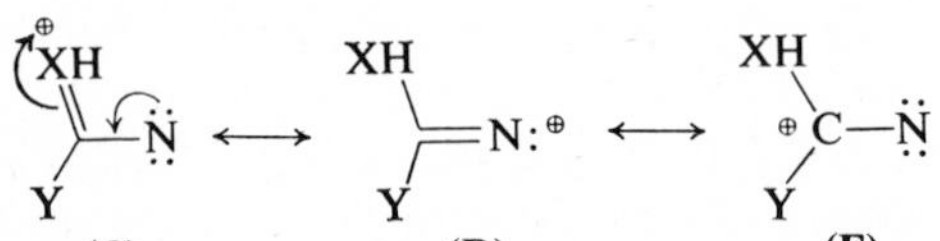

the electron-deficient intermediate satisfactorily (**C** ⟷ **D**); only loss of an α-proton could produce a nitrene. Such a situation may pertain in the Neber rearrangement of oxime *p*-toluenesulfonates which is induced by strong bases (6).

Confining ourselves for the time being to X = Y = carbon, we may formulate a scheme (in part speculative at this stage) showing the conjugate acid–base relationship of a vinyl azide and its protonated form and their decomposition (groups other than molecular nitrogen can also be lost, as will be shown later) to produce nitrenes or nitrenium ions (see

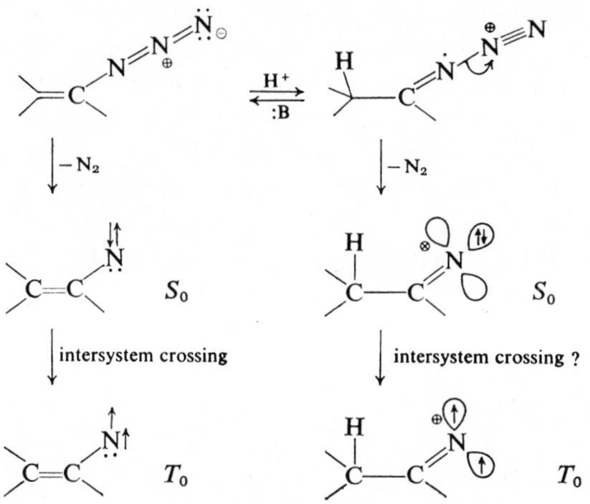

Scheme 2

Scheme 2). This hypothetical scheme, which can be applied to other nitrene precursors also, implies that nitrenium ions, like nitrenes, may have singlet and triplet electronic states, with perhaps the latter being more stable.

We may envision the singlet nitrenium ion to contain sp hybridized nitrogen with the non-bonding lone pair occupying a $2sp$ orbital and a vacant $2p$ orbital orthogonal to the two sp orbitals and the $2p$ π-bonding orbital. A similar bonding picture has been considered for vinyl cations (7)

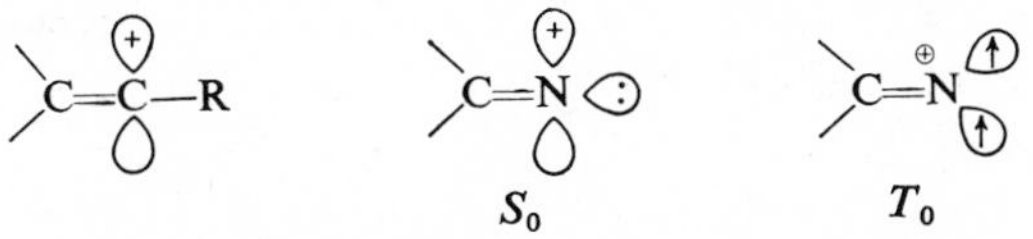

which are, of course, isoelectronic with nitrenium ions. The triplet state, on the other hand, may plausibly be represented as having sp^2 hybridization for nitrogen with two unpaired electrons of parallel spin.

How might we distinguish which nitrogen cation is at hand under a given set of experimental circumstances? Ideally, electron spin resonance spectroscopy could detect the triplet species if it could be generated under conditions where further reaction is slow. The Bell Telephone Laboratories group has carried out elegant studies (8) of triplet nitrenes generated photolytically in rigid matrices at 77°K and measured their ESR spectra.

Lacking experimental data, we may perhaps hypothesize that bond insertion will be a reaction characteristic of the singlet state, by analogy with
studies of carbethoxynitrene (9). In this latter case C—H bond insertion
occurs only with the singlet nitrene, paralleling the general trend in carbene
chemistry (10). Furthermore, a reaction (particularly if *intermolecular*)
suspected to involve a singlet nitrenium ion may be suppressed by collisional deactivation with inert solvent (9) or by a heavy atom solvent
effect (11) *if* the ground state is indeed the triplet.

In order to generate a nitrenium cation it would appear that the nature
of the leaving group (e.g., N_2, Cl^-, polyphosphate, etc.) is of secondary
importance, compared with the need to inhibit the ordinarily facile
migrations of groups from carbon to electrophilic nitrogen (3) (as
in the stereospecific rearrangements of oximes or *N*-chloroimines) or
oxime fragmentation (12) (see Scheme 3). These well-known oxime

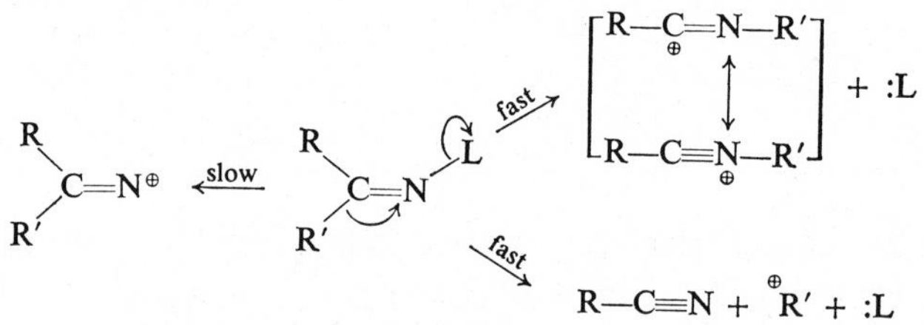

Scheme 3

reactions (L = OH) occur because synchronous migration (3) or fragmentation (12) of R′ is expected to delocalize any developing electron
deficiency on an electronegative nitrogen, no matter how "good" the
leaving group. In the transition state for aryl migration, charge is delocalized into the migrating group (13), perhaps with the formation of an
unstable intermediate. In an oxime fragmentation, R′ (Scheme 3) must be
a reasonably stable carbonium ion, e.g., as shown in equation 1. Thus, if

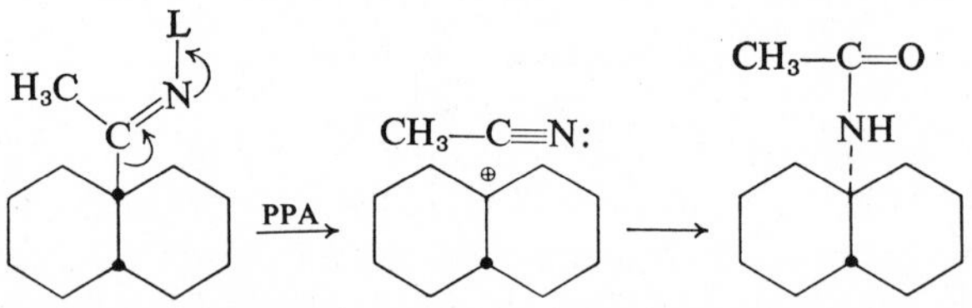

(1)

one can *prevent* a suitably chosen R′ from participating during ionization,
the desired generation of the reactive nitrenium ion may be possible.

To prevent R′ from participating during ionization, we initially chose
4-bromo-7-*t*-butyl-1-indanone oxime as the substrate (4), because its
fragmentation would involve a primary carbonium ion, and stereospecific

aryl migration would develop severe torsional strain in the transition state. Polyphosphate was chosen as the leaving group, because Hammett ρ values show it to require almost no aryl participation. We hoped to observe one or more of the reactions shown in Scheme 4. Perhaps other

Beckmann reactions	ρ	Ref.
Acetophenone oximes in PPA	-0.25	14
Acetophenone oximes in H_2SO_4	-2.0	15
Acetophenone oxime picrates	-4.1	16

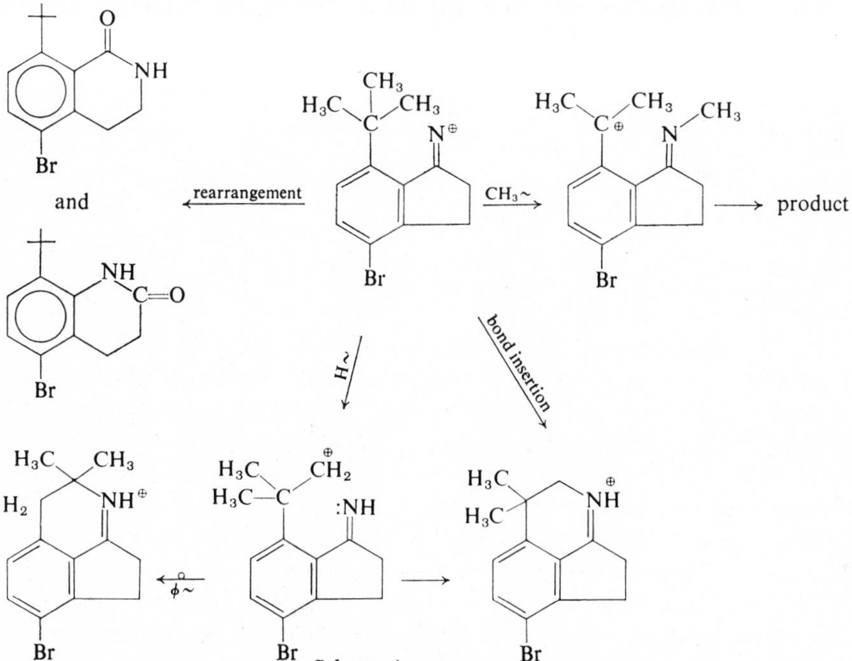

Scheme 4

model systems would also lead to primary products attributable to highly reactive nitrenium ions: these might include nonstereospecific lactam formation, transannular hydride or alkyl shifts, and bond insertions.

The Beckmann reaction of 4-bromo-7-t-butyl-1-indanone oxime (**1**) in polyphosphoric acid (PPA) gave **2**, **3**, and **4**, indicating that both bond-insertion and nonstereospecific rearrangement occur. Imine **2** was characterized tentatively by infrared and NMR spectroscopy (ruling out the isomeric enamine possibility) and, owing to its extreme lability toward air oxidation, was converted to the crystalline perchlorate. The latter was

(1)   (2)   (3)   (4)

(70–75%)   20% (ratio of **3:4** is *ca.* 4:1)

degraded to a 1,2,3,4-tetrahydroisoquinoline (**5**) that was unambiguously and independently synthesized (4) from the β-methylallylimine of 1-indanone (**6**). The major reaction product from **1** is thus *bond insertion*

2—ClO$_4^{\ominus}$ (5)   (6)

into a proximal *unactivated* C—H bond. The possibility that **2** arose from a vinyl nitrene intermediate (**7**) rather than the nitrenium ion (**8**) was dismissed by carrying out the reaction in fully deuterated PPA and observing essentially no incorporation of **D** into product (**4**). The absence of imine with the aryl migrated structure **9** is consistent with a *three-center bond insertion* process (as formulated above) involving a triangular transition state of the type invoked in other electrophilic bond insertions at saturated carbon (e.g., hydrolysis of diarylpyridine boranes (17), insertion by carbenes and nitrenes). The kinetic isotope effect for the insertion reaction, which was expected to be small (18), was determined by using **1** containing two CD$_3$ groups and one CH$_3$ group, and found to be $k_H/k_D = 1.45 \pm 0.1$ as determined both by NMR and mass-spectral methods (19). This isotope effect for nitrenium ion insertion is in good agreement with Lwowski's k_H/k_D value of 1.5 for singlet carbethoxynitrene insertion into cyclohexane (20) and suggests that our insertion proceeds via the singlet state. Such a postulate is consistent with results in nitrene chemistry, wherein singlet → triplet intersystem crossing proceeds quite slowly, judging from the requirement for extensive dilution to get triplet additions of carbethoxynitrene to olefins (9). Thus, in spite of the presence of a bromine atom in **1**, which might facilitate spin inversion (11), *intramolecular insertion* clearly occurs before conversion to triplet can compete (if, in fact, this is the stable ground state). The debrominated 7-*t*-butyl-1-indanone oxime-d_6 also gives $k_H/k_D = 1.45 \pm 0.1$ (average of 3 runs). If the triplet nitrenium cation had been the species leading to **2**, we might have observed a substantially larger k_H/k_D (perhaps 3–4) (see eq. 2),

based on Corey and Hertler's study (21) of the Löffler-Freytag reaction, for which $k_H/k_D = 3.54$ in the H *atom* abstraction step by nitrogen (see eq. 3). Summarizing, in a suitably constrained oxime polyphosphate, highly reactive nitrenium ions are generated and these insert into proximal C—H bonds. No C—C bond insertion occurred, but nonstereospecific Beckmann rearrangement accompanied the insertion reaction (5, 22).

$$(3)$$

The major lactam (3) resulting from rearrangement of 1 was that arising from net *cis*-migration (4)! Since the oxime configuration must be *anti*-aryl in 7-alkyl-1-indanone oximes (and cannot isomerize to *syn*-aryl), particularly where a 7-*t*-butyl group is involved (5), the high tendency for "*cis*-rearrangement," which is greatest for *t*-butyl and decreases on going to ethyl and then methyl, must be a consequence of the highly reactive nitrenium ion. The latter rearranges *without* stereospecificity, as predicted, the actual lactam ratios depending on how the 7-alkyl groups affect the respective rearrangement transition states *inter alia*. (See Table I). Thus

TABLE I
Lactam Ratios from Beckmann Rearrangement of 7-Alkyl-1-indanone Oximes in PPA (5)

Compound	% Aryl migration	% Alkyl migration	%R/%Ar
1-Indanone	90	10	0.1
4,7-Dimethyl-1-indanone	34	66	1.9
4,7-Diethyl-1-indanone	27	73	2.7
4-Bromo-7-*t*-butyl-1-indanone	19	81	4.3

a second characteristic of nitrenium ions is their *nonstereospecific* rearrangement to amides, a result in sharp contrast with conventional Beckmann rearrangements in which strict *trans*-stereochemistry is observed (3)!

In searching for possible extensions of nitrenium ion chemistry to more flexible substrates, particularly for synthetic purposes, the homolog of 1, 5-bromo-8-*t*-butyl-1-tetralone oxime (11), was synthesized and subjected to reaction conditions identical to those used with 1 (5). No basic imine (from bond insertion) was found; the sole product (97% yield) was 6-bromo-3,4-dihydrohomocarbostyril (12) resulting from normal stereospecific Beckmann rearrangement (and subsequent dealkylation). Several other 8-alkyl-1-tetralone oximes also rearranged only by aryl migration and without concomitant bond insertion (5), in contrast with related 7-alkyl-1-indanone oximes (see above).

Our previously stated feeling that inhibition of normal Beckmann rearrangement and fragmentation is the most important condition for

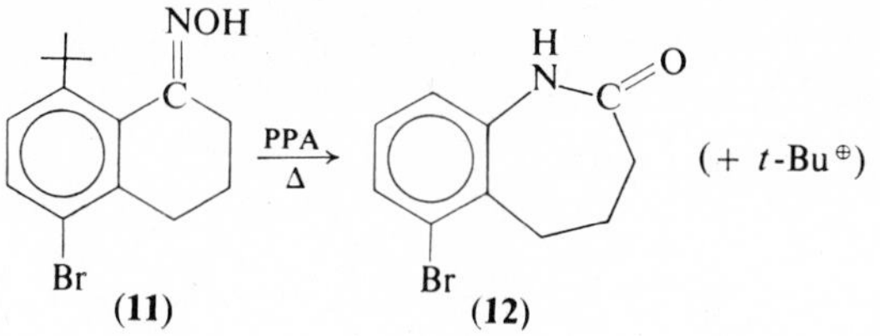

$(+ t\text{-Bu}^{\oplus})$

(11) **(12)**

generating nitrenium ions is now clearly vindicated. We have been unable as yet to uncover suitable substrates for generating nitrenium ions other than the 7-alkyl-1-indanone system. One such search for nitrenium ions involved the PPA reaction of *p*-phenylbenzophenone oxime, to give 8-methyl-6-phenylphenanthridine (23). By using *o*-methyl-benzophenone oxime, it was possible to show that normal carbon-to-nitrogen rearrangement precedes carbonium ion closure. Direct electrophilic aromatic substitution by a nitrenium ion would have given the isomeric 3-methyl-6-phenylphenanthridine which was not found (23) (see Scheme 5). Also, Loeppky and Rotman (24) encountered only stereospecific lactam formation from Ag⁺-assisted rearrangement of some diaryl and aralkyl N-chloroketimines. Although this method may lead to nitrenium ions with the right substrate, it is again clear that the synchronous reaction process is preferred.

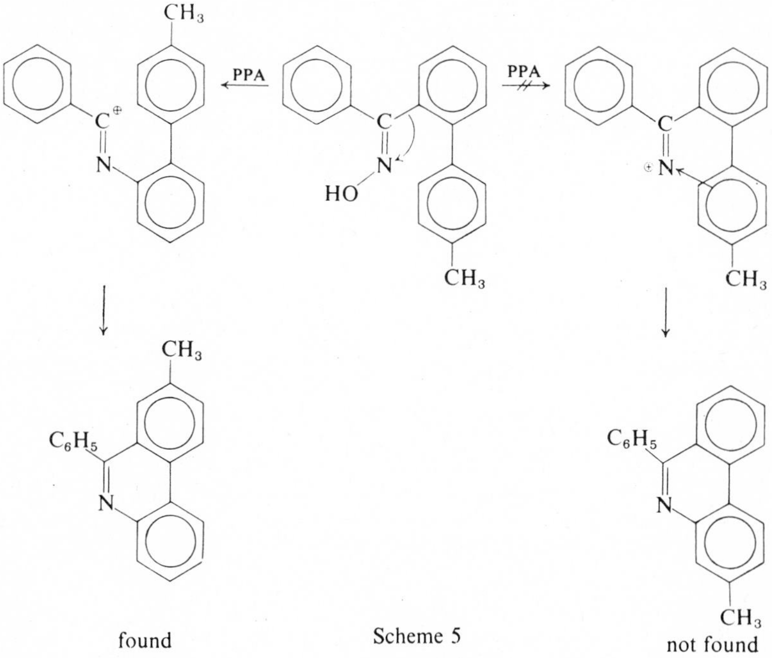

found   Scheme 5   not found

Other 7-alkyl-1-indanone oximes which gave lactam mixtures from Beckmann reactions in PPA also provide some C—H insertion product (5). Insertion into a proximal methyl group to give a strained five-membered ring occurs to a lesser extent than with **1** and the product is isolable only as the acetylated aminoketone resulting from imine hydrolysis (under conditions where **2** survived). The labile five-membered ring is also formed when an ethyl group with two possible sites for insertion is presented for nitrenium ion attack (5). One point worth noting, however, is that

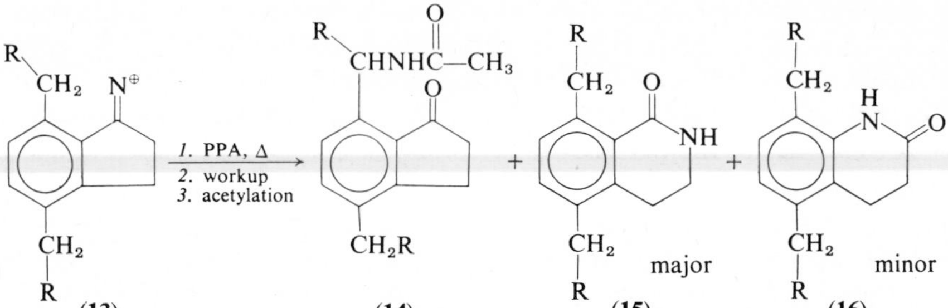

(13)   (14)   (15)   (16)

R = H, CH₃   R = H, CH₃   R = H   R = CH₃

these products resulting from formal bond insertion (**15** and **16**) may, in fact, arise from intramolecular 1,5-hydride shifts, followed by closure of imine nitrogen to the resulting carbonium ion since *benzylic* hydrogen is now available and the transition state for bond insertion may now be more strained than in **1**. This point remains to be clarified (see Scheme 6).

With the availability of suitable substrates for generating nitrenium ions, it is obviously desirable to study other leaving groups and some

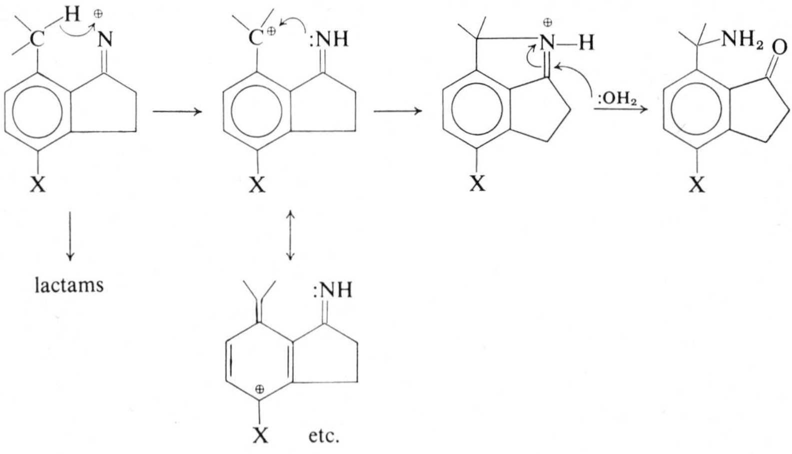

lactams

:NH

X etc.

Scheme 6

work along these lines has already been done in our laboratories (4,5). Schmidt reactions of 7-alkyl-1-indanones and diazotization of the derived hydrazones, both in PPA, produce insertion products and lactam mixtures in similar, but not necessarily identical ratios. Additional studies in this area are needed before conclusions can be drawn.

There is little doubt that nitrenium ions are not to be commonly expected as intermediates in organic chemical reactions, but what about transition states having "nitrenium" ion character? Since the Beckmann transition state involves substantial bond angle strain (3), a remote nucleophilic site situated in close proximity to the oxime nitrogen from which a group is departing may be able to displace the leaving group with much less strain. Although such a hypothesis needs ample experimental study, one apparent example was provided by Griot and Wagner-Jauregg (25), who observed the involvement of a remote alkene function in an oxime reaction leading to cyclization instead of rearrangement (eq. 4).

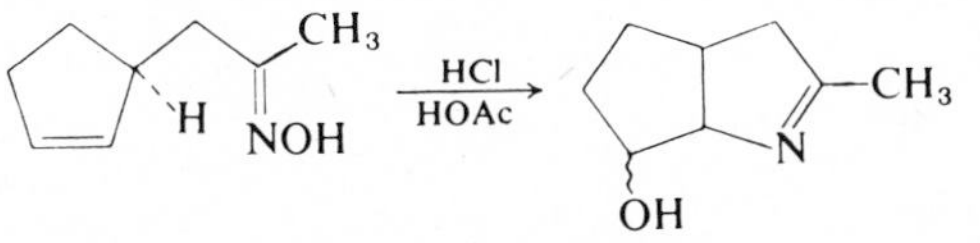

$$\text{(4)}$$

From inspection of models, it is clear that the alkene π orbital can attack nitrogen from the rear, as indicated below, in homoallylic fashion and without developing bond angle strain. Such a transition state involves nucleophilic attack along the axis of the developing vacant $2p$ orbital on nitrogen (see Scheme 7). Alternatively, the acetoxime side chain can

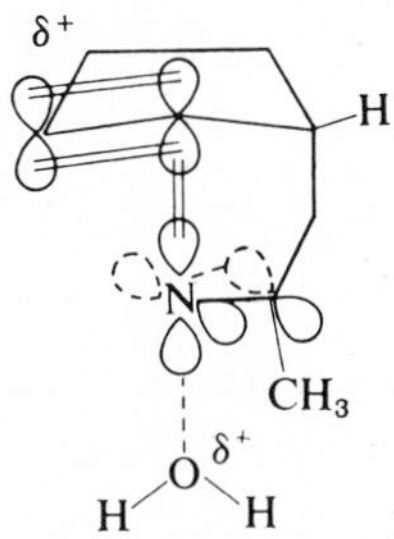

Scheme 7

readily assume a conformation such that double bond participation is

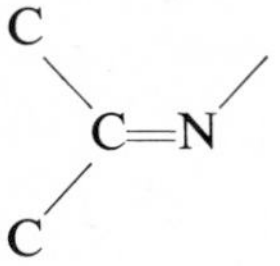

from the direction *perpendicular* to the C=N plane to produce an addition intermediate from which the imine π bond is regenerated by departure of the leaving group (eq. 5). This latter mechanism is more like

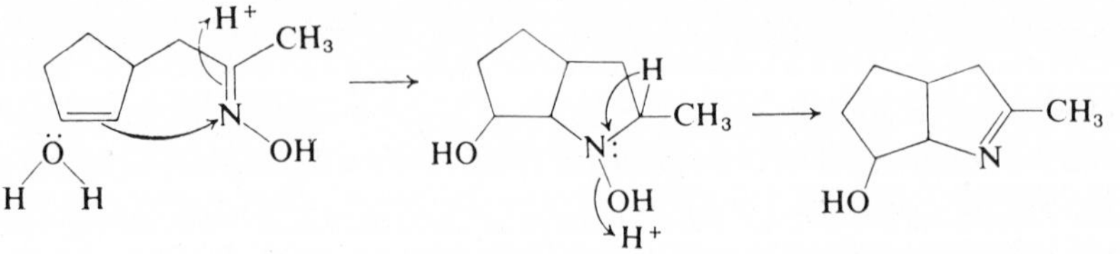

the addition–elimination sequence supposedly involved in many nucleophilic displacements on vinyl halides (26). It is of both synthetic and theoretical interest to assess the scope and mechanisms of such annelation reactions. One potentially illuminating example is illustrated below, in which a nucleophilic sulfoxide group is placed *cis* and rearward to a neighboring oxime *p*-toluenesulfonate (eq. 6). As indicated from models,

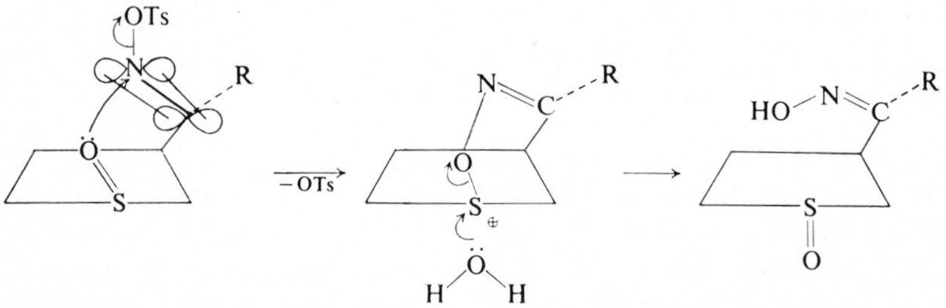

sulfoxide attack on nitrogen can only occur *anti* to the leaving group. The product of such a hypothetical reaction has inverted sulfoxide and oxime configurations. Other substrates where participation by some nucleophilic group can occur *only* perpendicular to the $\overset{\displaystyle C}{\underset{\displaystyle C}{\diagdown}}C{=}N\diagup$ plane would also be of obvious interest. If these oxime cyclizations prove to be of some generality, the already impressive array of oxime transformations will be further diversified!

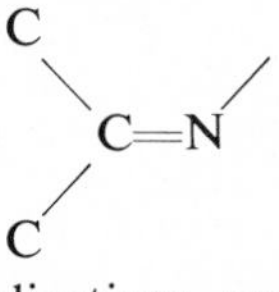

In restricting the present discussion to univalent nitrenium ions we can only briefly mention other interesting divalent nitrogen cation chemistry, such as Gassman's *N*-chloramine rearrangements (27) (eq. 7). The recent demonstration of internal return during the related silver ion-catalyzed reactions in several 2-azabicyclo-[2.2.1]heptenes makes it more likely that these are concerted processes, rather than initial heterolysis of the N—Cl bond with complete loss of chloride (28). Other reactions possibly involving

$$\text{(structure)} \xrightarrow[\text{CH}_3\text{OH}]{\text{Ag}^+} \text{(structure with N}^+\text{)} \longrightarrow \text{(structure with N, OCH}_3\text{)} \tag{7}$$

R_2N^+ and $R—\overset{+}{N}H$ intermediates (which are isoelectronic with carbenes) have been reviewed by Abramovitch and Davis (29) and Smith (3). We have also drawn heavily from our own experiences without meaning to pass over significant research of other groups. In fact, it is hoped that some of the speculations raised in this chapter will spur others to devise and carry out critical experiments which will provide the real proof of the pudding.

ACKNOWLEDGMENTS

It is a pleasure to acknowledge the enthusiastic collaboration of my co-workers on nitrenium ion chemistry whose names appear in the references and financial support of the U.S. Army Research Office (Durham) and the Petroleum Research Fund, administered by the American Chemical Society.

APPENDIX

Since the original manuscript was written, several highly relevant articles have appeared in the recent literature.

Hall and co-workers (30) have measured the deuterium isotope effect for the formation of N-t-butylaniline from triplet phenylnitrene $(C_6H_5—\overset{\uparrow}{\underset{\uparrow}{N}}\uparrow)$ and isobutane, obtaining $k_H/k_D = 4.1$ at 160°. This high value, which is similar to Corey's data on the Löffler-Freytag reaction (21), is consistent with an H atom abstraction reaction, as expected for a triplet species, and not a direct three-center bond insertion. The latter mechanism, proposed by us for the intramolecular insertion in 8 via the singlet state and by Lwowski for singlet carbethoxynitrene insertions (20) is characterized by lower k_H/k_D's (ca. 1.4–1.5), thus demonstrating the clear contrast between the two electronically different intermediates, and further validating the use of isotope effects as a mechanistic probe.

Additional examples of intramolecular nucleophilic attack at oxime nitrogen have been obtained. Autrey (31) has quoted unpublished results of Woodward which involve remote double bond participation in oxime tosylate solvolysis, much like Wagner-Jauregg's earlier work (25) (eq. 9). This result and additional evidence led Autrey to propose direct displacement by divalent sulfur ($\rightarrow$ **i**) during fragmentation of an α-methioxy-ketoxime, although formation of the observed product (**ii**) did not require

(9)

(i)   (ii)

such an intermediate. The case for **i** is strengthened by earlier work of Crawford (32), who had shown that certain *o*-substituents (e.g., —SCH$_3$, I) greatly enhance the transformation of *syn*-benzaldoxime derivatives to benzonitriles. Quite dramatically, the *o*-methioxy derivative (**iii** below) solvolyzes 11,000 times more rapidly than the *para* isomer, supposedly forming an isothiazolinium ion in the rate-determining step. Further

rigorous evidence for such heterocyclic intermediates would evolve from the appearance of products, e.g., isothiazolines, which absolutely require such precursors.

REFERENCES

1. T. Do Minh, H. E. Gunning, and O. P. Strausz, *J. Am. Chem. Soc.*, **89**, 6785 (1967).

2. H. Budzikiewicz, C. Djerassi, and D. H. Williams, *Interpretation of Mass Spectra of Organic Compounds*, Holden-Day, San Francisco, Calif., 1964, p. 111.
3. P. A. S. Smith, *Molecular Rearrangements*, Vol. 1, P. de Mayo, Ed., Interscience, New York, 1963, pp. 457–528.
4. P. T. Lansbury, J. G. Colson, and N. R. Mancuso, *J. Am. Chem. Soc.*, **86**, 5225 (1964).
5. P. T. Lansbury and N. R. Mancuso, *J. Am. Chem. Soc.*, **88**, 1205 (1966).
6. H. O. House and W. F. Berkowitz, *J. Org. Chem.*, **28**, 307, 2271 (1963).
7. H. Richey and J. Richey, *Carbonium Ions*, Vol. 2, G. Olah and P. Schleyer, Eds., Interscience, New York, 1969.
8. E. Wasserman, this volume, Chapter 13.
9. W. Lwowski, this volume, Chapter 6.
10. J. Hine, *Divalent Carbon*, Ronald Press, New York, 1964.
11. A. G. Anastassiou, *J. Am. Chem. Soc.*, **89**, 3184 (1967).
12. C. A. Grob and P. W. Schiess, *Angew. Chem., Intern. Ed.*, **6**, 1 (1967).
13. R. Huisgen, J. Witte, and I. Ugi, *Chem. Ber.*, **90**, 1844 (1957).
14. D. E. Pearson and R. M. Stone, *J. Am. Chem. Soc.*, **83**, 1716 (1961).
15. D. E. Pearson, J. F. Baxter, and J. C. Martin, *J. Org. Chem.*, **17**, 1511 (1952).
16. R. Huisgen, J. Witte, H. Walz, and W. Jira, *Ann.*, **604**, 191 (1957).
17. M. F. Hawthorne and E. S. Lewis, *J. Am. Chem. Soc.*, **80**, 4296 (1958).
18. F. Westheimer, *Chem. Rev.*, **61**, 265 (1961).
19. P. T. Lansbury and P. C. Briggs, unpublished results.
20. W. Lwowski and T. J. Maricich, *J. Am. Chem. Soc.*, **87**, 3630 (1965).
21. E. J. Corey and W. R. Hertler, *J. Am. Chem. Soc.*, **82**, 1657 (1960).
22. P. T. Lansbury and N. R. Mancuso, *Tetrahedron Letters*, **1965**, 2445.
23. P. T. Lansbury and R. P. Spitz, *J. Org. Chem.*, **32**, 2623 (1967).
24. R. Loeppky, and M. R. Rotman, *J. Org. Chem.*, **32**, 4010 (1967).
25. R. Griot and T. Wagner-Jaurreg, *Helv. Chim. Acta.*, **42**, 605 (1959).
26. E. L. Eliel, *Stereochemistry of Carbon Compounds*, McGraw-Hill, New York, 1962, pp. 368–369.
27. P. G. Gassman and B. L. Fox, *J. Am. Chem. Soc.*, **89**, 338 (1967).
28. P. G. Gassman and R. L. Cryberg, *J. Am. Chem. Soc.*, **90**, 1355 (1968).
29. R. A. Abramovitch and B. A. Davis, *Chem. Rev.*, **64**, 149 (1964).
30. J. H. Hall, J. W. Hill and J. M. Forgher, *J. Am. Chem. Soc.*, **90**, 5313 (1968).
31. R. L. Autrey and P. W. Scullard, *J. Am. Chem. Soc.*, **90**, 4924 (1968).
32. R. J. Crawford and C. Woo, *Can. J. Chem.*, **43**, 3178 (1965).

CHAPTER 12

Other Nitrenes

WALTER LWOWSKI

Research Center, New Mexico State University,
Las Cruces, New Mexico 88001

There are potential nitrenes which are not, or not fully, considered in the preceding chapters. Most of these potential nitrenes have been, or could be discussed as intermediates in the decomposition of metaloid azides, and some might become important in organic chemistry.

I. NITRENE, N—H

Chapter 2 of this book deals with the spectroscopy of NH, and with its reactions with olefins, but not with some other aspects (which turn out not to involve much organic chemistry). Abramovitch and Davis (1) have reviewed the formation and chemistry of NH in 1964. Most of this is inorganic chemistry. Chloroamine has been much discussed as a precursor for NH, especially in connection with the Raschig synthesis of hydrazine: $H_2N—Cl + 2NH_3 \rightarrow H_2N—NH_2 + NH_4Cl$. Raschig (2) proposed a nitrene mechanism (eq. 1), while Bodenstein (3) favored an S_N2 type of nucleophilic displacement on nitrogen (eq. 2). The evidence

$$H_2N—Cl + :NH_3 \rightleftharpoons H—\bar{N}—Cl + {}^+NH_4$$

$$-Cl^- \downarrow$$

$$H—N \xrightarrow{NH_3} H_2N—NH_2 \qquad (1)$$

$$H_3N: + H_2N—Cl \longrightarrow H_3\overset{+}{N}—NH_2 + Cl^-$$

$$\downarrow$$

$$H_2N—NH_2 + H_4N^+ \qquad\qquad (2)$$

now available strongly favors Bodenstein's mechanism (1,4). Chloroamine does not appear to be an efficient NH precursor. Hydroxylamine-O-sulfonic acid, $HO_3S—O—NH_2$, (5) has also been discussed as an NH precursor. Again, α-elimination to give NH and sulfate ion does not seem to be an important reaction path, compared to nucleophilic displacement on nitrogen (6). In the presence of base, hydroxylamine-O-sulfonic acid reacts with ketones to give oxaziridines, at a rate (disappearance of hydroxylamine-O-sulfonic acid) 1000 times faster than that in the presence of the base alone (7). This cannot be explained by a nitrene mechanism in which NH is in equilibrium with its precursor, because there is no uptake of external sulfate ion (^{35}S labeled) in the hydroxylamine-O-sulfonic acid (7). Base-induced decomposition of hydroxylamine-O-sulfonic acid in the presence of butadiene gave small amounts of pyrroline (8), perhaps due to the formation of a little NH, perhaps by another route. So far, no efficient α-elimination path to NH is known.

With the elimination of NH as an intermediate in the formation of oxazirane and diaziranes from ketones and azomethines, respectively (6), the organic chemistry of NH becomes rather meager. There are the reactions with olefins (see Chapter 2) in the gas phase, and the decomposition of hydrazoic acid in organic solvents at $-80°$ (9), and Milligan's (10,11) additions of NH to ethylene and carbon dioxide in solid matrices. As far as the solution experiments go, Koch (9) concludes that NH is not involved in the main path of the reactions.

II. AZIDES AND NITRENES OF ELEMENTS OTHER THAN
C, N, S, OR H

Nitrene functions bound to elements other than C, N, S, or H are not known with certainty. In some cases, such as boron, there are reactions that might be insertion reactions and thus indicate the intermediacy of a nitrene, in other cases there are just likely nitrene precursors, notably azides, which raise the prospect of future nitrene chemistry. A brief summary of such potential nitrene reactions, and of the parent azides, is given below. For a review on the azides, the reader should consult Thayer's paper (12).

A. Boron

The chemistry of boron azides has been studied mostly by Paetzold (13), who recently reviewed the field (14). He found a rearrangement, with loss of nitrogen and migration of various groups from boron to nitrogen (15–19), resembling in some ways the rearrangements of triarylmethyl azides (see Chapter 3). The rearrangement product is obtained in polymeric form:

$$nX_2B - N_3 \xrightarrow{180-260°} nN_2 + [XB - NX]_n$$

where $X = Cl$, OR, NR_2, R, Ar; and $n = 2$, 3 or higher. Among other mechanisms, Paetzold considered a nitrene intermediate for the rearrangement, just as carbonylnitrenes have been considered in the Curtius rearrangement (see Chapter 6, Section V). The suggestion of a boronitrene, without further experimental support, found its way into other publications (12,20), although Paetzold rejected it on the basis of his kinetic data and determinations of migratory aptitudes. The thermodynamic data (14) show wide variation both in the enthalpies and in the entropies of activation, a dilemma not entirely dissimilar to that in the Curtius rearrangement (Chapter 6, Section V). Paetzold did find indications for dehydrogenation and insertions into N—H bonds, which one could attribute to a diphenylboronitrene, $(C_6H_5)_2B$—N (14). It will be interesting to observe future developments in this sector. In view of Saunders's results with triarylmethyl azides (Chapter 3, Section III-A), it might be worthwhile to study in detail the mechanism of photolysis of boron azides.

B. Azides of Aluminum and Gallium

Azides of aluminum (21,22) and gallium (23), bearing organic groups on the metal, are known. Reactions likely to involve the corresponding nitrenes seem not to have been reported.

C. Silylazides

A substantial number of silyl azides has been prepared (12), and their chemistry has been studied (12,24). West observed the first example of what appears to be the insertion of a silylnitrene into an aromatic C—H bond (25)—a reaction similar to the formation of carbazoles from *ortho*-azidobiphenyls (cf. Chapter 4, Section I-c). The photolysis of dimethyl-(2-biphenyl)silyl azide gave the product of insertion into the 2'-position of the biphenyl system (eq. 3).

$$N_3—Si(CH_3)_2 \longrightarrow \underset{H}{N}—Si(CH_3)_2 + N_2 \qquad (3)$$

Scheiner (26) photolyzed trimethylsilyl azide in 2-methyl-butane and obtained, after hydrolysis, a complex mixture containing a 10% yield of

C_5-amines: The insertion product into the tertiary C—H bond, 2-amino-2-methylbutane, predominated, products of insertion into secondary and primary C—H bonds were present in lesser amounts. Photolysis of trimethylsilyl azide in cyclohexane gave, after hydrolysis, cyclohexylamine (26). In all these experiments, the major products were polymers (26, cf. 27). Other reactions of trimethylsilyl azide (which seem not to involve silylnitrene intermediates) have been reviewed (28).

D. Organotin Azides

A fair number of organotin azides are known (12,29–31), but reactions indicative of the intervention of Sn—N seem not to have been reported.

E. Organophosphorus Azides

A substantial number of organophosphorus azides are known (12, 32–39). Bock (40–42) has written the nitrene mechanism of equation 4, to explain the formation of an aziridine and of an amide from azido-diphenylphosphin-tosylimide. The intermediary formation of a triazoline

as the true precursor of the aziridine is not excluded by Bock's experiments, and the copper catalysis employed casts doubt on the intermediacy of a free nitrene. Franz and Osuch (43) found that diphenylphosphoryl azide reacts with strained olefins (such as norbornene) to give aziridines and imines; perhaps via a triazoline intermediate. Berlin and Wilson (44–46) found diethyl phosphorazidate, $(C_2H_5O)_2PO$—N_3, to add to norbornene to form a triazoline. This was stable up to 80°, at which temperature it decomposed to nitrogen and the imine.

Diphenylphosphinic azide, upon pyrolysis, undergoes a rearrangement with phenyl migration to nitrogen—presumably an analog of the boron azide rearrangement described above (47).

F. Halonitrenes

There is almost no organic chemistry of halonitrenes, but evidence for the existence of X—N has been obtained by spectroscopic and rate measurements.

Fluoronitrene, NF, has been implicated in the deamination (48) of organic amines by HNF_2, and the volume change of activation of the reaction of HNF_2 with base (49) indicated the intermediacy of NF (50).

Photolysis of halogen azides in frozen matrices, and the reaction of active nitrogen with halogens and hydrogen halides leads to species, the spectra of which indicate them to be halonitrenes (1,51–54).

ACKNOWLEDGMENTS

The author is greatly indebted to K. D. Berlin, P. I. Paetzold, and P. Scheiner for their help in writing this chapter.

REFERENCES

1. R. A. Abramovitch and B. A. Davis, *Chem. Rev.*, **64**, 149 (1964).
2. F. Raschig, *Schwefel- und Stickstoffstudien*, Verlag Chemie, Berlin, 1924, p. 76.
3. M. Bodenstein, *Z. Physik. Chem. A*, **139**, 397 (1928).
4. W. J. le Noble, *Tetrahedron Letters*, **1966**, 727.
5. F. Sommer, O. F. Schultz, and M. Nassau, *Z. Anorg. Allgem. Chem.*, **147**, 142 (1925).
6. E. Schmitz, *Dreiringe mit zwei Heteroatomen*, Springer, Berlin, 1967, p. 42.
7. E. Schmitz, R. Ohme, and S. Schramm, *Chem. Ber.*, **97**, 2521 (1964).
8. R. Appel and O. Büchner, *Angew. Chem.*, **74**, 430 (1962); *Angew. Chem., Intern. Ed.*, **1**, 332 (1962).
9. E. Koch, *Tetrahedron*, **23**, 1747 (1967).
10. M. E. Jacox and D. E. Milligan, *J. Am. Chem. Soc.*, **85**, 278 (1963).
11. D. E. Milligan, M. E. Jacox, S. W. Charles, and G. C. Pimentel, *J. Chem. Phys.*, **37**, 2303 (1962).
12. J. S. Thayer, *Organomet. Chem. Rev.*, **1**, 157 (1966).
13. P. I. Paetzold, *Angew Chem.*, **74**, 506 (1962); *Z. Anorg. Allgem. Chem.*, **326**, 53 (1963).
14. P. I. Paetzold, *Fortschr. Chem. Forsch.*, **8**, 437 (1967).
15. P. I. Paetzold, *Z. Anorg. Allgem. Chem.*, **326**, 58 (1963).
16. P. I. Paetzold and P. P. Habereder, *Angew. Chem.*, **76**, 598 (1964).
17. P. I. Paetzold, *Angew. Chem.*, **77**, 1035 (1965).
18. P. I. Paetzold, P. P. Habereder and R. Mullbayer, *J. Organomet. Chem.*, **7**, 45 (1967).
19. P. I. Paetzold and P. P. Habereder, *J. Organomet. Chem.*, **7**, 61 (1967).
20. J. B. Leach and J. H. Morris, *J. Organomet. Chem.*, **13**, 313 (1968).
21. J. Müller and K. Dehnicke, *Z. Anorg. Allgem. Chem.*, **348**, 261 (1966).
22. M. I. Prince and K. Weiss, *J. Organomet. Chem.*, **5**, 584 (1966).
23. J. Müller and K. Dehnicke, *J. Organomet. Chem.*, **7**, P1 (1967).
24. E. Ettenhuber and K. Rühlmann, *Ber.*, **101**, 743 (1968).

25. J. M. Gaidis and R. West, *J. Am. Chem. Soc.*, **86**, 5699 (1964).
26. P. Scheiner, private communication.
27. W. T. Reichle, *Inorg. Chem.*, **3**, 402 (1964).
28. L. Birkofer and A. Ritter, *Angew. Chem.*, **77**, 414 (1965); *Angew. Chem.*, *Intern. Ed.*, **4**, 417 (1965).
29. T. N. Srivastava and S. N. Bhattacharya, *J. Inorg. Nucl. Chem.*, **28**, 1480 (1966).
30. W. L. Lehn, *Inorg. Chem.*, **6**, 1061 (1967).
31. J. Lorberth, H. Krapf, and H. Noth, *Chem. Ber.*, **100**, 3511 (1967).
32. M. I. Kabachnik and V. A. Gilyarov, *Bull. Acad. Sci. USSR*, **1961**, 758.
33. R. A. Baldwin and R. M. Washburn, *J. Org. Chem.*, **30**, 3860 (1965).
34. R. A. Baldwin, *J. Org. Chem.*, **30**, 3866 (1965).
35. K. Utvary, *Inorg. Nucl. Chem. Letters*, **1**, 77 (1965).
36. R. M. Washburn, U.S. Pat. 3,212,844; *Chem. Abstr.*, **64**, 3601c (1966).
37. R. H. Kratzer and K. L. Paciorek, *J. Org. Chem.*, **32**, 853 (1967).
38. R. A. Baldwin, C. O. Wilson, Jr., and R. I. Wagner, *J. Org. Chem.*, **32**, 2172 (1967).
39. R. A. Baldwin and M. T. Cheng, *J. Org. Chem.*, **32**, 2636 (1967).
40. H. Bock and W. Wiegrabe, *Angew. Chem.*, **74**, 327 (1962).
41. H. Bock and W. Wiegrabe, *Angew. Chem.*, **75**, 1109 (1963).
42. H. Bock and W. Wiegrabe, *Chem. Ber.*, **99**, 1068 (1966).
43. J. E. Franz and C. Osuch, *Tetrahedron Letters*, **1963**, 837.
44. K. D. Berlin and L. A. Wilson, *Chem. Commun.*, **1965**, 280.
45. K. D. Berlin, L. A. Wilson, and L. M. Raff, *Abstracts*, 152nd Mtg. Am. Chem. Soc., Sept. 1966, *Abstr. S-97*.
46. L. A. Wilson, Ph.D. Thesis, Oklahoma State Univ., Stillwater, Okla., 1966.
47. W. T. Reichle, *Inorg. Chem.*, **3**, 402 (1964).
48. C. L. Bumgardner, K. J. Martin, and J. P. Freeman, *J. Am. Chem. Soc.*, **85**, 97 (1963).
49. G. A. Ward and C. M. Wright, *J. Am. Chem. Soc.*, **86**, 4333 (1964).
50. W. le Noble and D. Skulnik, *Tetrahedron Letters*, **1967**, 5217.
51. A. Elliott, *Proc. Roy. Soc. (London) Ser. A*, **169**, 469 (1939).
52. E. R. V. Milton and H. B. Dunford, *J. Chem. Phys.*, **34**, 51 (1961).
53. D. E. Milligan, *J. Chem. Phys.*, **35**, 372 (1961).
54. E. R. V. Milton, H. B. Dunford, and A. E. Douglas, *J. Chem. Phys.*, **35**, 1202 (1961).

CHAPTER 13

Electron Spin Resonance of Nitrenes

Due to unforeseen circumstances, Chapter 13 could not be included in this book. It is hoped that the material of this chapter will appear in the review literature in the near future.

WALTER LWOWSKI

Author Index

Numbers in parentheses are reference numbers and show that an author's work is referred to although his name is not mentioned in the text. Numbers in *italics* indicate the pages on which the full references appear.

Subject Index

In this index, chemical compounds are indexed under classes of compounds, unless special factors warrant a separate entry. Thus, for t-butyl hydroperoxide see under "hydroperoxides", etc.

58721